BIOFUNGICIDES

Eco-Safety and Future Trends

Volume 2: Novel Sources and Mechanisms

Editors

Kamel A. Abd-Elsalam
Plant Pathology Research Institute
Agricultural Research Center
Giza, Egypt

Mousa A. Alghuthaymi
Biology Department, Science and Humanities College
Shaqra University
Alquwayiyah, Saudi Arabia

Salah M. Abdel-Momen
Former Minister of Agriculture
Plant Pathology Research Institute
Agricultural Research Center
Giza, Egypt

CRC Press
Taylor & Francis Group
Boca Raton London New York

CRC Press is an imprint of the
Taylor & Francis Group, an **informa** business
A SCIENCE PUBLISHERS BOOK

Cover credit: Image provided by the first editor, Kamel A. Abd-Elsalam.

First edition published 2024
by CRC Press
2385 NW Executive Center Drive, Suite 320, Boca Raton FL 33431

and by CRC Press
4 Park Square, Milton Park, Abingdon, Oxon, OX14 4RN

CRC Press is an imprint of Taylor & Francis Group, LLC

Library of Congress Cataloging-in-Publication Data (applied for)

ISBN: 978-1-032-59013-4 (hbk)
ISBN: 978-1-032-59015-8 (pbk)
ISBN: 978-1-003-45257-7 (ebk)

DOI: 10.1201/9781003452577

Typeset in Times New Roman
by Radiant Productions

Dedication

I'd like to dedicate this book to Prof. Aly A. Aly, my father in agricultural research, for his guidance, direction, and pearls of wisdom, but more importantly, for putting up with my panic attacks and questions while providing amazingly timely feedback and encouragement precisely when needed, without which it would have been nearly impossible to produce this piece of work.

Kamel A. Abd-Elsalam

Preface

The creation of novel fungicides has gotten increasingly complex in recent years, as growing amounts of environmental and toxicological data are required to satisfy regulatory agencies. Furthermore, several present fungicides may be banned soon due to toxicological concerns. The widespread use of biofungicides will aid in crop protection against pest infestation at a cost that is both effective and reasonable. This, in turn, will increase agricultural productivity and assure food security. Trends in the biofungicides market show that there is an increasing demand for the use of biofungicides to boost agricultural yields. The book's 10 chapters emphasize the importance of developing novel antifungal medications and improving antifungal agent's regulations. The current volume entitled: BIOFUNGICIDES: Eco-Safety and Future Trends Volume 2: Novel Sources and Mechanisms investigate the novel resources of biofungicides in includings: rnai-based biofungicides, antifungal efficacy of algae and peptides/proteins, secondary metabolites antimicrobial, green synthesize nanoparticles as a source of nanofungicides to combat phytopathogens, potential prospects of trichoderma metabolites as biopesticides in managing plant health and diseases. In additions to the antifungal mechanisms such as exploring mechanisms of disease suppression using endophytic fungi as biocontrol agents, antifungal biomaterials and mechanisms, how to survey and select promising biofungicides, bioremediation of fungicide-contaminated environment. Crops treated with biofungicides will gain value since they are free of chemical residues, rot, and diseases that are hazardous to human health, supporting a global need for product quality. Finally, I'd like to state that the goal of this book is to (1) identify new botanical and microbiological fungicides (2) explains the role of biofungicides in the control of plant diseases (3) How should bio fungicides be surveyed and chosen (4) Regulations, toxicity, mechanism, and safety of biofungicides. In addition, to comprehend potentially effective biological control agents as well as the methodologies used to discover new biofungicides. There is no specific book on 'Recent Advances in Biofungicides: Eco-Safety and Future Trends,' so the suggested book will be a one-of-a-kind addition in the field with little competition. This book's multidisciplinary approach means that it will appeal to a wide range of readers, including those with an interest in organic farming; materials science; biology; chemistry; physics; plant science; chemical technology; microbiology; plant physiology; biotechnology; and industrial labour. Several different aspects of the topic are thoroughly covered, as seen by the breadth of the table of contents.

We are grateful to all the eminent authors who contributed chapters and contributed valuable thoughts and knowledge to this edited book. We are grateful to the Taylor & Francis team for their great collaboration throughout the book development process. We welcome any comments and criticism from topic experts and general readers of this work.

Kamel A. Abd-Elsalam
Plant Pathology Research Institute
Agricultural Research Center
Giza, Egypt

Mousa A. Alghuthaymi
Biology Department, Science and Humanities College
Shaqra University
Saudi Arabia

Salah M. Abdel-Momen
Plant Pathology Research Institute
Agricultural Research Center
Giza, Egypt

Contents

Novel Sources

Chapter 1

RNAi-Based Biofungicides

Siddhesh B. Ghag

1. Introduction

The ability of cultivating crop plants for food and feed source promoted the establishment of human civilization (Fuller and Stevens, 2019). Crop plants were selected from the wild and cultivated on a large scale. Thus the problem of crop plant destruction due to pests and pathogens was always threatening. Plant pests and pathogens are responsible for up to 40% of yield losses globally in major crops such as maize, potato, rice, soybean and wheat (Savary et al., 2019). Plant disease management strategies were constantly been practised and restructured based on available knowledge and technology. Fungal pathogens have always been a menace in an agricultural setup and are known to infect a range of crop plants leading to their destruction and threatening food security (Ristaino et al., 2021). Plant pathogenic fungi belonging to *Magnaporthe*, *Botrytis*, *Puccinia* and *Fusarium* are listed as the first four among the 10 most impactful plant pathogens with serious disease threats (Dean et al., 2012). Plant fungal pathogens have established agricultural history and proved their economic significance. In 1845, *Phytophthora infestans* destroyed the potato crop in Ireland that resulted in 2 million deaths and mass migration (Irish famine; Yoshida et al., 2013). Similarly, the brown spot on rice caused by *Cochliobolus miyabeanus* resulted in the Bengal famine with over 2 million deaths of Bengali people (Padmanabhan, 1973). The coffee rust caused by *Hemileia vastatrix* also caused large scale devastation in Central America (Avelino et al., 2015). Panama disease of banana caused by *Fusarium oxysporum* f. sp. *cubense* tropical race 4 is another example that is spreading rapidly and destroying plantations globally (Ghag et al., 2015). All these disease epidemics have resulted in hunger, poverty and increased migration. The spread of these fungal diseases that leads to huge devastation is most likely due to poor disease management practices, monoculture of susceptible

School of Biological Sciences, UM-DAE Centre for Excellence in Basic Sciences, University of Mumbai campus, Kalina, Santacruz, Mumbai-400 098, India.
Email: siddhesh.ghag@cbs.ac.in

cultivars and climate change (Corredor-Moreno and Saunders, 2020). Managing existing plant diseases has been a challenging task throughout, nonetheless, the emerging disease that occurs due to novel virulent strains and climate change pose further setbacks to the agricultural industry. Thus a fundamental understanding of the pathogen biology, spread and emergence of novel strains is required. With the available innovative technologies, it is possible for early detection of pathogens, modeling disease epidemics and providing quick and efficient management solutions. Host resistance achieved through traditional breeding, transgenesis or mutation breeding will help in cultivating resistant crop plants (Dong and Ronald, 2019; Mekapogu et al., 2021).

There are thousands of species of fungal pathogens that infect various crop plants and have diverse strategies to colonize and infect the host plants. Certain fungal pathogens called biotrophs (e.g., *Blumeria, Puccinia*) colonize the host plant and produce haustoria that penetrate the plant vascular system to receive nutrition (Glazebrook, 2005). In this interaction, the fungal pathogen does not kill the host plant but there is reduced growth and development. Necrotrophs (e.g., *Botrytis, Sclerotinia*) are another category of fungal pathogens that secretes a range of proteins and metabolites that result in the mass destruction of host plant cells. An intermediate category of fungal pathogens called hemibiotrophs (e.g., *Fusarium oxysporum, Colletotrichum*) possess both lifestyles during infection (Glazebrook, 2005). They have a biotrophic lifestyle at the onset of disease followed by necrotrophy resulting in the full-blown disease. Each plant-pathogen interaction system is unique and possesses diverse transcriptome, proteome and metabolome profiles. Each fungal pathogen has a particular infection cycle and infects specific plant tissues such as foliage, root, stem, fruit, etc. Fungal infections in plants are evident in the field during germination (Dieback), early maturity (wilt, rot, mildew and rust), maturity (wilt, rust, spots and mildew) and also during post-harvest storage (anthracnose, mold and spots). A sustainable strategy to manage all these fungal diseases in different crop plants that is equally effective and does not pose any health risks is needed. This chapter describes at length the impact of conventional fungicides and the potential of RNA interference (RNAi) technology in curbing plant fungal diseases.

2. Conventional synthetic fungicides and their limitations

A fungicide (natural or synthetic) is specifically used to control fungal disease by either inhibiting or killing the fungus causing the disease. The application of fungicides is an important component of the integrated disease management system and is widely used in traditional agriculture to control fungal diseases. Fungicide sales account for more than 40 and 35% of the total pesticide shares in the European Union and the world over respectively (Zubrod et al., 2019). Major classes of fungicides used in agriculture against fungal diseases include chloroalkylthiodicarboximides, carbamic acid derivatives, benzimidazoles, chloronitriles, demethylation inhibitors (DMIs) and strobilurins (Table 1.1). The fungicides are classified based on their mode of action into several groups. The most common ones include the inhibitors of mitochondrial electron transport chain (mitochondrial respiration inhibitors), sterol

Table 1.1: Major classes of fungicides used in agriculture for fungal disease management along with their mode of action, resistance mechanisms developed and the associated human health risks.

Synthetic fungicide	Commercial name	Mode of action	Resistance mechanism	Health risk	References
Chloroalkylthiodicarboximides	Captan, Folpet	Blocking the activity of thiol-containing enzymes involved in mitochondrial respiration	–	Genotoxic, mutagenic, carcinogen	Cohen et al., 2010
Carbamic acid derivatives	Thiram, Propamocarb, Mancozeb	Interfere with sulfydryl groups containing enzymes and inhibit biochemical processes in fungal cell cytoplasm and mitochondria	Over production of thiols for detoxification	Cholinergic effects, central nervous system toxicity	Lodovica et al., 2010
Benzimidazoles	Benomyl, Carbendazim	Target the fungal cytoskeleton by interfering with assembly of β-tubulin subunit	Mutations in β-tubulin genes	Neurotoxicity, oxidative stress	Hahn, 2014; Kara et al., 2020
Chloronitriles	Chlorothalonil, M05	Conjugate with thiols and deactivates glutathione, inhibits thiol dependent enzymes	Over production of thiols for detoxification	Carcinogen, hepatic and renal abnormalities	Tillman et al., 1973
Demethylation inhibitors (DMIs)	Tebuconazole, Epoxiconazole	Target cell membrane integrity through the inhibition of C14-demethylation during ergosterol biosynthesis	Mutations or overexpression of the cytochrome P450 sterol 14α-demethylase genes	Low toxicity	Mair et al., 2016; Zubrod et al., 2019
Strobilurins	Fluoxastrobin, Pyraclostrobin, Azoxystrobin	Binds to the cytochrome bc1 preventing the transfer of electrons from cytochrome b to cytochrome c1 thereby inhibiting mitochondrial respiration	Mutations in the cytochrome b gene	Genotoxic and cytotoxic	Ishii et al., 2001; Feng et al., 2020; Dodhia et al., 2021

biosynthesis inhibitors and multi-site enzyme, nucleic acid and protein synthesis inhibitors. The effectiveness of the fungicide depends on several factors including its uptake and distribution in the plant tissues, which is directly correlated with the hydrophilicity of the fungicide (Zubrod et al., 2019). Other than the synthetic ones, the multi-site action of inorganic sulfur (e.g., sulfur dust, Cosan) and divalent metals such as mercury (e.g., Merfusan, Cyclosan) and copper (e.g., Bordeaux mixture, Fungimar) are commonly used fungicides in fields. Application of fungicides in fields can be done either after infection (curative) or before infection (preventive). Fungicides can be applied to the seeds before sowing to prevent fungal pathogens from damaging the newly emerging plant or can be applied during pre-harvest (foliar sprays, soil drenching) and post-harvest practises (fruit sprays, fumigations).

Pesticides (primarily insecticides and fungicides) comprise an important module of the Green Revolution that increased overall agricultural productivity by reducing disease impacts (Pimentel, 1996). However extensive usage has led to two serious problems, toxicity to the environment and human health and the emergence of resistant pests and pathogens. Pesticide residues were found in a number of foodstuffs and water sources that are beyond safe limits and alarming (Carvalho, 2017). Pesticides can enter human or animal bodies through contact with the skin, ingestion and inhalation or via the food chain from plant or animal sources. Pesticides in the body can be metabolized or stored and can cause severe dermatological, neurological, gastrointestinal, respiratory, reproductive and endocrine defects (Nicolopoulou-Stamati et al., 2016). Many fungicides are carcinogenic and have shown to have damaging effects on the reproductive organs and disrupt hormonal balance (Axelstad et al., 2011; Bernabò et al., 2017). As a result, several governmental regulatory bodies across the world have either banned or restricted the use of these harmful fungicides. Some of the fungicides are found to be toxic even to the host plants leading to yield penalties (Shahid et al., 2018). Identification and synthesis of novel fungicides with low or negligible health risks are difficult and expensive. On the other hand, the injudicious use of fungicides has created selection pressure on the existing population of pathogenic fungi facilitating the proliferation of resistant strains (Hahn, 2014; Yang et al., 2019; Vielba-Fernández et al., 2020). The emergence of these novel multi-resistant strains poses a great challenge for the future application of fungicides. Therefore the focus has shifted towards an integrated disease management system that utilizes biological control either by using natural antifungal compounds or non-pathogenic strains of fungi or antagonistic organisms (biofungicides). Though biofungicides have a potential for future disease management there are a few shortfalls, such as reliability in open large fields and the population density effect on other life forms in the vicinity need to be evaluated thoroughly.

3. Biofungicides

Biofungicides are preparations of live organisms used for the management of plant diseases caused by phytopathogenic fungi. Biofungicides are those groups of organisms (bacteria, fungi, actinomycetes) that produce a wide range of antifungal compounds, compete with pathogenic fungi for resources, and are capable of eliciting an immune response (systemic acquired resistance response) in the host plants imparting protection against the invading fungal pathogen (Hedin, 1982). Most biocontrol organisms can colonize the rhizospheric region and plant root surfaces thereby restricting nutrients, space and attachment sites for the pathogenic fungi. Moreover, the exudates released by the plant roots that trigger the infection process through chemotaxis are metabolized by the biocontrol organisms (Köhl et al., 2019). The biocontrol organisms may show parasitism or produce molecules with antifungal activity inhibiting or killing fungal pathogens. Other than these inhibitory actions most biocontrol organisms promote plant growth by providing nitrogen sources, solubilization of minerals and production of plant hormones that improve plant health. A healthy plant is better equipped to mount a response against an invading pathogen resulting in a reduction of the disease severity index.

Various commercial formulations of biofungicides are been marketed by agro-based companies and supplied in bulk for sprays and soil drench applications. *Trichoderma* species have been widely acknowledged as the best biocontrol agents for different crop plants and are usually isolated from compost. In the soil, they are competitively dominant over other soil inhabitants and display mycoparasitism, competition and antibiosis (Sood et al., 2020). *Bacillus* and *actinomycete* species are other classes of biocontrol agents more frequently used for protection against root and foliar diseases. Actinovate® marketed by Novozymes contains *Streptomyces lydicus* that protects against Powdery mildew, Downy mildew, *Botrytis*, *Rhizoctonia*, *Pythium*, *Sclerotinia*, *Phytophthora* infections (*https://www.novozymes.com/en/advance-your-business/agriculture/crop-production1/actinovate*). Companion® Liquid (Growth Products, Ltd) contains the active ingredient *Bacillus subtilis* GB03 that protects plants from Black Root Rot, Early Blight, Crown Rot, Damping-off, Gray Mold, Leaf blight, Root Rot, Powdery Mildew, Fusarium Wilt, Bacterial Leaf spot and also possess plant growth growth promoting ability (*https://www3.epa.gov/pesticides/chem_search/ppls/071065-00003-20111102.pdf*). Double Nickel® (Certis Biologicals) contains the active ingredient *Bacillus amyloliquefaciens* D747 and protects against Powdery Mildew, *Botrytis*, *Alternaria*, Fire Blight and Bacterial Spot and Speck diseases of plants (*https://www.certisbio.com/products/biofungicides/double-nickel-55-wdg*). *Streptomyces griseoviridis* present in Mycostop (Verdera Oy) is efficacious against *Pythium*, *Fusarium*, *Botrytis*, *Alternaria* and *Phomopsis*. PlantShield HC (BioWorks) containing *Trichoderma harzianum* T-22 protects against root diseases caused by *Pythium*, *Rhizoctionia*, *Fusarium*, *Thielaviopsis* and *Cylindrocladium* (*https://www.planetnatural.com/product/mycostop/*). SoilGard (Olympic Horticultural Products Inc.) has *Gliocladium virens* GL-21 that protects *Rhizoctonia* and *Pythium* infection (*https://www.ohp.com/Products/soilgard.php*). Limitations on the use of chemical fungicides have resulted in an increased commercial interest in biocontrol organisms. The demand for biofungicides has become greater over time because of increased awareness of organic farming and the negative effect of chemical fungicides on health and the environment. Nevertheless, biofungicides are most suitable as a preventive rather than a curative strategy. Pathogens already present in the soil or infected plants cannot be eradicated using biofungicides. Moreover, the goal of a biofungicide is to keep a check on the pathogen size in soil and hinder smooth movement from one host to another. Thus the overall disease severity and disease incidence is reduced. Furthermore, the efficacy of a biofungicide depends on the formulation and storage conditions to retain the viability of the active microbial components (Miastkowska et al., 2020). Nonetheless, biofungicides have turned out to be a vital module in the integrated disease management program.

4. RNAi as a Biofungicide

RNAi is a natural gene silencing phenomenon observed in plants, animals and fungi (Zamore, 2001). This involves the formation of hairpin RNA structures post-transcription and processed by DICER (or DICER-like) proteins into fragments of 21–24 nucleotide sequence. These single-stranded DNA strands (small interfering RNA; siRNA) later get incorporated into the RNA-Induced Silencing Complex (RISC). The RISC with the help of the siRNA identifies complementary mRNA sequences in the cytosol and cleaves them bringing about Post-Transcriptional Gene Silencing (PTGS; Waterhouse et al., 2001). Gene silencing attained by inhibiting transcription activity is called Transcriptional Gene Silencing (TGS) and is associated with DNA methylation. Thus RNAi machinery fundamentally regulates gene expression, prevents transposon activity and hinders viral replication (Vaucheret and Fagard, 2001; Vaucheret et al., 2001; Bühler and Moazed, 2007). RNAi in plants is the prime antiviral defense strategy. This strategy has now been extensively studied to demonstrate its potential as a strong defense response against a multitude of plant pests and pathogens (Ghag, 2017).

The strategies for using RNAi for disease management include Host-Induced Gene Silencing (HIGS), Virus-Induced Gene Silencing (VIGS) and Spray-Induced Gene Silencing (SIGS; Fig. 1.1). HIGS encompass genetic engineering of host plants to express dsRNA and ultimately siRNAs that find entry into the pathogen during infection and target vital transcripts. This can be achieved by transforming the plants with a vector carrying vital fungal full-length genes or fragment/s arranged in opposite directions or driven by two promoters placed across it (Ghag et al., 2021; Fig. 1.2). VIGS make use of viral vectors carrying the pathogen gene sequences ranging from 300–800 base pairs (Lu et al., 2003). The virus used for infecting the host plants is modified so that that it retains infectivity but does not cause disease in plants. Several copies of RNAi signals are made during virus multiplication in the cell that is processed by the RNAi machinery which when ingested by fungal pathogens results in gene silencing. The response generated in VIGS is high, transient and spreads systemically along with the virus. Nevertheless, this response persists for a few weeks to months and then gradually decreases over time. SIGS involve spraying of dsRNAs and siRNAs that target essential pathogen genes on plant surfaces to provide protection. VIGS and SIGS strategies present short-lived resistance, however HIGS on the other hand endeavors durable resistance and does not require viral transmission (as in VIGS) or preparation of synthetic dsRNA or siRNAs (as in SIGS). In HIGS, host plants are engineered to produce dsRNAs or siRNAs and defend themselves against invading pathogens. To improve the efficacy and prevent the building up of resistance against this technology multiple

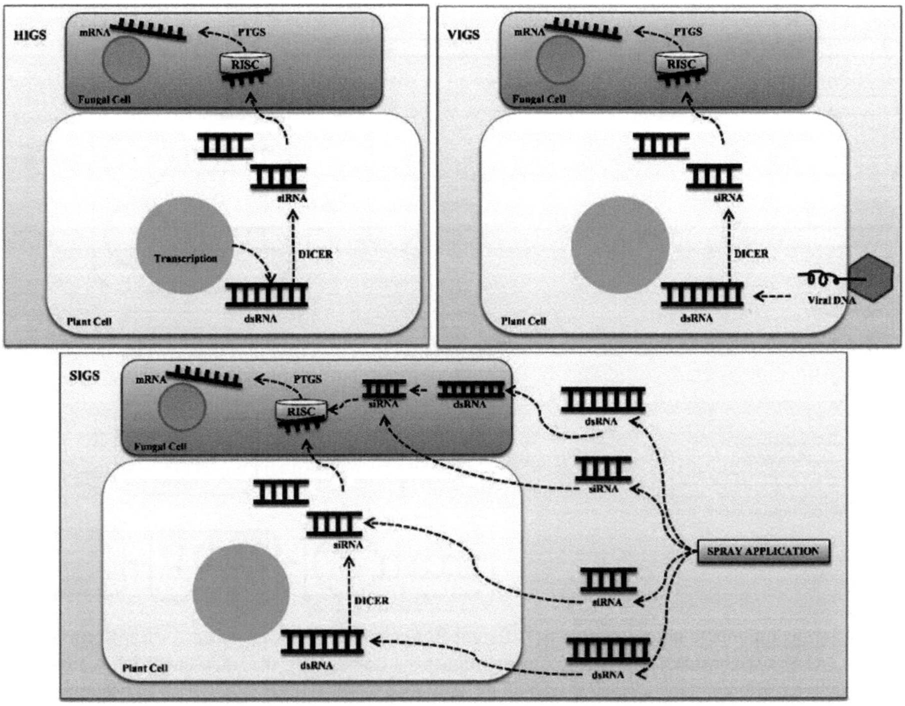

Fig. 1.1: Schematic representation of Host-induced gene silencing (HIGS), Viral-induced gene silencing (VIGS) and Spray-induced gene silencing (SIGS) employed in host plants for bringing about post-transcription gene silencing (PTGS) in fungal pathogen.

crucial pathogen genes can be targeted by stacking them together in the host plant. Furthermore, DNA fragments from different pathogens can be stacked together to offer protection against a multitude of pathogens (Fig. 1.2E).

Fungal phytopathogen (whether biotrophic or necrotroph) infects plants and receives nutrition from them. The nutrient material either in a simple or complex form is devoured by the fungi. Taking advantage of the conserved mechanism of RNAi in both plants and fungi, it is possible to deliver dsRNAs or siRNAs into fungal pathogens via plant tissues that target fungal-specific genes leading to its inhibition. The idea of RNAi as a defense strategy against fungi emerged when the uptake of dsRNA by nematodes leading to gene silencing was demonstrated (Urwin et al., 2002). The first evidence of the principle study establishing the use of RNAi to target fungal genes was carried out in transgenic tobacco plants expressing siRNAs to silence GUS gene expressed in the plant pathogenic fungi *Fusarium verticillioides* (Tinoco et al., 2010). Further Cai et al. (2018) reported naturally occurring siRNAs (e.g., ta1c-siR483, ta2-siR453) from *Arabidopsis* which are packed into extracellular vesicles (tetraspanin 8-associated exosomes) and silence *Botrytis cinerea* virulence genes. Targeting effector genes has been the choice as they are essential for pathogenesis and unique to the fungus. HIGS strategy was used to restrict powdery mildew fungus *Blumeria graminis* barley (*Hordeum vulgare*) and wheat (*Triticum*

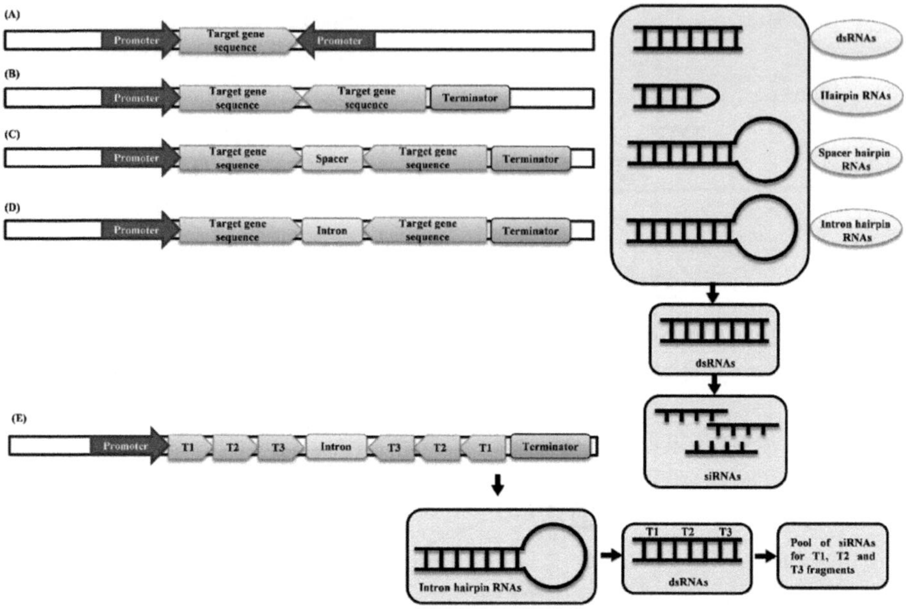

Fig. 1.2: Diagrammatic representation of different RNAi construct designs used for transforming host plants. (A) RNAi construct with two promoter sequences transcribing the intervening DNA from both the direction resulting in the formation of a double-stranded RNA. (B) Target DNA sequence arranged in tandem with opposite orientation transcribed by a single promoter to give hairpin RNA. (C) Construction of a hairpin RNA with an intervening spacer in between the target sequences. (D) RNAi constructs forming hairpin RNA structure with a splicable intron sequence in between. (E) Multiple target sequences for different genes from a given pathogen or different sequences targeting multiple genes from different pathogens aligned together to form a hairpin RNA. These hairpin RNAs formed are processed further to give double-stranded RNA (dsRNA) and small-inhibiting RNA (siRNA).

aestivum) by expressing dsRNA targeting the effector gene *Avra10*, *CSEP0081* and *CSEP0254* (Nowara, 2010; Ahmed et al., 2016). Fifty crucial effector proteins of powdery mildew fungus *Blumeria graminis* f. sp. *hordei* were identified using HIGS among which eight were found to be essential for infection. These include β-1,3 glucosyltransferases, metallo-proteases and microbial secreted ribonucleases that could be likely RNAi targets to generate transgenic barley resistance to powdery mildew (Pliego et al., 2013). HIGS was demonstrated to control white mold disease caused by *Sclerotinia sclerotiorum* by silencing essential effector genes *Ss-caF1* (putative Ca2+ binding protein), *SspG1d* (endopolygalacturonase) and *SsiTL* (integrin; Maximiano et al., 2022). Haustoria-specific genes in biotrophic fungus are unique and crucial for host-fungus communication. This is second in line as the best choice for targeting using RNAi. A number of haustoria specific genes were identified and shown to demonstrate loss of virulence when silenced. VIGS strategy was used to identify these genes and score the virulence response in suitable pathosystems (Yin et al., 2015). Chitinase and HXT1p hexose transporter gene was identified in *Puccinia striiformis* f. sp. *tritici* and silenced using *Barley stripe mosaic virus* (BSMV) in wheat (Yin et al., 2011). VIGS strategy (with BSMV) was used to develop resistance to stem rust or stripe rust fungus caused by *Puccinia triticina* in wheat plants engineered to silence pathogenicity genes such as mitogen-activated protein kinase 1 (*PtMAPK1*), cyclophilin (*PtCYC1*) or calcineurin B (*PtCNB*; Panwar et al., 2013). Pst_12806 is an essential haustorium-specific effector protein in *Puccinia striiformis* f. sp. *tritici* that is transported to the chloroplasts of wheat plants, interacts with the Rieske domain of TaISP and disturbs photosynthesis rate, ROS accumulation and defense response. Silencing either *Pst_12806* or *TaISP* in this pathosystem resulted in reduced *Puccinia* growth and disease development (Xu et al., 2019). Fungal-specific transcription factors, pathogenicity genes, toxin production genes and plant susceptibility factors are also likely targets for silencing to impart effective resistance in plants. Transgenic *Arabidopsis* and barley plants expressing a dsRNAs targeting fungal sterol cytochrome P450 lanosterol C-14α-demethylase (*CYP51*) gene exhibited resistance to Fusarium head blight disease (Koch et al., 2013). Similar results were obtained when *CYP51B* gene from *Fusarium oxysporum* was targeted by RNAi in transgenic soybean plants (Pérez et al., 2021). Transgenic wheat cultivars co-expressing three RNAi constructs with sequence mapped on three different regions of the chitin synthase (*Chs*) gene conferred resistance to both Fusarium head blight and Fusarium seedling blight disease (Cheng et al., 2015). Targeting crucial genes of *Fusarium oxysporum* f. sp. *cubense* namely velvet and Fusarium transcription factor 1 (*ftf1*) by HIGS strategy caused resistance in banana plants against Fusarium wilt disease (Ghag et al., 2014). Introduction of silencing constructs for genes namely, F-box protein Required for Pathogenicity1 (*FRP1*), *Fusarium oxysporum* Wilt 2 (*FOW2*) and homology to plant 12-oxophytodienoate-10, 11-reductase gene (*OPR*) in *Arabidopsis* resulted in enhanced resistance to *Fusarium oxysporum* f. sp. *conglutinans* (Hu et al., 2015). Transgenic wheat (*Triticum aestivum*) plants expressing dsRNA targeting a MAPK kinase (*PsFUZ7*) gene which is essential for regulating hyphal morphology and development in *Puccinia striiformis* f. sp. *tritici* exhibited protection against rust disease (Zhu et al., 2017). Other MAP kinase genes (RPMK1-1 and RPMK1-2) were targeted in *Rhizoctonia solani* to provide protection against sheath blight disease in transgenic rice (Tiwari et al., 2017). MoAP1, a bZIP transcription factor in *Magnaporthe oryzae* regulates downstream genes such as *MoAAT*, *MoSSADH* and *MoACT* that are essential for growth, development and pathogenicity. Transgenic rice lines expressing RNAi constructs targeting *MoAP1* gene showed improved blast disease resistance (Guo et al., 2019). Transgenic lettuce plants targeting cellulose synthase (CES1) genes of *Bremia lactucae*, the causative agent of downy mildew in lettuce, showed decreased *B. lactucae* growth and inhibited sporulation (Govindarajulu et al., 2015). Silencing a few plant genes (susceptibility factors) that are used by fungal pathogens for their infection strategy can protect the plants and hinder pathogen proliferation. *Triticum aestivum TaRac6* gene is known to get regulated under *Puccinia striiformis* f. sp. *tritici* infection. It displayed inhibition of Bax-induced cell death. Targeting *TaRac6* gene in wheat lines using VIGS provided resistance to *Puccinia* by modulating ROS response (Zhang et al., 2020). Transgenic tomato and potato plants expressing dsRNA targeting *Botrytis cinerea BcTOR* gene significantly inhibited gray mold disease (Xiong et al., 2019). Varied results were obtained when RNAi technology directed against oomycetes infection in plants was carried out. No GFP silencing was observed when GFP-tagged *Phytophthora parasitica* were allowed to colonize *Arabidopsis* lines expressing GFP RNAi constructs (Zhang et al., 2011). However, targeting the G protein β-subunit (*PiGPB1*) of *Phytophthora infestans* and glutathione-S-transferase gene of *Phytophthora parasitica* var. *nicotianae* in potato and tobacco plants respectively demonstrated reduced disease progression and increased resistance (Hernández et al., 2009; Jahan et al., 2015). In due course, several oomycetes effectors were identified that can suppress RNA silencing in plants and prevent the biogenesis of siRNAs (Qiao et al., 2013; Hou et al., 2019; Dunker et al., 2020).

Spray-induced gene silencing (SIGS) has become one of the prime technologies for crop disease management (Islam and Sherif, 2020). It is valued for its non-transgenic and environment-friendly nature. The other two strategies HIGS and VIGS have their limitations where SIGS has an advantage. HIGS and VIGS are restricted due to the transgenic character and available transformation methods. In addition, VIGS provide a transient resistance response and depends on the systemic movement of the virus. SIGS involve the exogenous application of dsRNAs or siRNAs targeting unique pathogen genes required for disease development presenting a unique potential as an innovative plant protection strategy. The efficacy of SIGS technology is dependent on several parameters that include uptake by plants, systemic translocation through the vascular system, amplification of RNAi signals and finally transmission to the interacting pathogen (Das and Sherif, 2020; Qiao et al., 2021). The sprayed dsRNAs are either taken up by the host plants or fungal pathogens and are processed by their respective RNAi machinery. An alternate mechanism was proposed by Zhao et al. (2021) wherein sRNAs can be transferred from host-associated beneficial microbes to the pathogens in the soil or on leaves, targeting virulence genes. This promotes an alternative non-transgenic approach for plant protection against pathogens. However, the RNAi signals processed by the fungal pathogen are found to be more effective and thus dsRNA application is more appropriate rather than siRNAs (Koch et al., 2016; Werner et al., 2020). The dsRNAs targeting cytochrome P450 lanosterol C-14α-demethylases genes (CYP3) of *Fusarium graminearum* when sprayed on barley leaves inhibited *Fusarium graminearum* growth (Koch et al., 2016). Multiple genes required for toxin biosynthesis, ROS response and cell cycle regulation in *Sclerotinia sclerotiorum* and *Botrytis cinerea* were used to make dsRNAs for topical applications and found to effectively control fungal growth on the leaves of *Arabidopsis thaliana* (McLoughlin et al., 2018). *Myo5* gene is involved in important cellular processes and virulence in *F. asiaticum*, *F. graminearum*, *F. tricinctum* and *F. oxysporum*, when targeted by spraying dsRNAs protected wheat coleoptiles against *F. asiaticum* (Song et al., 2018). SIGS was also potentially used for managing oomycetes infection in plants. Spraying of dsRNAs targeting *Phytophthora infestans* genes namely guanine-nucleotide binding protein b-subunit (*PiGPB1*), haustorial membrane protein (*PiHmp1*), cutinase (*PiCut3*) and endo-1,3(4)-b-glucanase (*PiEndo3*) on potato leaves with the pathogen resulted in inhibition of disease progression associated with sporulation defects and lower sporangia (Kalyandurg et al., 2021). Exogenous application of small RNAs (sRNAs) or dsRNAs targeting *Botrytis DCL1* and *DCL2* genes on the surfaces of tomato, grape, strawberry, lettuce, onion and rose decreased gray mold disease (Wang et al.,

2016). Application of dsRNA or antisense sRNA targeting cellulose synthase A3 (*CesA3*) gene of *Hyaloperonospora arabidopsidis* prevented downy mildew disease in *Arabidopsis* (Bilir et al., 2019). Recently SIGS was used to control sheath blight of rice by topical application of dsRNA targeting the vesicle trafficking pathway genes namely, DCTN1 (DYNACTIN1) and SAC1 (SUPPRESSOR OF ACTIN 1) or polygalacturonase genes of the basidiomycete fungus *Rhizoctonia solani* (Qiao et al., 2021). A combination of vesicle traffickings pathway genes such as vacuolar protein sorting 51 (VPS51), DCTN1 and SAC1 was targeted in *Botrytis cinerea* by spraying dsRNAs on lettuce leaves, tomato fruits, rose petals and grapefruits. All the treated plant tissues displayed a reduction in disease symptoms as compared to the controls (Qiao et al., 2021). Though SIGS is the most suitable biofungicide, especially for post-harvest protection, some technical glitches are waiting to be resolved. RNA in itself is very prone to degradation and has a short lifespan thus creating scope for the discovery of inert adjuvants that will hold and prevent RNAs from degradation. The use of nanomaterial or BioClay complexed with dsRNAs was found suitable in spray application and slow release of dsRNAs (Fletcher et al., 2020). To improve the efficiency and stability of RNAi and prevent degradation of RNAi signals, they can be encapsulated using nanomaterials which in turn also stimulate cellular uptake (de Oliveira Filho et al., 2021). Preparation of synthetic dsRNAs or sRNAs is expensive and demands the improvisation of cost-effective methods of preparation for field-scale application. For large-scale production and open-field applications, dsRNA was produced and encapsulated in *Escherichia coli*-derived anucleated minicells (ME-dsRNA). These ME-dsRNAs were designed to target chitin synthase class III (Chs3a, Chs3b) and DICER-like proteins (DCL1 and DCL2) genes of *Botryotinia fuckeliana* demonstrated significant fungal growth inhibition and prevented gray mold disease progression in strawberries (Islam et al., 2021). The effectiveness of RNAi depends on the uptake of RNAi signals into fungal cells, either from the environment or indirectly from plant cells (Šečić and Kogel, 2021).

5. Environment and human health associated risks

All technologies are advantageous to mankind but at the same time do carry a few risks. These risks can be managed appropriately or completely avoided by proper design and judicious application. Compared to other transgenic technologies RNAi-based technology provides clean and safe solutions for the management of plant diseases. In these plants only trans-RNA is either made or supplemented with no protein production, thereby reducing the trouble of GMO regulatory approvals. RNAs are quite unstable, incompetent to cross vascular and cellular barriers and are degraded in the gut to pose any risk associated with human or animal health (Chan and Snow, 2017; Chen et al., 2018). Moreover, even if dsRNA or sRNA find an entry into the human system it is less likely that it may find its complementary sequence to bind and bring about silencing. A study by Petrick et al. (2013) showed that even ingesting large amounts of dsRNAs and siRNAs that are completely homologous to the mouse *vATPase* gene (RNAi target for control of corn rootworms) did not affect the gene expression or physiological response. Plant sources also contain dsRNAs and siRNAs against several interacting viruses which may find entry into the human diet. Considerable amounts of dsRNA from virus-infected plant sources have been consumed for a long time without any detectable effects (Jensen et al., 2013).

The release of RNAi signals in the environment may pose risks to beneficial non-target species that are closely associated with the host plant or target organisms. If there is sequence homology in the target sequence between pathogen genes and genes from beneficial organisms it may lead to mortality. However, a few studies have demonstrated the off-target effects of RNAi in some different groups of the organism and ineffective against others even with high sequence homology. DsRNAs targeting *V-ATPase A* gene of Western Corn Rootworm (WCR) demonstrated ineffectiveness against Colorado potato beetles, cotton boll weevil, monarch butterfly and honey bees (Vélez et al., 2016; Pan et al., 2017; Fletcher et al., 2020). Thus the uptake and sensitivity to dsRNA in different species are complicated and require a case-by-case investigation approach. Moreover, depending on the sequence conservation the effect may be more pronounced and with reduced sequence homology, the effects are almost negligible. Thus, the thrust of targeting only pathogens without harming another organism in the vicinity lies in the unique sequence selection for RNAi. Further, unintended health or environmental effects of this technology are case-specific and needs independent investigations to show any potential effects of each of these RNA signals. The European Union (EU) and global sustainability policies are mainly aimed toward agricultural sustainability and food security with a significant reduction in harmful pesticides (Taning et al., 2021). Thus RNAi-based biofungicide is a most promising contender to support this policy with no risks and can substitute the presently used harmful pesticides. RNAi technology offers a clean and sustainable plant disease management solution with no significant human health or environmental impact.

6. Conclusion & future prospects

Plant fungal diseases are most devastating and result in huge economic losses. Conventionally fungal diseases were managed using chemical fungicides. However, these synthetic chemicals are detrimental to the health of different life forms including humans and are hazardous environmental pollutants (Fig. 1.3). Moreover, injudicious use of chemical fungicides has resulted in multiple mutant strains resistant to these chemicals making them ineffective. Innovative biocontrol strategies are warranted at this time that can effectively control fungal diseases and also pose no or minimal risk to health and the environment. Antagonistic organisms that compete, inhibit or improve plant potential for defense responses are used to reduce disease impact in the fields. RNAi technology is best-suited to generate host plants producing RNAi signals (dsRNA and siRNA) that have the potential to inhibit fungal invasion by silencing vital fungal-specific genes. RNAi technology (HIGS) has been proven highly effective against a range of fungal pathogens and prevented infections in crop plants. SIGS technology provides ready deployable topical application of RNAi molecules with negligible toxicity, species-specificity and nominal environmental

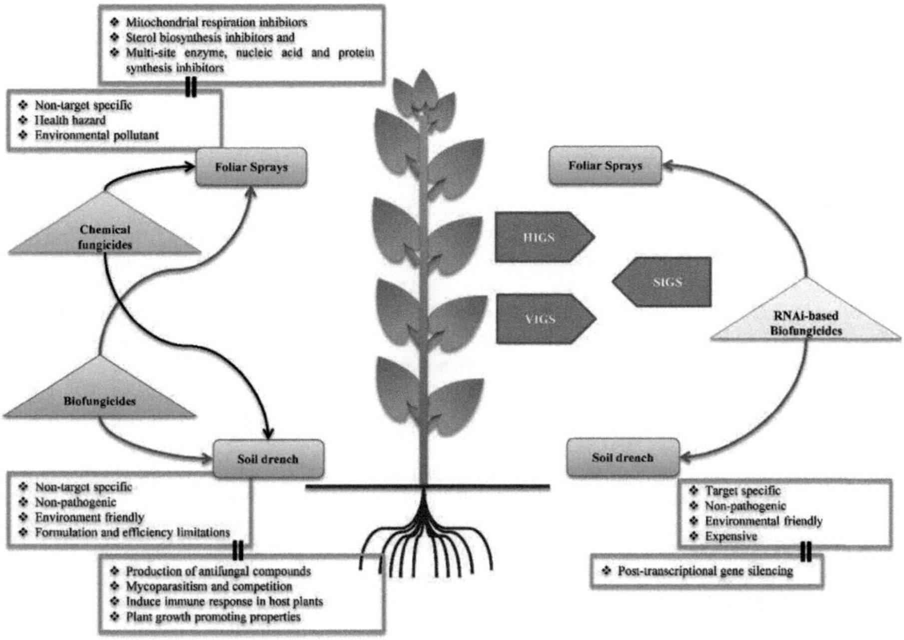

Fig. 1.3: Schematic representation depicting three methods of fungal disease management, namely chemical fungicides, biofungicides and RNAi-based biofungicides along with their mode of action, merits and demerits.

impact. In some countries, SIGS is considered as any other agricultural chemical product and is not subject to regulation by the GMO regulatory bodies. RNAi signals in the environment have a transient nature and are rapidly degraded to pose any severe impact on the off-target organisms. Due to its sequence-specific activity, it appears to be the best solution to target only a particular pathogen. Uptake of RNAi signals by humans through diet is poor and they are usually destroyed in the acidic gut, thereby preventing any significant deleterious health impacts. Closely related beneficial or off-target organisms may get affected if there appears to be a sequence similarity, but this may vary and needs further investigation. If the design and application are conducted meticulously, then RNAi can be the best solution to manage plant fungal diseases safely and effectively.

References

Ahmed, A.A., Pedersen, C. and Thordal-Christensen, H. (2016). The barley powdery mildew effector candidates CSEP0081 and CSEP0254 promote fungal infection success. PLoS One, 11(6): e0157586.

Avelino, J., Cristancho, M., Georgiou, S., Imbach, P., Aguilar, L., Bornemann, G., Läderach, P., Anzueto, F., Hruska, A.J. and Morales, C. (2015). The coffee rust crises in Colombia and Central America (2008–2013): Impacts, plausible causes and proposed solutions. Food Security, 7(2): 303–321.

Axelstad, M., Boberg, J., Nellemann, C., Kiersgaard, M., Jacobsen, P.R., Christiansen, S., Hougaard, K.S. and Hass, U. (2011). Exposure to the widely used fungicide mancozeb causes thyroid hormone disruption in rat dams but no behavioral effects in the offspring. Toxicological Sciences, 120(2): 439–446.

Bernabò, I., Guardia, A., Macirella, R., Tripepi, S. and Brunelli, E. (2017). Chronic exposures to fungicide pyrimethanil: Multi-organ effects on Italian tree frog (*Hyla intermedia*). Scientific Reports, 7(1): 1–6.

Bilir, Ö., Telli, O., Norman, C., Budak, H., Hong, Y. and Tör, M. (2019). Small RNA inhibits infection by downy mildew pathogen *Hyaloperonospora arabidopsidis*. Molecular Plant Pathology, 20(11): 1523–1534.

Bühler, M. and Moazed, D. (2007). Transcription and RNAi in heterochromatic gene silencing. Nature, 14: 1041–1048.

Cai, Q., Qiao, L., Wang, M., He, B., Lin, F.M., Palmquist, J., Huang, S.D. and Jin, H. (2018). Plants send small RNAs in extracellular vesicles to fungal pathogen to silence virulence genes. Science, 360(6393): 1126–1129.

Carvalho, F.P. (2017). Pesticides, environment, and food safety. Food and Energy Security, 6(2): 48–60.

Chan, S.Y. and Snow, J.W. (2017). Formidable challenges to the notion of biologically important roles for dietary small RNAs in ingesting mammals. Genes & Nutrition, 12(1): 13.

Chen, X., Mangala, L.S., Rodriguez-Aguayo, C., Kong, X., Lopez-Berestein, G. and Sood, A.K. (2018). RNA interference-based therapy and its delivery systems. Cancer and Metastasis Reviews, 37(1): 107–124.

Cheng, W., Song, X.S., Li, H.P., Cao, L.H., Sun, K., Qiu, X.L., Xu, Y.B., Yang, P., Huang, T., Zhang, J.B. and Qu, B. (2015). Host-induced gene silencing of an essential chitin synthase gene confers durable resistance to Fusarium head blight and seedling blight in wheat. Plant Biotechnology Journal, 13(9): 1335–1345.

Cohen, S.M., Gordon, E.B., Singh, P., Arce, G.T. and Nyska, A. (2010). Carcinogenic mode of action of folpet in mice and evaluation of its relevance to humans. Critical Reviews in Toxicology, 40(6): 531–545.

Corredor-Moreno, P. and Saunders, D.G. (2020). Expecting the unexpected: Factors influencing the emergence of fungal and oomycete plant pathogens. New Phytologist, 225(1): 118–125.

Das, P.R. and Sherif, S.M. (2020). Application of exogenous dsRNAs-induced RNAi in agriculture: Challenges and triumphs. Frontiers in Plant Science, 11: 946.

Dean, R., Van Kan, J.A., Pretorius, Z.A., Hammond-Kosack, K.E., Di Pietro, A., Spanu, P.D., Rudd, J.J., Dickman, M., Kahmann, R., Ellis, J. and Foster, G.D. (2012). The Top 10 fungal pathogens in molecular plant pathology. Molecular Plant Pathology, 13(4): 414–430.

de Oliveira Filho, J.G., Silva, G.D., Cipriano, L., Gomes, M. and Egea, M.B. (2021). Control of postharvest fungal diseases in fruits using external application of RNAi. Journal of Food Science, 86(8): 3341–3348.

Dodhia, K.N., Cox, B.A., Oliver, R.P. and Lopez-Ruiz, F.J. (2021). Rapid *in situ* quantification of the strobilurin resistance mutation G143A in the wheat pathogen *Blumeria graminis* f. sp. *tritici*. Scientific Reports, 11(1): 1–15.

Dong, O.X. and Ronald, P.C. (2019). Genetic engineering for disease resistance in plants: Recent progress and future perspectives. Plant Physiology, 180(1): 26–38.

Dunker, F., Trutzenberg, A., Rothenpieler, J.S., Kuhn, S., Pröls, R., Schreiber, T., Tissier, A., Kemen, A., Kemen, E., Hückelhoven, R. and Weiberg, A. (2020). Oomycete small RNAs bind to the plant RNA-induced silencing complex for virulence. Elife, 9: e56096.

Feng, Y., Huang, Y., Zhan, H., Bhatt, P. and Chen, S. (2020). An overview of strobilurin fungicide degradation: Current status and future perspective. Frontiers in Microbiology, 11: 389.

Fletcher, S.J., Reeves, P.T., Hoang, B.T. and Mitter, N. (2020). A perspective on RNAi-based biopesticides. Frontiers in Plant Science, 11: 51.

Fuller, D.Q. and Stevens, C.J. (2019). Between domestication and civilization: The role of agriculture and arboriculture in the emergence of the first urban societies. Vegetation History and Archaeobotany, 28(3): 263–282.

Ghag, S.B. (2017). Host induced gene silencing, an emerging science to engineer crop resistance against harmful plant pathogens. Physiological and Molecular Plant Pathology, 100: 242–254.

Ghag, S.B. (2021). RNAi strategy for management of phytopathogenic fungi. In CRISPR and RNAi Systems (pp. 535–550). Elsevier.

Ghag, S.B., Shekhawat, U.K. and Ganapathi, T.R. (2014). Host-induced post-transcriptional hairpin RNA-mediated gene silencing of vital fungal genes confers efficient resistance against Fusarium wilt in banana. Plant Biotechnology Journal, 12(5): 541–553.

Ghag, S.B., Shekhawat, U.K. and Ganapathi, T.R. (2015). Fusarium wilt of banana: Biology, epidemiology and management. International Journal of Pest Management, 61(3): 250–263.

Glazebrook, J. (2005). Contrasting mechanisms of defense against biotrophic and necrotrophic pathogens. Annual Review in Phytopathology, 43: 205–227.

Govindarajulu, M., Epstein, L., Wroblewski, T. and Michelmore, R.W. (2015). Host-induced gene silencing inhibits the biotrophic pathogen causing downy mildew of lettuce. Plant Biotechnology Journal, 13(7): 875–883.

Guo, X.Y., Li, Y., Fan, J., Xiong, H., Xu, F.X., Shi, J., Shi, Y., Zhao, J.Q., Wang, Y.F., Cao, X.L. and Wang, W.M. (2019). Host-induced gene silencing of *MoAP1* confers broad-spectrum resistance to *Magnaporthe oryzae*. Frontiers in Plant Science, 10: 433.

Hahn, M. (2014). The rising threat of fungicide resistance in plant pathogenic fungi: *Botrytis* as a case study. Journal of Chemical Biology, 7(4): 133–141.

Hedin, P.A. (1982). New concepts and trends in pesticide chemistry. Journal of Agricultural and Food Chemistry, 30(2): 201–215.

Hernández, I., Chacón, O., Rodriguez, R., Portieles, R., López, Y., Pujol, M. and Borrás-Hidalgo, O. (2009). Black shank resistant tobacco by silencing of glutathione S-transferase. Biochemical and Biophysical Research Communications, 387(2): 300–304.

Hou, Y., Zhai, Y.I., Feng, L.I., Karimi, H.Z., Rutter, B.D., Zeng, L., Choi, D.S., Zhang, B., Gu, W., Chen, X. and Ye, W. (2019). A Phytophthora effector suppresses trans-kingdom RNAi to promote disease susceptibility. Cell Host & Microbe, 25(1): 153–165.

Hu, Z., Parekh, U., Maruta, N., Trusov, Y. and Botella, J.R. (2015). Down-regulation of *Fusarium oxysporum* endogenous genes by host-delivered RNA interference enhances disease resistance. Frontiers in Chemistry, 3: 1.

Ishii, H., Fraaije, B.A., Sugiyama, T., Noguchi, K., Nishimura, K., Takeda, T., Amano, T. and Hollomon, D.W. (2001). Occurrence and molecular characterization of strobilurin resistance in cucumber powdery mildew and downy mildew. Phytopathology, 91(12): 1166–1171.

Islam, M.T. and Sherif, S.M. (2020). RNAi-based biofungicides as a promising next-generation strategy for controlling devastating gray mold diseases. International Journal of Molecular Sciences, 21: 2072.

Islam, M.T., Davis, Z., Chen, L., Englaender, J., Zomorodi, S., Frank, J., Bartlett, K., Somers, E., Carballo, S.M., Kester, M. and Shakeel, A. (2021) Minicell-based fungal RNAi delivery for sustainable crop protection. Microbial Biotechnology, 14(4): 1847–1856.

Jahan, S.N., Åsman, A.K., Corcoran, P., Fogelqvist, J., Vetukuri, R.R. and Dixelius, C. (2015). Plant-mediated gene silencing restricts growth of the potato late blight pathogen *Phytophthora infestans*. Journal of Experimental Botany, 66(9): 2785–2794.

Jensen, P.D., Zhang, Y., Wiggins, B.E., Petrick, J.S., Zhu, J., Kerstetter, R.A., Heck, G.R. and Ivashuta, S.I. (2013). Computational sequence analysis of predicted long dsRNA transcriptomes of major crops reveals sequence complementarity with human genes. GM Crops & Food, 4(2): 90–97.

Kalyandurg, P.B., Sundararajan, P., Dubey, M., Ghadamgahi, F., Zahid, M.A., Whisson, S.C. and Vetukuri, R.R. (2021). Spray-induced gene silencing as a potential tool to control potato late blight disease. Phytopathology, 111(12): 2168–2175.

Kara, M., Oztas, E., Ramazanoğulları, R., Kouretas, D., Nepka, C., Tsatsakis, A.M. and Veskoukis, A.S. (2020). Benomyl, a benzimidazole fungicide, induces oxidative stress and apoptosis in neural cells. Toxicology Reports, 7: 501–509.

Koch, A., Biedenkopf, D., Furch, A., Weber, L., Rossbach, O., Abdellatef, E., Linicus, L., Johannsmeier, J., Jelonek, L., Goesmann, A. and Cardoza, V. (2016). An RNAi-based control of *Fusarium graminearum* infections through spraying of long dsRNAs involves a plant passage and is controlled by the fungal silencing machinery. PLoS Pathogens, 12(10): e1005901.

Koch, A., Kumar, N., Weber, L., Keller, H., Imani, J. and Kogel, K.H. (2013). Host-induced gene silencing of cytochrome P450 lanosterol C14α-demethylase–encoding genes confers strong resistance to Fusarium species. Proceedings of the National Academy of Sciences, 110(48): 19324–19329.

Köhl, J., Kolnaar, R. and Ravensberg, W.J. (2019). Mode of action of microbial biological control agents against plant diseases: Relevance beyond efficacy. Frontiers in Plant Science, 10: 845.

Lodovica, M., Tinivella, F., Garibaldi, A., Kemmit, G.M., Bacci, L. and Sheppard, B. (2010). Mancozeb: Past, Present, and Future. Journal of Plant Disease, 94(9): 1076–1087.

Lu, R., Martin-Hernandez, A.M., Peart, J.R., Malcuit, I. and Baulcombe, D.C. (2003). Virus-induced gene silencing in plants. Methods, 30(4): 296–303.

Mair, W.J., Deng, W., Mullins, J.G., West, S., Wang, P., Besharat, N., Ellwood, S.R., Oliver, R.P. and Lopez-Ruiz, F.J. (2016). Demethylase inhibitor fungicide resistance in *Pyrenophora teres* f. sp. *teres* associated with target site modification and inducible overexpression of *Cyp51*. Frontiers in Microbiology, 7: 1279.

Maximiano, M.R., Santos, L.S., Santos, C., Aragão, F.J.L., Dias, S.C., Franco, O.L. and Mehta, A. (2022). Host induced gene silencing of *Sclerotinia sclerotiorum* effector genes for the control of white mold. Biocatalysis and Agricultural Biotechnology, 102302.

McLoughlin, A.G., Wytinck, N., Walker, P.L., Girard, I.J., Rashid, K.Y., de Kievit, T., Fernando, W.G., Whyard, S. and Belmonte, M.F. (2018). Identification and application of exogenous dsRNA confers plant protection against *Sclerotinia sclerotiorum* and *Botrytis cinerea*. Scientific Reports, 8(1): 7320.

Mekapogu, M., Jung, J.A., Kwon, O.K., Ahn, M.S., Song, H.Y. and Jang, S. (2021). Recent progress in enhancing fungal disease resistance in ornamental plants. International Journal of Molecular Sciences, 22(15): 7956.

Miastkowska, M., Michalczyk, A., Figacz, K. and Sikora, E. (2020). Nanoformulations as a modern form of biofungicide. Journal of Environmental Health Science and Engineering, 18(1): 119–128.

Nicolopoulou-Stamati, P., Maipas, S., Kotampasi, C., Stamatis, P. and Hens, L. (2016). Chemical pesticides and human health: The urgent need for a new concept in agriculture. Frontiers in Public Health, 4: 148.

Nowara, D., Gay, A., Lacomme, C., Shaw, J., Ridout, C., Douchkov, D., Hensel, G., Kumlehn, J. and Schweizer, P. (2010). HIGS: Host-induced gene silencing in the obligate biotrophic fungal pathogen *Blumeria graminis*. The Plant Cell, 22(9): 3130–3141.

Padmanabhan, S.Y. (1973). The great Bengal famine. Annual Review of Phytopathology, 11(1): 11–24.

Pan, H., Yang, X., Bidne, K., Hellmich, R.L., Siegfried, B.D. and Zhou, X. (2017). Dietary risk assessment of v-ATPase A dsRNAs on monarch butterfly larvae. Frontiers in Plant Science, 8: 242.

Panwar, V., McCallum, B. and Bakkeren, G. (2013). Host-induced gene silencing of wheat leaf rust fungus *Puccinia triticina* pathogenicity genes mediated by the *Barley stripe mosaic virus*. Plant Molecular Biology, 81(6): 595–608.

Pérez, C.E., Cabral, G.B. and Aragão, F.J. (2021). Host-induced gene silencing for engineering resistance to *Fusarium* in soybean. Plant Pathology, 70(2): 417–425.

Petrick, J.S., Moore, W.M., Heydens, W.F., Koch, M.S., Sherman, J.H. and Lemke, S.L. (2015). A 28-day oral toxicity evaluation of small interfering RNAs and a long double-stranded RNA targeting vacuolar ATPase in mice. Regulatory Toxicology and Pharmacology, 71(1): 8–23.

Pimentel, D. (1996). Green revolution agriculture and chemical hazards. Science of the Total Environment, 188: S86–S98.

Pliego, C., Nowara, D., Bonciani, G., Gheorghe, D.M., Xu, R., Surana, P., Whigham, E., Nettleton, D., Bogdanove, A.J., Wise, R.P. and Schweizer, P. (2013). Host-induced gene silencing in barley powdery mildew reveals a class of ribonuclease-like effectors. Molecular Plant Microbe Interactions, 26(6): 633–642.

Qiao, L., Lan, C., Capriotti, L., Ah-Fong, A., Sanchez, J.N., Hamby, R., Heller, J., Zhao, H., Glass, N.L., Judelson, H.S. and Mezzetti, B. (2021). Spray-induced gene silencing for disease control is dependent on the efficiency of pathogen RNA uptake. Plant Biotechnology Journal, 19(9): 1756.

Qiao, Y., Liu, L., Xiong, Q., Flores, C., Wong, J., Shi, J., Wang, X., Liu, X., Xiang, Q., Jiang, S. and Zhang, F. (2013). Oomycete pathogens encode RNA silencing suppressors. Nature Genetics, 45(3): 330–333.

Ristaino, J.B., Anderson, P.K., Bebber, D.P., Brauman, K.A., Cunniffe, N.J., Fedoroff, N.V., Finegold, C., Garrett, K.A., Gilligan, C.A., Jones, C.M. and Martin, M.D. (2021). The persistent threat of emerging plant disease pandemics to global food security. Proceedings of the National Academy of Sciences, 118(23): e2022239118.

Savary, S., Willocquet, L., Pethybridge, S.J., Esker, P., McRoberts, N. and Nelson, A. (2019). The global burden of pathogens and pests on major food crops. Nature Ecology & Evolution, 3(3): 430–439.

Šečić, E. and Kogel, K.H. (2021). Requirements for fungal uptake of dsRNA and gene silencing in RNAi-based crop protection strategies. Current Opinion in Biotechnology, 70: 136–142.

Shahid, M., Ahmed, B., Zaidi, A. and Khan, M.S. (2018). Toxicity of fungicides to *Pisum sativum*: A study of oxidative damage, growth suppression, cellular death and morpho-anatomical changes. RSC Advances, 8(67): 38483–38498.

Song, X.S., Gu, K.X., Duan, X.X., Xiao, X.M., Hou, Y.P., Duan, Y.B., Wang, J.X., Yu, N. and Zhou, M.G. (2018). Secondary amplification of siRNA machinery limits the application of spray-induced gene silencing. Molecular Plant Pathology, 19(12): 2543–2560.

Sood, M., Kapoor, D., Kumar, V., Sheteiwy, M.S., Ramakrishnan, M., Landi, M., Araniti, F. and Sharma, A. (2020). *Trichoderma*: The "secrets" of a multitalented biocontrol agent. Plants, 9(6): 762.

Taning, C.N., Mezzetti, B., Kleter, G., Smagghe, G. and Baraldi, E. (2021). Does RNAi-based technology fit within EU sustainability goals? Trends in Biotechnology, 39: 644–647.

Tillman, R.W., Siegel, M.R. and Long, J.W. (1973). Mechanism of action and fate of the fungicide chlorothalonil (2, 4, 5, 6-tetrachloroisophthalonitrile) in biological systems: I. Reactions with cells and subcellular components of *Saccharomyces pastorianus*. Pesticide Biochemistry and Physiology, 3(2): 160–167.

Tinoco, M.L.P., Dias, B., Dall'Astta, R.C., Pamphile, J.A. and Aragão, F.J. (2010). *In vivo* trans-specific gene silencing in fungal cells by in planta expression of a double-stranded RNA. BMC Biology, 8: 27.

Tiwari, I.M., Jesuraj, A., Kamboj, R., Devanna, B.N., Botella, J.R. and Sharma, T.R. (2017). Host delivered RNAi, an efficient approach to increase rice resistance to sheath blight pathogen (*Rhizoctonia solani*). Scientific Reports, 7(1): 7521.

Urwin, P.E., Lilley, C.J. and Atkinson, H.J. (2002). Ingestion of double-stranded RNA by preparasitic juvenile cyst nematodes leads to RNA interference. Molecular Plant Microbe Interaction, 15(8): 747–752.

Vaucheret, H. and Fagard, M. (2001). Transcriptional gene silencing in plants. Targets, inducers and regulators. Trends in Genetics, 17: 29–35.

Vaucheret, H., Beclin, C. and Fagard, M. (2001). Post-transcriptional gene silencing in plants. Journal of Cell Science, 114: 3083–3091.

Vélez, A.M., Jurzenski, J., Matz, N., Zhou, X., Wang, H., Ellis, M. and Siegfried, B.D. (2016), Developing an *in vivo* toxicity assay for RNAi risk assessment in honey bees, *Apis mellifera* L. Chemosphere, 144: 1083–1090.

Vielba-Fernández, A., Polonio, Á., Ruiz-Jiménez, L., de Vicente, A., Pérez-García, A. and Fernández-Ortuño, D. (2020). Fungicide resistance in powdery mildew fungi. Microorganisms, 8(9): 1431.

Wang, M., Weiberg, A., Lin, F.M., Thomma, B.P., Huang, H.D. and Jin, H. (2016). Bidirectional cross kingdom RNAi and fungal uptake of external RNAs confer plant protection. Nature Plants, 2(10): 16151.

Waterhouse, P.M., Wang, M.B. and Lough, T. (2001). Gene silencing as an adaptive defence against viruses. Nature, 411: 834–842.

Werner, B.T., Gaffar, F.Y., Schuemann, J., Biedenkopf, D. and Koch, A.M. (2020). RNA-spray-mediated silencing of *Fusarium graminearum AGO* and *DCL* genes improve barley disease resistance. Frontiers in Plant Science, 11: 476.

Xiong, F., Liu, M., Zhuo, F., Yin, H., Deng, K., Feng, S., Liu, Y., Luo, X., Feng, L., Zhang, S. and Li, Z. (2019). Host-induced gene silencing of *BcTOR* in *Botrytis cinerea* enhances plant resistance to grey mould. Molecular Plant Pathology, 20: 1722–1739.

Xu, Q., Tang, C., Wang, X., Sun, S., Zhao, J., Kang, Z. and Wang, X. (2019). An effector protein of the wheat stripe rust fungus targets chloroplasts and suppresses chloroplast function. Nature Communications, 10(1): 5571.

Yang, L.N., He, M.H., Ouyang, H.B., Zhu, W., Pan, Z.C., Sui, Q.J., Shang, L.P. and Zhan, J. (2019). Cross-resistance of the pathogenic fungus *Alternaria alternata* to fungicides with different modes of action. BMC Microbiology, 19(1): 205.

Yin, C., Downey, S.I., Klages-Mundt, N.L., Ramachandran, S., Chen, X., Szabo, L.J., Pumphrey, M. and Hulbert, S.H. (2015). Identification of promising host-induced silencing targets among genes preferentially transcribed in haustoria of *Puccinia*. BMC Genomics, 16(1): 579.

Yin, C., Jurgenson, J.E. and Hulbert, S.H. (2011). Development of a host-induced RNAi system in the wheat stripe rust fungus *Puccinia striiformis* f. sp. *tritici*. Molecular Plant-Microbe Interactions, 24: 554–561.

Yoshida, K., Schuenemann, V.J., Cano, L.M., Pais, M., Mishra, B., Sharma, R., Lanz, C., Martin, F.N., Kamoun, S., Krause, J. and Thines, M. (2013). The rise and fall of the *Phytophthora infestans* lineage that triggered the Irish potato famine. Elife, 2: e00731.

Zamore, P.D. (2001). RNA interference: listening to the sound of silence. Nature Structural Biology, 8(9): 746–750.

Zhang, M., Wang, Q., Xu, K., Meng, Y., Quan, J. and Shan, W. (2011). Production of dsRNA sequences in the host plant is not sufficient to initiate gene silencing in the colonizing oomycete pathogen Phytophthora parasitica. PLoS One, 6: e28114.

Zhang, Q., Zhang, X., Zhuang, R., Wei, Z., Shu, W., Wang, X. and Kang, Z. (2020). TaRac6 is a potential susceptibility factor by regulating the ROS burst negatively in the wheat-*Puccinia striiformis* f. sp. *tritici* interaction. Frontiers in Plant Science, 11: 716.

Zhao, J.H., Zhang, T., Liu, Q.Y. and Guo, H.S. (2021). Trans-kingdom RNAs and their fates in recipient cells: Advances, utilization, and perspectives. Plant Communications, 2: 100167.

Zhu, X., Qi, T., Yang, Q., He, F., Tan, C., Ma, W., Voegele, R.T., Kang, Z. and Guo, J. (2017). Host-induced gene silencing of the MAPKK gene PsFUZ7 confers stable resistance to wheat stripe rust. Plant Physiology, 175(4): 1853–1863.

Zubrod, J.P., Bundschuh, M., Arts, G., Brühl, C.A., Imfeld, G., Knäbel, A., Payraudeau, S., Rasmussen, J.J., Rohr, J., Scharmüller, A. and Smalling, K. (2019). Fungicides: An overlooked pesticide class? Environmental Science & Technology, 53(7): 3347–3365.

Chapter 2

Antifungal Efficacy of Algae

Aruna Jyothi Kora[1,2]

1. Introduction

Algae are eukaryotic, predominantly aquatic, photosynthetic organisms with sizes ranging from mm to m in diameter and length. They are the fastest growing organisms with higher photosynthetic efficiency than other plants. They occur in diverse habitats including, freshwater, seawater, desert, snow and on the surfaces of rocks, soil, stones, other plants and animals. These autotrophic organisms reproduce in sexual, asexual and vegetative modes (Ibrahim et al., 2017). Based on the structure, they are categorized under microalgae (unicellular) and macroalgae (multicellular) (Ahirwar et al., 2022). Depending on the type of coloring pigment, storage reserve product; flagellar details, cell wall biochemical composition and cell and thallus morphology; they are classified under various classes. These are chlorophyceae (green algae), rhodophyceae (red algae), phaeophyceae (brown algae), cyanophyceae (blue-green algae), xanthophyceae (yellow-green algae), chrysophyceae (golden algae), bacillariophyceae (yellow or golden brown algae), dinophyceae (dinoflagellates), cryptophyceae, chloromonadineae and euglenineae. The marine macroalgae are generally called seaweeds comprised of green algae, red algae and brown algae which are macroscopic and multicellular (Abdussalam, 1990; Ibrahim et al., 2017; Mandrekar et al., 2019; Mickymaray and Alturaiki, 2018; Righini et al., 2022).

Important marine green algae include, *Ulva* spp (*Ulva lactuca, U. prolifera, U. reticulata, U. rigida, U. fasciata*) (Barreto et al., 1997; Karthik et al., 2020; Kumari et al., 2017; Mickymaray and Alturaiki, 2018; Thavasi Alagan et al., 2020), *Caulerpa* spp (*C. serrulata, Caulerpa racemosa, C. scalpelliformis, C. filiformis*) (Barreto et al., 1997; Kumari et al., 2017; Lavanya and Veerappan, 2012; Liu et al., 2013), *Halimeda gracilis* (Bhagyaraj and Kunchithapatham, 2016), *Spirulina platensis* (Andreea et al., 2010), *Codium* spp (*C. decorticatum, C. tomentosum*) (Fayzi et al., 2012), etc.,

[1] National Centre for Compositional Characterisation of Materials (NCCCM), Bhabha Atomic Research Centre (BARC) ECIL PO, Hyderabad-500 062, India.

[2] Homi Bhabha National Institute (HBNI) Anushakti Nagar, Mumbai-400 094, India.

Email: koramaganti@gmail.com, koraaj@barc.gov.in

(Fig. 2.1) and are abundant in shallow water and tide pools. Some examples of marine red algae are *Hypnea* spp (*H. musciformis*, *H. spicifera*) (Barreto et al., 1997; Saim et al., 2021; Souza et al., 2018), *Porphyra* spp (*P. vietnamensis*, *P. umbilicalis*) (Bhatia et al., 2015; De Corato et al., 2017), *Laurencia* spp (*L. paniculata*, *L. catarinensis*) (Mickymaray and Alturaiki, 2018; Stein et al., 2011), *Kappaphycus alvarezii* (Banu et al., 2020), *Asparagopsis armata* (Messahli et al., 2021), *Osmundaria serrata* (Barreto et al., 1997), *Plocamium cartilagineum* (red algae) (Martorell et al., 2021), *Gelidium Sesquipedale* (El Wahidi et al., 2015), *Galaxaura rugosa*, *Liagora hawaiiana* (Al-Enazi et al., 2018), *Gracilariopsis* spp (*G. persica*, *G. cervicornis*) (Pourakbar et al., 2021; Sampaio et al., 2022), *Bostrychia tenella* (Kumar et al., 2020), *Gymnogongrus turquetii* (Martorell et al., 2021), *Chondracanthus teedei* (Soares et al., 2016), etc. (Fig. 2.2). While marine brown algae is composed of *Sargassum* spp (*S. wightii*, *S. tenerrimum*, *S. vulgare*, *S. muticum*) (Ammar et al., 2017; Banu et al., 2020; Kumar et al., 2020; Lavanya and Veerappan, 2012; Manivannan et al., 2011; Nofal et al., 2022; Veeragoni et al., 2018), *Turbinaria* spp (*T. ornata*, *T. conoides*) (Karthik et al., 2020; Lavanya and Veerappan, 2012; Manivannan et al., 2011), *Padina* spp (*P. tetrastromatica*, *P. gymnospora*, *P. boryana*) (Abhishek et al., 2021; Manivannan et al., 2011; Veeragoni et al., 2022), *Colpomenia sinuosa* (Monla et al., 2020), *Cystoseira* spp (*C. barbata*, *C. tamariscifolia*, *C. compressa*, *C. brachycarpa*)

Fig. 2.1: Photographs of some of the green marine algae, (a) *Caulerpa serrulata* and (b) *Halimeda gracilis.*

Fig. 2.2: Photographs of some of the red marine algae, (a) *Acanthophora spicifera* and (b) *Gracilaria edulis.*

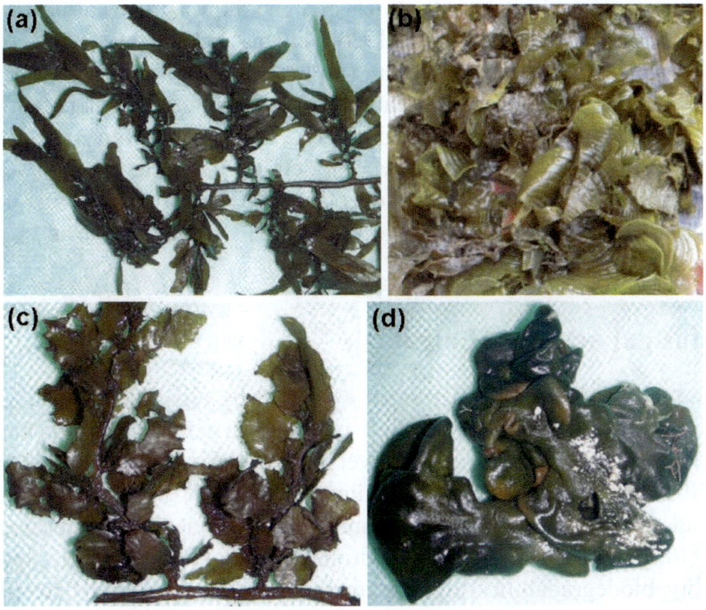

Fig. 2.3: Photographs of some of the brown marine algae, (a) *Sargassum wightii*, (b) *Padina tetrastromatica*, (c) *Turbinaria ornata* and (d) *Colpomenia sinuosa*.

(Bennamara et al., 1999; El Wahidi et al., 2015; Sellimi et al., 2017), *Laminaria digitata* (De Corato et al., 2017), *Gracilaria* spp (*G. edulis*, *G. foliifera*) (Algotiml et al., 2022; Kumari et al., 2017), *Ascophyllum nodosum* (Procházka et al., 2022), *Bifurcaria bifurcate* (El Wahidi et al., 2015), *Dictyopteris membranacea* (Akremi et al., 2017), *Undaria pinnatifida* (De Corato et al., 2017), *Alaria esculenta* (Andreea et al., 2010), *Zonaria toumefoni* (Barreto et al., 1997), etc. (Fig. 2.3).

2. Applications of algae

Algae and seaweeds are used for a range of applications in the fields of agriculture, food, pharmaceutical, animal husbandry, cosmetics, etc., due to the innate availability of diverse biomolecules. The extracts of seaweeds are used as biostimulants and liquid fertilizers in agriculture for enhancing plant nutrition, growth, productivity, crop quality, abiotic and biotic stress tolerance, plant disease resistance, etc. (Karthik et al., 2020; Kora, 2022b). Seaweed coats are utilized as biopreservatives for the shelf life extension of perishable fruits and vegetables (Banu et al., 2020). Edible seaweeds are employed as gelling agents, thickeners, stabilizers, potassium and iodine sources, etc., in food and pharma industries in addition to their uses as functional food/ nutraceuticals (Abdussalam, 1990; Mohamed et al., 2012). Green algae such as *Ulva* and other seaweeds are used as feeding supplements for livestock such as sheep, horses, cattle, rabbits, birds, etc., (Nofal et al., 2022; Thavasi Alagan et al., 2020). Algae are used in the manufacturing of cosmetics due to their photoprotective, skin whitening/antimelanogenic, antioxidant, antiacne, anti-wrinkling, antifungal and antiallergic activities (Senevirathne and Kim, 2013).

The aqueous and other solvent extracts of algae show a broad variety of biological properties including, antibacterial (Akremi et al., 2017), antifungal (Andreea et al., 2010), trypanocidal, antileishmanial (de Felicio et al., 2010), antiviral (Nabeta et al., 2021), antimalarial, antiulcer, anti-inflammatory, anticoagulant, antithrombin, antidiabetic, antiobesitic, antiangiogenic, antituberculotic, antialzheimer, hepatoprotective, radioprotective, hypolipidemic, antifeedant (Ahirwar et al., 2022; Ibrahim et al., 2017; Mohamed et al., 2012), wound healing, insecticidal (Abdussalam, 1990), antioxidant (Sellimi et al., 2017; Senevirathne and Kim, 2013), anticancer activities, etc. (Souza et al., 2018).

3. Antifungal activity of algae

Among the diverse biological activities exhibited by algal extracts, the antifungal property is of relevance owing to their potential in controlling and management of different plant diseases caused by diverse fungi, affecting the global food chain supply (El-Hossary et al., 2017; Ibrahim et al., 2017; Mogoşanu et al., 2016). Biofungicides from algae are gaining momentum as a sustainable alternative to chemical fungicides due to their environmental safety, minimal phytotoxicity, renewability, biodegradability, low cost, practical/field applicability, etc. (dos Santos Gomes et al., 2021; Vicente et al., 2021).

3.1 Important phytopathogenic fungi

Along with useful, edible fungi such as mushrooms (Kora, 2020), there are many pathogenic fungi which cause major diseases in various crops and plants. Some important phytopathogenic fungi are *Alternaria* spp (*A. alternata, A. solani, A. dauci, A. longipes*) (Andreea et al., 2010), *Fusarium* spp (*F. oxysporum, F. roseum, F. solani*) (Lavanya and Veerappan, 2012; Senousy et al., 2022), *Aspergillus* spp (*A. niger, A. flavus, A. parasiticus, A. fumigatus, A. furigate, A. tetreus)* (Manivannan et al., 2011), *Cladosporium* spp (*C. cladosporioides, C. sphaerospermum*) (Martorell et al., 2021), *Penicillium* spp (*P. digitatum, P. expansum*) (Bahammou et al., 2021; Manivannan et al., 2011), *Helminthosporium turcicum*, *Magnaporthe grisea* (Pourakbar et al., 2021), *Macrophomina phaseolina* (Karthik et al., 2020), *Rhizoctonia* spp (*R. solani, R. bataticola*) (Kora, 2019a, 2019b; Kora et al., 2020), *Sclerotium* spp (*S. rolfsii, S. oryzae*) (Karthik et al., 2020), *Sclerotinia sclerotiorum* (Kora, 2021b), *Verticillium alboatrum* (Bennamara et al., 1999), *Colletotrichum falcatum* (Kora, 2021b), *Syncephalastrum racemosum* (Al-Enazi et al., 2018), *Botrytis cinerea* (Boček et al., 2013; Kora, 2022a), *Monilinia laxa* (De Corato et al., 2017), *Pythium ultimum* (Kora, 2022b), *Pseudoperonospora humuli* (Procházka et al., 2022), *Phytophthora infestans* (Kora, 2022a), etc. (Fig. 2.4). Phytopathogenic fungi attack and cause various diseases such as blast, rust, blight, wilt, anthracnose, necrosis, rot, damping off, leaf spot, etc., on a myriad of crops such as cereals, millets, legumes, oil seeds, cash crops, fruits, vegetables, ornamental plants, trees, etc. (Karthik et al., 2020; Kora, 2022a, 2022b). Important human pathogenic fungi include *Candida albicans, C. tropicalis, C. glabrata, Geotricum candidum,*

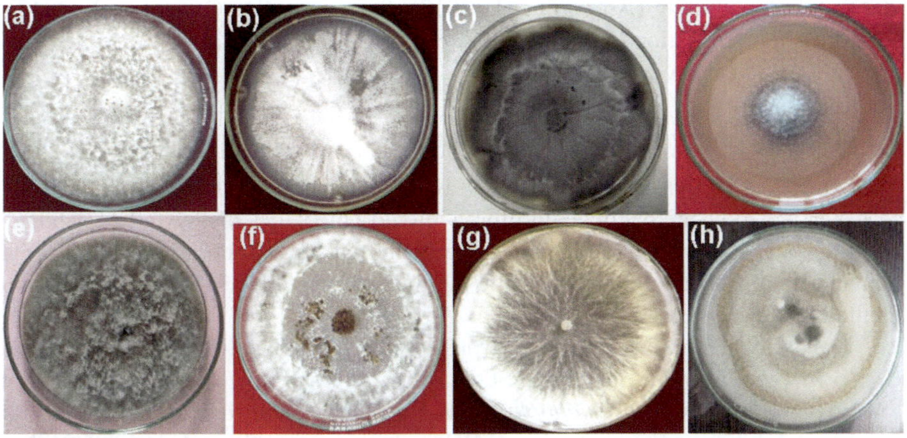

Fig. 2.4: Images of some important phytopathogenic fungi grown in petriplates, (a) *Alternaria alternata*, (b) *Fusarium oxysporum*, (c) *Helminthosporium* sp, (d) *Magnaporthe grisea*, (e) *Macrophomina* sp, (f) *Rhizoctonia solani*, (g) *Sclerotium rolfsii* and (h) *Colletotrichum falcatum*.

Cryptococcus neoformans, *A. niger*, *Mucor*, *Paecilomyces*, etc. (Kora, 2021b; Mickymaray and Alturaiki, 2018).

3.2 Methodologies used for antifungal evaluation

The fungistatic and fungicidal action of chemical and biological antifungal agents and fungicides is studied in a broad range of *in vitro* and *in vivo* assays. The *in vitro* methods include solid agar techniques such as disk diffusion (Messahli et al., 2021; Saim et al., 2021), well diffusion (Karthik et al., 2020), poison food technique (Kora, 2019a, 2019b; Kora et al., 2020) and the liquid broth assays consisting of micro broth dilution (Al-Enazi et al., 2018), broth dilution (Kora, 2022c; Kora et al., 2020); conidial/spore germination (Kora, 2022a), sclerotia germination (Kora, 2021b), optical microscopy-based micro culture technique (De Corato et al., 2017), TLC plate assay (de Felicio et al., 2010), detached leaf assay (Kora et al., 2020), etc. (Fig. 2.5). While *in vivo* assays include foliar spray (Kora, 2021a; Procházka et al.,

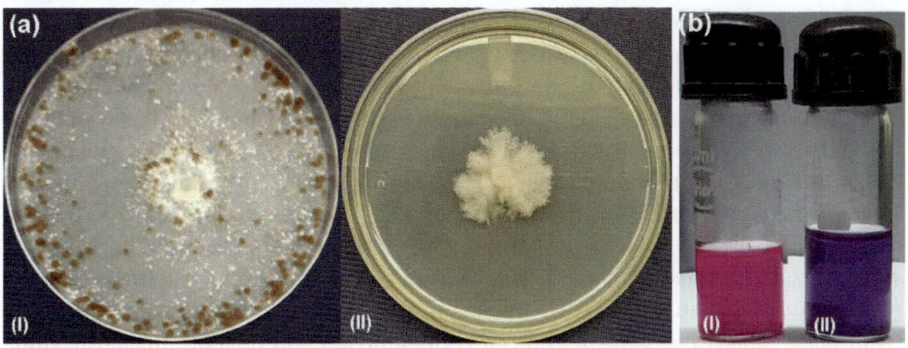

Fig. 2.5: *In vitro* antifungal evaluation methodologies used against *Rhizoctonia solani*, (a) poison food technique and (b) broth dilution method at (I) 0 and (II) 10 µg/mL of propiconazole.

2022), tuber wounding (Ammar et al., 2017), fruit spray (De Corato et al., 2017), seed dressing, fertigation (Kora, 2021a), seedling immersion (Kora, 2022b), etc., as preventive and curative treatments.

3.3 Antifungal evaluation of algal extracts

The extracts of various microalgae and seaweeds prepared in different solvents such as water, ethanol, methanol, chloroform, dichloromethane, diethyl ether, hexane, ethyl acetate, etc., were used for checking the antifungal action against diverse human pathogenic and phytopathogenic fungi, employing diverse *in vitro* and *in vivo* methods (Table 2.1).

The prenylated para-xylene, caulerprenylols B (2) isolated from the green algae *Caulerpa racemosa* of Zhanjiang St coastline, China showed antifungal action towards *Candida glabrata, Trichophyton rubrum* and *Cryptococcus neoformans* with Minimum Inhibitory Concentration (MIC) values ranging from 4–16 µg/mL (Liu et al., 2013). The dichloromethane extract of the red algae, *Asparagopsis armata* from the Mediterranean Coast of Algeria showed anticandidal action towards two strains of *C. albicans* using well and disk diffusion and micro dilution assays. The inhibition zones of 20–24 mm and 55 mm were observed at 3 mg/disk and 12 mg/well, respectively. The MIC values ranged from 0.58–2.34 mg/mL. The lipophilic extract abundant in brominated compounds and fatty acids was accountable for the antifungal action of *A. armata* (Messahli et al., 2021). The kappa carrageenan of *Hypnea musciformis* (red algae) collected from Fleicheiras beach, Trairí, Ceará, Brazil demonstrated antifungal action towards *C. albicans* in micro broth dilution assay and the IC_{50} value was 147.3 µg/mL (Souza et al., 2018). The sulfated polysaccharide of *Porphyra vietnamensis* (red algae) obtained from the Arabian Sea Coast of Maharashtra, India was studied for its antifungal action against *C. albicans* in well diffusion assay and the IC_{50} values ranged from 2.7–4.9 µg/mL and 4.2–5.3 µg/mL, respectively for cold ethanol and hot alkaline extracts (Bhatia et al., 2015). The fungicidal action of chloroform extract of *Laurencia catarinensis* (red algae) collected from Santo state, Southeastern Brazil was studied against three pathogenic fungi, *C. albicans* and *C. neoformans* by micro broth dilution method. The MIC values were 303.8 and 600.4 µg/mL, respectively (Stein et al., 2011). The methanol and dichloromethane extracts of *Dictyopteris membranacea* (brown algae) from the Mediterranean Sea, Monastir Coast of Tunisia exhibited antifungal activity towards *C. albicans* with an inhibition zone of 12 mm at 15 µg/disc in the disk diffusion method. While the micro broth dilution method demonstrated MIC and Minimum Fungicidal Concentration (MFC) values of 125 and 500 µg/mL, respectively. The extract was abundant in phenolics, tannins and flavonoids and also displayed inhibition zones of 10–18.3 mm and the MIC and Minimum Bactericidal Concentration (MBC) values of 62.5–500 µg/mL and 125–500 µg/mL, respectively towards *Staphylococcus aureus*, *Streptococcus agalactiae*, *Bacillus subtilis*, *Enterococcus faecium*, *E. faecalis*, *Escherichia coli* and *Salmonella typhimurium* (Akremi et al., 2017). The ethanolic extracts of *Bifurcaria bifurcate*, *Cystoseira compressa*, *C. brachycarpa* (brown algae) and dichloromethane extract of *Gelidium Sesquipedale* (red algae) collected from the Atlantic Coast of El-Jadida

Table 2.1: A comparative table showing the antifungal action of seaweeds in terms of the scientific name, native habitat, extraction solvent, test phytopathogenic fungi, selected method and fungal growth inhibition.

Seaweed	Native habitat	Extraction solvent	Test phytopathogenic fungi	Assay used	Fungal growth inhibition	Reference
Gracilariopsis persica (red algae)	Suru of Bandar Abbas, Iran	Methanol	*Botrytis cinerea* *Aspergillus niger* *Penicillium expansum* *Pyricularia oryzae*	Poison food technique	Mycelial growth inhibition (100%) at 12.5 mg/mL	(Pourakbar et al., 2021)
Sargassum vulgare (brown algae)	Ischia Island, Italy	Ethanol	*A. parasiticus* *Fusarium oxysporum* *P. expansum*	Disk diffusion method	Inhibition zone values ranged from 15–20, 13.5–17 and 18–23 mm at 7.5 mg/disc	(Kumar et al., 2020)
Plocamium cartilagineum (red algae)	King George Island, South Shetlands, Antarctica	Hexane	*Penicillium* sp *Cladosporium cladosporioides* *Phialocephala* sp *Antarctomyces psychrotrophicus*	Well diffusion method	Inhibition zone values of 10.6, 11.7, 7.2 and 10.5 mm at 1 mg/well	(Martorell et al., 2021)
Gymnogongrus turquetii (red algae)	King George Island, South Shetlands, Antarctica	Hexane	*Penicillium* sp *C. cladosporioides* *Phialocephala* sp *A. psychrotrophicus*	Well diffusion method	Inhibition zone values of 10.5, 17, 1 and 7.7 mm at 1 mg/well.	(Martorell et al., 2021)
Cystoseira tamariscifolia (brown algae)	Rose Marie, near Rabat, Morocco	Diethyl ether	*B. cinerea,* *F. oxysporum* sp *mycopersici* *Verticillium alboatrum*	Spot inoculation method	Inhibition zone values of 39, 36 and 38 mm at 20 µg/plate	(Bennamara et al., 1999)

Table 2.1 contd. ...

...Table 2.1 contd.

Seaweed	**Native habitat**	**Extraction solvent**	**Test phytopathogenic fungi**	**Assay used**	**Fungal growth inhibition**	**Reference**
C. barbata (brown algae)	Kerkennah Island, Tunisia	Water	*F. oxysporum* *Alternaria solani*	Well diffusion method	Inhibition zone values of 16 and 8 mm at 2.5 mg/well	(Sellimi et al., 2017)
C. barbata (brown algae)	Kerkennah Island, Tunisia	Water	*F. oxysporum* *A. solani*	Micro broth dilution method	MIC values of 80 and 160 mg/mL	(Sellimi et al., 2017)
S. vulgare (brown algae)	Tunisian Coast	Water	*Pythium aphanidermatum*	Poison food technique	Mycelial growth inhibition (28%) at 50 mg/mL	(Ammar et al., 2017)
S. vulgare (brown algae)	Tunisian Coast	Methanol	*P. aphanidermatum*	Poison food technique	Mycelial growth inhibition (51%) at 50 mg/mL	(Ammar et al., 2017)
S. vulgare (brown algae)	Tunisian Coast	Water	*P. aphanidermatum*	Tuber wounding	*Pythium* disease reduction in potato (34.3%) at 50 mg/mL	(Ammar et al., 2017)
S. vulgare (brown algae)	Tunisian Coast	Methanol	*P. aphanidermatum*	Tuber wounding	*Pythium* disease reduction in potato (82%) at 50 mg/mL	(Ammar et al., 2017)
Turbinaria ornata (brown algae)	Mandapam Coast, Gulf of Mannar, India	Water	*Rhizoctonia solani* *Macrophomina phaseolina* *Sclerotium rolfsii* *A. solani*	Well diffusion method	Inhibition zone values of 15, 19, 18 and 22 mm at 20 mg/well	(Karthik et al., 2020)
Ulva reticulata (green algae)	Mandapam Coast, Gulf of Mannar, India	Water	*R. solani* *M. phaseolina* *S. rolfsii* *A. solani*	Poison food technique	Mycelial growth inhibition (65.6, 59.2, 92.7 and 91.1%) at 40 mg/mL	(Karthik et al., 2020)

Alginure-composed of algal extract (24%) from *Ascophyllum nodosum* and *Laminaria* sp (brown algae)	Commercial, liquid, algal extract	Water	*Pseudoperonospora humuli*	Foliar spray (5 times) at 7–14 day interval	*Pseudoperonospora humuli* mediated downy mildew disease suppression (80–94%)at a dosage of 3.5 L/ha in 600 L/ha spraying volume at 1%	(Procházka et al., 2022)
Laminaria digitata *Undaria pinnatifida* (brown algae)	Marine biorefinery	Supercritical carbon dioxide	*B. cinerea* *Monilinia laxa*	Poison food technique	Mycelial growth inhibition (100%) at 30 g/L	(De Corato et al., 2017)
L. digitata *U. pinnatifida* (brown algae)	Marine biorefinery	Supercritical carbon dioxide	*B. cinerea* *M. laxa*	Microscopy based microculture technique	Conidial germination inhibition (100%) at 30 g/L	(De Corato et al., 2017)
L. digitata *U. pinnatifida* (brown algae)	Marine biorefinery	Supercritical carbon dioxide	*B. cinerea*	Fruit wounding	Strawberry gray mold disease inhibition (73–85, 45–66%) at a dose of 30 g/L in preventive and curative treatments	(De Corato et al., 2017)
L. digitata *U. pinnatifida* (brown algae)	Marine biorefinery	Supercritical carbon dioxide	*M. laxa*	Fruit wounding	Peach brown rot disease inhibition (66–82, 44–70%) at a dose of 30 g/L in preventive and curative treatments	(De Corato et al., 2017)
Spirulina platensis (green algae) *Alaria esculenta* (brown algae)	Commercial source	Water-ethanol	*F. roseum* *F. oxysporum* *A. alternata* *A. dauci* *A. longipes* *B. cinerea*	Poison food technique	EC_{90} value was 20 mg/mL	(Andreea et al., 2010)
Gracilaria edulis (brown algae)	Mandapam Coast, Tamil Nadu, India	Methanol	*Nomuraea rileyi*	Well diffusion method	Inhibition zone values of 16, 16 and 13.6 mm at 3 mg/mL	(Kumari et al., 2017)

Table 2.1 contd. ...

...Table 2.1 contd.

Seaweed	Native habitat	Extraction solvent	Test phytopathogenic fungi	Assay used	Fungal growth inhibition	Reference
C. filiformis, *U. rigida* (green algae) *Zonaria toumefoni* (brown algae) *Hypnea spicifera* and *Osmundaria serrata* (red algae)	KwaZulu-Natal Coast, South Africa	Ethanol	*R. solani*	Poison food technique	Mycelial growth inhibition (%) were 57.8, 59.1, 74.9, 69.4 and 71% at 1 mg/mL	(Barreto et al., 1997)
C. filiformis, *U. rigida* (green algae) *Z. toumefoni* (brown algae) *H. spicifera* and *O. serrata* (red algae)	KwaZulu-Natal Coast, South Africa	Ethanol	*Verticillium*	Poison food technique	Mycelial growth inhibition (%) were 29.5, 32.3, 64.9, 34.7 and 77.8% at 1 mg/mL	(Barreto et al., 1997)
Kappaphycus alvarezii (red algae) *S. tenerrimum* (brown algae)	Commercial source	Water	*A. flavus* *A. furigate*	Disk diffusion method	Inhibition zone values of 6 and 5 mm and 12 and 11 mm at 40 μg/mL	(Banu et al., 2020)
Codium decorticatum *C. scalpelliformis* (green algae) *S. wightii* and *T. conoides* (brown algae)	Mandapam Coast, Gulf of Mannar, Tamil Nadu, India	Acetone, methanol, chloroform, ethyl acetate	*R. solani* *B. cinerea* *F. udum*	Disk diffusion method	Inhibition zone values of 5–10 mm at 5 mg/mL	(Lavanya and Veerappan, 2012)

province, Morocco showed antifungal action towards *C. albicans* and *C. neoformans* in terms of inhibition zones (15–20 mm) in the disk diffusion method at 500 μg/disk (El Wahidi et al., 2015).

The antifungal action of *Galaxaura rugosa* and *Liagora hawaiiana* (red algae) obtained from Alharra, Umluj, Red Seashore, Kingdom of Saudi Arabia was tested against an array of fungal pathogens using well diffusion and micro broth dilution methods. The ethanolic extract of *G. rugosa* showed inhibition zone values of 21, 15, 18, 17, 20, 15, 14 and 14 mm at 5 mg/well against *Aspergillus fumigatus*, *A. niger*, *C. albicans*, *C. tropicalis*, *C. neoformans*, *Geotricum candidum*, *Penicillium expansum*, *Syncephalastrum racemosum*, respectively. The corresponding MIC values were 1.2, 5, 2.5, 2.5, 1.2, 5, 10 and 10 mg/mL. In the case of chloroform extract of *L. hawaiiana*, the respective inhibition zone and MIC values were 16, 16, 22, 27, 25, 21, 20 and 15 mm and 5, 10, 0.625, 0.078, 0.312, 0.625. 1.25 and 5 mg/mL against the tested fungi (Al-Enazi et al., 2018). The methanolic extract of *Gracilariopsis persica* (red algae) collected from Suru of Bandar Abbas, Iran caused complete mycelial growth inhibition (100%) of plant pathogenic fungi, *Botrytis cinerea*, *A. niger*, *P. expansum* and *Pyricularia oryzae* in the poison food technique at 12.5 mg/mL. The fungicidal action was attributed to phenolics (rosmarinic acid, quercetin) and fatty acids (palmitic acid, oleic acid) and other extracted compounds (Pourakbar et al., 2021). The hexane and dichloromethane fractions of *Bostrychia tenella* (red algae) collected from the Coast of Ilha das Couves, Ubatuba, São Paulo, Brazil inhibited the growth of phytopathogenic fungi *Cladosporium cladosporioides* and *C. sphaerospermum* in TLC plate assay. These fractions were composed of fatty acids, low molecular weight hydrocarbons, esters and steroids (de Felicio et al., 2010). The ethanolic extract of *Sargassum vulgare* (brown algae) collected from Ischia Island, Italy was studied against *A. parasiticus*, *Fusarium oxysporum*, *P. expansum*, *Trichoderma harzianum* and *C. albicans* in the disk diffusion assay at 7.5 mg/disk. The inhibition zone values ranged from 15–20, 13.5–17, 18–23, 6–6.5 and 14–15 mm, respectively (Kumar et al., 2020).

The hexane extracts of *Plocamium cartilagineum* and *Gymnogongrus turquetii* (red algae) collected from King George Island, South Shetlands, Antarctica were studied for their antifungal action against *Penicillium* sp, *C. cladosporioides*, *Phialocephala* sp and *Antarctomyces psychrotrophicus*, in the well diffusion method at 1 mg/well. The *P. cartilagineum* showed respective inhibition zone values of 10.6, 11.7, 7.2 and 10.5 mm. The *G. turquetii* extract showed corresponding values of 10.5, 17, 1 and 7.7 mm (Martorell et al., 2021). The carrageenans extracted from *Chondracanthus teedei* var *lusitanicus* (red algae) caused swelling, branching and shortening of hyphal segments and reduction in chitin content of the cell wall in *Alternaria infectoria* and *A. fumigatus* on exposure at 100–150 μg/mL, confirming the strong antifungal action (Soares et al., 2016). The ethanolic extracts of *Ulva prolifera* (green algae) and *Laurencia paniculata* (red algae) obtained from the Red Sea Coast of Yanbu, Saudi Arabia were studied towards fungi isolated from bronchial asthmatic patients; *C. albicans*, *A. niger*, *Mucor* sp and *Paecilomyces* sp. The corresponding inhibition zone values of 17.3, 16, 17 and 16.6 mm; and 17.6, 16.3, 17 and 17.3 mm were obtained from the well diffusion assay, respectively. The MIC and

MFC values were 125–500 μg/mL and 125–1000 μg/mL, respectively based on the broth dilution method. The extracts were composed of terpene alcohol, diterpene, steroids, sesquiterpene and sesquiterpene alcohols (Mickymaray and Alturaiki, 2018).

The ether extract of *C. tamariscifolia* (brown algae) from Rose Marie, near Rabat, Morocco exhibited antifungal action towards tomato pathogenic fungi, *B. cinerea*, *F. oxysporum* sp *mycopersici* and *Verticillium alboatrum*. The inhibition zone diameters were 39, 36 and 38 mm, respectively at 20 μg/plate in the spot inoculation method and the active meroditerpenoid in the extract was identified as methoxybifurcarenone. The methoxybifurcarenon at 15 μg/disk also exhibited antibacterial action towards *Agrobacterium tumefaciens* and *E. coli* with corresponding inhibition zone values of 17 and 15 mm in the disk diffusion assay (Bennamara et al., 1999). The methanolic extract of *Turbinaria conoides*, *Padina gymnospora* and *S. tenerrimum* (brown algae) from Vedalai coastal waters, Gulf of Mannar, India displayed antifungal action in the disk diffusion method against the fungi, *A. niger*, *A. flavus*, *A. tetreus*, *Penicillium* sp, *C. albicans*, *C. glabrata* and *C. neoformans* at 6 μL/disk. The corresponding inhibition zones for *T. conoides* extract were 4.6, 14.6, 17, 18, 18, 8 and 13.3 mm. For *P. gymnospora* extract, the corresponding inhibition zones were 13.6, 12, 18, 16.3, 14.3, 14 and 20 mm. In the case of *S. tenerrimum* extract, the recorded inhibition zones were 5.6, 12.3, 11.6, 13.6, 15, 12.3 and 14 mm (Manivannan et al., 2011). The polyphenolic-protein-polysaccharide ternary conjugate of *Cystoseira barbata* (brown algae) collected from Kerkennah Island, Tunisia demonstrated antifungal action against *F. oxysporum* and *A. solani* in good diffusion and micro broth dilution methods. The inhibition zone and MIC values were 16 and 8 mm at 2.5 mg/well; and 80 and 160 mg/mL, respectively for the tested fungal strains (Sellimi et al., 2017). The aqueous and methanol extracts of *S. vulgare* (brown algae) obtained from the Tunisian Coast were tested towards *Pythium aphanidermatum* at 50 mg/mL. The noted growth inhibitions (%) in the poison food technique were 28 and 51%, respectively and the antifungal action was attributed to phenolic acids and flavonoids such as ascorbic acid, chlorogenic acid, rosmarinic acid, resorcinol, isorhamnetin and apigenin. The *Pythium* leak severity/disease in potatoes was reduced by 34.3 and 82% by the respective extracts (Ammar et al., 2017).

The 40% aqueous extract of *Turbinaria ornata* (brown algae) and *Ulva reticulata* (green algae) from the Mandapam Coast, Gulf of Mannar, India demonstrated antifungal action towards plant pathogenic fungi including, *Rhizoctonia solani*, *Macrophomina phaseolina*, *Sclerotium rolfsii* and *A. solani* in well diffusion and poison food techniques. The noted inhibition zones were 15, 19, 18 and 22 mm and the mycelial growth inhibition was 65.6, 59.2, 92.7 and 91.1% for the respective fungal strains at 20 mg/well and at 40 mg/mL of extract concentrations. In addition, the extract served as a liquid fertilizer in increasing the seed germination (%) of *Raphanus sativus* (Radish), *Phaseolus vulgaris* (Green Pea) and *Vigna radiata* (Mung). Further, the foliar spray enhanced the root and shoots length; and levels of carbohydrate, protein, amino acid, phenolics, chlorophyll and carotenoid pigments in leaves (Karthik et al., 2020). The repeated foliar spray (five times) of Alginure at 1% on *Humulus lupulus* (Hop) at 7–14 d intervals in the dosage of 3.5 L/ha in 600 L/ha spraying volume showed the preventive effect on downy mildew caused by *Pseudoperonospora humuli* and substantially suppressed (80–94%) the disease development. Alginure is a commercial, liquid, algal preparation composed of algal extract (24%) from *Ascophyllum nodosum* and *Laminaria* sp (brown algae); plant amino acids (7%) and phosphates (20%) (Procházka et al., 2022).

The antifungal action of crude extracts produced by supercritical carbon dioxide extraction technique from brown (*Laminaria digitata, Undaria pinnatifida*) and red (*Porphyra umbilicalis*) algae against the fruit postharvest pathogen fungi, *B. cinerea, Monilinia laxa, P. digitatum* was studied. In strawberry fruit, the causative agent for gray mold is *B. cinerea*. The brown rot in the peach fruit is induced by *M. laxa*. In the case of lemon fruit, the green mold is caused by *P. digitatum*. The extracts of *L. digitata* and *U. pinnatifida* caused complete (100%) inhibition of mycelial growth and conidial germination of *B. cinerea* and *M. laxa* at 30 g/L *in vitro* assays such as poison food and optical microscopy-based microculture techniques, respectively. For *P. digitatum*, the mycelial growth inhibition (%) and conidial germination ratio (%) ranged from 50–88 and 12–50%, respectively by the three extracts. In *in vivo* studies, the algal extracts strongly diminished the gray mold on strawberries (73–85, 45–66%), brown rot on peaches (66–82, 44–70%) and green mold on lemons (51–60, 34–53%) at a concentration of 30 g/L in preventive and curative treatments, in terms of fruit decay inhibition and disease severity reduction, respectively. The inhibition could be attributed to combined effects of fatty acid toxicity and polysaccharide elicited peroxidase mediated systemic resistance (De Corato et al., 2017).

The hydroalcoholic extracts of *Spirulina platensis* (green algae) and *Alaria esculenta* (brown algae) which were obtained from commercial sources showed inhibitory action against *F. roseum, F. oxysporum, A. alternata, A. dauci, A. longipes, B. cinerea* in the agar dilution method and the EC_{90} value was around 20 mg/mL (Andreea et al., 2010). The biocidal action of methanolic extracts of *Gracilaria edulis* (brown algae) and *U. fasciata, Caulerpa serrulata* (green algae) collected from the Mandapam Coast, Tamil Nadu, India was studied for silkworm pathogenic fungus, *Nomuraea rileyi* at 3 mg/mL in the well diffusion assay. The inhibition zones were 16, 16 and 13.6 mm and the fungal growth was inhibited *in vitro* with efficient silkworm rearing of 79, 75 and 69%, respectively (Kumari et al., 2017). The ethanolic extracts of *C. filiformis, U. rigida* (green algae), *Zonaria toumefoni* (brown algae) and *Hypnea spicifera* and *Osmundaria serrata* (red algae) collected from KwaZulu-Natal Coast; South Africa hindered the hyphal growth of *R. solani* and *Verticillium* in the poison food technique. The growth inhibitions (%) for *R. solani* were 57.8, 59.1, 74.9, 69.4 and 71% and for *Verticillium* the values were 29.5, 32.3, 64.9, 34.7 and 77.8% at 1 mg/mL (Barreto et al., 1997). The aqueous seaweed extracts of *Kappaphycus alvarezii* (red algae) and *S. tenerrimum* (brown algae) from commercial sources showed inhibitory action towards *A. flavus, A. furigate* and *C. albicans* in the disk diffusion method at 40 µg/mL. The inhibition zone values were 6, 5 and 6 mm and 12, 11 and 8 mm, respectively. The antifungal activity was attributed to seaweed phytonutrients such as flavonoids, steroids, saponins and tannins. Further, the coating of *K. alvarezii* extract at 3% maintained the fruit quality of tomato in terms of weight loss, texture, total acidity, ascorbic acid and juice contents for a storage period of 30 d. Thus, the seaweed extract coating serves as a natural preservative for the shelf life extension of fruits and vegetables (Banu et al., 2020). The different solvent (acetone, methanol, chloroform, ethyl acetate) extracts of *Codium decorticatum*, *C. scalpelliformis* (green algae), *S. wightii* and *Turbinaria conoides* (brown algae) obtained from the Mandapam Coast, Gulf of Mannar, Tamil Nadu, India were tested against *C. albicans*, *C. krusei*, *R. solani*, *B. cinerea* and *F. udum* employing the disk diffusion method. The inhibition zone values ranged from 5–10 mm at 5 mg/mL (Lavanya and Veerappan, 2012).

4. Biochemical composition of algae and mechanism of antifungal action

Algae are enriched with various bioactive compounds and secondary metabolites due to their exposure to different environmental and climatic conditions such as salinity, light, temperature and diverse marine chemicals. They include sulfated polysaccharides (ulvans, alginate, agar, carrageenan), polyphenolics, flavonoids, cardiac glycosides, phlorotannins, phycobiliproteins, sterols, polyhydroxyalkonates, alkaloids, mono/poly unsaturated fatty acids, bromophenol derivatives, anthraquinones, sesquiterpenoid quinines, terpenes, diterpenoids, triterpenoids, tannins, saponins, alkaloids, amino acids, proteins, minerals, vitamins, alcohols, lipids, steroids, carotenoids, chlorophylls, phycobilins, cyclic depsipeptides, hydrocarbons, oxygen heterocycles, carbonyl compounds, etc. (Ahirwar et al., 2022; Boukhatem et al., 2021; Ibrahim et al., 2017; Mandrekar et al., 2019; Mickymaray and Alturaiki, 2018; Mogoşanu et al., 2016; Reis et al., 2018; Vicente et al., 2021). The algal extracts are commercially available in the market as liquid and dry powder formulations (Boček et al., 2013; de Borba et al., 2022; Vicente et al., 2021).

The different mechanisms involved in the antifungal action of various algal extracts against diverse phytopathogenic fungi are complex and not entirely understood (dos Santos Gomes et al., 2021; Vicente et al., 2021). The biocontrol modes of fungal phytopathogens can be categorized as direct and indirect differences between the antagonistic agent and the phytopathogenic fungi. The important direct antagonistic routes include antibiosis, enzyme, phytohormone and secondary metabolite production; parasitism, nutrient and space competition, etc. The indirect route is the induction of Systemic Acquired Resistance (SAR) in plants via gene expression, enzymatic activity, metabolite accumulation, lignin biosynthesis, etc. (Fig. 2.6) (dos Santos Gomes et al., 2021; Kora, 2022b; Righini et al., 2022).

The important targets for any antifungal agent are the external barriers such as the cell wall, cell membrane; and intracellular organelles including, nucleic acids and mitochondria. After crossing the external barriers, the fungicides inhibit protein synthesis, replication and respiratory pathways. In addition, different algal compounds stimulate natural plant defense mechanisms against microbes (dos Santos Gomes et al., 2021; Vicente et al., 2021). Algal extracts are abundant sources of hydrolytic enzymes such as chitinases, glucanase, xylanases, amylase, cellulase, etc.,

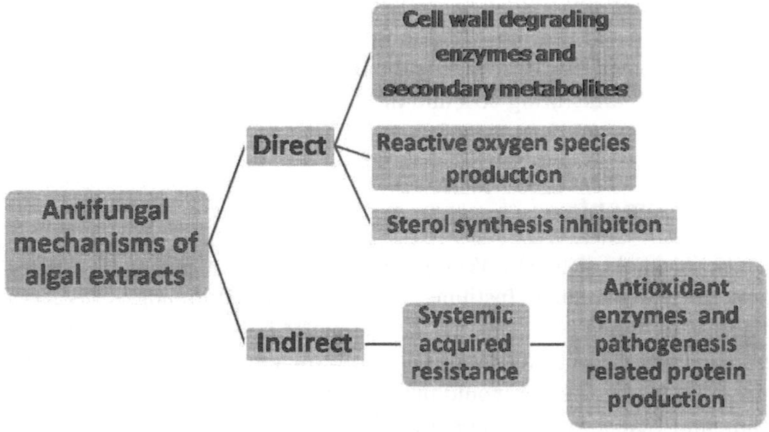

Fig. 2.6: The different mechanisms involved in the antifungal action of various algal extracts.

and can degrade the cell wall (Senousy et al., 2022). The carrageenans in red algae *Chondracanthus teedei* induced swelling of fungal hyphal filaments by acting on fungal cell wall by decreasing the chitin content in *A. fumigatus* (Soares et al., 2016). The antiviral red algal lectin caused cell death in *C. albicans* through disruption of cell wall integrity and induction of Reactive Oxygen Species (ROS) (Nabeta et al., 2021). Some of the algal extracts are known to inhibit the synthesis of sterol, the backbone component of the fungal membrane; enhance the membrane fluidity/permeability via incorporation of fungal fatty acids into the fungal membrane, trigger oxidative stress and lead to ultimate cell death (dos Santos Gomes et al., 2021; Vicente et al., 2021). The laminarin based antifungal formulation, Vacciplant elicited SAR in wheat against septoria *tritrici* leaf blotch fungus, *Zymoseptoria tritici* through the expression of plant defense-associated pathways such as phenylpropanoid (phenylalanine ammonia lyase, chalcone synthase), octadecanoid (lipoxygenase, allene oxide synthase) and pathogenesis-related proteins (de Borba et al., 2022). The important antioxidant enzymes such as Phenylalanine Ammonia Lyase (PAL), phosphoenol pyruvate carboxylase (PEP), peroxidase (PO), polyphenol oxidase (PPO); and Pathogenesis-Related Proteins (PRP) (endoglucanase, chitinase) are involved in the SAR (Righini et al., 2022). The antifungal action of fungicides is dependent on the class type, cell wall and membrane composition of fungi; and composition of algal extract (Kora, 2022a).

5. Conclusions

Various solvent extracts of microalgae and seaweeds are the candidates of choice as fungicides for the control and management of different devastating plant diseases caused by diverse phytopathogenic fungi. In plant disease management, they are sustainable and economical alternatives to chemical fungicides due to their merits including, renewability, biodegradability, environmental safety, minimal phytotoxicity and field applicability. Further, extensive studies are needed on formulation stability, mode of action, field efficacy, ecological interactions, environmental risk and impacts using morphological, biochemical and omics-based methodologies. In addition, the commercial viability/success of algal fungicides also depends on the manufacturing cost and their compliance with mandatory regulatory laws and guidelines (dos Santos Gomes et al., 2021).

Acknowledgements

The author is thankful to Dr. Veeragoni Dileepkumar, Council of Scientific and Industrial Research-Indian Institute of Chemical Technology, Hyderabad for providing photographs of seaweeds. Dr. Penumatsa Kishore Varma, Acharya N.G. Ranga Agricultural University, Guntur are acknowledged for petriplate photographs of fungal cultures. The author would like to thank Dr. M.V. Balarama Krishna, Head, Environmental Science and Nanomaterials Section, and Dr. Sanjiv Kumar, Head, NCCCM/BARC, for their continuous support and encouragement throughout the work.

References

Abdussalam, S. (1990). Drugs from seaweeds. Medical Hypotheses, 32(1): 33–35. doi: 10.1016/0306-9877(90)90064-l.

Abhishek, D., Sanjay, S. and Jadeja, B.A. (2021). Cytotoxicity, antioxidant and antimicrobial activity of marine macro algae (*Iyengaria stellata* and *Padina boryana*) from the Gujarat Coast. Journal of the Maharaja Sayajirao University of Baroda, 55(1): 130–140.

Ahirwar, A., Kesharwani, K., Deka, R., Muthukumar, S., Khan, M.J., Rai, A., Vinayak, V., Varjani, S., Joshi, K.B. and Morjaria, S. (2022). Microalgal drugs: A promising therapeutic reserve for the future. J. Biotechnol., 349: 32–46. doi: 10.1016/j.jbiotec.2022.03.012.

Akremi, N., Cappoen, D., Anthonissen, R., Verschaeve, L. and Bouraoui, A. (2017). Phytochemical and *in vitro* antimicrobial and genotoxic activity in the brown algae *Dictyopteris membranacea*. South African Journal of Botany, 108: 308–314. doi: 10.1016/j.sajb.2016.08.009.

Al-Enazi, N.M., Awaad, A.S., Alqasoumi, S.I. and Alwethairi, M.F. (2018). Biological activities of the red algae *Galaxaura rugosa* and *Liagora hawaiiana* butters. Saudi Pharmaceutical Journal, 26(1): 25–32. doi: 10.1016/j.jsps.2017.11.003.

Algotiml, R., Gab-Alla, A., Seoudi, R., Abulreesh, H.H., El-Readi, M.Z. and Elbanna, K. (2022). Anticancer and antimicrobial activity of biosynthesized Red Sea marine algal silver nanoparticles. Scientific Reports, 12(1): 2421. doi: 10.1038/s41598-022-06412-3.

Ammar, N., Jabnoun-Khiareddine, H., Mejdoub-Trabelsi, B., Nefzi, A., Mahjoub, M.A. and Daami-Remadi, M. (2017). *Pythium* leak control in potato using aqueous and organic extracts from the brown alga *Sargassum vulgare* (C. Agardh, 1820). Postharvest Biology and Technology, 130: 81–93. doi: 10.1016/j.postharvbio.2017.04.010.

Andreea, C., Axine, O. and Iacomi, B. (2010). Antifungal activity of macroalgae extracts. Scientific Papers, UASVM Bucharest, Series A, LIII: 442–447.

Bahammou, N., Raja, R.S. Carvalho, I., Cherifi, K., Bouamama, H. and Cherifi, O. (2021). Assessment of the antifungal and antioxidant activities of the seaweeds collected from the Coast of Atlantic Ocean, Morocco. 2021: 9(4). doi: 10.48317/IMIST.PRSM/morjchem-v9i3.25910.

Banu, A.T., Ramani, P.S. and Murugan, A. (2020). Effect of seaweed coating on quality characteristics and shelf life of tomato (*Lycopersicon esculentum* mill). Food Science and Human Wellness, 9(2): 176–183. doi: 10.1016/j.fshw.2020.03.002.

Barreto, M., Straker, C.J. and Critchley, A.T. (1997). Short note on the effects of ethanolic extracts of selected South African seaweeds on the growth of commercially important plant pathogens,

Rhizoctonia solani Kühn and *Verticillium* sp. South African Journal of Botany, 63(6): 521–523. doi: 10.1016/s0254-6299(15)30808-5.

Bennamara, A., Abourriche, A., Berrada, M., Charrouf, M.h., Chaib, N., Boudouma, M. and Garneau, F.X. (1999). Methoxybifurcarenone: An antifungal and antibacterial meroditerpenoid from the brown alga *Cystoseira tamariscifolia*. Phytochemistry, 52(1): 37–40. doi: 10.1016/S0031-9422(99)00040-0.

Bhagyaraj, I. and Kunchithapatham, V.R. (2016). Diversity and distribution of seaweeds in the shores and water lagoons of Chennai and Rameshwaram coastal areas, South-Eastern coast of India. Biodiversity Journal, 7(4): 923–934.

Bhatia, S., Sharma, K., Nagpal, K. and Bera, T. (2015). Investigation of the factors influencing the molecular weight of porphyran and its associated antifungal activity. Bioactive Carbohydrates and Dietary Fibre, 5(2): 153–168. doi: 10.1016/j.bcdf.2015.03.005.

Boček, S., Salaš, P., Sasková, H. and Mokričková, J. (2013). Effect of Alginure® (seaweed extract), Myco-Sin®VIN (sulfuric clay) and Polyversum® (Pythium oligandrum Drechs.) on yield and disease control in organic strawberries. Acta Universitatis Agriculturae et Silviculturae Mendelianae Brunensis, 60(8): 19–28. doi: 10.11118/actaun201260080019.

Boukhatem, T., Chadli, R. and Berrahal, N. (2021). Fatty acid profile and isolation of bioactive compounds from the acetone extract of *Cystoseira stricta* L., harvested from Western Algeria. Egyptian Journal of Microbiology, 56(1): 99–108. doi: 10.21608/ejm.2021.59732.1186.

de Borba, M.C., Velho, A.C., de Freitas, M.B., Holvoet, M., Maia-Grondard, A., Baltenweck, R., Magnin-Robert, M., Randoux, B., Hilbert, J.-L., Reignault, P., Hugueney, P., Siah, A. and Stadnik, M.J. (2022). A laminarin-based formulation protects wheat against *Zymoseptoria tritici* via direct antifungal activity and elicitation of host defense-related genes. Plant Disease, 106(5): 1408–1418. doi: 10.1094/pdis-08-21-1675-re.

De Corato, U., Salimbeni, R., De Pretis, A., Avella, N. and Patruno, G. (2017). Antifungal activity of crude extracts from brown and red seaweeds by a supercritical carbon dioxide technique against fruit postharvest fungal diseases. Postharvest Biology and Technology, 131: 16–30. doi: 10.1016/j.postharvbio.2017.04.011.

de Felicio, R., de Albuquerque, S., Young, M.C., Yokoya, N.S. and Debonsi, H.M. (2010). Trypanocidal, leishmanicidal and antifungal potential from marine red alga *Bostrychia tenella* J. Agardh (Rhodomelaceae, Ceramiales). J. Pharm. Biomed. Anal., 52(5): 763–769. doi: 10.1016/j.jpba.2010.02.018.

dos Santos Gomes, A.C., da Silva, R.R., Moreira, S.I., Vicentini, S.N.C. and Ceresini, P.C. (2021). Biofungicides: An eco-friendly approach for plant disease management. pp. 641–649. *In*: Ó. Zaragoza and A. Casadevall (eds.). Encyclopedia of Mycology. Oxford: Elsevier.

El-Hossary, E.M., Cheng, C., Hamed, M.M., El-Sayed Hamed, A.N., Ohlsen, K., Hentschel, U. and Abdelmohsen, U.R. (2017). Antifungal potential of marine natural products. Eur. J. Med. Chem., 126: 631–651. doi: 10.1016/j.ejmech.2016.11.022.

El Wahidi, M., El Amraoui, B., El Amraoui, M. and Bamhaoud, T. (2015). Screening of antimicrobial activity of macroalgae extracts from the Moroccan Atlantic coast. Ann. Pharm. Fr., 73(3): 190–196. doi: 10.1016/j.pharma.2014.12.005.

Fayzi, L., Askarne, L., Boufous, E.H., Cherifi, O. and Cherifi, K. (2022). Antioxidant and antifungal activity of some Moroccan seaweeds against three postharvest fungal pathogens. Asian Journal of Plant Sciences, 21(2): 328–338. doi: 10.3923/ajps.2022.328.338.

Ibrahim, M., Salman, M., Kamal, S., Rehman, S., Razzaq, A. and Akash, S.H. (2017). Algae-based biologically active compounds. Algae Based Polymers, Blends, and Composites (pp. 155–271), Elsevier.

Karthik, T., Sarkar, G., Babu, S., Amalraj, L.D. and Jayasri, M.A. (2020). Preparation and evaluation of liquid fertilizer from *Turbinaria ornata* and *Ulva reticulata*. Biocatalysis and Agricultural Biotechnology, 28: 101712. doi: 10.1016/j.bcab.2020.101712.

Kora, A.J. (2019a). Multifaceted activities of plant gum synthesised platinum nanoparticles: Catalytic, peroxidase, PCR enhancing and antioxidant activities. IET Nanobiotechnol., 13(6): 602–608. doi: 10.1049/iet-nbt.2018.5407.

Kora, A.J. (2019b). Plant arabinogalactan gum synthesized palladium nanoparticles: Characterization and properties. Journal of Inorganic and Organometallic Polymers and Materials, 29(6): 2054–2063. doi: 10.1007/s10904-019-01164-6.

Kora, A.J. (2020). Nutritional and antioxidant significance of selenium-enriched mushrooms. Bulletin of the National Research Centre, 44(1). doi: 10.1186/s42269-020-00289-w.

Kora, A.J. (2021a). Applications of waste decomposer in plant health protection, crop productivity and soil health management. pp. 609–624. *In*: Inamuddin (ed.). Application of Microbes in Environmental and Microbial Biotechnology. Singapore: Springer Verlag.

Kora, A.J. (2021b). Chapter 18—Applications of biogenic silver nanocrystals or nanoparticles as bactericide and fungicide. pp. 335–352. *In*: S. Mallakpour and C.M. Hussain (eds.). Industrial Applications of Nanocrystals, Elsevier.

Kora, A.J. (2022a). Chapter 6—Copper-based nanopesticides. pp. 133–153. *In*: K.A. Abd-Elsalam (ed.). Copper Nanostructures: Next-generation of Agrochemicals for Sustainable Agroecosystems, Elsevier.

Kora, A.J. (2022b). Commercial bacterial and fungal microbial biostimulants used for agriculture in India: An overview. pp. 159–175. *In*: Inamuddin, Charles Oluwaseun Adetunji, Mohd Imran Ahamed and T. Altalhi (eds.). Microbial Biostimulants for Sustainable Agriculture and Environmental Bioremediation (1st Edition ed.). USA: CRC Press.

Kora, A.J. (2022c). Synthesis, characterization and *in vitro* antifungal action of gum ghatti capped copper oxide nanoparticles. Biointerface Research in Applied Chemistry, 13(2): 138. doi: 10.33263/briac132.138.

Kora, A.J., Mounika, J. and Jagadeeshwar, R. (2020). Rice leaf extract synthesized silver nanoparticles: An *in vitro* fungicidal evaluation against *Rhizoctonia solani,* the causative agent of sheath blight disease in rice. Fungal Biol., 124(7): 671–681. doi: 10.1016/j.funbio.2020.03.012.

Kumar, A., Buia, M.C., Palumbo, A., Mohany, M., Wadaan, M.A.M., Hozzein, W.N., Beemster, G.T.S. and AbdElgawad, H. (2020). Ocean acidification affects biological activities of seaweeds: A case study of *Sargassum vulgare* from Ischia volcanic CO_2 vents. Environ. Pollut., 259: 113765. doi: 10.1016/j.envpol.2019.113765.

Kumari, S.S., Kumar, V.D. and Priyanka, B. (2017). Antifungal efficacy of seaweed extracts against fungal pathogen of silkworm, *Bombyx mori* L. International Journal of Agricultural Research, 12(3): 123–129. doi: 10.3923/ijar.2017.123.129.

Lavanya, R. and Veerappan, N. (2012). Pharmaceutical properties of marine macroalgal communities from Gulf of Mannar against human fungal pathogens. Asian Pacific Journal of Tropical Disease, 2: S320–S323. doi: 10.1016/s2222-1808(12)60174-1.

Liu, A.H., Liu, D.Q., Liang, T.J., Yu, X.Q., Feng, M.T., Yao, L.G., Fang, Y., Wang, B., Feng, L.H., Zhang, M.X. and Mao, S.C. (2013). Caulerprenylols A and B, two rare antifungal prenylated para-xylenes from the green alga *Caulerpa racemosa*. Bioorg. Med. Chem. Lett., 23(9): 2491–2494. doi: 10.1016/j.bmcl.2013.03.038.

Mandrekar, V.K., Gawas, U.B. and Majik, M.S. (2019). Brominated molecules from marine algae and their pharmacological importance. Studies in Natural Products Chemistry, 61: 461–490. doi: 10.1016/b978-0-444-64183-0.00013-0.

Manivannan, K., Karthikai devi, G., Anantharaman, P. and Balasubramanian, T. (2011). Antimicrobial potential of selected brown seaweeds from Vedalai coastal waters, Gulf of Mannar. Asian Pacific Journal of Tropical Biomedicine, 1(2): 114–120. doi: 10.1016/s2221-1691(11)60007-5.

Martorell, M.M., Lannert, M., Matula, C.V., Quartino, M.L., de Figueroa, L.I.C., Cormack, W.M. and Ruberto, L.A.M. (2021). Studies toward the comprehension of fungal-macroalgae interaction in cold marine regions from a biotechnological perspective. Fungal Biol., 125(3): 218–230. doi: 10.1016/j.funbio.2020.11.003.

Messahli, I., Gouzi, H., Sifi, I., Chaibi, R., Rezzoug, A. and Rouari, L. (2021). Anticandidal activity of dichloromethane extract obtained from the red algae *A. armata* of the Algerian coast. Acta Ecologica Sinica. doi: 10.1016/j.chnaes.2021.08.005.

Mickymaray, S. and Alturaiki, W. (2018). Antifungal efficacy of marine macroalgae against fungal isolates from bronchial asthmatic cases. Molecules, 23(11): 3032. doi: 10.3390/molecules23113032.

Mogoşanu, G.D., Grumezescu, A.M., Bejenaru, L.E. and Bejenaru, C. (2016). Marine natural products in fighting microbial infections. pp. 351–375. *In*: K. Kon and M. Rai (eds.). Antibiotic Resistance: Mechanisms and New Antimicrobial Approaches, Academic Press.

Mohamed, S., Hashim, S.N. and Rahman, H.A. (2012). Seaweeds: A sustainable functional food for complementary and alternative therapy. Trends in Food Science & Technology, 23(2): 83–96. doi: 10.1016/j.tifs.2011.09.001.

Monla, R.M.A., Dassouki, Z.T., Gali-Muhtasib, H. and Mawlawi, H.R. (2020). Chemical analysis and biological potentials of extracts from *Colpomenia sinuosa*. Pharmacognosy Research, 12(3): 272–278.

Nabeta, H.W., Kouokam, J.C., Lasnik, A.B., Fuqua, J.L. and Palmer, K.E. (2021). Novel antifungal activity of Q-Griffithsin, a broad-spectrum antiviral lectin. Microbiology Spectrum, 9(2): e00957–00921. doi: doi:10.1128/Spectrum.00957-21.

Nofal, A., Azzazy, M., Ayyad, S., Abdelsalm, E., Abousekken, M.S. and Tammam, O. (2022). Evaluation of the brown alga, *Sargassum muticum* extract as an antimicrobial and feeding additives. Braz. J. Biol., 84: e259721. doi: 10.1590/1519-6984.259721.

Pourakbar, L., Moghaddam, S.S., Enshasy, H.A.E. and Sayyed, R.Z. (2021). Antifungal activity of the extract of a macroalgae, *Gracilariopsis persica*, against four plant pathogenic fungi. Plants (Basel), 10(9). doi: 10.3390/plants10091781.

Procházka, P., Řehoř, J., Vostřel, J. and Fraňková, A. (2022). Use of botanicals to protect early stage growth of hop plants against *Pseudoperonospora humuli*. Crop Protection, 157: 105978. doi: 10.1016/j.cropro.2022.105978.

Reis, R.P., Carvalho Junior, A.A.d., Facchinei, A.P., Calheiros, A.C.d.S. and Castelar, B. (2018). Direct effects of ulvan and a flour produced from the green alga *Ulva fasciata* Delile on the fungus *Stemphylium solani* Weber. Algal Research, 30: 23–27. doi: 10.1016/j.algal.2017.12.007.

Righini, H., Francioso, O., Martel Quintana, A. and Roberti, R. (2022). Cyanobacteria: A natural source for controlling agricultural plant diseases caused by fungi and oomycetes and improving plant growth. *Horticulturae*, 8(1): 58. doi: 10.3390/horticulturae8010058.

Saim, S., Sahnouni, F., Bouhadi, D. and Kharbouche, S. (2021). The antimicrobial activity of two marine red algae collected from Algerian West Coast. Trends in Pharmaceutical Sciences, 7(4): 233–242. doi: 10.30476/tips.2021.89827.1078.

Sampaio, T.M.M., Lucas dos Santos, A.T., de Freitas, M.A., Almeida-Bezerra, J.W., Fonseca, V.J.A., Coutinho, H.D.M., Morais-Braga, M.F.B., de Miranda, G.E.C. and Andrade-Pinheiro, J.C. (2022). Antifungal activity of *Gracilaria cervicornis* (Turner) J. Agardh against *Candida* spp. South African Journal of Botany, 150: 146–152. doi: 10.1016/j.sajb.2022.06.057.

Sellimi, S., Benslima, A., Barragan-Montero, V., Hajji, M. and Nasri, M. (2017). Polyphenolic-protein-polysaccharide ternary conjugates from *Cystoseira barbata* Tunisian seaweed as potential biopreservatives: Chemical, antioxidant and antimicrobial properties. Int. J. Biol. Macromol., 105(Pt 2): 1375–1383. doi: 10.1016/j.ijbiomac.2017.08.007.

Senevirathne, W.S.M. and Kim, S.K. (2013). Cosmeceuticals from algae. pp. 694–713. *In*: H. Domínguez (ed.). Functional Ingredients from Algae for Foods and Nutraceuticals. Woodhead Publishing.

Senousy, H.H., El-Sheekh, M.M., Saber, A.A., Khairy, H.M., Said, H.A., Alhoqail, W.A. and Abu-Elsaoud, A.M. (2022). Biochemical analyses of ten cyanobacterial and microalgal strains isolated from Egyptian habitats, and screening for their potential against some selected phytopathogenic fungal strains. Agronomy, 12(6): 1340. doi: 10.3390/agronomy12061340.

Soares, F., Fernandes, C., Silva, P., Pereira, L. and Gonçalves, T. (2016). Antifungal activity of carrageenan extracts from the red alga *Chondracanthus teedei* var. *lusitanicus*. Journal of Applied Phycology, 28(5): 2991–2998. doi: 10.1007/s10811-016-0849-9.

Souza, R.B., Frota, A.F., Silva, J., Alves, C., Neugebauer, A.Z., Pinteus, S., Rodrigues, J.A.G., Cordeiro, E.M.S., de Almeida, R.R., Pedrosa, R. and Benevides, N.M.B. (2018). *In vitro* activities of kappa-carrageenan isolated from red marine alga *Hypnea musciformis*: Antimicrobial, anticancer and neuroprotective potential. Int. J. Biol. Macromol., 112: 1248–1256. doi: 10.1016/j.ijbiomac.2018.02.029.

Stein, E.M., Colepicolo, P., Afonso, F.A.K. and Fujii, M.T. (2011). Screening for antifungal activities of extracts of the Brazilian seaweed genus *Laurencia* (Ceramiales, Rhodophyta). Revista Brasileira de Farmacognosia, 21(2): 290–295. doi: 10.1590/s0102-695x2011005000085.

Thavasi Alagan, V., Nakulan Vatsala, R., Sagadevan, I., Subbiah, V. and Ragothaman, V. (2020). Effect of dietary supplementation of seaweed (*Ulva lactuca*) and *Azolla* on growth performance, haematological and serum biochemical parameters of Aseel chicken. Beni-Suef University Journal of Basic and Applied Sciences, 9: 58. doi: 10.1186/s43088-020-00087-3.

Veeragoni, D., Deshpande, S., Rachamalla, H.K., Ande, A., Misra, S. and Mutheneni, S.R. (2022). *In vitro* and *in vivo* anticancer and genotoxicity profiles of green synthesized and chemically synthesized silver nanoparticles. ACS Applied Bio Materials, 5(5): 2324–2339. doi: 10.1021/acsabm.2c00149.

Veeragoni, D., Mutheneni, S.R., Misra, S. and Savarapu, S.K. (2018). Evaluation of different solvent extracts of *Sargassum wightii* (brown algae) for its antifungal efficacy against silkworm pathogens. Journal of Entomology and Zoology Studies, 6(3): 1125–1130.

Vicente, T.F.L., Lemos, M.F.L., Felix, R., Valentao, P. and Felix, C. (2021). Marine macroalgae, a source of natural inhibitors of fungal phytopathogens. J. Fungi (Basel), 7(12): 1006. doi: 10.3390/jof7121006.

Chapter 3

A Natural Approach to Fungal Control

Peptides and Proteins as Antifungal Agents

Ahya Abdi Ali,[1,2] *Parisa Mohammadi*[1,2] and *Parinaz Ghadam*[3,*]

1. Introduction

Fungi are remarkable and ubiquitously life forms. These eukaryotes range from large structures such as mushrooms to tiny unicellular molds and yeasts. These organisms have been recognized as microbiota at different points of our body (Enaud et al., 2018; Huffnagle and Noverr, 2013; Kapitan et al., 2018). Fungi have been used in food industries for years (Campbell-Platt and Cook, 1989) as well as in numerous industries, including peptides' production, chemicals, enzymes, antibiotics and organic acids (Money, 2016; Mukherjee et al., 2018).

Nowadays, fungi are responsible for a wide range of diseases in people. It is assessed that fungal infections impact many individuals around the world, 150 million of whom experience extreme diseases (Bongomin et al., 2017). Fungal disease includes superficial and subcutaneous damages which affect the keratinous tissues, skin and mucous tissue (Kaushik et al., 2015) as well as dangerous systemic diseases influencing the heart, brain, liver, lungs, kidneys and spleen (Rautemaa-Richardson and Richardson, 2017). It is particularly important for patients with autoimmune disorders and immunocompromised conditions, and those going through organ transplantation or anticancer chemotherapy. The crucial human pathogens are *Cryptococcus neoformans*, *Aspergillus fumigatus*, *Candida albicans* and *C. auris* and other fungal species like *Malassezia furfur* or *Histoplasma capsulatum*, which are emerging (Fernández de Ullivarri et al., 2020; Roemer and Krysan, 2014).

[1] Department of Microbiology, Faculty of Biological Sciences, Alzahra University, Tehran, Iran.
[2] Research Center for Applied Microbiology and Microbial Biotechnology, Alzahra University, Tehran, Iran.
[3] Department of Biotechnology, Faculty of Biological Sciences, Alzahra University, Tehran, Iran.
* Corresponding author: pghadam@alzahra.ac.ir

The increased occurrence of mycological contagions is a result of the improvement in therapeutic technologies and rehabilitations, applied to critical diseases (Pfaller and Diekema, 2007). Increasing immunocompromised people due to multi-organ failure or severe disease can affect the occurrence of Invasive Fungal Infections (IFI) (Badiee and Hashemizadeh, 2014; Lum, Kah Yean et al., 2015; Pappas et al., 2016; Pluim et al., 2012; Yapar, 2014). The morbidity and mortality of IFIs are of great concern (Brown et al., 2012).

At present, existing antifungals can be categorized into five groups: azoles, polyenes, pyrimidine analogs, echinocandins and allylamines (Healey et al., 2016; Hope et al., 2004; Kołaczkowska and Kołaczkowski, 2016; Petrikkos and Skiada, 2007; Posteraro et al., 2003; Sanglard, 2016; Szymański et al., 2022). Among them, allylamines are used on superficial infectious diseases, and the other four drug groups for invasive fungal diseases (Sanglard, 2016). Two groups of these antifungals are non-toxic: triazoles and echinocandins (Healey et al., 2016). Extensive usage of antifungals in treatments, as well as lengthy treatment, results in the increase of resistance (Sanglard, 2016). The mechanisms of resistance have been reported for all commercially accessible antifungals (Bondaryk et al., 2017) (Table 3.1).

The necessity for new, safe and more effective antifungals appears in parallel with the growing number of resistant fungal isolates (Lewis, 2011; Pfaller et al., 2012. Studies (Chen et al., 2011) confirmed the effectiveness of the combination of antifungal treatment procedures in which novel groups of antifungal agents are applied (third generation of lipopeptides-echinocandins and azoles) against severe fungal diseases. In addition, caspofungin and other echinocandins have not been a correct remedy for *Histoplasma capsulatum* (Espinel-Ingroff, 1998; Kohler et al.,

Table 3.1: Antifungal agents.

Name	Category	Resistance mechanisms	References
Fluorinated pyrimidine analogs	Flucytosine (5-FC)	Pyrimidine salvage enzyme deficiency and 5-FC metabolism _ Mutations in *FCA1*, *FUR1*, *FCY21*, *FCY22*	(Hope et al., 2004; Vandeputte et al., 2012)
Polyenes	Nystatin Natamycin Amphotericin B	Decreased ergosterol in the cell membrane Membrane composition alteration Capsule enlargement	(Posteraro et al., 2003; Vandeputte et al., 2012)
Echinocandins	Caspofungin Micafungin	Mutation in the *FKS1* and *FKS2* genes Lack of 1,3-_-glucan in the cell wall	(Kołaczkowska and Kołaczkowski, 2016; Sanglard, 2016)
Allylamines	Terbinafine Nattifine	Drug target modification The naphthalene ring degradation	(Kołaczkowska and Kołaczkowski, 2016; Sanglard, 2016)
Azoles	Voriconazole Posoconazole Fluconazole	*ERG11* mutation or overexpression, Decreased azoles inside the fungal cell, ERG3 mutation	(Kołaczkowska and Kołaczkowski, 2016; Sanglard, 2016; Shahi et al., 2022; Vandeputte et al., 2012)

2000), especially for immunosuppressed people since this fungal infection shows significant morbidity and mortality in immunosuppressed people (Bondaryk et al., 2017; Stringer et al., 2002).

On the other hand, a large number of food constituents are subject to fungal pollution. The main problems are of yield, production of food spoilage as well as food contact with invasive fungal toxins. The resistance of fungal species to communal preservatives is another problem increasing the need for novel methods to make the shelf lifetime of materials and crops longer (Thery et al., 2019). The attempt worldwide to avoid spoilage and fungal toxins on food as well as invasive fungal infections necessitates the expansion of novel antifungal methods. AMPs with antifungal action have attracted considerable attention because of their characteristics such as a variety of structures and functions, antifungal range, mechanism of action, great strength and the simple accessibility of their biotechnological production. Due to their multistep activities, the increase of fungal resistance to AMPs is slower or deleted in comparison to routinely used fungicides. Furthermore, AMPs select significant new biological targets such as inhibition of mycotoxin biosynthesis that provides new features for their use in farming and food production contaminated by mycotoxin (Martínez-Culebras et al., 2021). AMPs can be found in the immune system of most organisms. Nevertheless, despite their rising reports in food applications, the number of permitted peptides is low which can be attributed to the low solubility, toxicity and time-consuming extraction. It appears that a deeper understanding of their modes of action can compensate these disadvantages (Thery et al., 2019). In comparison to usual antibiotics, AMPs can display a wide range of antimicrobial actions against fungi, bacteria, protozoa, viruses and some cancer cells (Hancock and Chapple, 1999). AMP uses diverse mechanisms and also hampers forming resistance in bacteria and fungi. Furthermore, AMPs help body defense through the homeostasis of the gut and inflammatory responses of the host.

Issues restricting peptide use, like low steadiness, conceivable accumulation and reduced activity in the presence of salts and metal ions could be overwhelmed by compound enhancement and new conveyance techniques (Vlieghe et al., 2010). The rise of resistence to antifungal medication reinforces the requirement for improvement of novel antifungal particles (Denning and Bromley, 2015). The Antimicrobial Peptide Data set (APD) contains more than 3000 arrangements, among which 1084 antifungal peptides can be found (Váradi et al., 2018). In contrast with regular antimicrobials, which for the most part are focused against microorganisms, AMPs can show wide antimicrobial and anticancer activity (Hancock and Chapple, 1999). In this chapter, an outline of the activity and order of AFPs, as well as their method of activity and benefits over momentum antifungal medications (Fernández de Ullivarri et al., 2020) are given.

2. AMPs characteristics

AMPs are small cationic bioactive compounds that are produced by almost all living organisms. AMPs are principally innate immune systems and the first defense boundary against inffective microorganism attacks on other living organisms.

Moreover, AMPs might limit the growth of other microorganisms by competition (Magana et al., 2020; Moretta et al., 2021). As mentioned above, AMPs are produced in all microorganisms, plants and animals (Thery et al., 2019; Van der Weerden et al., 2013; Wang et al., 2016). Interestingly, some AMPs are effective on fungal cells (Buda De Cesare et al., 2020; Struyfs et al., 2020; Thery et al., 2019) some of which display antifungal activity on toxicogenic fungi. AMPs are mainly produced by two biosynthetic paths. Most of them are ribosomally encoded and other AMPs are produced through Non-Ribosomal Peptide Synthases (NRPSs). The latter are mostly observed in bacteria, particularly in Actinomycetes and Bacilli (Finking and Marahiel, 2004; Martínez-Culebras et al., 2021). NRPS-produced AMPs are recognized via the combination of nonproteinogenic amino acids in structure. They are frequently produced through glycosylation, hydroxylation, cyclization and lipidation (Wang et al., 2016).

Being aware of the mode of action of AMPs is important to increase their activity as well as avoiding the expansion of resistance. Furthermore, it speeds up their applications in therapeutics or food preservatives. There are a lot of reports existing on AMPs mechanisms (Fig. 3.1) (Brogden, 2005; Marcos and Gandía, 2009; Nguyen et al., 2011; Nicolas, 2009; Rautenbach et al., 2016). Overall, AMPs can act on several targets in the cell. Cationic AMPs act on negatively charged microbial envelopes (Zasloff, 2002). Electrostatic ligands cannot completely report other activities of AMPs and it seems other components of envelopes complete their actions. In fungi, the cell wall has a crucial role. Diverse AMPs have been described to destroy the fungal cell walls by preventing chitin or glucan synthesis and targeting mannoproteins of the cell wall in sensitive fungi (Buda De Cesare et al., 2020; Martínez-Culebras et al., 2021).

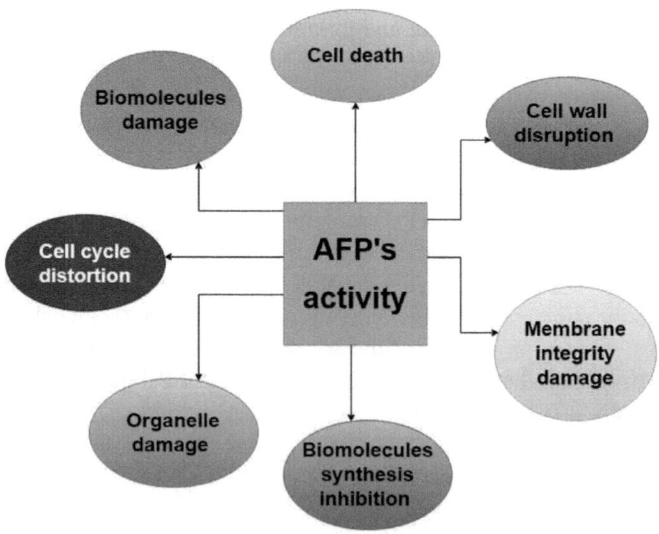

Fig. 3.1: Antifungal peptides/proteins (AFPs) include damaging fungal cell membranes, interfering with fungal cell wall formation, generating oxidative stress, and targeting specific fungal enzymes. Some peptides and proteins have immunomodulatory properties, which stimulate the host's defense mechanisms against fungal infections.

2.1 Unnatural AFPs

Unnatural peptides are made with the end goal of further developing dissolvability, steadiness, antifungal range, adequacy, cost, pharmacological properties, diminishing incidental effects as well as bringing down the immunogenicity of normal peptides. Normal peptides are a rich source of motivation for the plan of new unnatural peptides with bioactivity properties and are reasonable for various applications (Bondaryk et al., 2017). There are a few properties that can determine antifungal action, such as stereospecificity, charge, amphipathicity, hydrophobicity, optional design and the length of peptide, related with a part of these qualities. Amphipathicity and hydrophobicity are fundamental elements for peptide film associations and layer permeabilization (Lum et al., 2015); furthermore, these are significant factors in the plan of manufactured peptides.

Unnatural peptides can be received from the normal cleavage of proteins and *in vitro* proteolysis (Iavarone et al., 2018). Besides the recombination is a practical and time-proficient method for getting bigger peptides. Notwithstanding these regular-based subsidiaries, numerous manufactured peptides have been built using a new synthesis. Cyclization of peptides expanded specific poisonousness and steadiness against proteases (Kamysz et al., 2012) while short peptide lipidation designs could upgrade antifungal activity, furthermore, they develop cell selectivity by diminishing their cytotoxicity (Arnusch et al., 2012). New drug details, such as a consideration in a biocompatible hydrogel, nanofibers, liposome vesicles or conveyance frameworks, and the addition of strange side chains and unnatural amino acids could resolve the issue connected with the responsiveness of AFPs to proteolytic debasement, safeguarding their action and working on their bioavailability (Kong et al., 2016). A typical element found in most AFPs is that distribution and amount of the net charge correlate with biological activity. A few examinations showed that Trp buildups have areas of strength for embedding in lipid bilayers, impacting hemolytic and antimicrobial exercises (Schibli et al., 2002), while Cys deposits give expanded hemolytic potential to certain peptides (Duncan and O'Neil, 2013).

2.1.1 Synthetic and semisynthetic AFPs

The advance of synthetic peptides has become of interest due to their averting from the normal downsides related to regular peptides, including low antifungal activity, harmfulness or precariousness. The late examination has zeroed in on the use of different strategies yielding new atomic elements, with upgraded antimicrobial action, diminished poisonousness and lower expenses of production. Accordingly, new AFPs ought to be dynamic against cells inside biofilms. As per the last study, quantities of positively charged buildup improve the collaboration of peptides with the fungal cell wall (Bondaryk et al., 2017). The examination of different peptide arrangements and designs plays a critical part in a few boundaries in the antifungal power, for example, the cationic, the hydrophobicity and the dispersion of the buildups and related construction, amphipathicity and length (Sagaram et al., 2011). Numerous manufactured peptides have been built using once more blending a gathering of peptides, named PAFs, which have been planned utilizing a combinatorial library (Munoz et al., 2006). Peptides and regular proteins can be used for the plan of novel manufactured bioactive peptides that are stronger than the first ones. Many exploration groups planned different Trp-rich manufactured peptides that work on antifungal action. Preferably, remedial AFPs ought to perform their activity in physiological situations; nonetheless, numerous peptide efficiencies are significantly decreased under high salt concentration (Yu et al., 2011). Subsequently, the quantity of Trp buildups alongside amphipathic construction might improve the antifungal action of peptides (Lum et al., 2015). Arrangement changes of regular AFPs intend to improve their action and decrease harmfulness. For example, a replacement of hydrophobic amino acids in PMAP-23, antimicrobial peptide secluded from porcine, with Trp improved anticandidal capability of made analogs (Lee et al., 2002). The report on uperin antibiofilm activity analogs showed that upgraded antibiofilm action connects with the expanded cationicity of peptides (Lum et al., 2015). Trp-rich peptides, for example, tritrpticin and indolicidin are intense antifungal specialists (Benincasa et al., 2006). Again peptide configuration might help lower production costs, likely harmfulness and liability and expand the *in vivo* action (Steckbeck et al., 2014). There are a few elements which are fundamental in improving synthetic peptides including cationicity, hydrophobicity, amphipathicity and length.

Expanding the number of fundamental deposits in AFPs frequently works on their antifungal activity (Jin et al., 2016). Although the presence of basic amino acids is significant, the selection among Lys and Arg buildups is possibly dangerous (Guo et al., 2013). Due to their scattered charge and related solidness to Lys buildups, the association of Arg with layer surface through H-bonds is essential. It is clear that the general charge, in addition the area of the fundamental buildups in the peptide is significant.

Hydrophobicity is perceived as one more important boundary, as the AFP is not supposed to tie to the layers without any hydrophobic deposits. An expansion in hydrophobicity can also be accomplished by replacements including Ile, Val and Leu buildups, bringing about an increment of antifungal action. Nonetheless, the hydrophobicity alone does not direct the antifungal activity of these peptides (Fernández-Vidal et al., 2007). The inclusion of a Trp on the hydrophilic site of a helix decreases its activity, though the inclusion of hydrophobic deposits on the hydrophobic substance of the peptide rich in His, P-113, does not influence its candidacidal action (Rothstein et al., 2001). Although hydrophobicity deposits display a significant role in working on the antifungal activity, the nature and position of these buildups in the peptide are critical. Trp and Phe buildups are used to provide a hydrophobic activity on the peptide or to expand protection from salt and serum (Sonesson et al., 2011). Additionally, the substitution of hydrophobic buildups (Gopal et al., 2013) is sometimes used to work on the beneficial list of a peptide. A deliberate expansion in hydrophobicity (especially on the nonpolar face) has an adverse consequence prompting self-relationship of the peptide (López-Abarrategui et al., 2015).

Hence, the antifungal activity of AFPs is thought to have corresponded with both amphipathicity and hydrophobicity. Helicity is generally more connected with peptide harmfulness than antimicrobial activity. The decrease of hemolytic peptides helicity, through joining of D-amino acids, saves the antimicrobial properties and brings down the hemolytic action.

The position and nature of the deposits directly impact factors like hydrophobicity, cationicity and peptide action. Another likely unfavorable consequence of a very large peptide are the improvement of unfortunate dissolvability (Gopal et al., 2013). A peptide should be made of no less than seven to eight amino acids to guarantee a decent amphipathicity. Notwithstanding, present-day methods and a constant expansion in information have featured different boundaries that could be regulated to work on both the activity and the strength of recently planned peptides. In any case, shortening peptides might be a method for expanding antifungal activities. Subsequently, an effective α-helical peptide needs 22 amino acids to have the option to cross the plasma layer. Moreover, the length of a peptide should be important in the design of the peptide. The improvement of PC-based techniques and devices for the expectation of activity have prompted a quicker recognizable proof of strong AFPs (Thery et al., 2019).

2.1.2 Cryptides

Recently, various examinations have concentrated on the activity of peptides obtained from proteins (cryptides) supplied with natural activities unique or comparative, yet not indistinguishable from that of the precursor protein. Cryptides could be partitioned into three gatherings as their action and presence *in vivo* are normally handled sections or proteolysis production *in vitro* (Samir and Link, 2011). A part of these is described by different other natural capacities like antimicrobial, anticancer and immunomodulatory. Numerous regular cryptides, such as lactoferricin (Lfcin) have antifungal activities (Thery et al., 2019).

2.2 Natural AFPs

AMPs are made by various macro and microorganisms, which have been isolated from different natural habitats (De Lucca and Walsh, 1999) (Fig. 3.2). Most AMPs are effective on pathogenic fungi *in vitro* (Freitas and Franco, 2016; McNair et al., 2018). Moreover, there are novel methodologies for the prediction and recognition of different AMPs through *in silico* tools, that have attempted to decease the connected costs (Agrawal et al., 2018; Fjell et al., 2007; Garrigues et al., 2017; Robinson, 2011; Schneider and Fechner, 2005).

Biological AMPs are classified based on their sources (Table 3.2). AMPs made by microorganisms were of increased study for medicinal applications (Essig et al., 2014). Other different sources, such as plant sources which are rich in AMPs, were used for various purposes, like the control of phytopathogens (De Lucca, 2000). Moreover, AMP from archaea with antifungal and antibiofilm activities was introduced (Roscetto et al., 2018). Generally, AMPs have an a-helix structure, β- sheet with two cysteines or a mixture of two structures in close contact with membranes. Some AMPs have a lot of some amino acids; therefore, they are named Pro-rich, Gly-rich, His-rich, Arg-rich and Trp-rich (Bondaryk et al., 2017). However, the

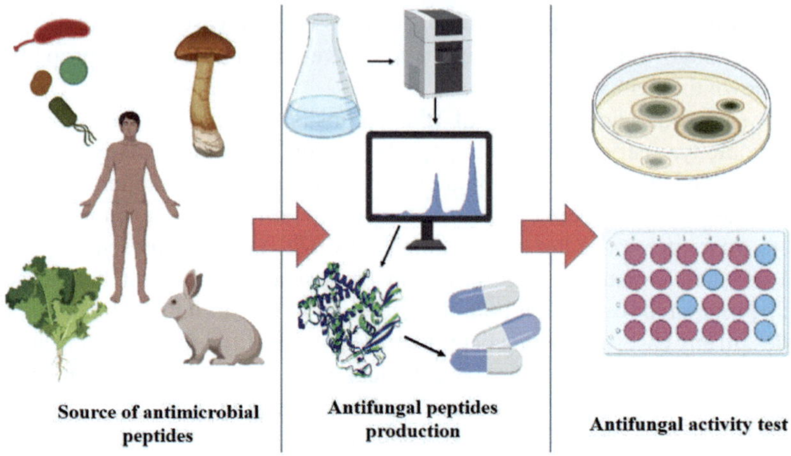

Fig. 3.2: Antifungal peptides/proteins can be found in animals, plants, and microbes, among other places. Sources of animal products: - Mammals: Anti-fungal peptides such as defensins and cathelicidins are found in the milk, saliva, and skin secretions of numerous mammalian species. - Insects: Anti-fungal peptides such as attacin and cecrin are found in insect hemolymph (blood) and cuticle. Plant sources include: - Seeds: Anti-fungal peptides such as thionins, defensins, and lipid transfer proteins are found in many plant seeds. - Leaves: Some plant leaves, such as tobacco leaves, contain anti-fungal peptides such snakin-1. - Fruits: Anti-fungal peptides such as citrusin are found in certain fruits such as grapefruit. Microbiological sources: *Bacillus subtilis* and *Lactobacillus plantarum*, for example, produce antifungal peptides such as subtilosin and plantaricin. - Fungus: Anti-fungal peptides such as gigasin and harzianin are produced by fungi such as *Aspergillus giganteus* and *Trichoderma harzianum*. These natural sources of anti-fungal peptides could be useful in the development of new anti-fungal medicines.

majority of AMPs have not been recognized as yet. Making alterations after translation plays a crucial role in three-dimension structures and AMPs action, which cannot be predicted by usual *in silico* tools (Agrawal et al., 2018).

3. Mechanisms of AFPs activity

Although a few peptides such as His-rich or lipopeptides (e.g., echinocandins) have mainly antifungal impacts, peptides with membrane-lysis impacts (e.g., protegrins) have more extensive antimicrobial activity, which encompasses a wide range of microorganisms. There is an explanation of how AFPs affect fungal pathogens in this section (Fernández de Ullivarri et al., 2020) (Table 3.3 and Fig. 3.3).

3.1 Glucan synthesis inhibition

The β-glucan synthase is an enzyme with activity in cell wall integrity. This enzyme can be inhibited by lipids with cyclic structure, resulting in cell wall destabilization, prompting its susceptibility to osmotic pressure and lysis of the cell. 1,3-β-glucans are used in the division of septum and cell wall assembly, which can be stopped by inhibitors. This enzyme has been reported in fungi such as Candida, Aspergillus, Cryptococcus and Pneumocystis species (Guyard et al., 2002; Matejuk et al., 2010; van der Weerden et al., 2013).

Table 3.2: Natural AFPs families.

Source	Name	Characteristics	Activity	References
Archaea	Cryptic CAMP-like	Cationic	Targets cell wall	(Roscetto et al., 2018)
Bacteria	Iturin	Small cyclic peptidolipids with a lipid-soluble B-amino acid linked to a peptide with D- and L-amino acids	Pore formation in membranes	(Alvarez et al., 2012; De Lucca, 2000; De Lucca and Walsh, 1999; Landy et al., 1948)
	Syringomycins	Small cyclic lipodepsipeptides	Forms voltage-sensitive Candida spp. ion channels	(De Lucca, 2000; De Lucca and Walsh, 1999; Sinden et al., 1971)
Fungi	Nikkomycins	Peptide nucleosides	Inhibit chitin biosynthesis	(De Lucca, 2000; De Lucca and Walsh, 1999; Hector et al., 1990; McCarthy et al., 1985)
	Polyoxins	Peptide nucleosides	Inhibit chitin biosynthesis	(De Lucca, 2000; De Lucca and Walsh, 1999; Hori et al., 1974; Suzuki et al., 1965)
	Echinocandins	Cyclic hexapeptides with N-linked acyl lipid side chains	Inhibit glucan synthesis	(De Lucca, 2000; De Lucca and Walsh, 1999; Emri et al., 2013; Eschenauer et al., 2007)
	Aureobasidin	Cyclic depsipeptide	Lysis by altering actin assembly and delocalizing chitin in fungal walls/ sphingolipid synthesis inhibition	(De Lucca, 2000; Endo et al., 1997)
	Leucinostatins	With five unusual amino acids, 4-methylproline (MePro), 2-amino-6-hydroxy-4-methyl-8-oxodecanoic acid (AHMOD), threo-b-hydroxyleucine (HyLeu), three 2-aminoisobutyric acids (Aib), and b-alanine (b-Ala)	Uncouplers mitochondria	(Abe et al., 2018; De Lucca, 2000; De Lucca and Walsh, 1999)
Plants	Defensins	Small, highly stable, cysteine-rich peptides	Membrane pore, ion efflux, induction of ROS and Apoptosis	(Bondaryk et al., 2017; Sher Khan et al., 2019)

Table 3.2 contd. ...

...Table 3.2 contd.

Source	Name	Characteristics	Activity	References
Mammals	α-defensins	β-sheet with Cys forming intramolecular S-S bonds	Cell lysis	(De Lucca and Walsh, 1999; De Lucca, 2000)
	β-defensins	β-sheet with Cys with an S-S motif different from a-defensins. Amino termini are blocked with a pyroglutamyl residue	Cell lysis	(De Lucca and Walsh, 1999; De Lucca, 2000)
	Protegrins Cathelicidins	Cationic, cysteine-rich β-defensins	Pore formations and Lysis	(De Lucca and Walsh, 1999; De Lucca, 2000; Bondaryk et al., 2017)
	Histatins	Basic and neutral helical peptides	Induction of cell death, osmosis stress	(Bondaryk et al., 2017; De Lucca, 2000; Koshlukova et al., 1999)

Table 3.3: Different groups of antifungal peptides.

Name of the group	Representative peptides	Origin	Structure	References
β-glucan synthase inhibitors	Echinocandins Pneumocandins Aculeacins	Fungi	Cyclic lipopeptides	(Epstein et al., 2018; Fromtling and Abruzzo, 1989; Iwata et al., 1982)
	Mulundocandins		Threonine is substituted by serine; and lineoyl with 12-methylmyristoyl	(Hawser et al., 1999; Roy et al., 1987)
Cell wall chitin inhibitors	Nikkomycin polyoxins	Bacteria	Nucleoside peptide antibiotic	(Hector et al., 1990; McCarthy et al., 1985)
	Aureobacidins	Fungi	Cyclic lipophilic depsipeptide with 8 amino acids and anhydroxyacid	(Endo et al., 1997; Ikai et al., 1991; Nagiec et al., 1997)
Membrane targeting inhibitors	Rs-ARF2	plant	α-helical 3-stranded β-sheets, 4 disulphide bridges	(Rautenbach et al., 2016; Schaaper et al., 2001; Thevissen et al., 2007)
	Drosomycin	Insects	α-helical 3-stranded β-sheets, 4 disulphide bridges	(Bulet and Stocklin, 2005; Cohen et al., 2009; Fehlbaum et al., 1994; Yuan et al., 2007)
	Bacillomycin F Iturin A	Bacteria	Cyclic with lipid-soluble β-amino acid linked to the D/L aa	(Besson and Michel, 1984; Mhammedi et al., 1982)
Histatins	Histatin 1–12	Primates	α-helical in hydrophobic environment; increase histidine content	(Baev et al., 2004; Helmerhorst et al., 1999b; Mochon and Liu, 2008; Pollock et al., 1984; Troxler et al., 1990; Tsai and Bobek, 1997)

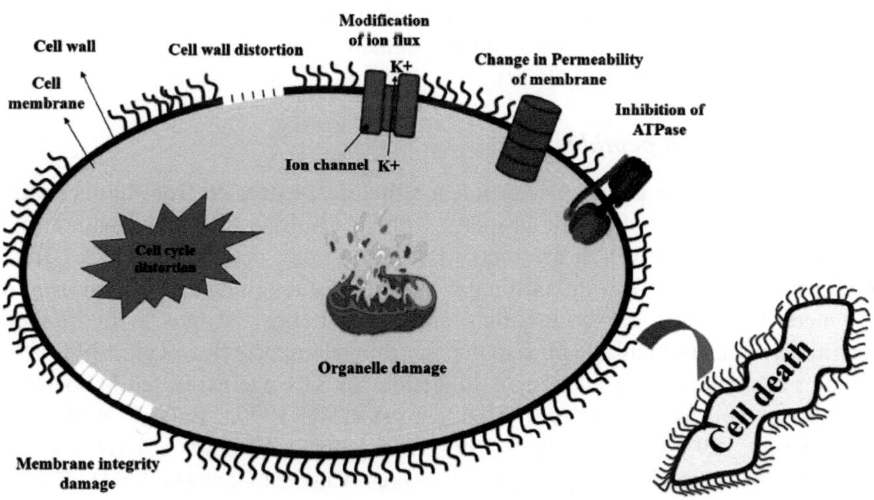

Fig. 3.3: Schematic presentation of antifungal peptide mechanisms.

3.2 Chitin biosynthesis inhibition

Chitin, a complex compound of fungal cell walls, is fundamental for keeping the integrity of the cell wall. Aureobasidins are cyclic lipophilic compounds with 8-mer depsipeptides and α-hydroxy acid, that show two mechanisms: the disturbance of the cell wall/membrane through changing assembly of actin and chitin (Endo et al., 1997) and sphingolipids synthesis interruption (Nagiec et al., 1997).

A few derivations of the aureobasidin are effective on Candida spp. Nikkomycins are primary analogs of uridine diphosphate N-acetylglucosamine, which is a significant component of chitin. These peptides displayed inhibition of chitin synthesis in *C. albicans* (McCarthy et al., 1985) whereas human cells are not impacted. Besides they display considerable action against, *B. dermatitidis*, *C. immitis* and a moderate impact on *H. capsulatum* (Hector et al., 1990; Clemons and Stevens, 1997) and mold; however, they are not affected by these compounds.

3.3 Specific activity on membranes

The plant defensin, Rs-ARF2 focuses on glucosylceramide of the fungal membrane, initiating permeability, causing Ca^{2+} take-up, K^+ efflux and moderate alkalinization. Besides, defensin makes the synthesis of toxic ROS. This peptide and its analogs have less cytotoxicity against mammalian cells at an effective repressing dosage on their pathogens (Matejuk et al., 2010; Sher Khan et al., 2019).

Iturins result in the formation of a pore in the membrane and spillage of key ions (Besson and Michel, 1984). Their antimicrobial effect is restricted essentially to fungi. Antifungal iturins are toxic to the mammalian membrane. Unlike most AFPs which are cationic, iturins are neutral such as iturin A or anionic like bacillomycin L. One derivative from this group, bacillomycin F, actually represses *C. albicans*, *A. niger* and *C. tropicalis* growth. Histatin 5 encompasses 7 His, 4 Arg and 3 Lys, allowing the peptide to embrace α-helical conformity in non-aqueous conditions (Raj et al., 1998). Histatin 5 disorganizes the mitochondrial membrane integrity when the cell is in respiratory metabolisms (Helmerhorst et al., 1999a). The resulting ATP restricting surface P2X receptors induce signaling cascade, prompting cell death (Fernández de Ullivarri et al., 2020).

3.4 Histidine-rich peptides

Histatins are cationic peptides with linear structures, extracted from human saliva. They have a weak amphipathic property without disulfide bonds and with a lot of histidine amino acids which are recognized from other cationic peptides (Brewer et al., 1998) and show strong antifungal impacts. Although histatins form irregular polymers in aqueous conditions, they embrace α-helical designs in hydrophobic conditions which particularly present their antifungal impacts (Raj et al., 1998, 1994). A major characteristic of histatins is fungicidal activity against different species of Candida as well as *A. fumigatus* (Helmerhorst et al., 1999a; Pollock et al., 1984; Rayhan et al., 1992; Tsai and Bobek, 1997). This group of 12 short (3–4 kD) histidine-rich peptides are made by proteolytic cleavages (Castagnola et al., 2004). They are only detected in primates' saliva (Ahmad et al., 2004; MacKay et al., 1984; Wyandt et al., 1989). Among these peptides, histatin 5 displays the most fungicidal action on *C. albicans* (Raj et al., 1990). The fragments of P-113, a 12-mer amino acid of histatin 5, are the shortest peptides that show a complete anticandidal effect, compared to its parental peptide (Rothstein et al., 2001). Histatin 5 and two other synthetic fragments (Dhar4 and 5) repress drug-resistant *C. albicans* biofilm (Douglas, 2003; Prijck et al., 2009; Pusateri et al., 2009). Besides mammalian cells, arthropods produce histidine-rich peptides that are effective on specific fungi such as *C. albicans* although do not show any activity against bacteria (Iijima et al., 1993; Lee et al., 1995a; Lee et al., 1995b). In comparision to histatin 5, synthetic histidine-rich peptides with branched structures have essentially more inhibitory activity on *C. albicans* (Zhu et al., 2006). In addition, these branched histidine-rich peptides with antifungal activity show less toxicity on mammalian cells.

The reason for the use of these histidine-rich peptides is that they exert a particular fungicidal impact. Although there are few studies in this regard, one review demonstrated that histatins are connected to the energized membrane (Mochon and Liu, 2008). Most reports show that histidine-rich peptides should cross the membrane and make an interaction with an intracellular target (Baev et al., 2004; Helmerhorst et al., 2001, 1999a). Remarkably, histidine-rich peptides in our laboratory acted as nucleic acid transporters due to their potential to cross through the membrane and disturb acidic endosomes (Chen et al., 2001; Leng et al., 2005; Leng and Mixson, 2005). Although the mechanisms to kill fungi by histatins and branched HK polymers are connected, this speculation has not been proved as yet (Matejuk et al., 2010).

4. Advantages of AFPs

A few antifungal medications influence normal eukaryotic targets in human cells and pathogenic fungi (Rautenbach et al., 2016). AFPs have advantages in comparison with other antifungal medications. First and foremost, it is a more grounded communication between the charge of the fungal membrane and the peptide's positive charge, as opposed to mammalian cell layers which are overwhelmingly neutral to host mammals. Second , AFPs are especially encouraging because they can perceive different microbial targets; in this way, the chance of resistance improvement is decreased (Rautenbach et al., 2016). Third , a few AFPs target layer lipids novel to growth and missing from mammalian cells, which likewise diminishes harmfulness (Rautenbach et al., 2016). This converts into a large variety and microbe selectivity and reduces the likelihood of toxicity against mammal cells (Rautenbach et al., 2016).

The determination of proper antifungal treatment is problematic due to the presence of a few classes of antifungals and the rising antifungal resistance (Pappas et al., 2018; Sanguinetti et al., 2015). Ergosterol synthesis in the cell wall is the neutral sites of azoles, echinocandins and polyenes. The low variety of current drugs results in becoming multi-resistant to antifungal agents. Some antifungals influence eukaryotic cell targets, which are present in both pathogenic fungal and human cells

(Rautenbach et al., 2016). AFPs show different antifungal mechanisms and targets in the cell, subsequently lessening the probability of drug resistance (Rautenbach et al., 2016). These targets include cell walls, fungal membranes and macromolecules which affect the synthesis of RNA, DNA and protein and the cell cycle (Bondaryk et al., 2017; Van der Weerden et al., 2013).

One of the characteristics of some AFPs is targeting the conserved fungal molecules which cause selectivity for fungal cells, and decrease the toxicity in mammalian cells (Rautenbach et al., 2016). However, it does not guarantee the lack of cytotoxicity. For example, aculeacins are an antifungal of the echinocandin class that targets 1,3-β-glucan synthase of fungi, which shows hemolytic activity (Matejuk et al., 2010). Recently, three anidulafungin, echinocandins, micafungin and caspofungin have been approved (Beyda et al., 2012; Pound et al., 2011). These AFPs are substrates for cytochrome P450, which convert into less hepatotoxicity and reduce the risk for cessation of treatment (3.7–4.8%), compared to the other antifungal agents. It seems AFPs show diminished cytotoxicity because of two reasons (Matejuk et al., 2010). First, there is a stronger interaction between the negative charges of the fungal membrane due to many contents of phosphatidic acid and phosphatidylinositol and the positive charges of the peptides. The mammalian cell membranes are dominatingly neutral because of a lot of substances of phosphatidylcholine. Furthermore, some AFPs target special lipids of a membrane of fungi, that are not present in mammalian cells and lessen their toxicities (Nguyen et al., 2011; Rautenbach et al., 2016). All reports indicate not only the antimicrobial impacts of AFPs but also their effective multifactorial impacts.

5. Limitation of AFPs

Despite a lot of reports of identified AMPs from several macro and microorganisms, there is little success in the commercialization of AFPs due to the following problems:

- Long sequences of AFPs are both synthetically and economically difficult for medicinal scale production.
- AFPs molecules have disulfide bonds and CCKs compound which are difficult to be mimicked in industrial production.
- Folding of AMPs into α-helical or β-sheet as a secondary structure is necessary for their impacts, which may not be achieved in its production and prompt inactivity.
- Many AMPs that are formed in different organisms may exhibit toxicity and high immunogenicity when administrated in people making them therapeutically not suitable.
- AFPs have an inadequate serum half-life because of their susceptibility to protease activity as well as physiological salt concentration.
- Delivery of AMPs to the sites of infection is also challenging (Sarkar et al., 2021).

5.1 AFPs resistance

AMPs and AFPs often act through interaction with the membrane (Wu et al., 1999). Microorganisms quickly change and adjust rapidly when subject to antifungals, though the evolution of cell membranes is slower.

With the fast and strong impact on the membrane, combined with other inhibitory components by AFPs, in target microorganisms, *de novo* resistance is less likely to occur (Yeung et al., 2011).

The regular use of AFPs speeds up the resistance in fungi. The increased development of resistance to AFPs is a challenging subject, although it is less, compared with that notable for other antifungals; however, it completely depends on the way the antimicrobial peptide is administered. Furthermore, the lack or alteration in the fungal target through spontaneous mutation can prompt resistance, although such alteration of the conserved molecule could cause a decrease in the virulence of pathogens. What is more, treatment with a few drugs, including the utilization of AFPs with other peptides or antibiotics will diminish the increased rate of resistance (Fernández de Ullivarri et al., 2020; Matejuk et al., 2010; Rautenbach et al., 2016).

5.2 Large production of AFPs

Insufficient studies on AFPs and the lack of an optimized process for their production have limited their use. An increment in the degree of AMPs has been seen in plant species during improvement, in the wake of injury or in signaling molecules' presence (Thery et al., 2019). Since synthetic peptides production is becoming cheaper, it may from now on, be more feasible financially in comparision with normal production or extraction of recombinant (regular) peptides. Nonetheless, disregarding moderately modest unrefined components, the extraction and decontamination of normal AFPs remain a costly course. Seeds are helpful sources regularly utilized for the segregation of plant AFPs. The processes of extraction using seeds can be enhanced in different ways, through the prompt articulation of the plant AFPs in seedlings in light of a few abiotic and biotic boosts (Stotz et al., 2009). Potential answers for the issue of proteolysis incorporate the definition of the peptide to bear the cost of protection from proteases, including liposomal plans, as utilized for different medications (Steimbach et al., 2017), use of non-natural or D-amino acids (Oliva et al., 2018), design and development of peptidomimetics (Kuppusamy et al., 2019) and multivalent peptides (Lakshminarayanan et al., 2014). The definition and conveyance of AMP will assume key parts in viability results including decreasing debasement of susceptible AMP for protease, restricted to plasma and different macromolecules and proteins, controlling the parameters of dose-exposure and even possibly focusing on microorganisms straightforwardly (e.g., intracellular microbes or microorganisms in biofilms). The formulation that has been tried for AMP incorporates the use of hydrogels (Kong et al., 2016), liposomal formulations (Zambom et al., 2019), carbon nanotubes (Viraka Nellore et al., 2015), PEGylation (Nordström and Malmsten, 2017) and nanoparticles (Bajaj et al., 2017). In contrast to the extensive collection of exploration focusing on the revelation of AMPs and the advancement of their activity, impressively less exertion has been given to conveyance frameworks, definitions or courses of AMP organization. AMP charge (and its type and dispersion), size, solvency, hydrophobicity and construction can influence stacking and action as well as the properties of the transporter depending on pH, charge, ionic strength, network size and formation strategy (where fitting) (Deo et al., 2022; Mercer and O'Neil, 2020).

5.3 The effect of salt, pH, temperature and proteolysis on AFPs

As a general rule, the physical and compound properties of the antifungal substances are not entirely resolved. The necessity for environmental conditions is mostly based on the understanding of the application of these peptides.

Salt: The various associations that occur between the fungi, ions and AFPs and these examinations can be useful for AFP capacities. A diminished action with a focus of high salt has also been found for the manufactured [RLLR]5 peptides, that have been made by the interruption of the α-helix (Park et al., 2004). The antifungal action of a number of AFPs, including defensins (Berrocal-Lobo et al., 2002), is diminished at high salt focusing as well as on the presence of cations. Impact of salt as a food additive, a productive peptide should hold fractional inhibitory action in the presence of salt. Strategies, like start-to-finish cyclization (Yu et al., 2000) and the production of dimers or peptides hybridization can also improve salt resistance (Han et al., 2016). Various ions distinctively affect a similar AFP, and the impact of similar ions on various AFPs is not the same (Schmidt et al., 2019). Metal ions (K^+, Na^+, Mg^{2+}, Ca^{2+}, Zn^{2+}, Cu^{2+} and so on) additionally influence the activity of a few antifungal peptides (Li et al., 2021).

Temperature: Defensin-type peptides are additionally impervious to high temperature, although an activity of slight loss can occur at temperatures higher than 80 to 90°C when the structure is irreversibly different (Thery et al., 2019). The most notable temperature endured by most AFPs is around 100°C (Zhang et al., 2017). Investigations of the impact of low temperature on the activity of the peptides, particularly against growth, are very restricted, and effect is by all accounts peptide subordinate.

Proteolysis: Albeit a few AFPs follow up their activities after treatment with chymotrypsin or trypsin (Kim et al., 2014), the AFPs proteolytic resistance is intricate, and stable peptides can likewise be delicate (Hajji et al., 2010). Concerning protease strength, the antidegradation impact of the antifungal peptide on papain, protease K, trypsin, pepsin, etc. (Li et al., 2021).

pH: There is speculation in ongoing examinations revealing that high pH restrains the antifungal action of manufactured peptides (Walkenhorst et al., 2013), although some peptides display better activity at higher pH (Hajji et al., 2010). Many peptides are stable over a wide range of pH (Chan et al., 2012). To get a bigger pH range, dimers (P-113) can be made. Executioner poisons created by yeasts are typically dynamic inside a limited scope of pH (Soares and Sato, 2000). For example, AFP PcPAF is highly active in weakly acidic environments (Wen et al., 2014).

UV stability: AFP is illuminated with UV at different times and in various portions (Shokri et al., 2014). Researchers focus on AFPs that are resistant to these chemicals (Ramachandran et al., 2014).

5.4 Cell toxicity

Most AFPs are perceived as being nontoxic to mammalian cells due to their particular cooperation with microorganism explicit targets (Lohner and Leber, 2016). For instance, the higher substance of cholesterol in their layer represents mammalian cells' security against the poison of AFPs (Matsuzaki, 2009). In addition, contrasts in the organization of film lipids among mammalian and contagious cells are significant elements, without harmful impacts because of human use. The addition of PTMs, the alteration of peptide spine and disturbance of α-helical designs are normal techniques used to diminish the harmfulness of AFPs (Thery et al., 2019). When an antifungal species has been recognized and its activity stability determined its true capacity for additional examination and clinical application relies on its poison to the body. Other antifungal medications produce gentle unfriendly responses, including nearby phlebitis, fever, liver damage and gentle hemolysis (Denning, 2003). Today, the significantly unfriendly impact created by specific AFPs is hematotoxicity addressed by erythrocyte hemolysis, which occurs at various levels of seriousness. In addition, these unfavorable responses have continuously decreased in intensity as these antifungal medications have kept on being renewed and reached the next level (Hector and Bierer, 2011). Notwithstanding hemolysis, antifungal peptides can also harm DNA, e.g., actinomycin D (an individual from the chromopeptids) (Yamamoto et al., 2011). For instance, the medications customarily used to treat profound intrusive growths (azole or polyene drugs) can prompt hepatotoxicity (Tverdek et al., 2016) or nephrotoxicity (Hamill, 2013) creating great harm to different parts of the body under the influence of numerous variables. In this unique situation, the decrease or end of the poisonousness of antifungal substances has been a significant issue for some time. Generally one of the fundamental burdens of a few existing antifungal medications is that they are exceptionally poisonous to the liver, kidneys as well as blood (Li et al., 2021).

6. Future perspectives

Chemoinformatics and bioinformatics devices have become the key for antimicrobial medication research diminishing the expense of high-throughput screening and further developing the achievement rate (Danishuddin and Khan, 2015). New objectives can be accomplished with blended treatment, in which AFPs or their analogs could synergistically build the activity of traditional antifungals (Ferreira et al., 2015). AFPs from regular sources, described by strong and wide-range exercises against both planktonic and biofilm contagious microorganisms, could be a significant asset for the control of fungal diseases (Kang et al., 2014). *In silico* sub-atomic displaying strategies could be used to make this cycle more effective and will speed up drug revelation (Medina Marrero et al., 2015). Proceeding with enhancements in normal products research is required to keep up this kind of examination determinedly along with other medication revelation strategies. Structure-based plan innovation has advanced and today, these procedures are by and large broadly utilized and credited for the revelation and plan of the majority of the new medication materials on the lookout. Combinatorial science has also allowed the distinguishing proof of AFPs utilizing nonsupport bound and blend-based synthetic combinatorial libraries. Soon, normal/changed AFPs or their analogs/mirrors will likely lead to a class of promising novel medications that are suitable for safe microbe development. In addition, programmatic experience strategies could probably be used to investigate their fungicidal system at the nuclear level and sub-atomic element techniques to explore connections among peptides and lipid bilayers (Matejuk et al., 2010). At long last, expanding information about the construction work connection of AFPs brings to light new chances to work on their capacity for the production of mycotoxin and their (specific) antifungal action, for instance, by the replacement of single amino acids or the union of non-normal peptides. Peptides production by heterologous articulation frameworks has turned into a rapidly growing area of examination, and pertinent instances of AFPs delivered in suitable amount in microorganisms, yeast, filamentous fungi or plants have been accounted for (Martínez-Culebras et al., 2021). These methodologies allow the improvement of synthetic peptides with more prominent antifungals and are hostile to mycotoxin actions and qualities, for example, decreased harmfulness and steadiness which are fundamental for their application as food additives (Meyer, 2008). The advancement in the improvement of new production frameworks and aging cycles will ensure the development of steady, unadulterated and practical AFPs in amounts expected for effective commercialization. Notwithstanding the developing number of logical reports on antimicrobial mixtures to be applied in horticulture, postharvest and food, the concentration on genuine use of AFPs is still very restricted. The commercial utilization of these materials is hampered by the difficulties in their production. In future, qualities like dissolvability, strength or cytotoxicity should be improved; finally, AFPs could be considered to be dynamic over the long haul in complex grids like food varieties (Thery et al., 2019). The activity of AFPs will be compromised by various qualities of food grids like high convergence of salts (Thery et al., 2020). Additionally, the adequacy of AFPs can be improved through structure adjustment, peptide concatemerization, as well as the age of peptide half breed combinations (Kerenga et al., 2019).

7. Conclusion

AMPs exhibit extra capacities for decreasing mycotoxin production which makes them integral assets to the conflict against the mycotoxin pollution of food. In this condition, AFPs are regular options for post harvesting and food as bioprotective devices to battle fungal decay and mycotoxin defilement. The instruments by which

AFPs collaborate with mycotoxin production are changed and complicated, going from the association of oxidative pressure and the hindrance of substrate fermentation to the particular restraint of enzymatic parts of the pathways of mycotoxin biosynthetic (Martínez-Culebras et al., 2021). Notwithstanding the excess gaps in information, particularly in regards to their system of activity, a comprehension of key determinants of AFPs' action as powerful synthetic peptides that have a minimal expense, low poisonousness and high steadiness is a strong starting point for numerous reports. Regardless of its promising outcomes, the utilization of these manufactured peptides as food additives should challenge the rising consumers' interest in green-named normal additives and the difficulties related to acquiring administrative endorsement. The rising use of computational-based strategies has smoothed out the plan and enhanced novel peptides which are consistent for the rising number of peptides with biotechnology applications. AFPs are financially appealing to applicants regarding fabricating costs, choices, expanding administrative acknowledgement of peptide therapeutics and so on. Additionally, dependability of the definitions, conveyance procedures and the general remedial productivity along with creation costs at a modern scale and administrative hindrances still need to be settled. At this point, compound amalgamation is financially reasonable just for short peptides and high-worth applications, yet a clever blend and decontamination advancements. Research on AFPs has been profoundly dynamic and nearly 1,000 peptides have been depicted. Endurance rates remain unsuitably low and no new antifungals have been presented over a long time for echinochandins and pneumocandins. AFPs have clear potential as more efficient and safer therapeutic agents than usual antifungal drugs (Fernández de Ullivarri et al., 2020). Normal peptides could act as layouts for the plan of new antifungal medications by substance change, derivatization or blended once more. Peptides could address a limitless source as driving designs for the normal plan of another class of remedial mixtures to be utilized alone, in collaboration with existing medications or formed into different mixtures, as self-gathered or formed nanoparticles. Normal antifungal peptides add to intrinsic insusceptibility and address a deep-rooted instrument of the host to safeguard against fungi. Peptides have various benefits over small particles albeit a few issues are related to their use as antifungal medications. Peptides with antifungal activity are broadly dispersed in nature and have high selectivity and power against growths. Taking into account, the examination introduced by this represents the potential for the utilization of AFPs to control mycotoxin-delivering organisms and mycotoxin production. In this regard, AFPs from various starting points and designs have been also reported in this chapter to show the antagonist of mycotoxin capacities. With the improvement of resistance against fungicides, experts have been encouraged to use them. AFPs are considered potential answers for the constant presence of food decay and mycotoxin defilement, which are subjects of the central issue. Due to their power, a wide range of action, various wellsprings of accessibility in nature, absence of quick advancement of obstruction, low harmfulness and quick killing activity, these proteins and peptides reveal a few benefits over traditionally utilized fungicides and additives. There are a few issues restricting peptide use, for example, hemolytic action, low bioavailability, conceivable collection, shakiness, the significant expense of production, fast turnover in the body of humans, unfortunate capacity to cross physiological boundaries and loss of activity in high salt concentration (Ciociola et al., 2016). As underscored of late in a few examinations, for these peptides and peptide-impersonates to be used, their methods of antifungal activity need to be known. Arising contagious protection from regular treatments requires the improvement of novel antifungal strategies (Bondaryk et al., 2017). These variables are without a doubt, a part of the explanation for there not being more advance of AFP through the medication improvement cycle as well as AFP gaining importance for the status of skin treatment as conveyance frameworks, definition, courses of organization and span of treatment for AFP have not been sufficiently enhanced. Now is the ideal opportunity for more notable double-dealing of AFP and another immunotherapeutic as antifungal medication competitors as one gains a comprehension of how best to test these medications gaining prominence *in vitro* and how administrative pathways and clinical examinations can be more amenable for peptides. Unfortunately, one does not witness sufficient medications likely to advance enduring the medication progression pipeline, as *in vitro* and *in vivo* testing approaches are not generally suitable or potentially streamlined for AFP (Mercer et al., 2020). AFPs are promising applicants as therapeutics for contagious disease treatment and are necessary for clinical practise because of the restricted collection of treatment choices and expanding protection from existing antifungals. As the worldwide antifungal resistance emergency declines and the increase of contagious diseases , the capability of these medication competitors should be satisfied as soon as possible.

References

Abe, H., Kawada, M., Sakashita, C., Watanabe, T. and Shibasaki, M. (2018). Structure-activity relationship study of leucinostatin A, a modulator of tumor-stroma interaction. Tetrahedron, 74: 5129–5137.

Agrawal, P., Bhalla, S., Chaudhary, K., Kumar, R., Sharma, M. and Raghava, G.P.S. (2018). *In silico* approach for prediction of antifungal peptides. Front. Microbiol., 9: 323.

Ahmad, M., Piludu, M., Oppenheim, F.G., Helmerhorst, E.J. and Hand, A.R. (2004). Immunocytochemical localization of histatins in human salivary glands. J. Histochem. Cytochem., 52: 361–370.

Alvarez, F., Castro, M., Príncipe, A., Borioli, G., Fischer, S., Mori, G. and Jofré, E. (2012). The plant-associated Bacillus amyloliquefaciens strains MEP 218 and ARP 23 capable of producing the cyclic lipopeptides iturin or surfactin and fengycin are effective in biocontrol of sclerotinia stem rot disease. J. Appl. Microbiol., 112: 159–174.

Arnusch, C.J., Ulm, H., Josten, M., Shadkchan, Y., Osherov, N., Sahl, H.-G. and Shai, Y. (2012). Ultrashort peptide bioconjugates are exclusively antifungal agents and synergize with cyclodextrin and amphotericin B. Antimicrob. Agents Chemother., 56: 1–9.

Badiee, P. and Hashemizadeh, Z. (2014). Opportunistic invasive fungal infections: Diagnosis & clinical management. Indian J. Med. Res., 139: 195.

Baev, D., Rivetta, A., Vylkova, S., Sun, J.N., Zeng, G.-F., Slayman, C.L. and Edgerton, M. (2004). The TRK1 potassium transporter is the critical effector for killing of Candida albicans by the cationic protein, Histatin 5. J. Biol. Chem., 279: 55060–55072.

Bajaj, M., Pandey, S.K., Nain, T., Brar, S.K., Singh, P., Singh, S., Wangoo, N. and Sharma, R.K. (2017). Stabilized cationic dipeptide capped gold/silver nanohybrids: Towards enhanced antibacterial and antifungal efficacy. Colloids Surfaces B Biointerfaces, 158: 397–407.

Benincasa, M., Scocchi, M., Pacor, S., Tossi, A., Nobili, D., Basaglia, G., Busetti, M. and Gennaro, R. (2006). Fungicidal activity of five cathelicidin peptides against clinically isolated yeasts. J. Antimicrob. Chemother., 58: 950–959.

Berrocal-Lobo, M., Segura, A., Moreno, M., López, G., Garcıa-Olmedo, F. and Molina, A. (2002). Snakin-2, an antimicrobial peptide from potato whose gene is locally induced by wounding and responds to pathogen infection. Plant Physiol., 128: 951–961.

Besson, F. and Michel, G. (1984). Action of the antibiotics of the iturin group on artificial membranes. J. Antibiot. (Tokyo), 37: 646–651.

Beyda, N.D., Lewis, R.E. and Garey, K.W. (2012). Echinocandin resistance in Candida species: Mechanisms of reduced susceptibility and therapeutic approaches. Ann. Pharmacother., 46: 1086–1096.

Bondaryk, M., Staniszewska, M., Zielińska, P. and Urbańczyk-Lipkowska, Z. (2017). Natural antimicrobial peptides as inspiration for design of a new generation antifungal compounds. J. Fungi, 3: 46.

Bongomin, F., Gago, S., Oladele, R.O. and Denning, D.W. (2017). Global and multi-national prevalence of fungal diseases—estimate precision. J. Fungi, 3: 57.

Brewer, D., Hunter, H. and Lajoie, G. (1998). NMR studies of the antimicrobial salivary peptides histatin 3 and histatin 5 in aqueous and nonaqueous solutions. Biochem. Cell Biol., 76: 247–256.

Brogden, K.A. (2005). Antimicrobial peptides: Pore formers or metabolic inhibitors in bacteria? Nat. Rev. Microbiol., 3: 238–250.

Brown, G.D., Denning, D.W., Gow, N.A.R., Levitz, S.M., Netea, M.G. and White, T.C. (2012). Hidden killers: Human fungal infections. Sci. Transl. Med., 4: 165rv13–165rv13.

Buda De Cesare, G., Cristy, S.A., Garsin, D.A. and Lorenz, M.C. (2020). Antimicrobial peptides: A new frontier in antifungal therapy. MBio, 11: e02123–20.

Bulet, P. and Stocklin, R. (2005). Insect antimicrobial peptides: structures, properties and gene regulation. Protein Pept. Lett., 12: 3–11.

Campbell-Platt, G. and Cook, P.E. (1989). Fungi in the production of foods and food ingredients. J. Appl. Bacteriol., 67: 117s–131s.

Castagnola, M., Inzitari, R., Rossetti, D.V., Olmi, C., Cabras, T., Piras, V., Nicolussi, P., Sanna, M.T., Pellegrini, M. and Giardina, B. (2004). A cascade of 24 Histatins (Histatin 3 fragments) in human saliva: suggestions for a pre-secretory sequential cleavage pathway. J. Biol. Chem., 279: 41436–41443.

Chan, Y.S., Wong, J.H., Fang, E.F., Pan, W.L. and Ng, T.B. (2012). An antifungal peptide from Phaseolus vulgaris cv. brown kidney bean. Acta Biochim. Biophys. Sin., 44: 307–315.

Chen, Q.-R., Zhang, L., Stass, S.A. and Mixson, A.J. (2001). Branched co-polymers of histidine and lysine are efficient carriers of plasmids. Nucleic Acids Res., 29: 1334–1340.

Chen, S.C.-A., Slavin, M.A. and Sorrell, T.C. (2011). Echinocandin antifungal drugs in fungal infections. Drugs, 71: 11–41.

Ciociola, T., Giovati, L., Conti, S., Magliani, W., Santinoli, C. and Polonelli, L. (2016). Natural and synthetic peptides with antifungal activity. Future Med. Chem., 8: 1413–1433.

Clemons, K.V. and Stevens, D.A. (1997). Efficacy of nikkomycin Z against experimental pulmonary blastomycosis. Antimicrobial Agents and Chemotherapy, 41(9): 2026–2028.

Cohen, L., Moran, Y., Sharon, A., Segal, D., Gordon, D. and Gurevitz, M. (2009). Drosomycin, an innate immunity peptide of Drosophila melanogaster, interacts with the fly voltage-gated sodium channel. J. Biol. Chem., 284: 23558–23563.

Danishuddin, M. and Khan, A.U. (2015). Structure based virtual screening to discover putative drug candidates: necessary considerations and successful case studies. Methods, 71: 135–145.

De Lucca, A.J. (2000). Antifungal peptides: potential candidates for the treatment of fungal infections. Expert Opin. Investig. Drugs, 9: 273–299.

De Lucca, A.J. and Walsh, T.J. (1999). Antifungal peptides: novel therapeutic compounds against emerging pathogens. Antimicrob. Agents Chemother., 43: 1–11.

Denning, D.W. (2003). Echinocandin antifungal drugs. Lancet, 362: 1142–1151.

Denning, D.W. and Bromley, M.J. (2015). How to bolster the antifungal pipeline. Science, (80-). 347: 1414–1416.

Deo, S., Turton, K.L., Kainth, T., Kumar, A. and Wieden, H.-J. (2022). Strategies for improving antimicrobial peptide production. Biotechnol. Adv., 107968.

Douglas, L.J. (2003). Candida biofilms and their role in infection. Trends Microbiol. 11: 30–36.

Duncan, V.M.S. and O'Neil, D.A. (2013). Commercialization of antifungal peptides. Fungal Biol. Rev., 26: 156–165.

Emri, T., Majoros, L., Tóth, V. and Pócsi, I. (2013). Echinocandins: Production and applications. Appl. Microbiol. Biotechnol., 97: 3267–3284.

Enaud, R., Vandenborght, L.-E., Coron, N., Bazin, T., Prevel, R., Schaeverbeke, T., Berger, P., Fayon, M., Lamireau, T. and Delhaes, L. (2018). The mycobiome: A neglected component in the microbiota-gut-brain axis. Microorganisms, 6: 22.

Endo, M., Takesako, K., Kato, I. and Yamaguchi, H. (1997). Fungicidal action of aureobasidin A, a cyclic depsipeptide antifungal antibiotic, against Saccharomyces cerevisiae. Antimicrob. Agents Chemother., 41: 672–676.

Epstein, D.J., Seo, S.K., Brown, J.M. and Papanicolaou, G.A. (2018). Echinocandin prophylaxis in patients undergoing haematopoietic cell transplantation and other treatments for haematological malignancies. J. Antimicrob. Chemother., 73: i60–i72.

Eschenauer, G., DePestel, D.D. and Carver, P.L. (2007). Comparison of echinocandin antifungals. Ther. Clin. Risk Manag., 3: 71.

Espinel-Ingroff, A. (1998). Comparison of *in vitro* activities of the new triazole SCH56592 and the echinocandins MK-0991 (L-743,872) and LY303366 against opportunistic filamentous and dimorphic fungi and yeasts. J. Clin. Microbiol., 36: 2950–2956.

Essig, A., Hofmann, D., Münch, D., Gayathri, S., Künzler, M., Kallio, P. T. and Aebi, M. (2014). Copsin, a novel peptide-based fungal antibiotic interfering with the peptidoglycan synthesis. Journal of Biological Chemistry, 289(50): 34953–34964.

Falla, T.J. and Hancock, R.E. (1997). Improved activity of a synthetic indolicidin analog. Antimicrob. Agents Chemother., 41: 771–775.

Fehlbaum, P., Bulet, P., Michaut, L., Lagueux, M., Broekaert, W.F., Hetru, C. and Hoffmann, J.A. (1994). Insect immunity. Septic injury of Drosophila induces the synthesis of a potent antifungal peptide with sequence homology to plant antifungal peptides. J. Biol. Chem., 269: 33159–33163.

Fernández-Vidal, M., Jayasinghe, S., Ladokhin, A.S. and White, S.H. (2007). Folding amphipathic helices into membranes: Amphiphilicity trumps hydrophobicity. J. Mol. Biol., 370: 459–470.

Fernández de Ullivarri, M., Arbulu, S., Garcia-Gutierrez, E. and Cotter, P.D. (2020). Antifungal peptides as therapeutic agents. Front. Cell. Infect. Microbiol., 10: 105.

Ferreira, S.Z., Carneiro, H.C., Lara, H.A., Alves, R.B., Resende, J.M., Oliveira, H.M., Silva, L.M., Santos, D.A. and Freitas, R.P. (2015). Synthesis of a new peptide–coumarin conjugate: A potential agent against cryptococcosis. ACS Med. Chem. Lett., 6: 271–275.

Finking, R. and Marahiel, M.A. (2004). Biosynthesis of nonribosomal peptides. Annu. Rev. Microbiol., 58: 453–488.

Fjell, C.D., Hancock, R.E.W. and Cherkasov, A. (2007). AMPer: A database and an automated discovery tool for antimicrobial peptides. Bioinformatics, 23: 1148–1155.

Freitas, C.G. and Franco, O.L. (2016). Antifungal peptides with potential against pathogenic fungi BT - recent trends in antifungal agents and antifungal therapy. pp. 75–95. *In*: A. Basak, R. Chakraborty and S.M. Mandal (eds.). Springer India, New Delhi.

Fromtling, R.A. and Abruzzo, G.K. (1989). L-671, 329, A new antifungal agent III. *In vitro* activity, toxicity and efficacy in comparison to aculeacin. J. Antibiot. (Tokyo), 42: 174–178.

Garrigues, S., Gandía, M., Borics, A., Marx, F., Manzanares, P. and Marcos, J.F. (2017). Mapping and identification of antifungal peptides in the putative antifungal protein AfpB from the filamentous fungus Penicillium digitatum. Front. Microbiol., 8: 592.

Gopal, R., Park, S., Ha, K., Cho, S.J., Kim, S.W., Song, P.I., Nah, J., Park, Y. and Hahm, K. (2009). Effect of leucine and lysine substitution on the antimicrobial activity and evaluation of the mechanism of the HPA3NT3 analog peptide. J. Pept. Sci. an Off. Publ. Eur. Pept. Soc., 15: 589–594.

Gopal, R., Seo, C.H., Song, P.I. and Park, Y. (2013). Effect of repetitive lysine–tryptophan motifs on the bactericidal activity of antimicrobial peptides. Amino Acids, 44: 645–660.

Guo, C., Pan, L., Xiao, S., Chen, H. and Jiang, Z. (2013). Short simple linear peptides mimic antimicrobial complex cyclodecapeptides based on the putative pharmacophore. Med. Chem. (Los. Angeles), 4: 322–329.

Guyard, C., Dehecq, E., Tissier, J.P., Polonelli, L., Dei-Cas, E., Cailliez, J.C. and Menozzi, F.D. (2002). Involvement of β-glucans in the wide-spectrum antimicrobial activity of Williopsis saturnus var. mrakii MUCL 41968 killer toxin. Molecular Medicine, 8: 686–694.

Hajji, M., Jellouli, K., Hmidet, N., Balti, R., Sellami-Kamoun, A. and Nasri, M. (2010). A highly thermostable antimicrobial peptide from Aspergillus clavatus ES1: Biochemical and molecular characterization. J. Ind. Microbiol. Biotechnol., 37: 805–813.

Hamill, R.J. (2013). Amphotericin B formulations: a comparative review of efficacy and toxicity. Drugs, 73: 919–934.

Han, J., Jyoti, M.A., Song, H.-Y. and Jang, W.S. (2016). Antifungal activity and action mechanism of histatin 5-halocidin hybrid peptides against Candida ssp. PLoS One, 11: e0150196.

Hancock, R.E.W. and Chapple, D.S. (1999). Peptide antibiotics. Antimicrob. Agents Chemother., 43: 1317–1323.

Hawser, S., Borgonovi, M., Markus, A. and Isert, D. (1999). Mulundocandm, an echinocandin-like lipopeptide antifungal agent: Biological activities *in vitro*. J. Antibiot. (Tokyo), 52: 305–310.

Healey, K.R., Zhao, Y., Perez, W.B., Lockhart, S.R., Sobel, J.D., Farmakiotis, D., Kontoyiannis, D.P., Sanglard, D., Taj-Aldeen, S.J. and Alexander, B.D. (2016). Prevalent mutator genotype identified in fungal pathogen Candida glabrata promotes multi-drug resistance. Nat. Commun., 7: 1–10.

Hector, R.F. and Bierer, D.E. (2011). New β-glucan inhibitors as antifungal drugs. Expert. Opin. Ther. Pat., 21: 1597–1610.

Hector, R.F., Zimmer, B.L. and Pappagianis, D. (1990). Evaluation of nikkomycins X and Z in murine models of coccidioidomycosis, histoplasmosis, and blastomycosis. Antimicrob. Agents Chemother., 34: 587–593.

Helmerhorst, E.J., Breeuwer, P., van't Hof, W., Walgreen-Weterings, E., Oomen, L.C.J.M., Veerman, E.C.I., Amerongen, A.V.N. and Abee, T. (1999a). The cellular target of histatin 5 on *Candida albicans* is the energized mitochondrion. J. Biol. Chem., 274: 7286–7291.

Helmerhorst, E.J., Reijnders, I.M., van't Hof, W., Simoons-Smit, I., Veerman, E.C.I. and Amerongen, A.V.N. (1999b). Amphotericin B-and fluconazole-resistant Candida spp., Aspergillus fumigatus, and other newly emerging pathogenic fungi are susceptible to basic antifungal peptides. Antimicrob. Agents Chemother., 43: 702–704.

Helmerhorst, E.J., Van't Hof, W., Breeuwer, P., Veerman, E.I., Abee, T., Troxler, R.F., Amerongen, A.V.N. and Oppenheim, F.G. (2001). Characterization of histatin 5 with respect to amphipathicity, hydrophobicity, and effects on cell and mitochondrial membrane integrity excludes a candidacidal mechanism of pore formation. J. Biol. Chem., 276: 5643–5649.

Hope, W.W., Tabernero, L., Denning, D.W. and Anderson, M.J. (2004). Molecular mechanisms of primary resistance to flucytosine in Candida albicans. Antimicrob. Agents Chemother., 48: 4377–4386.

Hori, M., Kakiki, K. and Misato, T. (1974). Interaction between polyoxin and active center of chitin synthetase. Agric. Biol. Chem., 38: 699–705.

Huffnagle, G.B. and Noverr, M.C. (2013). The emerging world of the fungal microbiome. Trends in Microbiology, 21(7): 334–341.

Iavarone, F., Desiderio, C., Vitali, A., Messana, I., Martelli, C., Castagnola, M. and Cabras, T. (2018). Cryptides: Latent peptides everywhere. Crit. Rev. Biochem. Mol. Biol., 53: 246–263.

Iijima, R., Kurata, S. and Natori, S. (1993). Purification, characterization, and cDNA cloning of an antifungal protein from the hemolymph of Sarcophaga peregrina (flesh fly) larvae. J. Biol. Chem., 268: 12055–12061.

Ikai, K., Takesako, K., Shiomi, K., Moriguchi, M., Umeda, Y., Yamamoto, J., Kato, I. and Naganawa, H. (1991). Structure of aureobasidin AJ Antibiot., 44: 925–933.

Iwata, K., Yamamoto, Y., Yamaguchi, H. and Hiratani, T. (1982). *In vitro* studies of aculeacin A, a new antifungal antibiotic. J. Antibiot. (Tokyo), 35: 203–209.

Jin, L., Bai, X., Luan, N., Yao, H., Zhang, Z., Liu, W., Chen, Y., Yan, X., Rong, M. and Lai, R. (2016). A designed tryptophan-and lysine/arginine-rich antimicrobial peptide with therapeutic potential for clinical antibiotic-resistant Candida albicans vaginitis. J. Med. Chem., 59: 1791–1799.

Kamysz, E., Sikorska, E., Karafova, A. and Dawgul, M. 2012. Synthesis, biological activity and conformational analysis of head-to-tail cyclic analogues of LL37 and histatin 5. J. Pept. Sci., 18: 560–566.

Kang, S.-J., Park, S.J., Mishig-Ochir, T. and Lee, B.-J. (2014). Antimicrobial peptides: Therapeutic potentials. Expert Rev. Anti. Infect. Ther., 12: 1477–1486.

Kapitan, M., Niemiec, M.J., Steimle, A., Frick, J.S. and Jacobsen, I.D. (2018). Fungi as part of the microbiota and interactions with intestinal bacteria. Fungal Physiol. Immunopathogenes, 265–301.

Kaushik, N., Pujalte, G.G.A. and Reese, S.T. (2015). Superficial fungal infections. Prim. care Clin. Off. Pract., 42: 501–516.

Kerenga, B.K., McKenna, J.A., Harvey, P.J., Quimbar, P., Garcia-Ceron, D., Lay, F.T., Phan, T.K., Veneer, P.K., Vasa, S. and Parisi, K. (2019). Salt-tolerant antifungal and antibacterial activities of the corn defensin ZmD32. Front. Microbiol., 10: 795.

Kim, H., Jang, J.H., Kim, S.C. and Cho, J.H. (2014). *De novo* generation of short antimicrobial peptides with enhanced stability and cell specificity. J. Antimicrob. Chemother., 69: 121–132.

Kohler, S., Wheat, L.J., Connolly, P., Schnizlein-Bick, C., Durkin, M., Smedema, M., Goldberg, J. and Brizendine, E. (2000). Comparison of the echinocandin caspofungin with amphotericin B for treatment of histoplasmosis following pulmonary challenge in a murine model. Antimicrob. Agents Chemother., 44: 1850–1854.

Kołaczkowska, A. and Kołaczkowski, M. (2016). Drug resistance mechanisms and their regulation in non-albicans Candida species. J. Antimicrob. Chemother., 71: 1438–1450.

Kong, E.F., Tsui, C., Boyce, H., Ibrahim, A., Hoag, S.W., Karlsson, A.J., Meiller, T.F. and Jabra-Rizk, M.A. (2016). Development and *in vivo* evaluation of a novel histatin-5 bioadhesive hydrogel formulation against oral candidiasis. Antimicrob. Agents Chemother., 60: 881–889.

Koshlukova, S.E., Lloyd, T.L., Araujo, M.W.B. and Edgerton, M. (1999). Salivary histatin 5 induces Non-lytic release of ATP fromCandida albicans leading to cell death. J. Biol. Chem., 274: 18872–18879.

Kuppusamy, R., Willcox, M., Black, D.S. and Kumar, N. (2019). Short cationic peptidomimetic antimicrobials. Antibiotics, 8: 44.

Lakshminarayanan, R., Liu, S., Li, J., Nandhakumar, M., Aung, T.T., Goh, E., Chang, J.Y.T., Saraswathi, P., Tang, C. and Safie, S.R.B. (2014). Synthetic multivalent antifungal peptides effective against fungi. PLoS One, 9: e87730.

Landy, M., Warren, G.H., RosenmanM, S.B. and Colio, L.G. (1948). Bacillomycin: An antibiotic from Bacillus subtilis active against pathogenic fungi. Proc. Soc. Exp. Biol. Med., 67: 539–541.

Lee, D.G., Kim, P.Il, Park, Y., Woo, E.-R., Choi, J.S., Choi, C.-H. and Hahm, K.-S. (2002). Design of novel peptide analogs with potent fungicidal activity, based on PMAP-23 antimicrobial peptide isolated from porcine myeloid. Biochem. Biophys. Res. Commun., 293: 231–238.

Lee, S., Moon, H., Kawabata, S., Kurata, S., Natori, S. and Lee, B. (1995a). A sapecin homologue of Holotrichia diomphalia: Purification, sequencing and determination of disulfide pairs. Biol. Pharm. Bull., 18: 457–459.

Lee, S., Moon, H., Kurata, S., Natori, S. and Lee, B. (1995b). Purification and cDNA cloning of an antifungal protein from the hemolymph of Holotrichia diomphalia larvae. Biol. Pharm. Bull., 18: 1049–1052.

Leng, Q. and Mixson, A.J. (2005). Modified branched peptides with a histidine-rich tail enhance *in vitro* gene transfection. Nucleic Acids Res., 33: e40–e40.

Leng, Q., Scaria, P., Zhu, J., Ambulos, N., Campbell, P. and Mixson, A.J. (2005). Highly branched HK peptides are effective carriers of siRNA. J. Gene Med. A cross-disciplinary J. Res. Sci. gene Transf. its Clin. Appl., 7: 977–986.

Lewis, R.E. (2011). Current concepts in antifungal pharmacology. *In*: Mayo Clinic Proceedings. Elsevier, pp. 805–817.

Li, T., Li, L., Du, F., Sun, L., Shi, J., Long, M. and Chen, Z. (2021). Activity and mechanism of action of antifungal peptides from microorganisms: A review. Molecules, 26: 3438.

Lohner, K. and Leber, R. (2016). Antifungal host defense peptides. *In*: Host Defense Peptides and Their Potential as Therapeutic Agents. Springer, pp. 27–55.

López-Abarrategui, C., McBeth, C., Mandai, S.M., Sun, Z.J., Heffron, G., Alba-Menéndez, A., Migliolo, L., Reyes-Acosta, O., García-Villarino, M. and Nolasco, D.O. (2015). Cm-p5: An antifungal hydrophilic peptide derived from the coastal mollusk Cenchritis muricatus (Gastropoda: Littorinidae). FASEB J., 29: 3315–3325.

Lum, Kah Yean, Tay, S.T., Le, C.F., Lee, V.S., Sabri, N.H., Velayuthan, R.D., Hassan, H. and Sekaran, S.D. (2015). Activity of novel synthetic peptides against Candida albicans. Sci. Rep., 5: 1–12.

Lum, K.Y., Tay, S.T., Le, C.F., Lee, V.S., Sabri, N.H., Velayuthan, R.D., Hassan, H. and Sekaran, S.D. (2015). Activity of novel synthetic peptides against Candida albicans. Sci. Rep. 5: 9657.

MacKay, B.J., Pollock, J.J., Iacono, V.J. and Baum, B.J. (1984). Isolation of milligram quantities of a group of histidine-rich polypeptides from human parotid saliva. Infect. Immun., 44: 688–694.

Magana, M., Pushpanathan, M., Santos, A.L., Leanse, L., Fernandez, M., Ioannidis, A., Giulianotti, M.A., Apidianakis, Y., Bradfute, S. and Ferguson, A.L. (2020). The value of antimicrobial peptides in the age of resistance. Lancet Infect. Dis., 20: e216–e230.

Marcos, J.F. and Gandía, M. (2009). Antimicrobial peptides: to membranes and beyond. Expert Opin. Drug Discov., 4: 659–671.

Martínez-Culebras, P.V., Gandía, M., Garrigues, S., Marcos, J.F. and Manzanares, P. (2021). Antifungal peptides and proteins to control toxigenic fungi and mycotoxin biosynthesis. Int. J. Mol. Sci., 22: 13261.

Matejuk, A., Leng, Q., Begum, M.D., Woodle, M.C., Scaria, P., Chou, S.T. and Mixson, A.J. (2010). Peptide-based antifungal therapies against emerging infections. Drugs Future, 35: 197.

Matsuzaki, K. (2009). Control of cell selectivity of antimicrobial peptides. Biochim. Biophys. Acta (BBA)-Biomembranes, 1788: 1687–1692.

McCarthy, P.J., Troke, P.F. and Gull, K. (1985). Mechanism of action of nikkomycin and the peptide transport system of Candida albicans. Microbiology, 131: 775–780.

McNair, L.K.F., Siedler, S., Vinther, J.M.O., Hansen, A.M., Neves, A.R., Garrigues, C., Jäger, A.K., Franzyk, H. and Staerk, D. (2018). Identification and characterization of a new antifungal peptide in fermented milk product containing bioprotective Lactobacillus cultures. FEMS Yeast Res., 18: foy094.

Medina Marrero, R., Marrero-Ponce, Y., Barigye, S.J., Echeverria Diaz, Y., Acevedo-Barrios, R., Casanola-Martin, G.M., Garcia Bernal, M., Torrens, F. and Perez-Gimenez, F. (2015). QuBiLs-MAS method in early drug discovery and rational drug identification of antifungal agents. SAR QSAR Environ. Res., 26: 943–958.

Mercer, D.K. and O'Neil, D.A. (2020). Innate inspiration: Antifungal peptides and other immunotherapeutics from the host immune response. Front. Immunol., 2177.

Mercer, D.K., Torres, M.D.T., Duay, S.S., Lovie, E., Simpson, L., von Köckritz-Blickwede, M., De la Fuente-Nunez, C., O'Neil, D.A. and Angeles-Boza, A.M. (2020). Antimicrobial susceptibility testing of antimicrobial peptides to better predict efficacy. Front. Cell. Infect. Microbiol., 326.

Meyer, V. (2008). A small protein that fights fungi: AFP as a new promising antifungal agent of biotechnological value. Appl. Microbiol. Biotechnol., 78: 17–28.

Mhammedi, A., Peypoux, F., Besson, F. and Michel, G. (1982). Bacillomycin F, a new antibiotic of iturin group: Isolation and characterization. J. Antibiot. (Tokyo), 35: 306–311.

Mochon, A.B. and Liu, H. (2008). The antimicrobial peptide histatin-5 causes a spatially restricted disruption on the Candida albicans surface, allowing rapid entry of the peptide into the cytoplasm. PLoS Pathog., 4: e1000190.

Money, N.P. (2016). Fungi and biotechnology. In: The Fungi. Elsevier, pp. 401–424.

Moretta, A., Scieuzo, C., Petrone, A.M., Salvia, R., Manniello, M.D., Franco, A., Lucchetti, D., Vassallo, A., Vogel, H. and Sgambato, A. (2021). Antimicrobial peptides: A new hope in biomedical and pharmaceutical fields. Front. Cell. Infect. Microbiol., 11: 453.

Mukherjee, D., Singh, S., Kumar, M., Kumar, V., Datta, S. and Dhanjal, D.S. (2018). Fungal biotechnology: Role and aspects. In: Fungi and their Role in Sustainable Development: Current Perspectives. Springer, pp. 91–103.

Munoz, A., López-García, B. and Marcos, J.F. (2006). Studies on the mode of action of the antifungal hexapeptide PAF26. Antimicrob. Agents Chemother., 50: 3847–3855.

Nagiec, M.M., Nagiec, E.E., Baltisberger, J.A., Wells, G.B., Lester, R.L. and Dickson, R.C. (1997). Sphingolipid synthesis as a target for antifungal drugs: Complementation of the inositol phosphorylceramide synthase defect in a mutant strain of Saccharomyces cerevisiae by the AUR1 gene. J. Biol. Chem., 272: 9809–9817.

Nguyen, L.T., Haney, E.F. and Vogel, H.J. (2011). The expanding scope of antimicrobial peptide structures and their modes of action. Trends Biotechnol., 29: 464–472.

Nicolas, P. (2009). Multifunctional host defense peptides: Intracellular-targeting antimicrobial peptides. FEBS J., 276: 6483–6496.

Nordström, R. and Malmsten, M. (2017). Delivery systems for antimicrobial peptides. Adv. Colloid Interface Sci., 242: 17–34.

Oliva, R., Chino, M., Pane, K., Pistorio, V., De Santis, A., Pizzo, E., D'Errico, G., Pavone, V., Lombardi, A. and Del Vecchio, P. (2018). Exploring the role of unnatural amino acids in antimicrobial peptides. Sci. Rep., 8: 1–16.

Pappas, P.G., Kauffman, C.A., Andes, D.R., Clancy, C.J., Marr, K.A., Ostrosky-Zeichner, L., Reboli, A.C., Schuster, M.G., Vazquez, J.A. and Walsh, T.J. (2016). Clinical practice guideline for the management of candidiasis: 2016 update by the Infectious Diseases Society of America. Clin. Infect. Dis., 62: e1–e50.

Pappas, P.G., Lionakis, M.S., Arendrup, M.C., Ostrosky-Zeichner, L. and Kullberg, B.J. (2018). Invasive candidiasis. Nat. Rev. Dis. Prim., 4: 1–20.

Park, Y., Lee, D.G. and Hahm, K. (2004). HP (2–9)-magainin 2 (1–12), a synthetic hybrid peptide, exerts its antifungal effect on Candida albicans by damaging the plasma membrane. J. Pept. Sci. an Off. Publ. Eur. Pept. Soc., 10: 204–209.

Petrikkos, G. and Skiada, A. (2007). Recent advances in antifungal chemotherapy. Int. J. Antimicrob. Agents, 30: 108–117.

Pfaller, M.A. and Diekema, D. (2007). Epidemiology of invasive candidiasis: A persistent public health problem. Clin. Microbiol. Rev., 20: 133–163.

Pfaller, M.A., Castanheira, M., Lockhart, S.R., Ahlquist, A.M., Messer, S.A. and Jones, R.N. (2012). Frequency of decreased susceptibility and resistance to echinocandins among fluconazole-resistant bloodstream isolates of Candida glabrata. J. Clin. Microbiol., 50: 1199–1203.

Pluim, T., Halasa, N., Phillips, S.E. and Fleming, G. (2012). The morbidity and mortality of patients with fungal infections before and during extracorporeal membrane oxygenation support*. Pediatr. Crit. Care Med., 13.

Pollock, J.J., Denepitiya, L., MacKay, B.J. and Iacono, V.J. (1984). Fungistatic and fungicidal activity of human parotid salivary histidine-rich polypeptides on Candida albicans. Infect. Immun., 44: 702–707.

Posteraro, B., Sanguinetti, M., Sanglard, D., La Sorda, M., Boccia, S., Romano, L., Morace, G. and Fadda, G. (2003). Identification and characterization of a Cryptococcus neoformans ATP binding cassette (ABC) transporter-encoding gene, CnAFR1, involved in the resistance to fluconazole. Mol. Microbiol., 47: 357–371.

Pound, M.W., Townsend, M.L., Dimondi, V., Wilson, D. and Drew, R.H. (2011). Overview of treatment options for invasive fungal infections. Med. Mycol., 49: 561–580.

Prijck, K. De, Smet, N. De, Rymarczyk-Machal, M., Driessche, G. Van Devreese, B., Coenye, T., Schacht, E. and Nelis, H.J. (2009). Candida albicans biofilm formation on peptide functionalized polydimethylsiloxane. Biofouling, 26: 269–275.

Pusateri, C.R., Monaco, E.A. and Edgerton, M. (2009). Sensitivity of Candida albicans biofilm cells grown on denture acrylic to antifungal proteins and chlorhexidine. Arch. Oral Biol., 54: 588–594.

Raj, P.A., Edgerton, M. and Levine, M.J. (1990). Salivary histatin 5: Dependence of sequence, chain length, and helical conformation for candidacidal activity. J. Biol. Chem., 265: 3898–3905.

Raj, P.A., Marcus, E. and Sukumaran, D.K. (1998). Structure of human salivary histatin 5 in aqueous and nonaqueous solutions. Biopolym. Orig. Res. Biomol., 45: 51–67.

Raj, P.A., Soni, S.-D. and Levine, M.J. (1994). Membrane-induced helical conformation of an active candidacidal fragment of salivary histatins. J. Biol. Chem., 269: 9610–9619.

Ramachandran, R., Chalasani, A.G., Lal, R. and Roy, U. (2014). A broad-spectrum antimicrobial activity of Bacillus subtilis RLID 12.1. Sci. World J., 2014.

Rautemaa-Richardson, R. and Richardson, M.D. (2017). Systemic fungal infections. Medicine (Baltimore), 45: 757–762.

Rautenbach, M., Troskie, A.M. and Vosloo, J.A. (2016). Antifungal peptides: To be or not to be membrane active. Biochimie, 130: 132–145.

Rayhan, R., Xu, L., Santarpia III, R.P., Tylenda, C.A. and Pollock, J.J. (1992). Antifungal activities of salivary histidine-rich polypeptides against Candida albicans and other oral yeast isolates. Oral Microbiol. Immunol., 7: 51–52.

Robinson, J.A. (2011). Protein epitope mimetics as anti-infectives. Curr. Opin. Chem. Biol., 15: 379–386.

Roemer, T. and Krysan, D.J. (2014). Antifungal drug development: Challenges, unmet clinical needs, and new approaches. Cold Spring Harb. Perspect. Med., 4: a019703.

Roscetto, E., Contursi, P., Vollaro, A., Fusco, S., Notomista, E. and Catania, M.R. (2018). Antifungal and anti-biofilm activity of the first cryptic antimicrobial peptide from an archaeal protein against Candida spp. clinical isolates. Sci. Rep., 8: 1–11.

Rothstein, D.M., Spacciapoli, P., Tran, L.T., Xu, T., Roberts, F.D., Dalla Serra, M., Buxton, D.K., Oppenheim, F.G. and Friden, P. (2001). Anticandida activity is retained in P-113, a 12-amino-acid fragment of histatin 5. Antimicrob. Agents Chemother., 45: 1367–1373.

Roy, K., Mukhopadhyay, T., Reddy, G.C.S., Desikan, K.R. and Ganguli, B.N. (1987). Mulundocandin, a new lipopeptide antibiotic I. Taxonomy, fermentation, isolation and characterization. J. Antibiot. (Tokyo), 40: 275–280.

Sagaram, U.S., Pandurangi, R., Kaur, J., Smith, T.J. and Shah, D.M. (2011). Structure-activity determinants in antifungal plant defensins MsDef1 and MtDef4 with different modes of action against Fusarium graminearum. PLoS One, 6: e18550.

Samir, P. and Link, A.J. (2011). Analyzing the cryptome: Uncovering secret sequences. AAPS J., 13: 152–158.

Sanglard, D. (2016). Emerging threats in antifungal-resistant fungal pathogens. Front. Med., 3: 11.

Sanguinetti, M., Posteraro, B. and Lass-Flörl, C. (2015). Antifungal drug resistance among Candida species: Mechanisms and clinical impact. Mycoses, 58: 2–13.

Sarkar, T., Chetia, M. and Chatterjee, S. (2021). Antimicrobial peptides and proteins: From nature's reservoir to the laboratory and beyond. Frontiers in Chemistry, 9: 691532.

Schaaper, W.M.M., Posthuma, G.A., Meloen, R.H., Plasman, H.H., Sijtsma, L., Van Amerongen, A., Fant, F., Borremans, F.A.M., Thevissen, K. and Broekaert, W.F. (2001). Synthetic peptides derived from the $β$2− $β$3 loop of Raphanus sativus antifungal protein 2 that mimic the active site. J. Pept. Res., 57: 409–418.

Schibli, D.J., Epand, R.F., Vogel, H.J. and Epand, R.M. (2002). Tryptophan-rich antimicrobial peptides: Comparative properties and membrane interactions. Biochem. Cell Biol., 80: 667–677.

Schmidt, M., Arendt, E.K. and Thery, T.L.C. (2019). Isolation and characterisation of the antifungal activity of the cowpea defensin Cp-thionin II. Food Microbiol., 82: 504–514.

Schneider, G. and Fechner, U. (2005). Computer-based de novo design of drug-like molecules. Nat. Rev. Drug Discov., 4: 649–663.

Shahi, G., Kumar, M., Skwarecki, A.S., Edmondson, M., Banerjee, A., Usher, J., Gow, N.A.R., Milewski, S. and Prasad, R. (2022). Fluconazole resistant Candida auris clinical isolates have increased levels of cell wall chitin and increased susceptibility to a glucosamine-6-phosphate synthase inhibitor. Cell Surf., 8: 100076.

Sher Khan, R., Iqbal, A., Malak, R., Shehryar, K., Attia, S., Ahmed, T., Ali Khan, M., Arif, M. and Mii, M. (2019). Plant defensins: Types, mechanism of action and prospects of genetic engineering for enhanced disease resistance in plants. 3 Biotech., 9: 1–12.

Shokri, D., Zaghian, S., Khodabakhsh, F., Fazeli, H., Mobasherizadeh, S. and Ataei, B. (2014). Antimicrobial activity of a UV-stable bacteriocin-like inhibitory substance (BLIS) produced by Enterococcus faecium strain DSH20 against vancomycin-resistant Enterococcus (VRE) strains. J. Microbiol. Immunol. Infect., 47: 371–376.

Sinden, S.L., DeVay, J.E. and Backman, P.A. (1971). Properties of syringomycin, a wide spectrum antibiotic and phytotoxin produced by Pseudomonas syringae, and its role in the bacterial canker disease of peach trees. Physiol. Plant Pathol., 1: 199–213.

Soares, G.A.M. and Sato, H.H. (2000). Characterization of the Saccharomyces cerevisae Y500-4L killer toxin. Brazilian J. Microbiol., 31: 291–297.

Sonesson, A., Nordahl, E.A., Malmsten, M. and Schmidtchen, A. (2011). Antifungal activities of peptides derived from domain 5 of high-molecular-weight kininogen. Int. J. Pept., 2011.

Steckbeck, J.D., Deslouches, B. and Montelaro, R.C. (2014). Antimicrobial peptides: New drugs for bad bugs? Expert. Opin. Biol. Ther., 14: 11–14.

Steimbach, L.M., Tonin, F.S., Virtuoso, S., Borba, H.H.L., Sanches, A.C.C., Wiens, A., Fernandez-Llimós, F. and Pontarolo, R. (2017). Efficacy and safety of amphotericin B lipid-based formulations—A systematic review and meta-analysis. Mycoses, 60: 146–154.

Stotz, H.U., Thomson, J. and Wang, Y. (2009). Plant defensins: Defense, development and application. Plant Signal. Behav., 4: 1010–1012.

Stringer, J.R., Beard, C.B., Miller, R.F. and Wakefield, A.E. (2002). A New Name for <em>Pneumocystis</em> from humans and new perspectives on the host-pathogen relationship. Emerg. Infect. Dis. J., 8: 891.

Struyfs, C., Cools, T.L., De Cremer, K., Sampaio-Marques, B., Ludovico, P., Wasko, B.M., Kaeberlein, M., Cammue, B.P.A. and Thevissen, K. (2020). The antifungal plant defensin HsAFP1 induces autophagy, vacuolar dysfunction and cell cycle impairment in yeast. Biochim. Biophys. Acta (BBA)-Biomembranes, 1862: 183255.

Suzuki, S., Isono, K., Nagatsu, J., Mizutani, T., Kawashima, Y. and Mizuno, T. (1965). A new antibiotic, polyoxin A. J. Antibiot. Ser. A, 18: 131.

Szymański, M., Chmielewska, S., Czyżewska, U., Malinowska, M. and Tylicki, A. (2022). Echinocandins–structure, mechanism of action and use in antifungal therapy. J. Enzyme Inhib. Med. Chem., 37: 876–894.

Thery, T., Lynch, K.M. and Arendt, E.K. (2019). Natural antifungal peptides/proteins as model for novel food preservatives. Compr. Rev. Food Sci. Food Saf., 18: 1327–1360.

Thery, T., Lynch, K.M., Zannini, E. and Arendt, E.K. (2020). Isolation, characterisation and application of a new antifungal protein from broccoli seeds–New food preservative with great potential. Food Control, 117: 107356.

Thevissen, K., Kristensen, H.-H., Thomma, B.P.H.J., Cammue, B.P.A. and Francois, I.E.J.A. (2007). Therapeutic potential of antifungal plant and insect defensins. Drug Discov. Today, 12: 966–971.

Troxler, R.F., Offner, G.D., Xu, T., Vanderspek, J.C. and Oppenheim, F.G. (1990). Structural relationship between human salivary histatins. J. Dent. Res., 69: 2–6.

Tsai, H. and Bobek, L.A. (1997). Human salivary histatin-5 exerts potent fungicidal activity against Cryptococcus neoformans. Biochim. Biophys. Acta (BBA)-General Subj., 1336: 367–369.

Tverdek, F.P., Kofteridis, D. and Kontoyiannis, D.P. (2016). Antifungal agents and liver toxicity: A complex interaction. Expert. Rev. Anti. Infect. Ther., 14: 765–776.

Van der Weerden, N.L., Bleackley, M.R. and Anderson, M.A. (2013). Properties and mechanisms of action of naturally occurring antifungal peptides. Cell. Mol. life Sci., 70: 3545–3570.

Vandeputte, P., Ferrari, S. and Coste, A.T. (2012). Antifungal resistance and new strategies to control fungal infections. Int. J. Microbiol., 2012.

Váradi, G., Tóth, G.K. and Batta, G. (2018). Structure and synthesis of antifungal disulfide β-strand proteins from filamentous fungi. Microorganisms, 7: 5.

Viraka Nellore, B.P., Kanchanapally, R., Pedraza, F., Sinha, S.S., Pramanik, A., Hamme, A.T., Arslan, Z., Sardar, D. and Ray, P.C. (2015). Bio-conjugated CNT-bridged 3D porous graphene oxide membrane for highly efficient disinfection of pathogenic bacteria and removal of toxic metals from water. ACS Appl. Mater. Interfaces, 7: 19210–19218.

Vlieghe, P., Lisowski, V., Martinez, J. and Khrestchatisky, M. (2010). Synthetic therapeutic peptides: Science and market. Drug Discov. Today, 15: 40–56.

Walkenhorst, W.F., Klein, J.W., Vo, P. and Wimley, W.C. (2013). pH dependence of microbe sterilization by cationic antimicrobial peptides. Antimicrob. Agents Chemother., 57: 3312–3320.

Wang, G., Li, X. and Wang, Z. (2016). APD3: The antimicrobial peptide database as a tool for research and education. Nucleic Acids Res., 44: D1087–D1093.

Wen, C., Guo, W. and Chen, X. (2014). Purification and identification of a novel antifungal protein secreted by Penicillium citrinum from the Southwest Indian Ocean. J. Microbiol. Biotechnol., 24: 1337–1345.

Wu, M., Maier, E., Benz, R. and Hancock, R.E. (1999). Mechanism of interaction of different classes of cationic antimicrobial peptides with planar bilayers and with the cytoplasmic membrane of Escherichia coli. Biochemistry, 38: 7235–7242.

Wyandt, H.E., Skare, J.C., Milunsky, A., Oppenheim, F.G. and Troxler, R.F. (1989). Localization of the genes for histatins to human chromosome 4q13 and tissue distribution of the mRNAs. Am. J. Hum. Genet., 45: 381.

Yamamoto, K.N., Hirota, K., Kono, K., Takeda, S., Sakamuru, S., Xia, M., Huang, R., Austin, C.P., Witt, K.L. and Tice, R.R. (2011). Characterization of environmental chemicals with potential for DNA damage using isogenic DNA repair-deficient chicken DT40 cell lines. Environ. Mol. Mutagen., 52: 547–561.

Yapar, N. (2014). Epidemiology and risk factors for invasive candidiasis. Ther. Clin. Risk Manag., 10: 95.

Yeung, A.T., Gellatly, S.L. and Hancock, R.E. (2011). Multifunctional cationic host defence peptides and their clinical applications. Cellular and Molecular Life Sciences, 68: 2161–2176.

Yu, H.-Y., Tu, C.-H., Yip, B.-S., Chen, H.-L., Cheng, H.-T., Huang, K.-C., Lo, H.-J. and Cheng, J.-W. (2011). Easy strategy to increase salt resistance of antimicrobial peptides. Antimicrob. Agents Chemother., 55: 4918–4921.

Yu, Q., Lehrer, R.I. and Tam, J.P. (2000). Engineered salt-insensitive α-defensins with end-to-end circularized structures. J. Biol. Chem., 275: 3943–3949.

Yuan, Y., Gao, B. and Zhu, S. (2007). Functional expression of a Drosophila antifungal peptide in Escherichia coli. Protein Expr. Purif., 52: 457–462.

Zambom, C.R., da Fonseca, F.H., Crusca Jr, E., da Silva, P.B., Pavan, F.R., Chorilli, M. and Garrido, S.S. (2019). A novel antifungal system with potential for prolonged delivery of histatin 5 to limit growth of Candida albicans. Front. Microbiol. 10: 1667.

Zasloff, M. (2002). Antimicrobial peptides of multicellular organisms. Nature, 415: 389–395.

Zhang, Q.-X., Zhang, Y., Shan, H.-H., Tong, Y.-H., Chen, X.-J. and Liu, F.-Q. (2017). Isolation and identification of antifungal peptides from Bacillus amyloliquefaciens W10. Environ. Sci. Pollut. Res., 24: 25000–25009.

Zhu, J., Luther, P.W., Leng, Q. and Mixson, A.J. (2006). Synthetic histidine-rich peptides inhibit Candida species and other fungi *in vitro*: Role of endocytosis and treatment implications. Antimicrob. Agents Chemother., 50: 2797–2805.

Chapter 4

Antimicrobial Secondary Metabolites as Antifungal Agents

Khayalethu Ntushelo,[1,*] *Vuyisile Samuel Thibane,*[2] *Udoka Vitus Ogugua,*[1] *Lesiba Klaas Ledwaba*[3] and *Chimdi Mang Kalu*[1]

1. Introduction

Secondary metabolites are small organic molecules which are produced by an organism and, unlike primary metabolites, are not essential for the growth and development of the organism. They are specialized compounds which generally mediate ecological interactions giving the organism a competitive advantage over other members of the ecosystem. To survive in the ecosystem, an organism may produce antimicrobial secondary metabolites to antagonize fungi. This antagonistic activity may be inhibition of growth, development and reproduction. Due to the recent wish to obtain environmentally friendly agents to control harmful fungi studies on antimicrobial secondary metabolites as antifungal agents have become significant. Fungi occupy diverse habitats from ocean floors to hot springs and in various organisms either as ectophytes or endophytes and can be saprophytes or pathogens of their resident organisms or hosts. In cases where fungi are parasites of living organisms like humans, animals and plants, treatment with antifungal agents becomes necessary to lessen the number of viable propagules and therefore the overall harmful effect of the fungus. There are ever-continuing searches for antifungal agents with a preference for naturally occurring antifungal agents. Some antifungal secondary metabolites may be produced by biological control agents and various plants which have medicinal properties. Plants are among the organisms which

[1] Department of Agriculture and Animal Health, Corner Christiaan De Wet and Pionner Avenue, University of South Africa, Florida, 1710, South Africa.
[2] Sefako Makgatho Health Sciences University, Department of Biochemistry and Biotechnology, Ga-Rankuwa 0204, South Africa.
[3] Agricultural Research Council, Plant Health and Protection, Roodeplaat, 0030, South Africa.
* Corresponding author: ntushk@unisa.ac.za

produce secondary metabolites with antifungal properties. Secondary metabolites which are produced by plants to ward off harmful fungi include flavonoids, phenols, phenolic glycosides, unsaturated lactones, sulfur compounds, sponins, cyanogenic glycosides and glucosinolates (Gómez et al., 1990; Bennett and Wallsgrove, 1994; Grayer and Harborne, 1994; Osbourn, 1996) and plant secondary metabolites can be classified under three major groups namely, (i) flavonoids and allied phenolic and polyphenolic compounds, (ii) terpenoids, (iii) nitrogen-containing alkaloids and sulfur-containing compounds (reviewed in Pusztahelyi et al., 2015). These secondary metabolites are regarded as defense compounds which trigger defense processes and or as signaling like hormones. They play a large role in plant fitness. Besides plants, various other organisms have been found to possess antifungal agents and therefore used as biological control agents against harmful fungi. Some of the antifungal agents are other fungi such as *Trichoderma*. Trichoderma is a well-known antifungal agent with various Trichoderma-based commercial products developed to limit the impact of harmful fungi. *Trichoderma* species are found in different habitats such as soil and root ecosystems and are effective against a wide range of economically important plant pathogenic fungi such as *Rhizoctonia solani*, *Pythium ultimum* and *Alternaria solani* (Mazrou et al., 2020), *Stemphylium vesicarium* (Zapata-Sarmiento, 2020), *Botrytis cinerea*, *Sclerotinia sclerotiorum*, *Fusarium solani* f. *cucurbitae*, *Pythium aphanidermatum*, *Rhizoctonia solani* and *Mycosphaerella melonis* (Sánchez-Montesinos et al., 2021). A plethora of other studies which involve a demonstrated efficacy of *Trichoderma* against plant pathogenic fungi exist. Secondary metabolites produced by *Trichoderma* species include esters, lactones, organic acids (Tchameni et al., 2020), anthraquinones, daucanes, simple pyrones, koninginins, trichodermanides, viridins, viridiofungins, nitrogen heterocyclic compounds, trichodenones and cyclopentenone derivatives, azaphilones, harzialactones and derivatives, butenolides, trichothecenes, isocyano metabolites, setin-like metabolites, bisorbicillinoids, diketopiperazines, ergosterol derivatives, peptaibols, cyclonerodiol derivatives, statins, heptelidic acid and derivatives, acoranes, miscellanea as reviewed in Reino et al., 2008, epipolythiodioxopiperazines, peptaibols, pyrones, butenolides, pyridones, azaphilones, koninginins, steroids, anthraquinones, lactones as reviewed in Khan et al., 2020. Among bacteria, *Bacillus* is one of the most studied biological control agent against harmful fungi. Bacillus acts against a myriad of fungi which include *Helminthosporium*, *Altenaria* and *Fusarium* (Matar et al., 2009), *Colletotrichum*, *Phomopsis*, *Dothiorella*, *Lasiodiplodia* (Korsten and De Jager, 1995). Secondary metabolites produced by *Bacillus* include surfactins (non-ribosomal peptide), fengycins (non-ribosomal peptide) and iturins (non-ribosomal peptide) (reviewed in Ntushelo et al., 2019). Secondary metabolites produced by Bacillus have been effective against a number of fungi which include *Fusarium* and *Alternaria* (Win et al., 2021), *Aspergillus* and *Penicillium* (Saleh et al., 2021), *Botrytis*, *Colletotrichum* and *Phytophthora* (Kwon et al., 2021) and many other fungi which are pathogens of plants.

Conclusions reached on antifungal secondary metabolites have been based on studies such as dual cultures and other confrontational assays. In dual cultures, the antagonist is cultured together with the target fungus (Fig. 4.1) and this is followed by

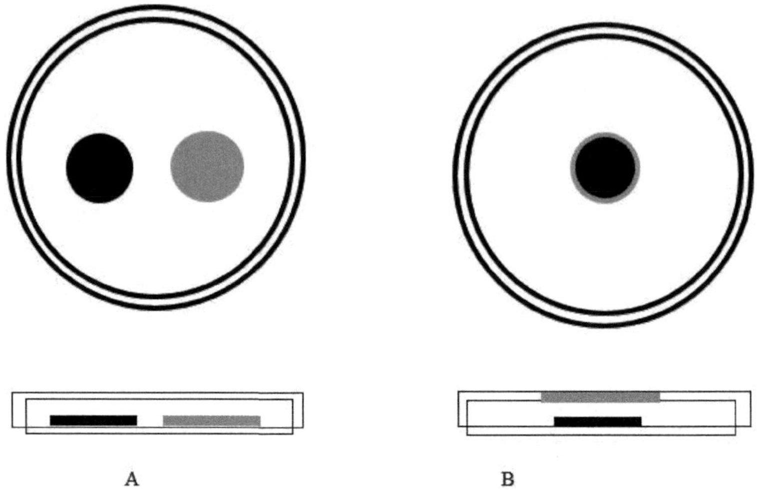

Fig. 4.1: An illustration of a dual culture to test antifungal activity (A) diffusible compounds (B) volatile compounds. To test the effect of diffusible compounds the cultures, the antagonist (gray) and the target (black) are plated alongside each other and to test for volatile compounds the organisms are covered separately, one on the plate and the other on the lid.

observations and measurements of growth to assess the impact of the antagonist on the target fungus. In agar plate assays zones of inhibition are often a good indicator of the effects of the antagonist against the target. In dual culture either diffusible antagonistic compounds act against the target fungus, this is shown by a zone of inhibition. The antifungal action can also be by volatile antagonistic compounds. To demonstrate the effect of volatile compounds the target fungus and the tested/candidate antifungal agent are covered in separate plates, usually a plate and its lid, and the two are closed to form a chamber (Fig. 4.1).

For higher throughput, multi-well plates are commonly used and generate a multitude of data than just agar Petri plates. Growth inhibition works are usually followed by studies to determine the active metabolites which cause the growth inhibition. Various tools to isolate and purify the secondary active metabolite and identify it often include the hyphenated techniques of chromatography and mass spectrometry. This chapter provides a background of secondary metabolites and methods to discover, identify and characterize them. Furthermore, cataloged antifungal secondary metabolites are described followed by basic methods to study antifungal activity. Finally, challenges and future perspectives are discussed.

2. What are metabolites?

Metabolites are the intermediates and end products of cellular metabolism responsible for the growth, development and reproduction of cells (Zaynab et al., 2019). Metabolites are small biomolecules strictly produced by living organisms and make up the basic molecular skeleton of living organisms. They can have several functions which can include energy conversion, signaling, epigenetic influence and act as co-factors (Govatati et al., 2021). The major groups of biomolecules found in living organisms include polysaccharides, proteins, nucleic acids and fatty acids. Polysaccharides are ubiquitous polymetric carbohydrates composed of simple sugars held together by glycosidic linkages (Delattre et al., 2011). Homopolysaccharides, which are those made up of similar monosaccharides and heteropolysaccharides, are those that are made up of different monosaccharides, exist in living organisms. The molecular structure of polysaccharides can either be linear or branched in complex formations depending on the absence or presence of double bonds. The biological function of polysaccharides is distinctly influenced by their physical structure brought about by differences in their physical and chemical composition (Ferreira et al., 2015; Montreuil, 1996). Amino acids are the building blocks of proteins and are characterized by the presence of a basic amino group ($–NH_2$), carboxylic group (–COOH) and a side chain that is unique to each amino acid. Proteins are large biomolecules made up of amino acid groups and are involved in almost all biological processes (Kadakeri et al., 2020). These include catalyzing chemical reactions, providing structural functions, protecting the cell against free radicals, serving as a transporter of molecules inside and outside the cell, functioning in cellular binding and facilitating joint movement as a cartilage, involved in gene expression and acting as antibodies (Rasheed et al., 2020). Nucleic acids are ubiquitous polymeric macromolecules serving as the primary information-carrying molecules in the cell. Nucleic acids are composed of nucleotides which assimilate into two main classes, deoxyribonucleic acid (DNA) and ribonucleic acid (RNA) (Minchin and Lodge, 2019). Fatty acids are carboxylic acids with long aliphatic chains which may be straight or branched, saturated or unsaturated and play several key roles in metabolism such as storage and transport of energy (Di Pasquale, 2009). Fatty acids further play a structural function as constituents of phospholipids which are the building blocks of the cellular membrane. Some of the biological functions of fatty acids also include their ability to inhibit microbial invasions in living cells (de Carvalho and Caramujo, 2018; Orhan et al., 2009). The focus of this chapter is on secondary metabolites which mediate lesser processes to equip the organism with fitness traits. Unlike primary metabolites, they are not primarily essential for the growth and development of the organism.

3. Analysis of secondary metabolites

3.1 Traditional identification and quantification methods of secondary metabolites

Plant secondary metabolites, derived from different chemical groups, are known to concentrate in a solution following solvent extraction. Traditional methods such as chromatography and spectrometry are routinely used to separate, identify and quantify these metabolic products (see basic workflows illustrated in Figs. 4.2 and 4.3). Separation techniques include Gas Chromatography (GC), Liquid Chromatography (LC), Thin Layer Chromatography (TLC) and High-Performance Liquid Chromatography (HPLC). Gas chromatographic (GC) techniques employ a principle where volatile compounds are separated and detected from complex mixtures. Liquid chromatographic (LC) techniques are more useful for analyzing non-volatile compounds into their parts using a mobile and a stationary phase (Poole, 2003). Detectable compounds comprise volatile aromatic compounds, sugars (mono-, di- and tri-saccharides), sugar alcohols/acids, amino and fatty acids, phosphorylated intermediates and many plants' secondary metabolites such as phenolics, terpenoids, steroids and alkaloids (Coskun, 2016). Spectrometric techniques such as Mass Spectrometry (MS), infrared (IR) spectrometry and Nuclear Magnetic Resonance (NMR) spectroscopy are used mainly for identification and quantification (Scheinmann, 2013). The coupling of chromatographic and spectrometric techniques provides largely robust and reliable analytical data, high specificity and handling of complex mixtures. This allows for the identification of compounds based on the use of a MS library and resources in combination with retention index data (Rohloff, 2015).

3.2 Technological advances in the identification and quantification of secondary metabolites

Chromatographic and spectrometric techniques to detect, identify and quantify metabolites in organisms have undergone major advances in recent years due to improvement in analysis time, detection limit and separation characteristics. The introduction of High-Performance Thin Layer Chromatography (HPTLC) is such an example, where the metabolite contents of the plant material could be analyzed both qualitatively and quantitatively (Attimarad et al., 2011). The traditional TLC plate technique can only provide qualitative data in a form of the Retention factor (Rf) of a plant's secondary metabolite. Analysis of secondary metabolites using the HPTLC technique has significantly enhanced data representation since correlation studies between the peak areas of present metabolites and biological activity (antimicrobial activity) can now be performed (Ansari et al., 2021). The Ultra-High-Performance Liquid Chromatography (UHPLC), an advancement of the traditional HPLC, has been another significant analysis technique for secondary metabolites. UHPLC techniques require fewer volume systems and enable operation at some high pressures and thus presenting a more robust analytical system. HPLC has been used extensively in the analysis of derivatives of the shikimic acid pathways, such as phenols, phenolic acids, tannins, coumarins, flavonoids and lignans (Waksmundzka-Hajnos and Sherma, 2011). Several hyphenated techniques have found increased application in the analysis of metabolites. These techniques combine and exploit the advantages of several analytical applications to provide a highly accurate analysis of the metabolite content of plant mixtures. Hyphenated techniques such as high-performance liquid chromatography photodiode array detection (HPLC-PDA), UHPLC-MS, high-performance liquid chromatography-electrospray ionization tandem (HPLC-ESI-MS/MS), LC-ESI-MS/MS and Ultra-Performance Liquid Chromatography-Quadrupole/Time-Of-Flight-Mass Spectrometry (UPLC-Q/TOF-MS) are routinely used in the analysis of secondary metabolites (Ali et al., 2021; de Vos et al., 2011; Saleem et al., 2020). Compared to a traditional method for quantification and detection of plants' secondary metabolites, the UPLC-Q/TOF-MS provides high detection sensitivity and short analysis time for those metabolites present in minute amounts in plants (Wang et al., 2015). The increased use of UPLC-Q/TOF-MS significantly advances plant metabolites analytical techniques specifically when targeting accurate molecular weights and structural conformation of metabolites (Cho et al., 2012). Epirodins from *Epicoccum nigrum*, are found in trace amounts and have been reported to possess antimicrobial potential (Alcock et al., 2015). Therefore, such techniques can be useful when exploring for tools to identify and quantify secondary metabolites with the desired activity but present at undetectable levels for traditional techniques. Most recently, the fate of different metabolites throughout the different metabolic pathways into their intended metabolic product could be tracked. The precise location of metabolite precursors, labeled with a heavy stable isotope, could be monitored by Nuclear Magnetic Resonance (NMR) in plants (Arroo et al., 2021). This coupling of stable isotope-labeling and NMR analysis techniques has future beneficial applications when targeting primary metabolites with inherent biological activity, such as antimicrobial potential, into their intended active secondary metabolites. The yield of the desired metabolite can henceforth be increased by selectively optimizing metabolic pathways responsible for their production. For example, the shikimic acid pathways play a critical role in the production of phenolic compounds which have antimicrobial potential (Lagrouh et al., 2017; Verma and Shukla, 2015). Coupling stable isotope-labeling and NMR analysis, precursor molecules of the shikimic acid pathway could be tracked and channeled, avoiding the aromatic amino acid metabolic pathway, to produce the desired metabolic compounds. The different analytical methods for secondary metabolites have enabled maximum utilization and exploitation of these beneficial plant derivatives. However, continuous improvement is required to improve the various aspects of secondary metabolite analysis. These include sensitivity and resolving power of the application, precision and accuracy, reliability, consistency, user-friendliness, simultaneous detection of secondary metabolites in a single sample, cost-effectiveness, throughput, off-site and rapid detection and environmental friendliness. The definitions provided are not necessarily standardized but are conveniently used to describe a suitable instrument or application for secondary metabolite analysis.

3.2.1 Sensitivity and resolution power

Chromatography and mass spectrometry have become versatile applications throughout the world and their use depends on their qualities as instruments which generate useful data for chemists, biochemists, industry personnel and the like. They are capable of (i) molecular mass, elemental and isotopic composition determinations, (ii) structural elucidation, and (iii) quantification. Sensitivity and resolution power are important parameters in chromatography and mass spectrometry to perform these important functions, namely, molecular mass, structure and quantity determination. Sensitivity means the instrument can detect and quantify minute quantities of the metabolite. Resolution means that the instrument can differentiate between two or more metabolites with very similar properties such as m/z value. For important applications like drug discovery, these two features are significant because a slight error may lead to different results with undesirable consequences. Similarly for identifying secondary metabolites with antifungal properties, instrument sensitivity and resolution are crucial to ensure that the metabolite is identified with the highest confidence possible.

3.2.2 Precision and accuracy

Equally important in chromatographs and mass spectrometers is precision which is the measure of deviation or closeness between two or more measurements of the same sample or replicated samples. Robust instruments are precise in their measuring of samples, especially technical replicates which are supposed to have the same measurements for a given parameter such as retention time or m/z ratio. Similarly, the accuracy of an instrument describes its ability to show an agreement between a measurement and a true value. This true value can be a standard sample which has been run and verified in many instruments multiple times. Serial dilutions of the samples should also be accurately measured in line with the dilution factors.

3.2.3 Reliability

Instrument measurements are prone to systematic, mechanical and human errors during routine operations. Reliability can be defined as the extent to which an instrument can yield consistent, reproducible and repeatable measurements which can be assessed using various experimental designs. Instrument reliability requires strict maintenance even when dealing with high loads of data and thus ensures high throughput. Instrument reliability also refers to the confidence an experimenter has in the instrument about giving correct measurements for a given sample.

3.2.4 Consistency

For consistency, all the effects and settings must yield the same effect. If settings such as injection volume, choice of the mobile phase, etc., are changed for a given sample on a particular application the instrument must consistently produce the same result. The instrument must be predictable in its output if the runs are correctly conducted by the application specialist.

3.2.5 User-friendliness

Often chromatographs and mass spectrometers are complicated to run. Either the in-built software packages are cumbersome and sometimes require users to be trained before they can competently use the instrument. For instruments that have less automation, the user may be required to spend time running the instrument. One of the qualities of a good instrument is user-friendliness. The tools should be spontaneous, user-friendly and open-source, allowing multiple users and parallel processes by the same instrument. User-friendly tools are needed a great deal especially to cope with the high volume of work handled by many laboratories that perform biomarker discovery work and general natural products research.

3.2.6 Simultenous detection

Simultaneous detection of compounds using chromatographic and spectrometric techniques provides an efficient system for metabolite detection and quantification. Hyphenated techniques combining both techniques have become favorable and popular in natural product research. These systems exploit simultaneous detection of metabolites while maintaining systems that are reliable with high throughput. Simultaneous detection means the instrument can detect a multitude of compounds in a single run. Most high-throughput and present-day state-of-the-art instruments have this function of picking a handful of metabolites and allow the user to take the few metabolites of interest for further characterization and finetuning of identification.

3.2.7 Cost-effectiveness

Various instruments for chromatography and mass spectrometry can be costly, especially when including the cost of shipping, the cost of commissioning and running costs. Coupled with this, these instruments require frequent calibration and servicing as well as repairs. Running costs include the price of consumables such as columns, solvents for metabolite extraction and phase mobility. Purchase or renewal of licences for software packages for data processing may also be expensive. All these costs limit the availability and use of the instruments in different laboratories that would otherwise contribute meaningful results. Preferable instruments are those with lower costs, either for purchase or for running when the instrument has been procured.

3.2.8 Throughput

Throughput can be defined as the amount of data that can be processed successfully within a specific period. Low throughput applications are often laborious and yield fewer data. High-throughput applications generate multitudes of data for a run usually saving the experimenter valuable laboratory time. Most high-end chromatographs and mass spectrometers are high-throughput and automated. This is usually for sample queuing, injection and filing of results. The experiment is, at times, hands-free and saves time. From multitudes of data, especially for untargeted experimental approaches, many metabolites of interest may be discovered and selected for further testing and verification.

3.2.9 Off-site and rapid detection

Samples of plants in the environment may sometimes be better tested off-site, generally preliminarily, before massive sampling and comprehensive work can be undertaken in the laboratory. Applications which are portable or can be set easily away from the laboratory are often handy when quick off-site work needs to be conducted. However, for such instruments, there may be trade-offs with other parameters required for a good instrument. Instruments which can be used off-site for rapid detection can therefore be used coupled with laboratory-based applications which offer the full range of advantages for analytical work for studying metabolites.

3.2.10 Environmental friendliness

Environmental friendliness is also known as green analytical chemistry. It means performing a reliable analysis while applying Quality Assurance/Quality Control (QA/QC) and abating any undesirable effects of the application in the environment. Green chromatography usually comes to mind when one thinks of chromatographic procedures that meet the standards of analytical chemistry. Chromatographic procedures are a sequence of operations involving sample collection, transportation, preparation, proper chromatographic separation and analysis. Green chromatography comprises all the stages of analysis focusing on three aspects (I) sample preparation, (II) reagents and solvents (III) instrumentation assessment. These were further subdivided into 15 subcategories (Kannaiah et al., 2021). Pressurized hot water extraction of metabolites is gaining popularity because it eliminates the need for solvents for metabolite extraction (Nuapia et al., 2021; Gbashi et al., 2020a; Gbashi et al., 2020b).

The main approach of green analytical chromatography includes reducing wide steps in analytical procedures, while real-time analysis of samples using in-line sample collection thereby eliminating toxic organic or inorganic chemicals in extraction or sampling procedures.

The use of various analytical suites is crucial for detecting and identifying antifungal secondary metabolites in organisms like plants and other microbes. However, the initial experiments are those that test antifungal activity in dual cultures of the antagonistic organism and the fungus or expose it to a candidate antifungal agent. Various organisms, plants, fungi and bacteria, among others, produce antifungal agents.

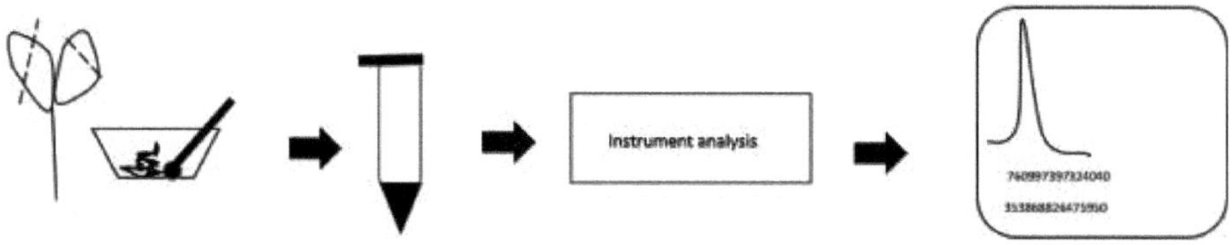

1 2 3 4

Fig. 4.2: An illustration of a simple workflow of plant secondary metabolite analysis. Plant tissue is sampled and ground (step 1), this is followed by metabolite extraction (step 2), then instrument analysis (step 3) and finally data analysis and interpretation (step 4).

4. Antifungal activity

Antifungal activity is described as an action caused by another organism that may lead to inhibition of the growth, development and reproduction of a fungus. Fungi often grow and proliferate in their desirable habitat and utilize the available

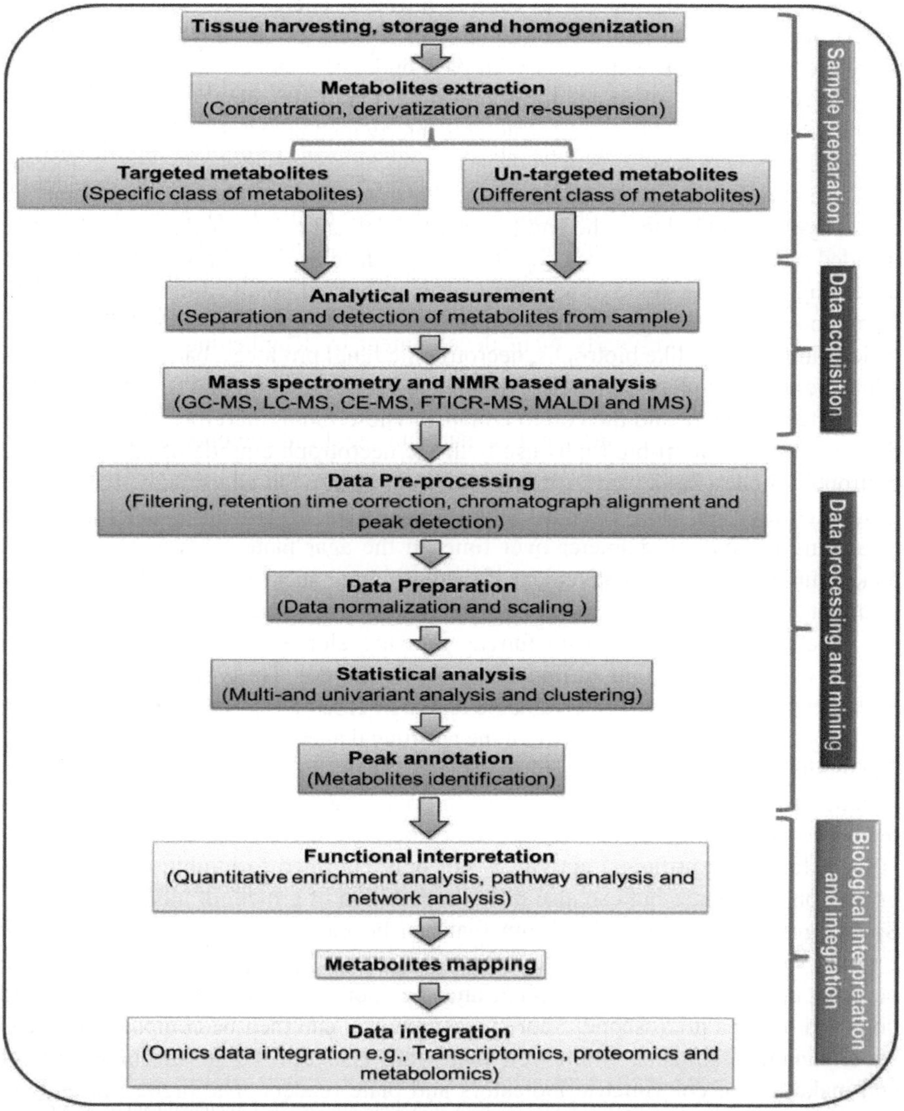

Fig. 4.3: Schematic representation of the multi-step workflow of a plant metabolomics study. Sample preparation, data acquisition, data processing and biological interpretation are key steps in plant metabolomics. These days, for data acquisition, different MS-based analytical tools (GC-MS, LC-MS CE-MS, FTICR-MS, MALDI and IMS) and NMR are available. The most important step in data processing and mining includes correction of baseline shifts, background noise reduction, chromatograph alignment and peaks detection. Biological interpretation and integration include enrichment analysis, networks and pathways analysis for a comprehensive scope of the metabolome. GC-MS, gas chromatography-mass spectrometry; IMS, Ion Mobility Spectrometry; LC-MS, Liquid Chromatography Mass-Spectroscopy; CE-MS, Capillary Electrophoresis-Mass Spectrometry; FTICR-MS, Fourier Transform Ion Cyclotron Resonance-Mass Spectrometry; MALDI, Matrix-Assisted Laser Desorption/Ionization; NMR, Nuclear Magnetic Resonance (Cited from Patel et al., 2021).

substrate for nutrition. Sources of carbon, nitrogen, energy and other nutrients for fungi vary from oil to hardwood and simpler sources of carbohydrates like crops. They metabolize many substrates with ease and can tolerate harsh conditions and can survive on these substrates for a long time. Fungi are complex manipulators of their host, they can form feeding structures intertwined in between the cell and draw nutrients for a very long time even decades causing serious losses to trees in forests. The structures of fungi create an intercellular network in the invaded plant are called hyphae and bulge at the end to form specialized cells called appressoria. The invaded plant tissue often does not die and therefore becomes a source of nutrition for a long time. The mode of attack which is of hyphal colonization, appressorium formation and a prolonged association between the fungal parasite and the plant is called biotrophy. Unlike biotrophs, necrotrophic fungi produce a barrage of enzymes which degrade the physical barriers of the plant such as pectin and cellulose causing cell contents to ooze and the cell to collapse. These fungi, therefore, draw nutrients from dead cells. Hemi-biotrophs use both the necrotrophic mode of attack and the biotrophic mode (Reviewed in Rauwane et al., 2020). On agar plates, fungi often grow laterally and form circular colonies. The rate of growth is determined by measuring the colony diameter over time on the agar plate. This circular growth makes inhibition studies convenient if the antifungal agent is cultured on the same agar plate as the fungus. The rate of fungal growth is measured over time to compare the experimental treatment (the fungus growing alongside the antifungal agent) and the controls (where no antifungal agent is cultured). These studies have gained prominence and continue to be valuable and their ease of use makes them favored in antifungal assays. In instances where the antifungal agent to be tested is a secondary metabolite (and not another organism) poison agar is formulated with the secondary metabolite as an ingredient of the agar. In poison agar studies growth inhibition is measured similarly to in co-culture studies because the area of the fungal colony is the reference. Antifungal activity is not only confined to inhibition of mycelial growth on agar plates but can also refer to inhibition of growth of individual spores. Spores growing in a liquid medium that can be incubated with an antagonist or a tested secondary metabolite can be an element of the liquid medium. This can be done in a time-course experiment and the spores are sampled at intervals and observed under a microscope. Spore measurements can then be compared between experimental treatments. Multi-well plates are also utilized in antifungal assays and make use of colorimetric indicators and plate readers. The higher throughput of multi-well plates allows various antifungal agents with different dosages to be tested simultaneously. Various antimicrobial and antifungal secondary metabolites exist. Due to their bioactivity against fungi, antifungal secondary metabolites can be useful in different industrial/commercial applications such as food preservation, pharmaceuticals, fungicides for use in agriculture. The focus of this chapter is antifungal secondary metabolites produced by plants, fungi and bacteria.

4.1 Antifungal secondary metabolites produced by plants

As part of the defense against invading fungi plants produce a multitude of secondary metabolites. Plants synthesize these secondary metabolites for their protection, as part of a battery of responses triggered when a pathogen enters the plant. On infection, the plant initiates defense by first recognizing conserved Ppathogen-Associated Molecular Patterns (PAMPs) by host pattern receptors (PPR) leading to PAMP-triggered immunity (PTI) (Jones and Dangl, 2006). PTI causes the plant to respond by callose deposition, generation of reactive oxygen species, Ca2+ signaling, protein phosphorylation cascades and accumulation of pathogenicity-related proteins, etc. (van Loon et al., 2006). Another defense method is systemic acquired resistance (SAR) which is an equally important component of PAMPs and provides long-distance protection through the production of plant hormones salicylic acid, ethylene and jasmonic acid. Plant secondary metabolites mediate plant defense to protect the plant against hazardous pathogens which include fungi.

The production of these secondary metabolites by plants emanates from the interaction of the plant with pathogens during millions of years of interaction with more than 100000 metabolites known to be involved in plant defense, however the situation is not clear (Wink, 2008). Well-known examples of plant-produced secondary metabolites with antifungal properties include flavonoids, phenols and phenolic glycosides, unsaturated lactones, sulfur compounds, saponins, cyanogenic glycosides and glucosinolates (Gómez Garibay et al., 1990; Bennett and Wallsgrove, 1994; Grayer and Harborne, 1994; Osbourn, 1996).

4.1.1 Flavonoids

Flavonoids are structurally-diverse secondary metabolites which mediate plant defense and involve differential pigmentation. Flavonoids cause various coloring patterns by the accumulation of anthocyanins and flavonols which are derived from the flavonoid biosynthesis pathways regulated by the MYB transcription factors in plants (Tian et al., 2015).

They have strong antioxidative properties and are derivatives of 2-phenyl-benzyl-γ-pyrone with over 9000 compounds known (Buer et al., 2010). The biosynthesis of flavonoids begins with the condensation of one p-coumaroyl-CoA molecule with three molecules of malonyl-CoA to yield chalcone (4',2',4',6'-tetrahydroxychalcone), catalyzed by chalcone synthase (CHS). This is followed by isomerization of chalcone to flavanone by chalcone synthase and then the formation of several different flavonoid classes including aurones, dihydrochalcones, flavanonols (dihydroflavonols), isoflavones, flavones, flavonols, leucoanthocyanidins, anthocyanins and proanthocyanidins (Mierziak et al., 2014). Further modifications can take place such as methylation by methyltransferases and glycosylation by specific glycosyltransferases and the effect can be changed in solubility, reactivity and stability. However, most flavonoids are present as glycosides under natural conditions (Bohm, 1998; Forkmann and Heller, 1999).

Their role in plant defense mediation has been observed in several plant-fungal interactions which include *Gymnosporangium yamadae* and apple (Lu et al., 2017), *Phytophthora sojae* and *Glycine max* (Subramanian et al., 2005), *Collectotrichum lindemuthianum* and *Phaseolus vulgaris* (Durango et al., 2013), *Plasmopara viticola* and *Vitis vinifera* (Ali et al., 2012), *Erysiphe pisi* and *Medicago truncatula*

(Foster-Hartnett et al., 2007), *Botrytis cineria* and *Cider bijugum* (Stevenson and Haware, 1999), *Cytonaema* and *Eucalyptus globules* (Eyles et al., 2003), *Neurospora crassa* and *Arabidopsis thaliana* (Parvez et al., 2004), *Cladosporium cucumerinum* and *Mariscus psilotachys* (Gafner et al., 1996). Another noteable involvement of flavonoids as antifungal agents is that of proanthocyanidins, and to some extent dihydroquercetin in the protection of barley against *Fusarium*. *Fusarium* flavonoids probably act against by cross-linking of microbial enzymes, inhibition of the cellulases of the fungus, xylanases and pectinases, chelation of metal ions, etc., (Skadhauge et al., 2004).

4.1.2 Phenols and phenol glycosides

Phenolics are chemical compounds with a hydroxyl group bonded to an aromatic hydrocarbon group. They can be polyphenolic which means many phenol units may be present in the same compound. They are ubiquitous secondary metabolites found in the plant and synthesized by both the shikimate and phenylpropanoid pathways. Phenol glycosides are some of the most abundant secondary metabolites in the tissues of plants. Phenol glycosides or phenolic glycosides are compounds which have a sugar unit bound to a phenol aglycone. The simplest phenol glycoside is salicin and variations of other phenol glycosides are built on the salicin core structure. The esterification of the sub-structure of salicin in complex phenol glycosides usually takes place at the primary alcohol function of the salicylic alcohol moiety at positions 2' and 6' of the glucose moiety.

Although the production of phenol glycosides is not clear, it is indicated that they are formed from cinnamic acid (Babst et al., 2010). Cinnamic acid is believed to be the precursor of the salicyl moiety of phenol glycosides (Zenk, 1967) with no clarity on the origin of the 6-hydroxy-2-cyclohexen-on-oyl (HCH) moiety found in certain phenol glycosides such as salicortin. Phenolics help the plant to evade pathogenic intrusions and protect plant tissues from the toxicity of reactive oxygen species and act as antioxidants and accumulate in the tissues especially when the invading fungus causes oxidative stress (Grassmann et al., 2002). Following a reaction with a radical when there is oxidative stress, hydroxyl groups attached to aromatic rings enable phenolics to become stable radical forms to terminate the production of new reactive species (Pereira et al., 2009). Invasion of the plant by a fungal pathogen causes upregulation of the phenylalanine ammonia-lyase gene and this takes place during PTI which is activated during recognition by the MAMPs of the plant (Pombo et al., 2014). Phenylalanine ammonia-lyase is a key enzyme in phenolic synthesis. The production of phenolic compounds to neutralize the effects of intrusive fungi in plants has been demonstrated in various pathosystems but, is demonstrated less for phenolic glycosides. These include *Colletotrichum lupini* and narrow leaf lupin (Wojakowska et al., 2013), *Olea europaea* and *Verticillium dahliae* (Báidez et al., 2007), *Galla rhois* and fungi *Magnaporthe grisea*, *Botrytis cinerea*, *Puccinia recondita*, *M. grisea* and *Erysiphe graminis* (Ahn et al., 2005) and some other plant-fungus interactions.

4.1.3 Unsaturated lactones

Lactones are volatile organic compounds which are derived from lipid metabolism (Kourist and Hilterhaus, 2015). Lactones are cyclic carboxylic esters which have a 1-oxacycloalkan-2-one structure or analogs with variations of atoms replacing one or more carbon atoms of the ring. Lactones are named by attaching a Greek prefix α, β, Ɣ, δ that specifies the number of carbon atoms in the ring. They are naturally present in fruits and vegetables and contribute to aroma and flavor (Gawdzik et al., 2015; Kourist and Hilterhaus, 2015; Perestrelo et al., 2006; Tahara et al., 1972; Labuda, 2009). Unsaturated lactones are lactones with unsaturation in one or more carbon atoms on the lactone ring. An unsaturated lactone, protoanemonin was found to be effective against *Epidermophyton floccosum*, a fungus which causes nail and skin infection in humans and against a yeast, *Rhodotorula glutinis* (Mares, 1987). Structural similarities between protoanemonin and other cytotoxic unsaturated lactones, and the reversal of the cysteine of the antifungal action suggest that this may be how this protoanemonin acts against fungi. It was earlier suggested that unsaturaturated lactones react with the sulfur in sulfur-containing amino acids present in enzyme proteins (Cavallito and Haskell, 1945).

4.1.4 Sulfur compounds

Sulfur compounds contain one or more sulfur elements. Sulfur is a component of various compounds which are used as antifungal agents. For instance, garlic juice which, through gas chromatography was shown to have 85.95% sulfur compounds, is effective against *Botrytis cinerea*, and *Penicillium expansum* (Daniel et al., 2015).

4.1.5 Saponins

Saponins occur naturally in legumes. They characteristically form stable, soaplike foams in aqueous solutions. They are complex and chemically diverse. Structurally they have a carbohydrate moiety attached to a triterpenoid or steroids. One of their roles in plants is defense against pathogens. Saponin rich-extracts from various plants, namely, *Balanites aegyptiaca* fruit mesocarp, *Quillja saponaria* bark and *Yucca schidigera* were proven to be effective against phytopathogenic fungi *Pythium ultimum*, *Fusarium oxysporum*, *Alternaria solani*, *Colletotrichum coccodes*, and *Verticillium dahliae* (Chapagain et al., 2007).

4.1.6 Cyanogenic glycosides

Cyanogenic glycosides are polar and water-soluble compounds that are described as glycosides of α-hydroxynitriles (cyanohydrins). They have been reported to be present in more than 200 plant species and these plant species can form hydrocyanic acid (HCN) in response to tissue damage (Davis, 1991; Nayik and Kour, 2022). The release of HCN is a defense mechanism of these cyanogenic plant species against herbivores and pathogens (Fry and Evans, 1977; Hughes, 1991). Davis (1991) reported the identification of over 300 cyanogenic glycosides with more than 50 distinct structures. They are biosynthesized from amino acid precursors through the formation of hydroxynitriles that are glycosylated to form cyanogenic glycosides (Hughes, 1991). Fry and Myers (1981) correlated the cyanogenic glycoside levels with resistance to fungal pathogens. The results obtained proved to be less conclusive on the impact of the compound on the fungal pathogens. From plant-fungus interactions, highly cyanogenic plant species are more susceptible to fungal attacks than low cyanogenic varieties (Lieberei et al., 1989). This implies that there is a linkage between the fungi's ability to infect cyanogenic plant species and their ability to tolerate HCN despite the lack of a consistent correlation between cyanogenic glycoside content and fungal disease resistance.

4.1.7 Glucosinolates

Glucosinolates are sulfur-containing glucosides. They are found among the members of the family Cruciferae such as *Brassica* spp., and *Arabidopsis* spp. These compounds play a crucial role in the feeding restraints of vertebrate and invertebrate pests (Chew, 1988; Giamoustaris and Mithen, 1995). Their biosynthesis is triggered by pest attacks on a plant (Burow and Halkier, 2017). Depending on their side chains glucosinolates are broken down into aliphatic, indolyl or aralkyl a-amino acids containing glucosinoates (Fenwick et al., 1983; Chew, 1988; Duncan, 1991). Hydrolysis products of glucosinolates are reported to be toxic to various fungal pathogens *in vitro* of Brassica spp., (Fenwick et al., 1983; Chew, 1988). Isothiocyanates are the major breakdown product of glucosinoates in leaves of *Brassica* spp., with allyl-(2-propenyl) and 3-butenyl isothiocyanate being the most fungitoxic products (Mithen et al., 1986). Indolyl glucosinolates are also considered to be toxic to fungi (Mithen et al., 1986).

4.2 Antifungal metabolites produced by fungi

Some of the antifungal secondary metabolites are produced by fungi to gain dominance over encroaching fungi which threaten them. For example, fungi which belong to the genus *Aspergillus* produce a diverse array of these secondary metabolites with *A. fumigatus* producing more than 226 secondary metabolites which include the commonly studied polyketides, such as cyclic peptides, alkaloids and sesquiterpenoids (Frisvad et al., 2009).

4.2.1 Polyketides

Polyketides are a large and diverse group of metabolites with various biological activities and their role as antifungal agents is documented in numerous experiments. PKSs show an extraordinary structural assortment, which are synthesized from simple acyl building blocks and are also distributed widely in fungi. PKSs are produced by polyketide synthases. Their main biological synthesis includes stepwise condensation of a starter unit which is characteristically acetyl-CoA or propionyl-CoA and also has an extender unit which is malonyl-CoA or methylmalonyl-CoA. The synthesis of PKSs is by the non-reducing group of iterative polyketide synthases

(NR-PKSs). The different structures of PKS appear from selective enzymatic variations of reactive enzyme-bound poly-β-keto intermediates (Crawford and Townsend, 2010; Miyanaga, 2017).

There are three groups of PKSs namely: Type I PKSs which are multifunctional enzymes that are organized into modules, which harbor a set of diverse catalytic domains. Modular type I PKSs comprises several modules that are each responsible for a single round of polyketide chain elongation, while iterative type I PKSs comprises a single module that acts iteratively for polyketide chain elongation. Type II PKSs consist of a complex of subunits such as ketosynthase (KS) and Acyl Carrier Protein (ACP). Type III PKSs have the simplest architecture, a homodimer of KS.

The fungal pigments produced as secondary metabolites are either known or unknown functions and can be broadly classified chemically as polyketides. Fungal polyketide pigments range in structure from tetraketides to octaketides, which have four or eight C2 units that contribute to the polyketide chain. Some of them are produced through mixed biosyntheses, which means that they involve other pathways (such as amino acid or terpenoid synthesis), in addition to the polyketide pathway. Typical classes include anthraquinones, hydroxyanthraquinones, naphthoquinones and azaphilone structures. Earlier, polyketide pigments of ascomycetous fungi were used mainly for identification and species differentiation. Cyclic peptides (also known as cyclopeptides) are polypeptide chains of cyclic compounds where the amino acid residues are covalently linked to generating the ring. They are primarily made through proteinogenic or non-proteinogenic amino acids linked together by either amide or peptide bonds. Cyclic peptides have been isolated from plants, bacteria, sponges, algae, mammals and fungi. Fungi amongst these organisms are well-known producers of several ranges of cyclic peptides and have interesting structures and biological activities. Mycotoxins are some of the fungal cyclic peptides which pose a threat to plants. Cyclic peptides have significant biological activities, such as antimicrobial, insecticidal, cytotoxic and anticancer, they have greatly important physiological and ecological functions. Fungal cyclic peptides largely comprise cyclic di-, tri-, tetra-, penta-, hexa-, hepta-, octa-, nona- and decapeptides. They are two categorized types of bonds within the ring of cyclic peptides. Homodetic cyclic peptides (cyclized from head-to-tail, e.g., cyclosporine A) are composed completely of standard peptide bonds. Heterodetic cyclic peptides are cyclized between side chains or between a side chain and one of the termini. Alkaloids are another important molecule of secondary metabolites with different biological properties. They are the largest groups of natural products and characterize a vastly different group of chemical entities. They are three central types of alkaloids namely: true alkaloids, protoalkaloids and pseudoalkaloids. The correct alkaloids and protoalkaloids are produced from amino acids, while the pseudoalkaloids are not obtained from these compounds. Alkaloids comprise a colossal class of roughly 12 000 natural products. There are several known alkaloids namely: morphine, strychnine, quinine, ephedrine and nicotine (Bribi, 2018). The presence of a basic nitrogen atom at any position in the molecule is the primary requirement for classification as an alkaloid, this does not include nitrogen in an amide or peptide bond. As implied by this exceptionally broad definition, alkaloids form a group of structurally diverse and biogenically unrelated molecules (Wink, 2010; Kaisa et al., 2011; Badri et al., 2019). Sesquiterpenes are a class of terpenes that comprises three isoprene units and often have the molecular formula $C_{15}H_{24}$. Sesquiterprenes may be cyclic or encompass rings and several distinctive combinations. Sesquiterpenes are a large group of secondary metabolites that are important for plant defense and development (Ma et al., 2019). The synthesis of terpene synthases (TPSs) is the key structure of terpenoids, it exploits long-chains prenyl pyrophosphate to make a range of acrylic and cyclic compounds (Tholl, 2006). Depending on the carbon number, geranyl pyrophosphate (GPP, C_{10}), farnesyl secondary metabolites pyrophosphate (FPP, C_{15}) or geranylgeranyl pyrophosphate (GGPP, C_{20}) can be used as long-chain precursors with corresponding products of mono-, sesqui- and di-terpene, respectively (Christianson, 2006). The substrate preference for individual TPSs would be determined by the protein structure, thus phylogenetic analysis becomes a convenient tool to evaluate TPSs function as either mono-, sesqui or di-terpene synthase (Chen et al., 2011). Conversely, exact substrate conformation requires crucial catalytic residues of enzymes around the substrate (Christianson, 2006). The TPSs share a conserved aspartate-rich DDxxD motif, that protonates prenyl pyrophosphate and the interactions between the catalytic residues and intermediate affect subsequent cyclization or deprotonation, and create a different folding mechanism in individual TPSs (Gennadios et al., 2009). The TPSs originate from a tri-domain, bifunctional ent-copalyl diphosphate/ent-kaurene synthase, which is considered to be the most ancestral form of TPS for gibberellin production in plants (Hayashi, 2006) and the family of TPSs has diverged into subfamilies with different substrate preferences and protein structures (Bohlmann et al., 1998). The polyketide bacillaene is antagonistic against fungi (Um et al., 2013; Sang et al., 2018), and so is macrolactin (Gao et al., 2021), difficidin (Sang et al., 2018). Compounds or agents with antifungal activity can be tested against fungi in various assays, the most notable being poison agar tests or microwell plates assays.

4.3 Antifungal metabolites produced by bacteria

Bacillus bacteria have become prominent for their production of antifungal secondary metabolites which are mainly soluble compounds such as lipopeptides. The most notable compounds produced by *Bacillus* include subtilin (protease), bacilysin (non-ribosomal peptide), mycobacillin (cyclic peptide), bacillomycin (polypeptide), mycosubtilin (lipopeptide of the iturin family), iturins (non-ribosomal peptide), fengycins (non-ribosomal peptide) and surfactins (non-ribosomal peptide), bacillibactin and bacillibactin-like (non-ribosomal peptide); bacilysin (non-ribosomal peptide), the less prevalent non-ribosomal peptides locillomycin (non-ribosomal peptide), xenocoumacin (non-ribosomal peptide), pelgipeptin (non-ribosomal peptide) and tridecaptin (Weber et al., 2015), lichenysin (non-ribosomal peptide), the less-studied aerobactin-like (non-ribosomal peptide), bacitracin (non-ribosomal peptide) and polyketides (bacillaene, macrolactin, difficidin, kalimantacin, chalcone-like). The discussion excludes proteases, which are few prevalent compounds and less studied compounds.

4.3.1 Mycobacillin

Mycobacillin is a cyclic tridecapeptide with antifungal properties. This antifungal metabolite binds to ATP transporter on the plasma membrane and causes excessive release of ATP and eventually cell malfunction (Chowdhury et al., 1998; Das et al., 1987). Mycobacillin is synthesized by an enzyme complex which is separated into three fractions. The first fraction is involved in the synthesis of the first pentapeptide, the second catalyzes the synthesis of the nonapeptide and the third synthesizes the final product (Mannanov et al., 2001). Similar to bacylicin, mycobacillin is also antagonistic against the rubberwood sapstain fungus, *Lasiodiplodia theobromae* (Sajitha and Dev, 2016).

4.3.2 Mycosubtilin

Mycosubtilin, a lipopeptide antifungal biosurfactant, has a β-amino fatty acid linked to a heptapeptide moiety and exhibits several isoforms of varying length (C-15 to C-18) and the isomery (linear, *iso* and *anteiso*) of the fatty acid (Ongena and Jacques, 2008; Béchet et al., 2013). Mycosubtilin also contains a hydrophobic tail and a hydrophilic head for foaming action (Razafindralambo et al., 1998).

4.3.3 Iturin

Iturins are non-ribosomal cyclic lipopeptides. Iturins are a family of seven residues of α and one β amino acid that are unique compared to other lipopeptide antibiotics. Structurally iturins possess a heptapeptide backbone connected to a C_{13}-to-C_{17} β-amino fatty acid chain (Ongena and Jacques, 2008; Aranda et al., 2005). They have strong antifungal properties. They disrupt the fungal membranes by forming ion-conduction pores on contact will fungal membranes. Iturins are effective against numerous fungi which include *Monilinia* and *Penicillium* (Calvo et al., 2019), *Rhizoctonia solani* (Zohora et al., 2016); *Colletotricum acutatum* (Arroyave-Toro et al., 2017), *Fusarium oxysporum* (Fujita and Yokota, 2019; Lee et al., 2017; Cao et al., 2018), and *F. graminearum* (Zalila-Kolsi, 2016; Gong et al., 2015). Iturin A was found to inhibit Monilia species and *Penicillium expansum* on fruit (Calvo et al., 2019) *Penicillium italicum, P. vindicatum, Aspergillus ochraceus* and *A. versicolor* (Klich et al., 1991). In testing *B. subtilis* against the fungal pathogen *R. solani* which causes damping-off disease in tomato Zohora et al., 2016 found a signicant amount of Iturin A in the root zone soil in which tomato seed treated with *B. subtilis* was planted, an indication of the activity of iturin A against *R. solani*. Similarly, Arroyave-Toro (2017) also discovered the inhibitory action of iturin A against *C. acutatum* which causes anthracnose in the tamarillo fruit. In *F. graminearum* iturin A displayed strong antifungal properties by killing conidia which had severe morphological distortions, conglobation, inhibition of hyphal branch formation and eventually leakage of cellular contents (Gong et al., 2015). The cases of the inhibition of fungi by iturin are among the plethora of cases of antifungal activity by iturin against fungi.

4.3.4 Fengycins

Fengycins are Bacillus-produced biologically active lipopeptides with strong antifungal properties. Fengycins have a *β*-hydroxy fatty acid linked to a peptide part which has 10 amino acids eight of which are organized in a cyclic structure. Fengycins permeate the plasma membrane of target fungi and render them permeable, causing entry of toxic agents and oozing of cell contents. Eventually, the fungal cell loses firmness and collapses. Details of the molecular intricacies of this membrane perturbation are not fully understood but various attempts have shed some light. Deleu et al., 2008 found that the mechanism of action of fengycins appears to be a transition depending on the lipopeptide concentration with one state being monomeric as not deeply anchored and nonpertubing lipopeptide, and the other as a buried aggregated form responsible for irritation and disruption of the membrane to cause leakage.

4.3.5 Surfactins

Surfactins are natural lipopeptides that have antifungal activity. They are amphilic with polar amino acid heads and hydrocarbon chains. Their antifungal activity was discovered against *Fusarium graminearum, F. oxysporum* and *F. moniliforme* (at present *F. verticillioides*). However, the effect of *F. graminearum* can be on iron concentration and the conditions of the culture (Reviewed in Ntushelo et al., 2019).

4.3.6 Bacillibactin

Bacillibactin is a tris-catechoyl siderophore framed on a cyclic tri-ester scaffold (Barghouthi et al., 1989, Bultreys et al., 2006) produced by various *Bacillus* species which include *Bacillus subtilis*, *B. cereus*, *B. anthracis*, *B. thuringiensis* and *B. amyloliquefaciens* (Wilson et al., 2006; Chen, 2009; May et al., 2001). Bacillibactin, similar to many other siderophores is produced nonribosomally by large, multidomain enzymes called nonribosomal peptide synthetases that can assemble structural diverse peptides with broad biological activity (Konz and Marahiel, 1999). Bacillibactin has been demonstrated against some fungi, namely, *Macrophomina phaseolina, Fusarium moniliforme* (Pal et al., 2001) and *Botrytis cinerea* (Nifakos et al., 2021) and a plethora of other cases of activity against fungi.

4.3.7 Bacilysin

Bacilysin is a non-ribosomally synthesized dipeptide which is active against a wide range of bacteria and fungi. It has an L-alanine residue at the N terminus attached to a non-proteinogenic amino acid, l-anticapsin. Proteins in the *bac* operon, also known as *bacABCDE* (ywfBCDEF) gene cluster synthesize bacilysin (l-alanine-[2,3-epoxycyclohexano-4]-l-alanine) in *B. subtilis*. Apart from the bac operon the synthesis of bacilysin also involves thyA (thymidylate synthase), ybgG (homocysteine methyl transferase) and the oligopeptide permease OppA (Inaoka et al., 2003; Steinborn et al., 2005). Bacilysin production is likely regulated by a quorum sensing pathway that guides sporulation, competence development and surfactant synthesis (Inaoka et al., 2003). Bacylicin I is effective against several fungi which include *Phytophthora sojae*. Against *P. sojae* bacylicin damages the hyphae and this leads to leakage of cellular contents. Bacylicin also serves as a plant protectant preventing soybean sprouts from being infected by *P. sojae* and this can be a broad spectrum protection against other *Phytophthora* species such as *P. palmivora*, *P. melonis*, *P. capsica*, *P. litchi* and *P. infestans* (Han et al., 2021). Furthermore, bacylicin was found to be effective against the sapstain fungus, *Lasioplodia theobromae*. Against *L. theobromae* bacylicin complements the action of fengycin in inhibiting *L. theobromae* (Sajitha and Dev, 2016).

4.3.8 Lichenysin

Lichenysin is a biosurfactant produced by *Bacillus licheniformis* (Qiu et al., 2014). Lichenysin is synthesized by a multienzyme complex called lichenysin synthetase (LchA/Lic) encoded by 32.4 (26.6 kb) lichenysin operon *lch*A (lic) similar to the non-ribosomal biosynthesis of another Bacillus-produced surfactant named surfactin (Anuradha, 2010). Testing of lichenysin against fungi has not been done extensively, however testing against *Candida* showed that this biosurfactant prevents biofilm formation (Nelson et al., 2020). Various fungi are targeted by secondary metabolites to control their proliferation in different environments. This limit in the growth of fungi is part of the natural ecological balance which keeps in check all organisms in nature. Studies to assess compounds against fungi are conducted by incorporating the compound into agar on which the tested fungus is plated. Agar with an amendment such as an antifungal compound is called poison agar.

5. *In vitro* Screening of antifungal activity

5.1 Use of poison agar to test antifungal activity

Fungi are grown artificially on agar media for laboratory experiments or to upscale it for commercial purposes. Nutrient content and the conditions under which fungi are incubated determine the rate at which fungi grow and the media can be manipulated for a selective culture of a particular fungus. Similarly, the media can be amended by a test agent to assess its effects on the growth of the fungus. This is what is commonly referred to as poison agar assays. Common poison agar assays incorporate the agent, the poison, in different concentrations in the agar medium and following incubation, the effect of the poison is assessed either by measuring the colony diameter of the fungi to quantify the effect on growth. Poison agar studies are also employed in studies that involve assessing the effect of the poison on the genetic and physiological processes of the fungus. Most antifungal assays are based on the principle of poison agar to assess the effects of the antifungal agent on the fungus. The landmark discovery of antagonism between microbes was that of Alexander Fleming of 1928. Flemming observed that bacteria in proximity to mold colonies on a Petri dish were dying because a clear zone of inhibition developed between the bacteria and the mold. The mold was *Penicillium* which is effective against all Gram-positive bacteria. This discovery of mold juice is called penicillin. It paved way for the era of antibiotics. Penicillium is co-cultured with bacteria antagonized to form halos which are the focus of some fungal growth inhibition studies. In the poison agar method or the poisoned food method, the antifungal agent is incorporated into hot and liquid agar at the desired concentration, mixed well and poured on plates to solidify. Mycelial disks are placed on the solid agar preferably at the center of the plate. At different time intervals or at the termination of the test, the diameters of the fungal colonies are measured against the control which is the colony that is placed on the agar unamended.

5.2 Use of microwell plates to test antifungal activity

A microwell plate also known as a multiwell plate is a flat plate with multiple "wells" that serve the purpose of a small test tube. In most analytical and clinical diagnostic testing laboratories, a microwell plate is used as a standard tool (Lindström et al., 2009). A microplate could comprise sample wells of about 6, 12, 24, 48, 96, 384 or 1536 wells that have a 2:3 rectangular matrix arrangement (May, 2007). The first microplate was made by Dr. Gyula Takátsy, a Hungarian scientist in 1951. The microplate consists of six rolls and 12 sample wells (Takatsy, 1950). In the late 1980s, the introduction of a modern version of microplate by John Liner enhanced its applications. In the determination and quantification of apoptosis, 96 well plates have been employed. Apoptosis is an active process that occurs in a normal developing and healthy multicellular organism that involves programmed cell death. The study of the programmed cell death is recognized as a relevant field of biological inquiry as a reduction or excessive apoptosis is known as the root cause of cancers and autoimmune disorders as well as other diseases (Ribble et al., 2005). Hence, the need for a quick and easy assay for the quantification of apoptosis. Ribble et al. (2005) investigated the use of 96 well plates in the quantification of apoptosis in comparison to the conventional method of quantification and observed that the use of 96 well plates achieved more quantification than conventional methods. The application of 96 well plates has been recorded in cell migration or wound healing assay. Cell migration is a vital process in various physiological processes such as embryogenesis, angiogenesis, wound healing, repairing of intestinal mucosal damage and immune defense (Friedl and Wolf, 2003). Yue et al. (2010) observed better uniform denudation of cell monolayers using 96 well plates in their investigation of the effects of growth factors on rat and human cells. The use of 96 well plates has been used in the quantification of bacterial cells in diverse samples. Colonies counting the unit method involved the counting of bacterial cells from the plate (Miller, 1972). The exclusion of dead bacterial cells in the count and the ability to count any number of bacteria using dilutions are the major advantages of this method. However, the clumping of bacterial cells, the tedious nature of the method, and the long-time taken to complete the experiment are the notable disadvantage of this method. Furthermore, the method has limitations for high throughput screening studies (Hazan et al., 2012). Hazan et al. (2012) reported the start growth time method of bacterial count by applying 96 well plates that permitted rapid and serial quantification of a relative number of live cells in a bacterial culture in a high throughput manner. The use of 96 well plates enhanced the effectiveness of the method of the bacterial quantification according to the authors. The application of 96 well plates cuts across many research areas where it forms part of different methodologies. For example, Béchon et al. (2022) adopted a 96-well plate crystal violet biofilm assay to determine how bile triggers the formation of biofilm in many *Bacteroides thetaiotaomicron* in the gut of humans. The use of 96 well plates provided a quick and easy access method for obtaining the expected results. In the synthesis of drugs, 96 well plates application has made their findings fast and reliable. Babu et al. (2020) employed 96 well plates in cell viability analysis in anticancer studies of bioengineered gold nanoparticles from marine seaweed *Acanthophora spicifera*. The use of the 96 well plates by the authors provided a quick evaluation of the bioengineered gold nanoparticles. Zheng et al. (2020) also indicated the usefulness of 96 well plates in the quantification of syncytium formation in syncytium formation assay. From all the documented research work done with the application of 96 well plates, it has been reported to be a quick, cost-effective and less time-consuming approach in various analyses. A multitude of studies using multi-well plates to discover and study antifungal secondary metabolites are documented and have proven their worth for continued use.

6. Challenges and future trends

The discovery and validation of antifungal metabolites require the use of state-of-the-art instruments which may not be available in common low-budget laboratories. This means that discovery and validation of antifungal metabolites are reserved for a few workers. Moreover, instruments which combine the important qualities of high sensitivity and resolution power of the application, reliability, user-friendliness, simultaneous detection of secondary metabolites in a single sample, accuracy, consistency, reliability, cost-effectiveness, throughput, off-site and rapid detection and environmental friendliness may not be easily obtained and therefore the discovery and validation of antifungal metabolites may require the use of multiple instruments in different laboratories by different workers. Furthermore, the skills in this area of study may also not be sufficient in many parts of the world. When antifungal metabolites have been discovered and validated, mass production and upscaling the production may not be easy and therefore commercialization of the product may not be feasible. The future in the discovery and validation of antifungal metabolites must ensure that repeated efforts of testing and validation, and measures for upscaling must be devised for metabolites which can be used to control harmful and undesirable fungi.

7. Conclusion

This chapter reviewed antifungal secondary metabolites, methods for studying, discovery and validation. The traditional techniques of confrontational cultures are still useful to discover antifungal secondary metabolites produced by fungi and bacteria. Culturing of fungi on poison agar is favorable for studying antifungal properties of plant extracts. Multiwell plates and the use of plate readers are more convenient and have higher throughput than the use of Petri dishes either in confrontational cultures or poison agars. Hyphenated analytical techniques of chromatography and mass spectrometry complete studies which seek to discover and validate antifungal secondary metabolites. This review captures these important aspects and highlights the various challenges in this area of study and proposes continued efforts to improve systems for antifungal secondary metabolite discovery and validation as well as upscaling of production for promising antifungal metabolites.

References

Ahn, Y.J., Lee, H.S., Oh, H.S., Kim, H.T. and Lee, Y.H. (2005). Antifungal activity and mode of action of Galla rhois-derived phenolics against phytopathogenic fungi. Pestic. Biochem. Phys., 81(2): 105–112.

Alcock, A., Elmer, P., Marsden, R. and Parry, F. (2015). Inhibition of *Botrytis cinerea* by Epirodin: A secondary metabolite from New Zealand isolates of *Epicoccum nigrum*. J. Phytopathol., 163(10): 841–852.

Ali, K., Ali, A., Khan, M.N., Rahman, S., Faizi, S., Ali, M.S., Khalifa, S.A.M., El-Seedi, H.R. and Musharraf, S.G. (2021). Rapid identification of common secondary metabolites of medicinal herbs using high-performance liquid chromatography with evaporative light scattering detector in extracts. Metabolites, 11(8): 489.

Ali, K., Maltese, F., Figueiredo, A., Rex, M., Fortes, A.M., Zyprian, E., Pais, M.S., Verpoorte, R. and Choi, Y.H. (2012). Alterations in grapevine leaf metabolism upon inoculation with *Plasmopara viticola* in different time-points. Plant Science: Int. J. Exp. Plant Biol., 191-192: 100–107.

Ansari, S., Maaz, M., Ahmad, I., Hasan, S.K., Bhat, S.A., Naqui, S.K. and Husain, M. (2021). Quality control, HPTLC analysis, antioxidant and antimicrobial activity of hydroalcoholic extract of roots of qust (Saussurea lappa, CB Clarke). Drug Metab. Pers. Ther., 36(2): 145–153.

Anuradha, S. (2010). Structural and molecular characteristics of lichenysin and its relationship with surface activity. Biosurfactants, 304–315.

Aranda, F.J., Teruel, J.A. and Ortiz, A. (2005). Further aspects on the hemolytic activity of the antibiotic lipopeptide iturin A. Biochim. Biophys. Acta - Biomembr., 1713(1): 51–56.

Arroo, R.R., Bhambra, A.S., Hano, C., Renda, G., Ruparelia, K.C. and Wang, M.F. (2021). Analysis of plant secondary metabolism using stable isotope-labelled precursors. Phytochem. Anal., 32(1): 62–68.

Arroyave-Toro, J.J., Mosquera, S. and Villegas-Escobar, V. (2017). Biocontrol activity of *Bacillus subtilis* EA-CB0015 cells and lipopeptides against postharvest fungal pathogens. Biol. Control., 114: 195–200.

Attimarad, M., Mueen Ahmed, K.K., Aldhubaib, B.E. and Harsha, S. (2011). High-performance thin layer chromatography: A powerful analytical technique in pharmaceutical drug discovery. Pharm. Methods, 2(2): 71–75.

Babst, B.A., Harding, S.A. and Tsai, C.J. (2010). Biosynthesis of phenolic glycosides from phenylpropanoid and benzenoid precursors in populous. J. Chem. Ecol., 36(3): 286–297.

Babu, B., Palanisamy, S., Vinosha, M., Anjali, R., Kumar, P., Pandi, B., Tabarsa, M., You, S.G. and Prabhu, N.M. (2020). Bioengineered gold nanoparticles from marine seaweed *Acanthophora spicifera* for pharmaceutical uses: antioxidant, antibacterial, and anticancer activities. Bioprocess Biosyst. Eng., 43: 2231–2242.

Badri, S., Basu, V.R., Chandra, K. and Anasuya, D. (2019). A review on pharmacological activities of alkaloids. WJCMPR, 230–234.

Báidez, A.G., Gómez, P., Del Río, J.A. and Ortuño, A. (2007). Dysfunctionality of the xylem in *Olea europaea* L. plants associated with the infection process by *Verticillium dahliae* Kleb. Role of phenolic compounds in plant defense mechanism. J. Agric. Food Chem., 55(9): 3373–3377.

Barghouthi, S., Young, R., Olson, M.O., Arceneaux, J.E., Clem, L.W. and Byers, B.R. (1989). Amonabactin, a novel tryptophan-or phenylalanine-containing phenolate siderophore in Aeromonas hydrophila. J. Bacteriol., 171(4): 1811–1816.

Béchet, M., Castéra-Guy, J., Guez, J.S., Chihib, N.E., Coucheney, F., Coutte, F., Fickers, P., Leclère, V., Wathelet, B. and Jacques, P. (2013). Production of a novel mixture of mycosubtilins by mutants of *Bacillus subtilis*. Bioresour. Technol., 145: 264–270.

Béchon, N., Mihajlovic J., Lopes, A.E., Fernández, S.V., Deschamps, J., Briandet, R., Sismeiro, O., Martin-Verstraete, I., Dupuy, B. and Ghigo, J.M. (2022). Bacteroides thetaiotaomicron uses a widespread extracellular DNase to promote bile-dependent biofilm formation. Proc. Natl. Acad. Sci. U.S.A., 119(7): e2111228119.

Bennett, R.N. and Wallsgrove, R.M. (1994). Secondary metabolites in plant defence mechanisms. New Phytol., 127(4): 617–633.

Bohlmann, J., Crock, J., Jetter, R. and Croteau, R. (1998). Terpenoid-based defenses in conifers: cDNA cloning, characterization, and functional expression of wound-inducible (E)-α-bisabolene synthase from grand fir (*Abies grandis*). Proc. Natl. Acad. Sci. U.S.A., 95(12): 6756–6761.

Bohm, B.A. (1998). Introduction of Flavonoids. Harwood Academic Publishers, Singapore.

Bribi, N. (2018). Pharmacological activity of alkaloids: A review. Asian J. Bot., 1(1): 1–6.

Buer, C.S., Imin, N. and Djordjevic, M.A. (2010). Flavonoids: New roles for old molecules. J. Integr. Plant Biol., 52(1): 98–111.

Bultreys, A., Gheysen, I. and de Hoffmann, E. (2006). Yersiniabactin production by *Pseudomonas syringae* and *Escherichia coli*, and description of a second yersiniabactin locus evolutionary group. Appl. Environ. Microbiol., 72(6): 3814–3825.

Burow, M. and Halkier, B.A. (2017). How does a plant orchestrate defense in time and space? Using glucosinolates in Arabidopsis as case study. Curr. Opin. Plant Biol., 38: 142–147.

Calvo, H., Mendiara, I., Arias, E., Blanco, D. and Venturini, M.E. (2019). The role of iturin A from *B. amyloliquefaciens* BUZ-14 in the inhibition of the most common postharvest fruit rots. Food Microbiol., 82: 62–69.

Cao, Y., Pi, H., Chandrangsu, P., Li, Y., Wang, Y., Zhou, H., Xiong, H., Helmann, J.D. and Cai, Y. (2018). Antagonism of two plant-growth promoting Bacillus velezensis isolates against Ralstonia solanacearum and Fusarium oxysporum. Sci. Rep., 8(1): 1–14.

Cavallito, C.J. and Haskell, T.H. (1945). The mechanism of action of antibiotics. The reaction of unsaturated lactones with cysteine and related compounds. J. Am. Chem. Soc., 67(11): 1991–1994.

Chapagain, B.P., Wiesman, Z. and Tsror, L. (2007). *In vitro* study of the antifungal activity of saponin-rich extracts against prevalent phytopathogenic fungi. Ind. Crops and Prod., 26(2): 109–115.

Chen, F., Tholl, D., Bohlmann, J. and Pichersky, E. (2011). The family of terpene synthases in plants: A mid-size family of genes for specialized metabolism that is highly diversified throughout the kingdom. Plant J., 66(1): 212–229.

Chen, X.H., Koumoutsi, A., Scholz, R. and Borriss, R. (2009). More than anticipated–production of antibiotics and other secondary metabolites by *Bacillus amyloliquefaciens* FZB42. Microb. Physiol., 16(1-2): 14–24.

Chew, F.S. (1988). Biological effects of glucosinolates. pp. 155–181. *In*: H.G. Cutler (ed.). Biologically Active Natural Products-Potential Use in Agriculture. Proceedings of the ACS Symposium 380 (Washington, DC: American Chemical Society).

Cho, K., Kim, Y., Wi, S.J., Seo, J.B., Kwon, J., Chung, J.H., Park, K.Y. and Nam, M.H. (2012). Nontargeted metabolite profiling in compatible pathogen-inoculated tobacco (*Nicotiana tabacum* L. cv. Wisconsin 38) using UPLC-Q-TOF/MS. J. Agric. Food Chem., 60(44): 11015–11028.

Chowdhury, B., Das, S.K. and Bose, S.K. (1998). Use of resistant mutants to characterize the target of mycobacillin in *Aspergillus niger* membranes. Microbiology, 144(4): 1123–1130.

Christianson, D.W. (2006). Structural biology and chemistry of the terpenoid cyclases. Chem. Rev., 106(8): 3412–3442.

Coskun, O. (2016). Separation techniques: Chromatography. Northern Clinics of Istanbul, 3(2): 156–160.

Crawford, J.M. and Townsend, C.A. (2010). New insights into the formation of fungal aromatic polyketides. Nature Rev. Microbiol., 8(12): 879–889.

Daniel, C.K., Lennox, C.L. and Vries, F.A. (2015). *In-vitro* effects of garlic extracts on pathogenic fungi *Botrytis cinerea, Penicillium expansum* and *Neofabraea alba*. S. Afr. J. Sci., 111(7-8): 1–8.

Das, S.K., Mukherjee, S., Majumdar, S., Basu, S. and Bose, S.K. (1987). Physico-chemical interaction of mycobacillin With *Aspergillus niger* protoplast membrane, the site of its action. J. Antibiot., 40(7): 1036–1043.

Davis, R.H. (1991). Glucosinolates. pp. 202–225. *In*: J.P. DMello, C.M. Duffus and J.H. Duffus (eds.). Toxic Substances in Crop Plants (Cambridge, UK: Royal Society of Chemistry).

de Carvalho, C.C.C.R. and Caramujo, M.J. (2018). The various roles of fatty acids. Molecules, 23(10): 2583.

de Vos, R.C.H., Schipper, B. and Hall, R.D. (2011). High-Performance Liquid Chromatography–Mass Spectrometry analysis of plant metabolites in *Brassicaceae*. Methods Mol. Biol., 860: 111–128.

Delattre, C., Fenoradosoa, T.A. and Michaud, P. (2011). Galactans: An overview of their most important sourcing and applications as natural polysaccharides. Braz. Arch. Biol. Technol., 54(6): 1075–1092.

Deleu, M., Paquot, M. and Nylander, T. (2008). Effect of fengycin, a lipopeptide produced by Bacillus subtilis, on model biomembranes. Biophys. J., 94(7): 2667–2679.

Di Pasquale, M.G. (2009). The essentials of essential fatty acids. J. Diet Suppl., 6(2): 143–161.

Duncan, A.J. (1991). Glucosinolates. pp. 126–147. *In*: J.P. DMello, C.M. Duffus and J.H. Duffus (eds.). Toxic Substances in Crop Plants (Cambridge, UK: Royal Society of Chemistry).

Durango, D., Pulgarin, N., Echeverri, F., Escobar, G. and Quiñones, W. (2013). Effect of salicylic acid and structurally related compounds in the accumulation of phytoalexins in cotyledons of common bean (*Phaseolus vulgaris* L.) cultivars. Molecules, 18(9): 10609–10628.

Eyles, A., Davies, N.W., Yuan, Z.Q. and Mohammed, C. (2003). Host response to natural infection by *Cytonaema* sp. in the aerial bark of *Eucalyptus globulus*. For Pathol., 33(5): 317–331.

Fenwick, G.R., Heaney, R.K. and Mullin, W.J. (1983). Glucosinolates and their breakdown products in food and food plants. Crit. Rev. Food Sci. Nutr., 18: 123–301.

Ferreira, I.C.F.R., Heleno, S.A., Reis, F.S., Stojkovic, D., Queiroz, M.J.R.P., Vasconcelos, M.H. and Sokovic, M. (2015). Chemical features of Ganoderma polysaccharides with antioxidant, antitumor and antimicrobial activities. Phytochemistry, 114: 38–55.

Forkmann, G. and Heller, W. (1999). Comprehensive natural products chemistry. Elsevier, Amsterdam, The Netherlands. Biosynthesis of Flavonoids, 713–748.

Foster-Hartnett, D., Danesh, D., Peñuela, S., Sharopova, N., Endre, G., Vandenbosch, K.A., Young, N.D. and Samac, D.A. (2007). Molecular and cytological responses of *Medicago truncatula* to *Erysiphe pisi*. Mol. Plant Pathol., 8(3): 307–319.

Friedl, P. and Wolf, K. (2003). Tumour-cell invasion and migration: Diversity and escape mechanisms. Nat. Rev. Cancer, 3(5): 362–374.

Frisvad, J.C., Rank, C., Nielsen, K.F. and Larsen, T.O. (2009). Metabolomics of *Aspergillus fumigatus*. Med. Mycol., 47(Supplement 1): S53–S71.

Fry, W.E. and Evans, P.H. (1977). Association of formamide hydrolyase with fungal pathogenicity to cyanogenic plants. Phytopathology, 67: 1001–1006.

Fry, W.E. and Myers, D.F. (1981). Hydrogen cyanide metabolism by fungal pathogens of cyanogenic plants. pp. 321–334. *In*: B. Vennesland, C.J. Knowles, E.E. Conn, J. Westley and F. Wissing (eds.). Cyanide in Biology (London: Academic Press).

Fujita, S. and Yokota, K. (2019). Disease suppression by the cyclic lipopeptides iturin A and surfactin from *Bacillus* spp. against Fusarium wilt of lettuce. J. Gen. Plant Pathol., 85(1): 44–48.

Gafner, S., Wolfender, J.L., Mavi, S. and Hostettmann, K. (1996). Antifungal and antibacterial chalcones from *Myrica serrata*. Planta Med. 62(01):67–69.

Gao, C., Chen, X., Yu, L., Jiang, L., Pan, D., Jiang, S., Gan, Y., Liu, Y. and Yi, X. (2021). New 24-membered macrolactins isolated from marine bacteria *Bacillus siamensis* as potent fungal inhibitors against sugarcane smut. J. Agric. Food Chem., 69(15): 4392–4401.

Gawdzik, B., Kamizela, A. and Szyszkowska, A. (2015). Lactones with a fragrance properties. Chemik., 69: 346–349.

Gbashi, S., Njobeh, P.B., De Saeger, S., De Boevre, M. and Madala, N.E. (2020a). Development, chemometric-assisted optimization and in-house validation of a modified pressurized hot water extraction methodology for multi-mycotoxins in maize. Food Chem., 307: 125526.

Gbashi, S., Njobeh, P.B., Madala, N.E., De Boevre, M., Kagot, V. and De Saeger, S. (2020b). Parallel validation of a green-solvent extraction method and quantitative estimation of multi-mycotoxins in staple cereals using LC-MS/MS. Sci Rep., 10(1): 1–16.

Gennadios, H.A., Gonzalez, V., Di Costanzo, L., Li, A., Yu, F., Miller, D.J., Allemann, R.K. and Christianson, D.W. (2009). Crystal structure of (+)-δ-cadinene synthase from Gossypium arboreum and evolutionary divergence of metal-binding motifs for catalysis. Biochemistry, 48(26): 6175–6183.

Giamoustaris, A. and Mithen, R. (1995). The effect of modifying the glucosinolate content of leaves of oilseed rape (Brassica napus ssp. oleifera) on its interaction with specialist and generalist pests. Ann. Appl. Biol., 126: 347–363.

Gómez, G.F., Reyes, C.R., Quijano, L., Calderón, P.J.S. and Ríos, C.T. (1990). Methoxifurans auranols with fungostatic activity from *Lonchocarpus castilloi*. Phytochemistry, 29: 459–463.

Gong, A.D., Li, H.P., Yuan, Q.S., Song, X.S., Yao, W., He, W.J., Zhang, J.B. and Liao, Y.C. (2015). Antagonistic mechanism of iturin A and plipastatin A from *Bacillus amyloliquefaciens* S76-3 from wheat spikes against *Fusarium graminearum*. PLoS One, 10(2): e0116871.

Govatati, S., Pichavaram, P., Mani, A.M., Kumar, R., Sharma, D., Dienel, A., Meena, S., Puchowicz, M.A., Park, E.A. and Rao, G.N. (2021). Novel role of xanthine oxidase-dependent H_2O_2 production in 12/15-lipoxygenase-mediated de novo lipogenesis, triglyceride biosynthesis and weight gain. Redox. Biol., 47: 102163.

Grassmann, J., Hippeli, S. and Elstner, E.F. (2002). Plant's defence and its benefits for animals and medicine: Role of phenolics and terpenoids in avoiding oxygen stress. Plant Physiol. Biochem., 40(6-8): 471–478.

Grayer, R.J. and Harborne, J.B. (1994). A survey of antifungal compounds from higher plants, 1982–1993. Phytochemistry, 37(1): 19–42.

Han, X., Shen, D., Xiong, Q., Bao, B., Zhang, W., Dai, T., Zhao, Y., Borriss, R. and Fan, B. (2021). The plant-beneficial rhizobacterium *Bacillus velezensis* FZB42 controls the soybean pathogen *Phytophthora sojae* due to bacilysin production. Appl. Environ. Microbiol., 87(23): e0160121.

Hayashi, K.I., Kawaide, H., Notomi, M., Sakigi, Y., Matsuo, A. and Nozaki, H. (2006). Identification and functional analysis of bifunctional ent-kaurene synthase from the moss *Physcomitrella patens*. FEBS Lett., 580(26): 6175–6181.

Hazan, R., Que, Y.A., Maura, D. and Rahme, L.G. (2012). A method for high throughput determination of viable bacteria cell counts in 96-well plates. BMC Microbiol., 12: 259.

Hughes, M.A. (1991). The cyanogenic polymorphism in *Trifolium repens* L. (white clover). Heredity, 66: 105–115.

Inaoka, T., Takahashi, K., Ohnishi-Kameyama, M., Yoshida, M. and Ochi, K. (2003). Guanine nucleotides guanosine 5′-diphosphate 3′-diphosphate and GTP co-operatively regulate the production of an antibiotic bacilysin in *Bacillus subtilis*. J. Biol. Chem., 278(4): 2169–2176.

Jones, J.D. and Dangl, J.L. (2006). The plant immune system. Nature, 444(7117): 323–329.

Kadakeri, S., Arul, M.R., Bordett, R., Duraisamy, N., Naik, H. and Rudraiah, S. (2020). Protein synthesis and characterization. pp. 121–161. *In*: Wei, G. and Kumbar, Sangamesh, G. (eds.). Artificial Protein and Peptide Nanofibers. Woodhead Publishing Series.

Kaisa A. Salminen, Achim Meyer, Lenka Jerabkova et al. (2011). Inhibition of human drug metabolizing cytochromeP450 enzymes by plant isoquinoline alkaloids. Phytomedicine, (18): 533–538.

Kannaiah, K.P., Sugumaran, A., Chanduluru, H.K. and Rathinam, S. 2021. Environmental impact of greenness assessment tools in liquid chromatography—A review. Microchemical Journal, 170: 106685.

Khan, R.A.A., Najeeb, S., Mao, Z., Ling, J., Yang, Y., Li, Y. and Xie, B. (2020). Bioactive secondary metabolites from *Trichoderma* spp. against phytopathogenic bacteria and root-knot nematode. Microorganisms, 8(3): 401.

Klich, M.A., Lax, A.R. and Bland, J.M. (1991). Inhibition of some mycotoxigenic fungi by iturin A, a peptidolipid produced by *Bacillus subtilis*. Mycopathologia, 116(2): 77–80.

Konz, D. and Marahiel, M.A. (1999). How do peptide synthetases generate structural diversity? Chem. Biol., 6(2): R39–R48.

Korsten, L. and De Jager, E.E. (1995). Mode of action of *Bacillus subtilis* for control of avocado postharvest pathogens. South African Avocado Growers' Association Yearbook, 18: 124–130.

Kourist, R. and Hilterhaus, L. (2015). Microbial lactone synthesis based on renewable resources. *In*: B. Kamm (ed.). Microorganisms in Biorefineries, Microbiology Monographs. Springer Berlin Heidelberg: Berlin/Heidelberg, Germany, 26: 275–301.

Kwon, J.H., Won, S.J., Moon, J.H., Lee, U., Park, Y.S., Maung, C.E.H., Ajuna, H.B. and Ahn, Y.S. (2021). *Bacillus licheniformis* PR2 controls fungal diseases and increases production of jujube fruit under field conditions. Horticulturae, 7(3): 49.

Labuda, I. (2009). Flavor compounds. pp. 305–321. *In*: M. Schaechter (ed.). Encyclopedia of Microbiology, 3rd ed., Amsterdam Academic Press.

Lagrouh, F., Dakka, N. and Bakri, Y. (2017). The antifungal activity of Moroccan plants and the mechanism of action of secondary metabolites from plants. J. Mycol. Med., 27: 303–311.

Lee, T., Park, D., Kim, K., Lim, S.M., Yu, N.H., Kim, S., Kim, H.Y., Jung, K.S., Jang, J.Y., Park, J.C., Ham, H., Lee, S., Hong, S.K. and Kim, J.C. (2017). Characterization of *Bacillus amyloliquefaciens* DA12 showing potent antifungal activity against mycotoxigenic *Fusarium* species. Plant Pathol. J., 33(5): 499–507.

Lieberei, R., Biehl, B., Giesemann, A. and Junqueira, N.T.V. (1989). Cyanogenesis inhibits active defense reactions in plants. Plant Physiol., 90: 33–36.

Lindström, S., Eriksson, M., Vazin, T., Sandberg, J., Lundeberg, J., Frisén, J. and Andersson-Svahn, Helene. (2009). High-density microwell chip for culture and analysis of stem cells. PLoS One, 4(9): e6997.

Lu, Y., Chen, Q., Bu, Y., Luo, R., Hao, S., Zhang, J., Tian, J. and Yao, Y. (2017). Flavonoid accumulation plays an important role in the rust resistance of *Malus* plant leaves. Front. Plant Sci., 8: 1286.

Ma, L.T., Lee, Y.R., Liu, P.L., Cheng, Y.T., Shiu, T.F., Tsao, N.W., Wang, S.Y. and Chu, F.H. (2019). Phylogenetically distant group of terpene synthases participates in cadinene and cedrane-type sesquiterpenes accumulation in *Taiwania cryptomerioides*. Plant Sci., 289: 110277.

Mannanov, R.N. and Sattarova, R.K. (2001). Antibiotics produced by *Bacillus* bacteria. Chem. Nat. Compd., 37(2): 117–123.

Mares, D. (1987). Antimicrobial activity of protoanemonin, a lactone from ranunculaceous plants. Mycopathologia, 98(3): 133–140.

Matar, S.M., El-Kazzaz, S.A., Wagih, E.E., El-Diwany, A.I., Moustafa, H.E., Abo-Zaid, G.A., Abd-Elsalam, H.E. and Hafez, E.E. (2009). Antagonistic and inhibitory effect of *Bacillus subtilis* against certain plant pathogenic fungi, I. Biotechnology, 8(1): 53–61.

May, E.M.D. (2007). < date>/url=http:// www.genengnews.com/articles/ chtitem.aspx? tid=2136 "Array Tape for Miniaturized Genotyping". Genetic Engineering & Biotechnology News. Mary Ann Liebert, Inc. p. 22. Archived from the original on 2007. Retrieved 2008-07-06. (subtitle) Processing hundreds of microplate equivalents without complex plate-handling equipment.

May, J.J., Wendrich, T.M. and Marahiel, M.A. (2001). The dhb operon of *Bacillus subtilis* encodes the biosynthetic template for the catecholic siderophore 2,3-dihydroxybenzoate-glycine-threonine trimeric ester bacillibactin. J. Biol. Chem., 276(10): 7209–7217.

Mazrou, Y.S.A., Makhlouf, A.H., Elseehy, M.M., Awad, M.F. and Hassan, M.M. (2020). Antagonistic activity and molecular characterization of biological control agent *Trichoderma harzianum* from Saudi Arabia. Egypt. J. Biol. Pest. Control., 30(1): 1–8.

Mierziak, J., Kostyn, K. and Kulma, A. (2014). Flavonoids as important molecules of plant interactions with the environment. Molecules, 19(10): 16240–16265.

Miller, J.H. (1972). Determination of viable cell counts: bacterial growth curves. pp. 31–36. *In*: J.H. Miller (ed.). Experiments in Molecular Genetics. New York: Cold Spring Harbor.

Minchin, S. and Lodge, J. (2019). Understanding biochemistry: Structure and function of nucleic acids. Essays Biochem., 63(4): 433–456.

Mithen, R., Lewis, B.G., Fenwick, G.R. and Heaney, R.K. (1986). *In vitro* activity of glucosinolates and their products against *Leptosphaeria maculans*. Trans. Br. Mycol. SOC., 87: 433–440.

Miyanaga, A. (2017). Structure and function of polyketide biosynthetic enzymes: Various strategies for production of structurally diverse polyketides. Biosci. Biotechnol. Biochem., 81(12): 2227–2236.

Montreuil, J. (1996). Structure and biosynthesis of glycoproteins. pp. 273–328. *In*: S. Dumitriu (ed.). Polysaccharides in Medicinal Applications. Routledge, New York.

Nayik, G.A. and Kour, J. (2022). Handbook of Plant and Animal Toxins in Food: Occurrence, Toxicity, and Prevention (1st ed.). CRC Press.

Nelson, J., El-Gendy, A.O., Mansy, M.S., Ramadan, M.A. and Aziz, R.K. (2020). The biosurfactants iturin, lichenysin and surfactin, from vaginally isolated lactobacilli, prevent biofilm formation by pathogenic *Candida*. FEMS Microbiol. Lett., 367(15): fnaa126.

Nifakos, K., Tsalgatidou, P.C., Thomloudi, E.E., Skagia, A., Kotopoulis, D., Baira, E., Delis, C., Papadimitriou, K., Markellou, E., Venieraki, A. and Katinakis, P. (2021). Genomic analysis and secondary metabolites production of the endophytic *Bacillus velezensis* Bvel1: A biocontrol agent against *Botrytis cinerea* causing bunch rot in post-harvest table grapes. Plants, 10(8): 1716.

Ntushelo, K., Ledwaba, L.K., Rauwane, M.E., Adebo, O.A. and Njobeh, P.B. (2019). The mode of action of *Bacillus* species against *Fusarium graminearum*, tools for investigation, and future prospects. Toxins, 11(10): 606.

Nuapia, Y., Maraba, K., Tutu, H., Chimuka, L. and Cukrowska, E. (2021). *In situ* decarboxylation-pressurized hot water extraction for selective extraction of cannabinoids from *Cannabis sativa*. Chemometric Approach. Molecules, 26(11): 3343.

Ongena, M. and Jacques, P. (2008). *Bacillus* lipopeptides: Versatile weapons for plant disease biocontrol. Trends Microbiol., 16(3): 115–124.

Orhan, I., Deliorman-Orhan, D. and Özçelik, B. (2009). Antiviral activity and cytotoxicity of the lipophilic extracts of various edible plants and their fatty acids. Food Chem., 115(2): 701–705.

Osbourn, A.E. (1996). Preformed antimicrobial compounds and plant defense against fungal attack. Plant Cell, 8: 1821–1831.

Pal, K.K., Tilak, K.V.B.R., Saxcna, A.K., Dey, R. and Singh, C.S. (2001). Suppression of maize root diseases caused by *Macrophomina phaseolina, Fusarium moniliforme* and *Fusarium graminearum* by plant growth promoting rhizobacteria. Microbiol. Res., 156(3): 209–223.

Parvez, M.M., Tomita-Yokotani, K., Fujii, Y., Konishi, T. and Iwashina, T. (2004). Effects of quercetin and its seven derivatives on the growth of *Arabidopsis thaliana* and *Neurospora crassa*. Biochem. Syst. Ecol., 32: 631–635.

Patel, M.K., Pandey, S., Kumar, M., Haque, M.I., Pal, S. and Yadav, N.S. (2021). Plants metabolome study: Emerging tools and techniques. Plants, 10(11): 2409.

Pereira, D.M., Valentão, P., Pereira, J.A. and Andrade, P. 2009. Phenolics: From chemistry to biology. Molecules, 14(6): 2202–2211.

Perestrelo, R., Fernandes, A., Albuquerque, F.F., Marques, J.C. and Câmara, J.S. (2006). Analytical characterization of the aroma of *Tinta negra* mole red wine: Identification of the main odorants compounds. Anal. Chim. Acta, 563(1-2): 154–164.

Pombo, M.A., Zheng, Y., Fernandez-Pozo, N., Dunham, D.M., Fei, Z. and Martin, G.B. (2014). Transcriptomic analysis reveals tomato genes whose expression is induced specifically during effector-triggered immunity and identifies the Epk1 protein kinase which is required for the host response to three bacterial effector proteins. Genome Biol., 15: 492.

Poole, C.F. (2003). The Essence of Chromatography. Elsevier, Amsterdam.

Pusztahelyi, T., Holb, I.J. and Pócsi, I. (2015). Secondary metabolites in fungus-plant interactions. Front. Plant Sci., 6: 573.

Qiu, Y., Xiao, F., Wei, X., Wen, Z. and Chen, S. (2014). Improvement of lichenysin production in *Bacillus licheniformis* by replacement of native promoter of lichenysin biosynthesis operon and medium optimization. Appl. Microbiol. Biotechnol., 98(21): 8895–8903.

Rasheed, F., Markgren, J., Hedenqvist, M. and Johansson, E. (2020). Modeling to understand plant protein structure-function relationships—Implications for seed storage proteins. Molecules, 25(4): 873.

Rauwane, M.E., Ogugua, U.V., Kalu, C.M., Ledwaba, L.K., Woldesemayat, A.A. and Ntushelo, K. (2020). Pathogenicity and virulence factors of *Fusarium graminearum* including factors discovered using next generation sequencing technologies and proteomics. Microorganisms, 8: 305.

Razafindralambo, H., Popineau, Y., Deleu, M., Hbid, C., Jacques, P., Thonart, P. and Paquot, M. (1998). Foaming properties of lipopeptides produced by *Bacillus subtilis*: effect of lipid and peptide structural attributes. J. Agric. Food Chem., 46(3): 911–916.

Reino, J.L., Guerrero, R.F., Hernández-Galán, R. and Collado, I.G. (2008). Secondary metabolites from species of the biocontrol agent *Trichoderma*. Phytochem. Rev., 7: 89–123.

Ribble, D., Goldstein, N.B., Norris, D.A. and Shellman, Y.G. (2005). A simple technique for quantifying apoptosis in 96-well plates. BMC Biotechnol., 5: 12.

Rohloff, J. (2015). Analysis of phenolic and cyclic compounds in plants using derivatization techniques in combination with GC-MS-based metabolite profiling. Molecules, 20(2): 3431–3462.

Sajitha, K.L. and Dev, S.A. (2016). Quantification of antifungal lipopeptide gene expression levels in *Bacillus subtilis* B1 during antagonism against sapstain fungus on rubberwood. Biol. Control, 96: 78–85.

Saleem, H., Htar, T.T., Naidu, R., Anwar, S., Zengin, G., Locatelli, M. and Ahemad, N. (2020). HPLC–PDA polyphenolic quantification, UHPLC–MS secondary metabolite composition, and *in vitro* enzyme inhibition potential of *Bougainvillea glabra*. Plants, 9(3): 388.

Saleh, A.E., Ul-Hassan, Z., Zeidan, R., Al-Shamary, N., Al-Yafei, T., Alnaimi, H., Higazy, N.S., Migheli, Q. and Jaoua, S. (2021). Biocontrol activity of *Bacillus megaterium* BM344-1 against toxigenic fungi. ACS Omega, 6(16): 10984–10990.

Sánchez-Montesinos, B., Santos, M., Moreno-Gavíra, A., Marín-Rodulfo, T., Gea, F.J. and Diánez, F. (2021). Biological control of fungal diseases by *Trichoderma aggressivum* f. *europaeum* and its compatibility with fungicides. J. Fungi, 7(8): 598.

Sang, J., Yang, Y., Chen, Y., Cai, J., Lu, C. and Huang, G. (2018). Antibacterial activity analysis of lipopeptide and polyketide compounds produced by endophytic bacteria *Bacillus amyloliquefaciens* BEB17. Acta Phytopathol. Sin., 48(3): 402–412.

Scheinmann, F. (2013). An introduction to spectroscopic methods for the identification of organic compounds. *In*: F. Scheinmann (eds.). Elsevier Science.

Skadhauge, B., Thomsen, K.K. and von Wettstein, D. (2004). The role of barley testa layer and its flavonoid content in resistance to *Fusarium* infections. Hereditas, 126(2): 147–160.

Steinborn, G., Hajirezaei, M.R. and Hofemeister, J. (2005). *bac* genes for recombinant bacilysin and anticapsin production in *Bacillus* host strains. Arch. Microbiol., 183(2): 71–79.

Stevenson, P.C. and Haware, M.P. (1999). Maackiain in *Cicer bijugum* Rech. f. associated with resistance to Botrytis grey mould. Biochem. Syst. Ecol., 27(8): 761–767.

Subramanian, S., Graham, M.Y., Yu, O. and Graham, T.L. (2005). RNA interference of soybean isoflavone synthase genes leads to silencing in tissues distal to the transformation site and to enhanced susceptibility to *Phytophthora sojae*. Plant Physiol., 137(4): 1345–1353.

Tahara, S., Fujiwara, K., Ishizaka, H., Mizutani, J. and Obata, Y. (1972). γ-Decalactone—One of constituents of volatiles in cultured broth of *Sporobolomyces odorus*. Agric. Biol. Chem., 36(13): 2585–2587.

Takatsy, G. (1950). Uj modszer sorozatos higitasok gyors es pontos elvegzesere (A rapid and accurate method for serial dilutions). Kiserl. Orvostud, 5: 393–397.

Tchameni, S.N., Cotârleț, M., Ghinea, I.O., Bedine, M.A.B., Sameza, M.L., Borda, D., Bahrim, G. and Dinică, R.M. (2020). Involvement of lytic enzymes and secondary metabolites produced by *Trichoderma* spp. in the biological control of *Pythium myriotylum*. Int. Microbiol., 23(2): 179–188.

Tholl, D. (2006). Terpene synthases and the regulation, diversity and biological roles of terpene metabolism. Curr. Opin. Plant Biol., 9(3): 297–304.

Tian, J., Peng, Z., Zhang, J., Song, T., Wan, H., Zhang, M. and Yao, Y. (2015). McMYB10 regulates coloration via activating *McF3'H* and later structural genes in ever-red leaf crabapple. Plant Biotechnol. J., 13(7): 948–961.

Um, S., Fraimout, A., Sapountzis, P., Oh, D.C. and Poulsen, M. (2013). The fungus-growing termite *Macrotermes natalensis* harbors bacillaene-producing *Bacillus* sp. that inhibit potentially antagonistic fungi. Sci. Rep., 3(1): 1–7.

van Loon, L.C., Rep, M. and Pieterse, C.M. (2006). Significance of inducible defense-related proteins in infected plants. Annu. Rev. Phytopathol., 44: 135–162.

Verma, N. and Shukla, S. (2015). Impact of various factors responsible for fluctuation in plant secondary metabolites. J. Appl. Res. Med. Aromatic Plants, 2(4): 105–113.

Waksmundzka-Hajnos, M. and Sherma, J. (2011). High Performance Liquid Chromatography in Phytochemical Analysis. Taylor & Francis Group, Boca Raton. New York.

Wang, Q., Liang, Z., Peng, Y., Hou, J.L., Wei, S.L., Zhao, Z.Z. and Wang, W.Q. (2015). Whole transverse section and specific-tissue analysis of secondary metabolites in seven different grades of root of *Paeonia lactiflora* using laser microdissection and liquid chromatography-quadrupole/time of flight-mass spectrometry. J. Pharm. Biomed. Anal., 103: 7–16.

Weber, T., Blin, K., Duddela, S., Krug, D., Kim, H.U., Bruccoleri, R., Lee, S.Y., Fischbach, M.A., Müller, R., Wohlleben, W. and Breitling, R. (2015). antiSMASH 3.0—a comprehensive resource for the genome mining of biosynthetic gene clusters. Nucleic Acids Res., 43(W1): W237–W243.

Wilson, M.K., Abergel, R.J., Raymond, K.N., Arceneaux, J.E. and Byers, B.R. (2006). Siderophores of *Bacillus anthracis*, *Bacillus cereus*, and *Bacillus thuringiensis*. Biochem. Biophys. Res. Commun., 348(1): 320–325.

Win, T.T., Bo, B., Malec, P. and Fu, P. (2021). The effect of a consortium of *Penicillium* sp. and *Bacillus* spp. in suppressing banana fungal diseases caused by *Fusarium* sp. and *Alternaria* sp. J. Appl. Microbiol., 131(4): 1890–1908.

Wink, M. (2008). Plant secondary metabolism: diversity, function and its evolution. Nat. Prod. Commun., 3(8): 1205–1216.

Wink, M. (2010). Biochemistry, physiology and ecological function of secondary metabolites. Chapter 1 Introduction. Ann. Plant Rev., 40: 1–19.

Wojakowska, A., Muth, D., Narożna, D., Mądrzak, C., Stobiecki, M. and Kachlicki, P. (2013). Changes of phenolic secondary metabolite profiles in the reaction of narrow leaf lupin (*Lupinus angustifolius*) plants to infections with *Colletotrichum lupini* fungus or treatment with its toxin. Metabolomics, 9(3): 575–589.

Yue, P.Y.K., Leung, E.P.Y., Mak, N.K. and Wong, R.N.S. (2010). A simplified method for quantifying cell migration/wound healing in 96-Well plates. J. Biomol. Screen., 15(4): 427–433.

Zalila-Kolsi, I., Mahmoud, A.B., Ali, H., Sellami, S., Nasfi, Z., Tounsi, S. and Jamoussi, K. (2016). Antagonist effects of *Bacillus* spp. strains against *Fusarium graminearum* for protection of durum wheat (*Triticum turgidum* L. subsp. *durum*). Microbiol. Res., 192: 148–158.

Zapata-Sarmiento, D.H., Palacios-Pala, E.F., Rodríguez-Hernández, A.A., Melchor, D.L.M., Rodríguez-Monroy, M. and Sepúlveda-Jiménez, G. (2020). *Trichoderma asperellum*, a potential biological control agent of *Stemphylium vesicarium*, on onion (*Allium cepa* L.). Biol. Control, 140: 104105.

Zaynab, M., Fatima, M., Sharif, Y., Zafar, M.H., Ali, H. and Khan, K.A. (2019). Role of primary metabolites in plant defense against pathogens. Microb. Pathog., 137: 103728.

Zenk, M.H. (1967). Pathways of salicyl alcohol and salicin formation in *Salix purpurea* L. Phytochemistry, 6: 245–252

Zheng, M., Zhao, X., Zheng, S., Chen, D., Du, P., Li, X., Jiang, D., Guo, J.T., Zeng, H. and Lin, H. (2020). Bat SARS-Like WIV1 coronavirus uses the ACE2 of multiple animal species as receptor and evades IFITM3 restriction via TMPRSS2 activation of membrane fusion. Emerg Microbes Infect., 9(1): 1567–1579.

Zohora, U.S., Ano, T. and Rahman, M.S. (2016). Biocontrol of *Rhizoctonia solani* K1 by iturin A producer *Bacillus subtilis* RB14 seed treatment in tomato plants. Adv. Microbiol., 6: 424–431.

Chapter 5

Bio-Nanoparticles as a Source of Nanofungicides to Combat Phytopathogens

Chikanshi Sharma,[1] *Madhu Kamle*[1] and *Pradeep Kumar*[1,2,*]

1. Introduction

Nanotechnology is characterized as molecular, macro-molecular and atomic size research that has advanced tremendously in recent decades as evidenced by a 25-fold rise in a number of components that comprise nanoparticles for their biosynthesis between 2005 and 2010 (Wilson, 2010). Their unique general features particularly particle geometry, charge, surface reactivity, area and size in comparison to their initial bulk or dissolved predecessors are believed to have aided this evolution which opens the door to many different applications and research, because of these advantages green-biosynthesis of NPs has recently received a lot of recognition in different fields (Semenzin et al., 2015; Ito et al., 2005; Mie et al., 2014). For nanoparticle synthesis, a wide range of biological assets such as plants/plant parts and microorganisms can be used (Mohanpuria et al., 2008). Although there are many ways of making green NPs, most of them involve simply reacting plant and microbe extracts with a metallic salt before executing a biological reduction to turn the metal into NPs (Iravani, 2011; Mittal et al., 2013). Plants extract are more convenient for green synthesis of nanoparticles as compounds and phytochemicals present in plant extracts alleviate metal ions, reducing/stabilizing NPs at a faster rate by satisfying different parts of plants in comparison to microbes which take more time to reduce metallic ions (Ghosh et al., 2012; Khan et al., 2013; Rai and Yadav, 2013; Mittal

[1] Applied Microbiology Lab., Department of Forestry, North Eastern Regional Institute of Science and Technology, Nirjuli 791109, India.
[2] Department of Botany, University of Lucknow, Lucknow-226007, Uttar Pradesh, India.
* Corresponding author: pkbiotech@gmail.com

et al., 2013; Ahmed et al., 2017; Rai and Ingle, 2012; Iravani, 2011; Thakkar et al., 2010). Microbes can synthesize NPs in both intracellular/extracellular environments, bacteria of different types can withstand varied stressors and evolve many defense mechanisms that make them ideal for nanoparticle synthesis whereas, fungi are more favorable for nanoparticle synthesis because they produce a large quantity of proteins and enzymes (Singh et al., 2016; Iravani, 2014). Biological synthesis processes have several distinct advantages over traditional chemical and physical methods including (a) small nanoparticles can be made even in large-scale manufacturing (Klaus et al., 1999) (b) no toxic chemicals are used, resulting in an environmentally friendly and clean method (Senapati et al., 2005) (c) no external experimental conditions, such as high pressure and energy are required, resulting in significant energy savings (Bansal et al., 2004) (d) the cost of the synthesis process is reduced overall attributed to the capping and lowering effects of the bioactive component, such as an enzyme. Green or biosynthesis of silver nanoparticles (AgNPs) using plants-plant parts and microbes extracts is a viable alternative synthesis process that has received a lot of interest and recommendation recently. Alkaloids, sugars, phenolic acids, polyphenols, proteins and terpenoids among other metabolites found in plants play a considerable portion in the bio-reduction of Ag ions to AgNPs (Makarov et al., 2014). In this chapter, the antifungal effects of NPs against several phytopathogens, as well as their mechanisms of action are briefly discussed.

2. Nanoparticles

Nanoparticles are the most fundamental building block in the expansion of nanostructures. They are significantly far bigger than an atom or a simple molecule, which is the subject of quantum mechanics, but much narrower than the everyday objects that are governed by Newton's laws of motion (Horikoshi and Serpone, 2013). Nanoparticles have unique capabilities that are determined by their morphology, form and size and can interact with plants, animals and microorganisms (Husen and Siddiqi, 2014). Ultrafine particles classified as nanoparticles have a length of more than one nanometer (1 nm) and less than 100 nanometers (100 nm) in 2- or 3-dimensional space that lack size-related depth features that can change the nanoparticles physical or morphological properties. They are environmentally favorable because of their use in limiting microbial growth, also several other variables such as pollution reduction, waste prevention and the use of safe renewable/non-toxic solvents, can thus be explained as some essential principles of their environmental friendliness (Singh et al., 2018). Agriculture, nanomedicine, food science, microelectronics, hydrogen storage ferrofluids, chemical nanosensors and catalytic systems are just a few of the domains where nanoparticles are used. The biosynthesis of green NPs is the primary goal of the nascent scientific field known as nanotechnology whereas, biological or chemical methods can be used to make nanoparticles. Currently, a broad range of metallic nanostructures composed of silver, gold, titanium, copper, magnesium, alginate and zinc are used for a range of applications (Dubchak et al., 2010). Due to the presence of dangerous compounds accumulated on the surface, chemical manufacturing practices have been connected to a wide range of adverse effects. Environmentally friendly equivalents to physical and chemical methods of nanoparticle (NPs) biosynthesis include biodegradable materials using enzymes, bacteria, fungi, plants or plant extracts (Klaus et al., 1999). A relevant aspect of nanotechnology is the progression of environment-friendly methods for creating nanoparticles, especially silver nanoparticles because it has many uses (Vigneshwaran et al., 2007, Kyriacou et al., 2004). Simple molecules need not make up nanoparticles, they are made of three layers: (1) the central or core part, which is the nanoparticle's central component and commonly refers to the nanoparticle (NP) itself (2) The shell or helix layer, which is physically/chemically distinct from the core (central part), and (3) the surface layer, which can be functionalized with metal ions, polymers, surfactants and small molecules (Shin et al., 2016). Due to their unique features, these materials have piqued the curiosity of researchers from various disciplines. As morphology influences most of the properties of nanoparticles, morphological features are always of great interest. For morphological research, various characterization techniques are used, so microscopic approaches including scanning electron microscopy followed by transmission electron microscopy and polarized optical microscopy, are the most common. Scanning electron microscopy is a technology that uses electron scanning to offer all relevant information on nanoparticles at the nanoscale level. Researchers have used this technique in bulk or matrix to scrutinize not only the aesthetics of their nanomaterials but also the dispersion of nanoparticles in the matrix. The nature and composition of bonding materials are primarily studied using structural features. It provides a wealth of information regarding the subject material's bulk qualities. Inorganic nanoparticles, whether simple or complex, have unique properties and are becoming a more important material in the creation of innovative nanodevices for a variety of biological, physical, pharmacological and medicinal services (Loureiro et al., 2016; Nikalje, 2015; Martis et al., 2012). NPs have been employed in a variety of domains, including the management of plant pathogens and their related diseases (Masum et al., 2019, Oves et al., 2018). Pesticides are usually applied selectively to crop plant pests, with no adverse effects on the plants, wildlife, symbiotic flora or human beings (Hassall, 1965). However, indiscriminate pesticide applications are a major source of environmental impacts and public issues (Kutz et al., 1991). Furthermore, the emergence and growth of new pathogenic strains is a constant challenge and chemical treatment is both costly and ineffective. Recent advances in the field of green synthesis have prompted scientists and researchers to explore how they can use it to combat harmful microbes. Biosynthesized metallic nanoparticles have inhibitory effects against phytopathogenic fungi such as *F. oxysporum*, *A. fumigatus* and *A. niger*, as well as other pathogenic microorganisms, these nanoparticles have been shown to inhibit Gram-negative (–) and Gram positive (+) bacteria such as *E. coli*, *Staphylococcus aureu*s and *B. subtilis* (Nisar et al., 2019). Due to "green synthesis" by plants, fungus, yeasts or bacteria, silver NPs have received increased attention (Rafique et al., 2017). The biological biosynthesis of different nanoparticles is depicted in Fig. 5.1.

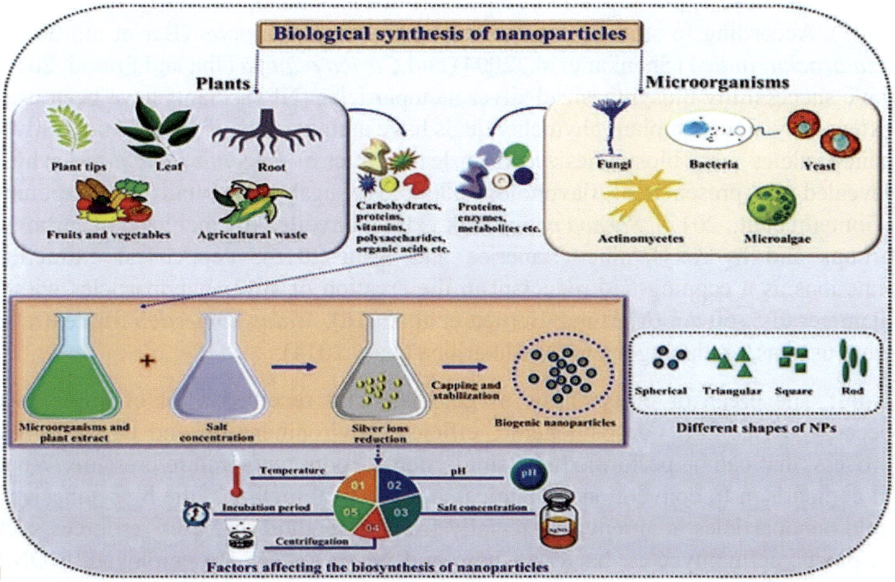

Fig. 5.1: Green synthesis of metallic nanoparticles (NPs) refers to the use of environmentally benign and renewable resources as reducing agents, stabilizers, and capping agents in the synthesis of metallic NPs. Several methods are available for the green synthesis of metallic NPs, including, plant extracts, microorganisms,biomolecules, and green solvents (Ali et al., 2020).

3. Types of nanoparticles

3.1 Metallic NPs

Faraday-1857 was the first to discover metallic nanoparticles in solution (Daniel and Astruc, 2004), while Mie,1908 was the first to describe the altering color behavior of NPs in the solution quantitatively (Link and El-Sayed, 2003). Large surface energies, a large ratio of surface to volume, specific electronic structure made available by their transition between metallic and molecular states, quantum confinement and Plasmon excitation are all important characteristics of metallic NPs. Silver (Saravanakumar et al., 2017), gold (Bollella et al., 2017), copper (Keihan et al., 2017), palladium (Lebaschi et al., 2017) and platinum (Aday et al., 2016) are among the metal NPs that have been created. Due to its unique antifungal and antibacterial characteristic and ease of reduction from monovalent silver to metallic silver, Ag NP is the most often favored and manufactured by green synthesis (Liu et al., 2017). Silver also has the highest plasmon excitation efficiency (Bastús et al., 2016). Similarly, gold NPs are important, but in a lesser amount than silver NPs. Other metallic NPs have been documented in the literature, although only in small quantities.

Silver: These nanoparticles are among the most effective antifungal and antibacterial agents, against bacteria, fungi, viruses and other pathogenic microorganisms (Gong et al., 2007). They are without a doubt the most frequently used nanomaterials, with a lot more applications in comparison to other nanoparticles (Sharma et al., 2009). According to studies, plants such as *Capsicum annuum* (Bar et al., 2009), *Azadirachta indica* (Shankar et al., 2004) and *Carica papaya* (Jha and Prasad, 2010) have successfully biosynthesized silver nanoparticles (NPs). Plants have been used extensively since the plant phytochemicals have a higher rate of stabilization. Silver nanoparticles were biosynthesized from leaf extract of *Eugenia jambolana*, which revealed the presence of flavonoids, alkaloids, sugar compounds and saponins (Gomathi et al., 2017). *Saraca asoca* bark extract unveiled the inclusion of carboxyl groups and hydroxylamine (Banerjee and Nath 2015). Pepper-leaf extraction functions as a capping and reductant in the creation of silver nanoparticles with a diameter of 5–60 nm (Martínez-Bernett et al., 2016). *Malus domestica* fruit extracts were used as a reducing agent (Mallikarjuna et al., 2014).

Gold: The green or biosynthesis of gold NPs has received a lot of importance because of its fast, non-pathogenic, efficient environmentally and cost-effective process, that can be performed in a single step at room temperature/pressure. When in comparison to conventional chemical and physical methods, the biosynthesis of gold nanoparticles is an environmentally friendly method that uses resources such as plants, actinomycetes, bacteria, fungi and yeast. AuNPs are employed in DNA fingerprinting as a laboratory tracer to detect the presence of DNA in a sample and identify protein interactions in immunochemical research. Amino glycoside drugs yielding gentamycin, neomycin and streptomycin are also observed using Au nanoparticles. Gold nanorods are being utilized to detect cancer stem cells, useful for cancer detection in humans and identifying distinct bacteria classes (Baban and Seymour, 1998). Extract from *Abelmoschus esculentus* produced Au nanoparticles with antifungal efficacy against *Aspergillus flavus*, *Puccinia graminis tritci*, *Candida albicans* and *A. niger* (Jayaseelan et al., 2013).

3.2 Metal oxide NPs

To make metal oxide nanoparticles (NPs), hydroxo (M-OH-M)/oxo (M-O-M) bridges are used to connect the metal centers, resulting in metal oxo as well as metal hydroxo composites in solution (Fernández-Garcia and Rodgriguez, 2007). Plant-based synthesized metal oxide nanoparticles using garlic and ginger has been recently reported (Haider et al., 2020).

3.3 Alloy NPs

Compared to their bulk predecessors, these nanoparticles exhibit different structural parameters (Ceylan et al., 2006). As Ag flakes have the greatest electrical conductivity of any metal filler and, unlike many other metals, their oxides have considerably higher conductivity, they are the most frequently utilized metal filler (Yun et al., 2008). Bimetallic alloy nanoparticles' characteristic is regulated by both metals, and they have greater advantages than typical metallic nanoparticles (Mohl et al., 2011). Combining multiple elements to make alloys can considerably expand the spectrum of properties of metallic nanoparticles (NPs). In most cases, alloying's synergistic effects result in improved specific qualities (Ferrando et al., 2008). The chemical, as well as physical properties of alloy NPs, can be easily modified by changing the size of the clusters, as well as the composition and atomic ordering, that has sparked a lot of attention. Nanoalloys can have properties that are considerably different, which can lead to one with a variety of applications in engineering, electronics and catalysis (Munoz-Flores et al., 2011).

3.4 Magnetic NPs

These NPs have distinct kinds of components: (a) Chemical elements with a specific purpose and (b) a component that is magnetic, like cobalt, iron or nickel. These nanoparticles are easily magnetically manipulated due to the prevalence of the magnetic component . They have been used in tissue-specific targeting, medical diagnostics and catalysis because they offer a variety of appealing features (Gubin et al., 2005). Nanoparticles, such as Fe_2O_3 and Fe_3O_4, are bio-compatible, in treating cancer treatment, magnetic hyperthermia, guided medication delivery, modification of stem cells and their sorting, also in DNA analysis, magnetic resonance imaging and gene therapy (Fan et al., 2009).

4. Biological synthesis of silver nanoparticle

Gardea-Torresdey et al. (2003) made the initial report on the biosynthesis of metallic nanoparticles utilizing plant extracts, including the synthesis of silver bio-nanoparticles using *Alfalfa sprouts*. The biological conceptualization uses green chemistry principles to substitute potentially hazardous chemicals like sodium borohydride (NaBH4) during the synthesis of silver nanoparticles by using natural products such as enzymes, biodegradable polymerase and phytochemicals capping/reducing agents (Dawadi et al., 2021). Extracting diverse portions of plants, microorganisms and biodegradable polymers might provide reducing agents for the synthesis. Plants can store heavy metals in numerous parts of their bodies as a result, plant extract-based biosynthesis approaches have provided a simple way to sustain efficient growth, as well as a cost-effective, simple synthesis process and classic nanoparticle preparation methods (Marchiol, 2012). As a consequence, their unusual biological, physical and chemical features, silver nanoparticles have been produced and researched extensively (Velmurugan et al., 2015). Plants have bio-molecules that can decrease metal salts such as carbohydrates, coenzymes and proteins. Other silver nanoparticles from herbal extract manufacturing approaches use extract-assisted biological synthesis, which has numerous advantages over approaches for creating physical and chemical nanoparticles. These approaches are also environmentally favorable, cost-effective and straightforward for high production (Kumar et al., 2017). Preparation of metal nanoparticles from plant extracts, living plants and inactivated plant tissue is a significant aspect of biosynthesis. The propensity of plants to eliminate metal ions on their exterior as well as in tissues and organs far away from the ion penetration site has long been recognized (Makarov et al., 2014). Proteins, enzymes, vitamins, amino acids and organic acids like citrates and polysaccharides are all biomolecules found in plant extracts that can decrease metal ions. *In vitro* techniques, in which plant and plant extracts are employed for bioreduction of ions to create nanoparticles have been effectively developed in recent years. Plants produce a family of organic polymers referred to as terpenoids, which have potent antioxidant capabilities, from 5-carbon isoprene units (Makarov et al., 2014). Flavonoids are a broad class of polyphenolic chemicals that can deliberately chelate and minimize metal ions into NPs and flavonoids with different functional groups can produce nanoparticles. Sugar can also be used to make metal nanoparticles in plant extracts. Monosaccharides, such as glucose, are known to act as reducing agents due to their free aldehyde group. Furthermore, polysaccharide and disaccharide reduction capacity is largely reliant on the concentration and type of specific monosaccharide components. Proteins containing different amino acids can decrease a variety of metal ions, resulting in NP formation. To minimize their Gibbs free energy, the resultant nanoparticles consolidate and promote to develop a more stable form, such as a truncated triangle (Makarov et al., 2014). Nanoparticle manufacturing frequently uses stabilizer chemicals to control their dispersion stability and formation (Ahmad et al., 2011). In this regard, plant hydrocarbons such as heptacosane and nonacosane have been hypothesized to have a favorable influence on silver nanoparticles stability (Roopan et al., 2013). Amino acids carbonyl component residues like cysteine, lysine, methionine and proteins as well as arginine can bind metal ions to form nanoparticles (NPs) for instance, sliver nanoparticle capping while preventing agglomeration and therefore stabilizing the medium. This implies that biological molecules may have a constitutive dual role in the stability and production of silver nanoparticles (Tran et al., 2013).

5. Antifungal activity of bionanoparticles

The outstanding anti-fungal properties of nanoparticles make them a very riveting choice for use in various domains due to their very small size and different surface arrangement, which increases their surface-to-volume ratio (Thirumalai Arasu et al., 2010). Plant pathogenic fungi, on the other hand, develop a resistance to various commercial fungicides subsequent to economic losses (Ishii and Holloman, 2015; Choudhury and Goswami, 2017). These issues have prompted researchers to look for new antifungal medicines, with plant-based nanoparticle-mediated formulations being a viable option. Nanotechnology has revolutionized the agriculture, food and medical industries by providing new methods for breaking phytopathogenic fungus resistance barriers and improving plants' ability to absorb nutrients (Del et al., 2016). Fungi have developed a variety of resistance strategies, including plasma membrane alteration to limit fungicide permeability, the production of lytic enzymes that destroy fungicide molecules and efflux transporters that excrete drug molecules to the extracellular environment. Antifungal characteristics of silver nanoparticles generated by green chemistry from *Trifolium resupinatum* plant seed transude and their effect on plant pathogenic fungus *Neofusicoccum parvum* and *Rhizoctonia solani* were investigated (Khatami et al., 2016). The fungal resistance activities of gold nanoparticles produced using *Mentha piperita* EOs against the fungus

Aspergillus niger were described (Thanighaiarassu et al., 2014). Nanoparticles like copper and copper oxide made from *Citrus medica* and *Stachys lavandulifolia*, have shown to have effective antifungal activity (Shende et al., 2015; Hernández-Díaz et al., 2021). Titanium dioxide (TiO_2) nanoparticles have also been shown to exhibit anti-fungal properties against pathogens of plants such as *Penicillium expansum* and *Alternaria brassicae* (Palmqvis et al., 2015; Maneerat and Hayata, 2006). Using plant extracts for NP synthesis is a viable way to create antimicrobial nanoparticles, utilizing the presence of a source of bioactive compounds in plants with a variety of biological functions, including antifungal capabilities. Different other plants and microbe-based NPs against plant pathogens are listed below in Table 5.1 and Table 5.2 respectively.

6. Mechanisms of action in opposition to phytopathogens of nanoparticles

Although the potential of green-produced metallic nanoparticles against phytopathogens has been investigated, the specific mode of action of nanoparticles is yet unknown (Hajipour et al., 2012). Genotoxicity, protein dysfunction (for instance, destruction of Fe–S cluster, oxidization of cysteine amino acid in Fe-binding site, exchange of structural metal and interchange of catalytic metal), impaired membrane function (deficit of membrane potential, membrane damage), disruption with nutrient retention and antioxidant depletion and fabrication of reactive oxygen species are some of the mechanisms that have been reported, so these systems may not work alone, however they do work together against diverse phytopathogens (Lemire et al., 2013). Nanoparticles with minimal negative (–) or positive (+) charges are electrostatically attracted to the negatively charged cell membrane of bacteria, that causes them to cling to the wall. The nanoparticle affects the membrane morphological structures and membrane depolarization disrupts respiratory functions and membrane permeability, ultimately damaging cell structures and leading to cell death. Internal cell content such as proteins, metabolites, DNA and enzymes seep out due to the rupture of the cell structure. Furthermore, nanoparticles may generate irregular pits in the microbial cell wall, allowing nanoparticles to enter the periplasmic space and inside the cells more easily. SEM/TEM can be handed-down to study the effects of nanoparticles on membrane damage and pit development on the surface of the cell (Gahlawat and Choudhury, 2019). Masum et al. (2019) used TEM to examine the impact of green-produced silver nanoparticles on the strain *Acidovorax oryzae* (RS-2) and found significant leakage of cytoplasmic and damaged cell walls nucleic contents and a bloated structure that led to bacterial mortality (Elbeshehy et al., 2015). Biosynthesized Ag NPs were administered to *Fusarium graminearum* strain (PH-1) in another investigation, and actions of antifungal activity such as cell wall damage and hyphae deformation were observed using TEM and SEM respectively (Ibrahim et al., 2020). Silver nanoparticles had similar effects on numerous other phytopathogenic fungi including *Botrytis cinera* and *Alternaria alternata* (Xia et al., 2016). The proffering of reactive oxygen species may be the cause of nanoparticle toxicity (Vankar and Shukla, 2012). Free radicals can harm the cell wall as well as

Table 5.1: Plant-based synthesized metallic nanoparticles against plant pathogen.

Plants-plant parts used	Metal NPs	Pathogen	Host	Shape	Size (nm)	References
Piper nigrum-Stem	Ag	*Erwinia cacticida*	Watermelon	crystalline	9–30	Paulkumar et al. (2014)
Citrus maxim-Fruits	Ag	*Acidovorax oryzae*	Rice	Spherical	11–13	Ali et al. (2020)
Artemisia absinthium-Leaves	Ag	*Phytophthora parasitica*	Citrus	Spherical	5–100	Ali et al. (2015)
Phyllanthu semblica-Fruits	Ag	*Dickeya dadantii*	Sweet potato	Polymorphic	20–200	Hossain et al. (2019)
Abelmoschus Esculentus-Seed	Au	*Puccinia graminis* pv. *Tritci*	Wheat	Spherical	45–75	Jayaseelan et al. (2013)
Matricaria Chamomilla- Flowers	ZnO	*Xanthomonas oryzae* pv. *oryzae*	Rice	Crystalline	50–192	Ogunyemi et al. (2019)
Lycopersicon Esculentum- Fruits	ZnO	*Xanthomonas oryzae* pv. *oryzae*	Rice	Crystalline	66–133	Ogunyemi et al. (2019)
Olea europaea-Leaves	ZnO	*Xanthomonas oryzae* pv. *oryzae*	Rice	Crystalline	41–124	Ogunyemi et al. (2019)
Parthenium Hysterophorus-Leaves	ZnO	*Fusarium culmorum*	Barley	Spherical and Hexagonal	28–84	Rajiv et al. (2013)
Syzygium Aromaticum-Bud	Cu	*Aspergillus niger*, *Aspergillus flavus*, and *Penicillium* spp.	Multiple crops	Spherical	15	Rajesh et al. (2018)
Rosmarinus Officinalis-Flowers	MgO	*Xanthomonas oryzae* pv. *oryzae*	Rice	Flower	< 20	Abdallah et al. (2019)
Matricaria Chamomilla-Flowers	MgO and MnO_2	*Acidovorax oryzae*	Rice	Disk-shaped Spherical	9–112	Ogunyemi et al. (2019)
Trachyspermum Ammi-Leaves	Ni	*Colletotrichum musae*	Banana	–	68	Jagana et al. (2017)

Table 5.2: Microbe-based synthesized metallic nanoparticles against plant pathogens.

Fungi	**Metal NPs**	**Host**	**Pathogen**	**Shape**	**Size (nm)**	**References**
Fusarium solani-Wheat grain	Ag	wheat, barley and corn	*Fusarium* spp., *Aspergillus* spp., *Alternaria* spp., and *Rhizopus Stolonifer*	Spherical	5–30	Abd et al. (2015)
Guignardia mangiferae-Leaves of medicinal plants	Ag	Rice	*Rhizoctonia solani*	Spherical	5–30	Balakumaran et al. (2015)
Trichiderma hazarium-Tomato	Ag	Multiple crops	*Helminthosporium* sp., *Alternaria alternata, Phytophthora arenaria,* and *Botrytis* sp.	Spherical	11–13	El-Moslamy et al. (2017)
Penicillium duclauxii-Corn seeds	Ag	Sorghum	*Bipolaris sorghicola*	Spherical	3–32	Almaary et al. (2020)
Aspergillus niger-Grape	Ag	Multiple crops	*Penicillin digitatum*, *Aspergillus flavus*, and *Fusarium oxysporum*	Spherical	10–100	Al-Zubaidi et al. (2019)
Trichoderma Longibrachiatum-Cucumber	Ag	Multiple crops	*Alternaria alternata, Pyricularia grisea, Fusarium verticillioides, Helminthosporium oryzae* and *Penicillium glabrum*	Spherical	1–25	Elamawi et al. (2018)
Setosphaeria rostrata- Solanum nigrum leaves	Ag	Multiple crops	*Aspergillus niger*, *Rhizoctonia solani*, *Fusarium graminearum,* and *Fusarium udum*	Spherical	2–50	Akther and Hemalatha (2019)
Bacteria	**Metal NPs**	**Host**	**Pathogen**	**Shape**	**Size (nm)**	**References**
Bacillus siamensis-Coriandrum sativum	Ag	Rice	*Xanthomonas oryzae* pv. *Oryzae*	Spherical	25–50	Ibrahim et al. (2019)
Bacillus cereu-Soil contaminated with waste water	Ag	Rice	*Xanthomonas oryzae* pv. *Oryzae*	Spherical	18–39	Ahmed et al. (2020)
Bacillus sp.-Soil	Ag	Tomato	*Fusarium oxysporum*	Spherical	7–21	Gopinath and Velusamy (2013)
Pseudomonas poae-Garlic	Ag	Wheat	*Fusarium graminearum*	Spherical	20–45	Ibrahim et al. (2020)

Table 5.2 contd. ...

...Table 5.2 contd.

Bacteria	**Metal NPs**	**Host**	**Pathogen**	**Shape**	**Size (nm)**	**References**
Stenotrophomonas sp-Soil	Ag	Chickpea	*Sclerotium rolfsii*	Spherical	12	Mishra et al. (2017)
Serratia sp.-Soil	Ag	Wheat	*Bipolaris sorokiniana*	Spherical	10–20	Mishra et al. (2014)
Pseudomonas sp., and *Achromobacter* sp	Ag	Chickpea	*Fusarium oxysporum* f. sp. *ciceri*	Spherical	20–50	Kaur et al. (2018)
Bacillus licheniformis-Soil	Ag	Faba bean	Bean yellow mosaic virus	Polymorphic	77–92	Elbeshehy et al. (2015)
Bacillus thuringensis-Soil	Ag	Cluster bean	Sun hemp rosette virus	Polymorphic	10–20	Jain and Kothari (2014)
Aeromonas hydrophila	ZnO	Maize	*Aspergillus flflavus*	Crystalline	57–72	Jayaseelan et al. (2012)
Streptomyces griseus-Rhizospheric soil of tea	Cu	Tea	*Poria hypolateritia*	Spherical	5–50	Ponmurugan et al. (2016)
Streptomyces Capillispiralis-*Convolvulus arvensis* leaves	Cu	Multiple crops	*Alternaria* spp., *Aspergillus niger*, *Pythium* spp., and *Fusarium* spp.	Spherical	4–59	Hassan et al. (2018)
Streptomyces spp.-*Oxalis corniculata* leaves	CuO	Multiple crops	*Alternaria alternata*, *Fusarium oxysporum*, *Pythium ultimum*, and *Aspergillus niger*	Spherical	78–80	Hassan et al. (2019)

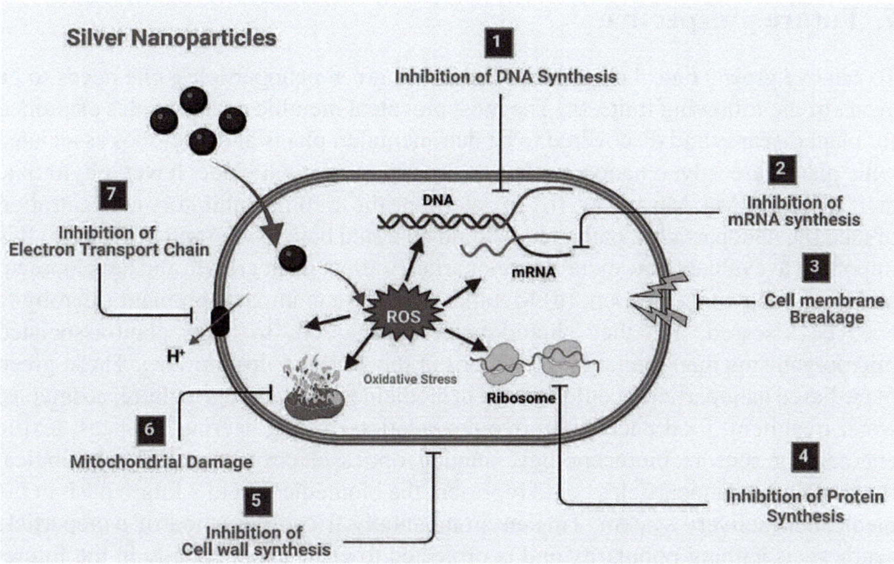

Fig. 5.2: Silver nanoparticles' antimicrobial mechanism includes the following steps: (1) inhibiting DNA synthesis; (2) inhibiting mRNA synthesis; (3) rupturing cell membranes and allowing cell components to leak out; (4) inhibiting protein synthesis; (5) inhibiting cell wall synthesis; (6) inflicting damage on the mitochondria; and (7) inhibiting the electron transport chain (Jain et al., 2021).

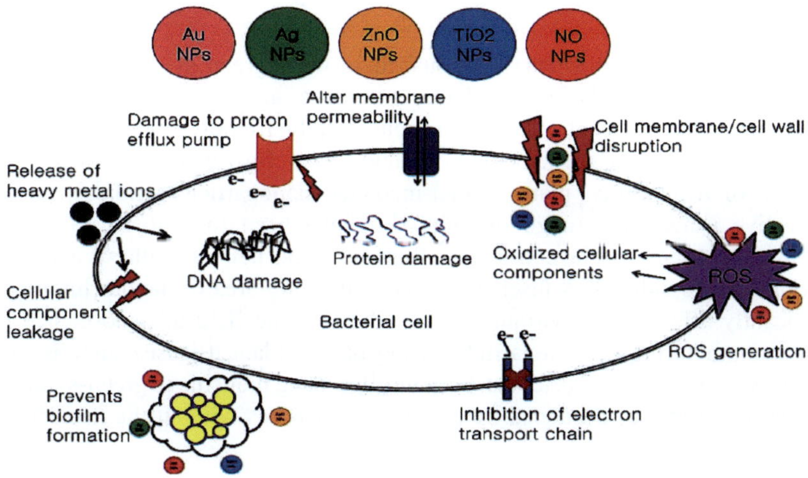

Fig. 5.3: Nanoparticles' mechanisms of action against phytopathogens (Shaikh et al., 2019).

other macromolecules like lipids, proteins and DNA. Deletions, mutations, double-strand breaks and single-strand breaks and crosslinking with proteins are all possible DNA damages (Soenen et al., 2011). The antimicrobial and antifungal mechanism of action of NPs is shown in Figs. 5.2 and 5.3.

7. Future prospective

To achieve green control of plant diseases with green nanoparticles, one needs to be aware of the following traits. (a) The most prevalent metallic nanoparticles examined for plant diseases and discovered to be detrimental to plants and microbes associated with plants are silver nanoparticles. It is important to include fewer phytotoxic metals like Zn, Mg, Mn and Fe. (b) To determine the minimal inhibitory concentration of metallic nanoparticles, pathogens should be tested both *in vivo* and *in vitro*. (c) It is important to evaluate how metallic nanoparticles affect plant growth and development at the working concentration. (d) Metallic nanoparticle impacts on plant microbiota need be assessed, and they should be less dangerous to other plant-associated microorganisms than the target pathogens at the working dose *in vivo*. These green plant-based nanoparticles could be used in medicine, catalysis, agriculture, cosmetics, water treatment, food packaging, dye degradation, bioengineering sciences, textile engineering, sensors, biotechnology, imaging, optics, electronics and other biological sectors. These nanoparticles could represent the biomedical field's future push in the medication delivery system. This environmentally friendly method of nanoparticle synthesis is gaining popularity and is projected to grow exponentially in the future; nevertheless, long-term effects on animals and people, as well as the accumulation of these nanoparticles in the environment and their influence, will need to be addressed in the future.

8. Conclusion

The desire for green chemistry has pushed forward green synthetic methods for producing nanomaterials using plants, microbes and other natural resources. Researchers have concentrated on environmentally friendly green production of nanoparticles. Due to their benign method, cost-effectiveness, eco-friendliness and ease of availability, plant extract-mediated nanoparticle synthesis and their prospective usage in a range of industries have been the subject of extensive investigation. Many special compounds found in plants help in synthesis and quicken the kinetics of synthesis. A fascinating and expanding area of nanotechnology that significantly affects the environment and advances the field of nanoscience is the green manufacturing of nanoparticles using plants. Plant diseases such as fungus, bacteria, viruses and oomycetes can be controlled using biogenic or green-synthesized metallic nanoparticles produced by plants and microbes without the need for toxic chemicals.

Acknowledgement

The authors are grateful to their respective authorities, departments, institutions and university for their support and cooperation. Authors (PK & MK) would like to thank DBT (BT/PR39789/NER/95/1664/2020), the Government of India.

Author contribution

Pradeep Kumar: Conceptualization; Chikanshi Sharma: Writing-original draft; Madhu Kamle: Writing-support & editing; Pradeep Kumar: critical review & final editing.

References

Abd El-Aziz, A.R.M., Al-Othman, M.R., Mahmoud, M.A. and Metwaly, H.A. (2015). Biosynthesis of silver nanoparticles using *Fusarium solani* and its impact on grain borne fungi. Digest Journal of Nanomaterials and Biostructures, 10(2): 655–662.

Abdallah, Y., Ogunyemi, S.O., Abdelazez, A., Zhang, M., Hong, X., Ibrahim, E., Hossain, A., Fouad, H., Li, B. and Chen, J. (2019). The green synthesis of MgO nano-flowers using *Rosmarinus officinalis* L. (Rosemary) and the antibacterial activities against *Xanthomonas oryzae pv. oryzae*. BioMed Research International, 2019.

Aday, B., Yıldız, Y., Ulus, R., Eris, S., Sen, F. and Kaya, M. (2016). One-pot, efficient and green synthesis of acridinedione derivatives using highly monodisperse platinum nanoparticles supported with reduced graphene oxide. New Journal of Chemistry, 40(1): 748–754.

Ahmad, N., Sharma, S., Singh, V.N., Shamsi, S.F., Fatma, A. and Mehta, B.R. (2011). Biosynthesis of silver nanoparticles from *Desmodium triflorum*: a novel approach towards weed utilization. Biotechnology Research International, 2011.

Ahmed, S., Chaudhry, S.A. and Ikram, S. (2017). A review on biogenic synthesis of ZnO nanoparticles using plant extracts and microbes: A prospect towards green chemistry. Journal of Photochemistry and Photobiology B: Biology, 166: 272–284.

Ahmed, T., Shahid, M., Noman, M., Niazi, M.B.K., Mahmood, F., Manzoor, I., Zhang, Y., Li, B., Yang, Y., Yan, C. and Chen, J. (2020). Silver nanoparticles synthesized by using Bacillus cereus SZT1 ameliorated the damage of bacterial leaf blight pathogen in rice. Pathogens, 9(3): 160.

Akther, T. and Hemalatha, S. (2019). Mycosilver nanoparticles: Synthesis, characterization and its efficacy against plant pathogenic fungi. BioNanoScience, 9(2): 296–301.

Ali, K.A., Yao, R., Wu, W., Masum, M.M.I., Luo, J., Wang, Y., Zhang, Y., An, Q., Sun, G. and Li, B. (2020). Biosynthesis of silver nanoparticle from pomelo (*Citrus Maxima*) and their antibacterial activity against *acidovorax oryzae* RS-2. Materials Research Express, 7(1): 015097.

Ali, M.A., Ahmed, T., Wu, W., Hossain, A., Hafeez, R., Islam, Masum, M.M., Wang, Y., An, Q., Sun, G. and Li, B. (2020). Advancements in plant and microbe-based synthesis of metallic nanoparticles and their antimicrobial activity against plant pathogens. Nanomaterials (Basel)., 10(6): 1146.

Ali, M., Kim, B., Belfield, K.D., Norman, D., Brennan, M. and Ali, G.S. (2015). Inhibition of *Phytophthora parasitica* and *P. capsici* by silver nanoparticles synthesized using aqueous extract of *Artemisia absinthium*. Phytopathology, 105(9): 1183–1190.

Almaary, K.S., Sayed, S.R., Abd-Elkader, O.H., Dawoud, T.M., El Orabi, N.F. and Elgorban, A.M. (2020). Complete green synthesis of silver-nanoparticles applying seed-borne *Penicillium duclauxii*. Saudi Journal of Biological Sciences, 27(5): 1333–1339.

Al-Zubaidi, S., Al-Ayafi, A. and Abdelkader, H. (2019). Biosynthesis, characterization and antifungal activity of silver nanoparticles by *Aspergillus niger* isolate. Journal of Nanotechnology Research, 1(1): 23–36.

Baban, D.F. and Seymour, L.W. (1998). Control of tumour vascular permeability. Advanced Drug Delivery Reviews, 34(1): 109–119.

Balakumaran, M.D., Ramachandran, R. and Kalaichelvan, P.T. (2015). Exploitation of endophytic fungus, *Guignardia mangiferae* for extracellular synthesis of silver nanoparticles and their *in vitro* biological activities. Microbiological Research, 178: 9–17.

Banerjee, P. and Nath, D. (2015). A phytochemical approach to synthesize silver nanoparticles for non-toxic biomedical application and study on their antibacterial efficacy. Nanosci. Technol., 2(1): 1–14.

Bansal, V., Rautaray, D., Ahmad, A. and Sastry, M. (2004). Biosynthesis of zirconia nanoparticles using the fungus *Fusarium oxysporum*. Journal of Materials Chemistry, 14(22): 3303–3305.

Bar, H., Bhui, D.K., Sahoo, G.P., Sarkar, P., De, S.P. and Misra, A. (2009). Green synthesis of silver nanoparticles using latex of *Jatropha curcas*. Colloids and Surfaces A: Physicochemical and Engineering Aspects, 339(1-3): 134–139.

Bastús, N.G., Piella, J. and Puntes, V. (2016). Quantifying the sensitivity of multipolar (dipolar, quadrupolar, and octapolar) surface plasmon resonances in silver nanoparticles: The effect of size, composition, and surface coating. Langmuir, 32(1): 290–300.

Bollella, P., Gorton, L., Ludwig, R. and Antiochia, R. (2017). A third generation glucose biosensor based on cellobiose dehydrogenase immobilized on a glassy carbon electrode decorated with electro deposited gold nanoparticles: Characterization and application in human saliva. Sensors, 17(8): 1912.

Ceylan, A., Jastrzembski, K. and Shah, S.I. (2006). Enhanced solubility Ag-Cu nanoparticles and their thermal transport properties. Metallurgical and Materials Transactions A, 37(7): 2033–2038.

Choudhury, S.R. and Goswami, A. (2017). Non-metallic nanoparticles & their biological implications. Basic Appl. Aspects, Nanobiotechnology, 61.

Daniel, M.C. and Astruc, D. (2004). Gold nanoparticles: Assembly, supramolecular chemistry, quantum-size-related properties, and applications toward biology, catalysis, and nanotechnology. Chemical Reviews, 104(1): 293–346.

Dawadi, S., Katuwal, S., Gupta, A., Lamichhane, U., Thapa, R., Jaisi, S., Lamichhane, G., Bhattarai, D.P. and Parajuli, N. (2021). Current research on silver nanoparticles: Synthesis, characterization, and applications. Journal of Nanomaterials, 2021.

Del Serrone, P., Buttazzoni, L. and Nicoletti, M. (2016). Nutrition and multiresistance alert. EC Nutrition, 4(1): 772–783.

Dubchak, S., Ogar, A., Mietelski, J.W. and Turnau, K. (2010). Influence of silver and titanium nanoparticles on *Arbuscular mycorrhiza* colonization and accumulation of radiocaesium in *Helianthus annuus*. Spanish Journal of Agricultural Research, (1): 103–108.

Elamawi, R.M., Al-Harbi, R.E. and Hendi, A.A. (2018). Biosynthesis and characterization of silver nanoparticles using Trichoderma longibrachiatum and their effect on phytopathogenic fungi. Egyptian Journal of Biological Pest Control, 28(1): 1–11.

Elbeshehy, E.K., Elazzazy, A.M. and Aggelis, G. (2015). Silver nanoparticles synthesis mediated by new isolates of Bacillus spp., nanoparticle characterization and their activity against Bean Yellow Mosaic Virus and human pathogens. Frontiers in Microbiology, 6: 453.

El-Moslamy, S.H., Elkady, M.F., Rezk, A.H. and Abdel-Fattah, Y.R. (2017). Applying Taguchi design and large-scale strategy for mycosynthesis of nano-silver from endophytic *Trichoderma harzianum* SYA. F4 and its application against phytopathogens. Scientific Reports, 7(1): 1–22.

Fan, T.X., Chow, S.K. and Zhang, D. (2009). Biomorphic mineralization: From biology to materials. Progress in Materials Science, 54(5): 542–659.

Fernández-Garcia, M. and Rodgriguez, J. (2007). Metal oxide nanoparticles (No. BNL-79479-2007-BC). Brookhaven National Lab.(BNL), Upton, NY (United States).

Ferrando, R., Jellinek, J. and Johnston, R.L. (2008). Nanoalloys: From theory to applications of alloy clusters and nanoparticles. Chemical Reviews, 108(3): 845–910.

Gahlawat, G. and Choudhury, A.R. (2019). A review on the biosynthesis of metal and metal salt nanoparticles by microbes. RSC Advances, 9(23): 12944–12967.

Gardea-Torresdey, J.L., Gomez, E., Peralta-Videa, J.R., Parsons, J.G., Troiani, H. and Jose-Yacaman, M. (2003). *Alfalfa sprouts*: A natural source for the synthesis of silver nanoparticles. Langmuir, 19(4): 1357–1361.

Ghosh, S., Patil, S., Ahire, M., Kitture, R., Gurav, D.D., Jabgunde, A.M., Kale, S., Pardesi, K., Shinde, V., Bellare, J. and Dhavale, D.D. (2012). *Gnidia glauca* flower extract mediated synthesis of gold nanoparticles and evaluation of its chemocatalytic potential. Journal of Nanobiotechnology, 10(1): 1–9.

Gomathi, S., Firdous, J. and Bharathi, V. (2017). Phytochemical screening of silver nanoparticles extract of *Eugenia jambolana* using Fourier infrared spectroscopy. Int. J. Res. Pharm. Sci., 8(3): 383–387.

Gong, P., Li, H., He, X., Wang, K., Hu, J., Tan, W., Zhang, S. and Yang, X. (2007). Preparation and antibacterial activity of Fe_3O_4@ Ag nanoparticles. Nanotechnology, 18(28): 285604.

Gopinath, V. and Velusamy, P. (2013). Extracellular biosynthesis of silver nanoparticles using Bacillus sp. GP-23 and evaluation of their antifungal activity towards *Fusarium oxysporum*. Spectrochimica Acta Part A: Molecular and Biomolecular Spectroscopy, 106: 170–174.

Gubin, S.P., Koksharov, Y.A., Khomutov, G.B. and Yurkov, G.Y. (2005). Magnetic nanoparticles: Preparation, structure and properties. Russian Chemical Reviews, 74(6): 489.

Haider, A., Ijaz, M., Ali, S., Haider, J., Imran, M., Majeed, H., Shahzadi, I., Ali, M.M., Khan, J.A. and Ikram, M. (2020). Green synthesized phytochemically (*Zingiber officinale* and *Allium sativum*) reduced nickel oxide nanoparticles confirmed bactericidal and catalytic potential. Nanoscale Research Letters, 15(1): 1–11.

Hajipour, M.J., Fromm, K.M., Ashkarran, A.A., de Aberasturi, D.J., de Larramendi, I.R., Rojo, T., Serpooshan, V., Parak, W.J. and Mahmoudi, M. (2012). Antibacterial properties of nanoparticles. Trends in Biotechnology, 30(10): 499–511.

Hassall, K.A. (1965). Pesticides: Their properties, uses and disadvantages. 1. General introduction; insecticides and related compounds. British Veterinary Journal, 121: 105–118.

Hassan, S.E.D., Fouda, A., Radwan, A.A., Salem, S.S., Barghoth, M.G., Awad, M.A., Abdo, A.M. and El-Gamal, M.S. (2019). Endophytic actinomycetes Streptomyces spp mediated biosynthesis of copper oxide nanoparticles as a promising tool for biotechnological applications. JBIC Journal of Biological Inorganic Chemistry, 24(3): 377–393.

Hassan, S.E.D., Salem, S.S., Fouda, A., Awad, M.A., El-Gamal, M.S. and Abdo, A.M. (2018). New approach for antimicrobial activity and bio-control of various pathogens by biosynthesized copper nanoparticles using endophytic actinomycetes. Journal of Radiation Research and Applied Sciences, 11(3): 262–270.

Hernández-Díaz, J.A., Garza-García, J.J., Zamudio-Ojeda, A., León-Morales, J.M., López-Velázquez, J.C. and García-Morales, S. (2021). Plant-mediated synthesis of nanoparticles and their antimicrobial activity against phytopathogens. Journal of the Science of Food and Agriculture, 101(4): 1270–1287.

Horikoshi, S. and Serpone, N. (ed.). (2013). Microwaves in Nanoparticle Synthesis: Fundamentals and Applications. John Wiley & Sons.

Hossain, A., Abdallah, Y., Ali, M., Masum, M., Islam, M., Li, B., Sun, G., Meng, Y., Wang, Y. and An, Q. (2019). Lemon-fruit-based green synthesis of zinc oxide nanoparticles and titanium dioxide nanoparticles against soft rot bacterial pathogen *Dickeya dadantii*. Biomolecules, 9(12): 863.

Husen, A. and Siddiqi, K.S. (2014). Phytosynthesis of nanoparticles: Concept, controversy and application. Nanoscale Research Letters, 9(1): 1–24.

Ibrahim, E., Fouad, H., Zhang, M., Zhang, Y., Qiu, W., Yan, C., Li, B., Mo, J. and Chen, J. (2019). Biosynthesis of silver nanoparticles using endophytic bacteria and their role in inhibition of rice pathogenic bacteria and plant growth promotion. RSC Advances, 9(50): 29293–29299.

Ibrahim, E., Zhang, M., Zhang, Y., Hossain, A., Qiu, W., Chen, Y., Wang, Y., Wu, W., Sun, G. and Li, B. (2020). Green-synthesization of silver nanoparticles using endophytic bacteria isolated from garlic and its antifungal activity against wheat Fusarium head blight pathogen *Fusarium graminearum*. Nanomaterials, 10(2): 219.

Iravani, S. (2011). Green synthesis of metal nanoparticles using plants. Green Chemistry, 13(10): 2638–2650.

Iravani, S. (2014). Bacteria in nanoparticle synthesis: Current status and future prospects. International Scholarly Research Notices, 2014.

Ishii, H. and Holloman, D.W. (2015). Fungicide Resistance in Plant Pathogens. Tokyo: Springer, doi, 10: 978–4.

Ito, A., Shinkai, M., Honda, H. and Kobayashi, T. (2005). Medical application of functionalized magnetic nanoparticles. Journal of Bioscience and Bioengineering, 100(1): 1–11.

Jagana, D., Hegde, Y.R. and Lella, R. (2017). Green nanoparticles: A novel approach for the management of banana anthracnose caused by *Colletotrichum musae*. Int. J. Curr. Microbiol. Appl. Sci., 6(10): 1749–56.

Jain, D. and Kothari, S.L. (2014). Green synthesis of silver nanoparticles and their application in plant virus inhibition. J. Mycol. Plant Pathol., 44(1): 21–24.

Jain, Ashvi S., Pranita S. Pawar, Aira Sarkar, Vijayabhaskarreddy Junnuthula and Sathish Dyawanapelly. (2021). Bionanofactories for green synthesis of silver nanoparticles: toward antimicrobial applications. International Journal of Molecular Sciences, 22(21): 11993.

Jayaseelan, C., Rahuman, A.A., Kirthi, A.V., Marimuthu, S., Santhoshkumar, T., Bagavan, A., Gaurav, K., Karthik, L. and Rao, K.B. (2012). Novel microbial route to synthesize ZnO nanoparticles using *Aeromonas hydrophila* and their activity against pathogenic bacteria and fungi. Spectrochimica Acta Part A: Molecular and Biomolecular Spectroscopy, 90: 78–84.

Jayaseelan, C., Ramkumar, R., Rahuman, A.A. and Perumal, P. (2013). Green synthesis of gold nanoparticles using seed aqueous extract of *Abelmoschus esculentus* and its antifungal activity. Industrial Crops and Products, 45: 423–429.

Jha, A.K. and Prasad, K. (2010). Green synthesis of silver nanoparticles using Cycas leaf. International Journal of Green Nanotechnology: Physics and Chemistry, 1(2): P110–P117.

Kaur, P., Thakur, R., Duhan, J.S. and Chaudhury, A. (2018). Management of wilt disease of chickpea in vivo by silver nanoparticles biosynthesized by rhizospheric microflora of chickpea (*Cicer arietinum*). Journal of Chemical Technology & Biotechnology, 93(11): 3233–3243.

Keihan, A.H., Veisi, H. and Veasi, H. (2017). Green synthesis and characterization of spherical copper nanoparticles as organometallic antibacterial agent. Applied Organometallic Chemistry, 31(7): e3642.

Khan, M., Khan, M., Adil, S.F., Tahir, M.N., Tremel, W., Alkhathlan, H.Z., Al-Warthan, A. and Siddiqui, M.R.H. (2013). Green synthesis of silver nanoparticles mediated by *Pulicaria glutinosa* extract. International Journal of Nanomedicine, 8: 1507.

Khatami, M., Nejad, M.S., Salari, S. and Almani, P.G.N. (2016). Plant-mediated green synthesis of silver nanoparticles using *Trifolium resupinatum* seed exudate and their antifungal efficacy on *Neofusicoccum parvum* and *Rhizoctonia solani*. IET Nanobiotechnology, 10(4): 237–243.

Klaus, T., Joerger, R., Olsson, E. and Granqvist, C.G. (1999). Silver-based crystalline nanoparticles, microbially fabricated. Proceedings of the National Academy of Sciences, 96(24): 13611–13614.

Kumar, B., Smita, K., Cumbal, L. and Debut, A. (2017). Green synthesis of silver nanoparticles using *Andean blackberry* fruit extract. Saudi Journal of Biological Sciences, 24(1): 45–50.

Kutz, F.W., Wood, P.H. and Bottimore, D.P. (1991). Organochlorine pesticides and polychlorinated biphenyls in human adipose tissue. Reviews of Environmental Contamination and Toxicology, 1–82.

Kyriacou, S.V., Brownlow, W.J. and Xu, X.H.N. (2004). Using nanoparticle optics assay for direct observation of the function of antimicrobial agents in single live bacterial cells. Biochemistry, 43(1): 140–147.

Lebaschi, S., Hekmati, M. and Veisi, H. (2017). Green synthesis of palladium nanoparticles mediated by black tea leaves (*Camellia sinensis*) extract: Catalytic activity in the reduction of 4-nitrophenol and Suzuki-Miyaura coupling reaction under ligand-free conditions. Journal of Colloid and Interface Science, 485: 223–231.

Lemire, J.A., Harrison, J.J. and Turner, R.J. (2013). Antimicrobial activity of metals: Mechanisms, molecular targets and applications. Nature Reviews Microbiology, 11(6): 371–384.

Link, S. and El-Sayed, M.A. (2003). Optical properties and ultrafast dynamics of metallic nanocrystals. Annual Review of Physical Chemistry, 54(1): 331–366.

Liu, Z., Qi, L., An, X., Liu, C. and Hu, Y. (2017). Surface engineering of thin film composite polyamide membranes with silver nanoparticles through layer-by-layer interfacial polymerization for antibacterial properties. ACS Applied Materials & Interfaces, 9(46): 40987–40997.

Loureiro, A.G., Azoia, N.C., Gomes, A. and Cavaco-Paulo, A. (2016). Albumin-based nanodevices as drug carriers. Current Pharmaceutical Design, 22(10): 1371–1390.

Makarov, V.V., Love, A.J., Sinitsyna, O.V., Makarova, S.S., Yaminsky, I.V., Taliansky, M.E. and Kalinina, N.O. (2014). "Green" nanotechnologies: Synthesis of metal nanoparticles using plants. Acta Naturae (англоязычная версия), 6(1(20)): 35–44.

Mallikarjuna, K., Sushma, N.J., Narasimha, G., Manoj, L. and Raju, B.D.P. (2014). Phytochemical fabrication and characterization of silver nanoparticles by using Pepper leaf broth. Arabian Journal of Chemistry, 7(6): 1099–1103.

Maneerat, C. and Hayata, Y. (2006). Antifungal activity of TiO_2 photocatalysis against *Penicillium expansum in vitro* and in fruit tests. International Journal of Food Microbiology, 107(2): 99–103.

Marchiol, L. (2012). Synthesis of metal nanoparticles in living plants. Italian Journal of Agronomy, 7(3): e37–e37.

Martínez-Bernett, D., Silva-Granados, A., Correa-Torres, S.N. and Herrera, A. (2016, February). Chromatographic analysis of phytochemicals components present in *Mangifera indica* leaves for the synthesis of silver nanoparticles by $AgNO_3$ reduction. In Journal of Physics: Conference Series 687(1): 012033. IOP Publishing.

Martis, E., Badve, R. and Degwekar, M. (2012). Nanotechnology based devices and applications in medicine: An overview. Chronicles of Young Scientists, 3(1): 68–68.

Masum, M., Islam, M., Siddiqa, M., Ali, K.A., Zhang, Y., Abdallah, Y., Ibrahim, E., Qiu, W., Yan, C. and Li, B. (2019). Biogenic synthesis of silver nanoparticles using *Phyllanthus emblica* fruit extract and its inhibitory action against the pathogen *Acidovorax oryzae* strain RS-2 of rice bacterial brown stripe. *Frontiers in Microbiology*, 10: 820.

Mie, R., Samsudin, M.W., Din, L.B., Ahmad, A., Ibrahim, N. and Adnan, S.N.A. (2014). Synthesis of silver nanoparticles with antibacterial activity using the lichen *Parmotrema praesorediosum*. International Journal of Nanomedicine, 9: 121.

Mishra, S., Singh, B.R., Naqvi, A.H. and Singh, H.B. (2017). Potential of biosynthesized silver nanoparticles using Stenotrophomonas sp. BHU-S7 (MTCC 5978) for management of soil-borne and foliar phytopathogens. Scientific Reports, 7(1): 1–15.

Mishra, S., Singh, B.R., Singh, A., Keswani, C., Naqvi, A.H. and Singh, H.B. (2014). Biofabricated silver nanoparticles act as a strong fungicide against *Bipolaris sorokiniana* causing spot blotch disease in wheat. Plos One, 9(5): e97881.

Mittal, A.K., Chisti, Y. and Banerjee, U.C. (2013). Synthesis of metallic nanoparticles using plant extracts. Biotechnology Advances, 31(2): 346–356.

Mohanpuria, P., Rana, N.K. and Yadav, S.K. (2008). Biosynthesis of nanoparticles: Technological concepts and future applications. Journal of Nanoparticle Research, 10(3): 507–517.

Mohl, M., Dobo, D., Kukovecz, A., Konya, Z., Kordas, K., Wei, J., Vajtai, R. and Ajayan, P.M. (2011). Formation of CuPd and CuPt bimetallic nanotubes by galvanic replacement reaction. The Journal of Physical Chemistry C, 115(19): 9403–9409.

Munoz-Flores, B.M., Kharisov, B.I., Jiménez-Pérez, V.M., Elizondo Martínez, P. and Lopez, S.T. (2011). Recent advances in the synthesis and main applications of metallic nanoalloys. Industrial & Engineering Chemistry Research, 50(13): 7705–7721.

Nikalje, A.P. (2015). Nanotechnology and its applications in medicine. Med. Chem., 5(2): 081–089.

Nisar, P., Ali, N., Rahman, L., Ali, M. and Shinwari, Z.K. (2019). Antimicrobial activities of biologically synthesized metal nanoparticles: An insight into the mechanism of action. JBIC Journal of Biological Inorganic Chemistry, 24(7): 929–941.

Ogunyemi, S.O., Abdallah, Y., Zhang, M., Fouad, H., Hong, X., Ibrahim, E., Masum, M.M.I., Hossain, A., Mo, J. and Li, B. (2019). Green synthesis of zinc oxide nanoparticles using different plant extracts and their antibacterial activity against *Xanthomonas oryzae pv. oryzae*. Artificial Cells, Nanomedicine, and Biotechnology, 47(1): 341–352.

Ogunyemi, S.O., Zhang, F., Abdallah, Y., Zhang, M., Wang, Y., Sun, G., Qiu, W. and Li, B. (2019). Biosynthesis and characterization of magnesium oxide and manganese dioxide nanoparticles using *Matricaria chamomilla* L. extract and its inhibitory effect on *Acidovorax oryzae* strain RS-2. Artificial Cells, Nanomedicine, and Biotechnology, 47(1): 2230–2239.

Oves, M., Aslam, M., Rauf, M.A., Qayyum, S., Qari, H.A., Khan, M.S., Alam, M.Z., Tabrez, S., Pugazhendhi, A. and Ismail, I.M. (2018). Antimicrobial and anticancer activities of silver nanoparticles synthesized from the root hair extract of *Phoenix dactylifera*. Materials Science and Engineering: C, 89: 429–443.

Palmqvist, N., Bejai, S., Meijer, J., Seisenbaeva, G.A. and Kessleral, V.G. (2015). Nano titania aided clustering and adhesion of beneficial bacteria to plant roots to enhance crop growth and stress management. Sci. Rep., 5: 10146.

Paulkumar, K., Gnanajobitha, G., Vanaja, M., Rajeshkumar, S., Malarkodi, C., Pandian, K. and Annadurai, G. (2014). *Piper nigrum* leaf and stem assisted green synthesis of silver nanoparticles and evaluation of its antibacterial activity against agricultural plant pathogens. The Scientific World Journal, 2014.

Ponmurugan, P., Manjukarunambika, K., Elango, V. and Gnanamangai, B.M. (2016). Antifungal activity of biosynthesised copper nanoparticles evaluated against red root-rot disease in tea plants. Journal of Experimental Nanoscience, 11(13): 1019–1031.

Rafique, M., Sadaf, I., Rafique, M.S. and Tahir, M.B. (2017). A review on green synthesis of silver nanoparticles and their applications. Artificial Cells, Nanomedicine, and Biotechnology, 45(7): 1272–1291.

Rai, M. and Ingle, A. (2012). Role of nanotechnology in agriculture with special reference to management of insect pests. Applied Microbiology and Biotechnology, 94(2): 287–293.

Rai, M. and Yadav, A. (2013). Plants as potential synthesiser of precious metal nanoparticles: Progress and prospects. IET Nanobiotechnology, 7(3): 117–124.

Rajesh, K.M., Ajitha, B., Reddy, Y.A.K., Suneetha, Y. and Reddy, P.S. (2018). Assisted green synthesis of copper nanoparticles using *Syzygium aromaticum* bud extract: Physical, optical and antimicrobial properties. Optik, 154: 593–600.

Rajiv, P., Rajeshwari, S. and Venckatesh, R. (2013). Bio-fabrication of zinc oxide nanoparticles using leaf extract of *Parthenium hysterophorus* L. and its size-dependent antifungal activity against plant fungal pathogens. Spectrochimica Acta Part A: Molecular and Biomolecular Spectroscopy, 112: 384–387.

Roopan, S.M., Madhumitha, G., Rahuman, A.A., Kamaraj, C., Bharathi, A. and Surendra, T.V. (2013). Low-cost and eco-friendly phyto-synthesis of silver nanoparticles using *Cocos nucifera* coir extract and its larvicidal activity. Industrial Crops and Products, 43: 631–635.

Saravanakumar, A., Peng, M.M., Ganesh, M., Jayaprakash, J., Mohankumar, M. and Jang, H.T. (2017). Low-cost and eco-friendly green synthesis of silver nanoparticles using *Prunus japonica* (Rosaceae) leaf extract and their antibacterial, antioxidant properties. Artificial Cells, Nanomedicine, and Biotechnology, 45(6): 1165–1171.

Semenzin, E., Lanzellotto, E., Hristozov, D., Critto, A., Zabeo, A., Giubilato, E. and Marcomini, A. (2015). Species sensitivity weighted distribution for ecological risk assessment of engineered nanomaterials: The n-TiO2 case study. Environmental Toxicology and Chemistry, 34(11): 2644–2659.

Senapati, S., Ahmad, A., Khan, M.I., Sastry, M. and Kumar, R. (2005). Extracellular biosynthesis of bimetallic Au–Ag alloy nanoparticles. Small, 1(5): 517–520.

Shaikh, S., Nazam, N., Rizvi, S.M.D., Ahmad, K., Baig, M.H., Lee, E.J. and Choi, I. (2019). Mechanistic insights into the antimicrobial actions of metallic nanoparticles and their implications for multidrug resistance. International Journal of Molecular Sciences, 20(10): 2468.

Shankar, S.S., Rai, A., Ankamwar, B., Singh, A., Ahmad, A. and Sastry, M. (2004). Biological synthesis of triangular gold nanoprisms. Nature Materials, 3(7): 482–488.

Sharma, V.K., Yngard, R.A. and Lin, Y. (2009). Silver nanoparticles: Green synthesis and their antimicrobial activities. Advances in Colloid and Interface Science, 145(1-2): 83–96.

Shende, S., Ingle, A.P., Gade, A. and Rai, M. (2015). Green synthesis of copper nanoparticles by *Citrus medica Linn.* (Idilimbu) juice and its antimicrobial activity. World Journal of Microbiology and Biotechnology, 31(6): 865–873.

Shin, W.K., Cho, J., Kannan, A.G., Lee, Y.S. and Kim, D.W. (2016). Cross-linked composite gel polymer electrolyte using mesoporous methacrylate-functionalized SiO2 nanoparticles for lithium-ion polymer batteries. Scientific Reports, 6(1): 1–10.

Singh, J., Dutta, T., Kim, K.H., Rawat, M., Samddar, P. and Kumar, P. (2018). 'Green' synthesis of metals and their oxide nanoparticles: Applications for environmental remediation. Journal of Nanobiotechnology, 16(1): 1–24.

Singh, P., Kim, Y.J., Zhang, D. and Yang, D.C. (2016). Biological synthesis of nanoparticles from plants and microorganisms. Trends in Biotechnology, 34(7): 588–599.

Soenen, S.J., Rivera-Gil, P., Montenegro, J.M., Parak, W.J., De Smedt, S.C. and Braeckmans, K. (2011). Cellular toxicity of inorganic nanoparticles: Common aspects and guidelines for improved nanotoxicity evaluation. Nano Today, 6(5): 446–465.

Thakkar, K.N., Mhatre, S.S. and Parikh, R.Y. (2010). Biological synthesis of metallic nanoparticles. Nanomedicine: Nanotechnology, Biology and Medicine, 6(2): 257–262.

Thanighaiarassu, R.R., Sivamai, P., Devika, R. and Nambikkairaj, B. (2014). Green synthesis of gold nanoparticles characterization by using plant essential oil Menthapiperita and their antifungal activity against human pathogenic fungi. J. Nanomed. Nanotechnol., 5(5): 1.

Thirumalai Arasu, V., Prabhu, D. and Soniya, M. (2010). Stable silver nanoparticle synthesizing methods and its applications. J. Bio. Sci. Res., 1: 259–270.

Tran, T.T.T., Vu, T.T.H. and Nguyen, T.H. (2013). Biosynthesis of silver nanoparticles using *Tithonia diversifolia* leaf extract and their antimicrobial activity. Materials Letters, 105: 220–223.

Vankar, P.S. and Shukla, D. (2012). Biosynthesis of silver nanoparticles using lemon leaves extract and its application for antimicrobial finish on fabric. Applied Nanoscience, 2(2): 163–168.

Velmurugan, P., Sivakumar, S., Young-Chae, S., Seong-Ho, J., Pyoung-In, Y., Jeong-Min, S. and Sung-Chul, H. (2015). Synthesis and characterization comparison of peanut shell extract silver nanoparticles with commercial silver nanoparticles and their antifungal activity. Journal of Industrial and Engineering Chemistry, 31: 51–54.

Vigneshwaran, N., Ashtaputre, N.M., Varadarajan, P.V., Nachane, R.P., Paralikar, K.M. and Balasubramanya, R.H. (2007). Biological synthesis of silver nanoparticles using the fungus *Aspergillus flavus*. Materials Letters, 61(6): 1413–1418.

Wilson, W. (2010). The project on emerging nanotechnologies. *International Center for Scholars. http:// www. nanotechproject. org/inventories/consumer/analysis_draft.*

Xia, Z.K., Ma, Q.H., Li, S.Y., Zhang, D.Q., Cong, L., Tian, Y.L. and Yang, R.Y. (2016). The antifungal effect of silver nanoparticles on *Trichosporon asahii*. Journal of Microbiology, Immunology and Infection, 49(2): 182–188.

Yun, J., Cho, K., Park, B., Kang, H.C., Ju, B.K. and Kim, S. (2008). Optical heating of ink-jet printable Ag and Ag–Cu nanoparticles. Japanese Journal of Applied Physics, 47(6S): 5070.

Zhang, D., Ma, X.L., Gu, Y., Huang, H. and Zhang, G.W. (2020). Green synthesis of metallic nanoparticles and their potential applications to treat cancer. Frontiers in Chemistry, 799.

Chapter 6

Potential Prospects of *Trichoderma* Metabolites as Biopesticides in Managing Plant Health and Diseases

Boregowda Nandini,[1,*] *Nagaraja Geetha*[2] and *Sanjay C. Jogigowda*[3]

1. Introduction

Pesticide resistance is an important challenge for agriculture today (Pittendrigh and Gaffney, 2001; Hahn et al., 2002; Gerhardson, 2002). In developing countries, pesticide use has progressively expanded to boost food production and manage pest-borne diseases, with detrimental impacts on human health and the biotic environment. When applying pesticides in agriculture and eating food with significant pesticide residues, consumers risk toxication effects (Hamilton et al., 2004; Maroni et al., 2006). Agronomists who handle pesticides remain unaware of their proper application or safeguards (Jørs et al., 2006; Deguine et al., 2021). Chronic respiratory illnesses, rashes and neurological abnormalities have also been described (Maroni et al., 2006; McCauley et al., 2006). Heavy metal deposition in the liver and kidney disrupts biochemical progressions, causing neurological, cardiovascular, bone and kidney problems (WHO, 1992; Aaseth et al., 2021). The overuse of pesticide affects soil, water and natural balance, harming the economy. Pesticide resistance is another concerning issue in this century. *Trichoderma* spp., produce Secondary Metabolites

[1] Department of Studies in Biotechnology, Post Graduate Wing of SBRR First Grade College (Autonomous), Affiliated to University of Mysore, Mysuru – 570016, Karnataka, India.
[2] Nanobiotechnology Laboratory, Department of Studies in Biotechnology, University of Mysore, Manasagangotri, Mysuru, 570 006, Karnataka, India.
[3] JSS Dental College and Hospital, JSS Academy of Higher Education and Research, SS Nagara, Mysuru-570015, Karnataka, India.
* Corresponding author: bnandini2010@gmail.com

(SMs) (volatile, nonvolatile, diffusable) that protect plants from pests, provide nutrients, solubilize minerals and have pharmacological effects. *Trichoderma* uses mycoparasitism, antibiosis and competition to fight pests. Secondary metabolism, mode of action and uses are useful for integrated pest and disease control (Patil et al., 2016; Henríquez-Urrutia et al., 2022).

The fungus *T. viride* is being developed as a biological control for the treatment of plant disease. Potential biocontrol chemicals produced by fungi vary in biochemical assembly structure and method of action, and include glycoproteins, polypeptides, amino acid byproducts, sterols, terpenoids and quinones (Kono et al., 1981; Stoessl, 1981; Metwally et al., 2022). Only one percent of the pesticide field is devoted to biopesticides, however they are becoming increasingly valued as an integral part of effective plant protection strategies (Copping and Menn, 2000). *Trichoderma* uses a variety of strategies to combat phytopathogenic fungi, including colonization, competition, direct mycoparasitism and antibiosis (Hjeljord and Tronsomo, 1988; Howell, 2003; Mahato, 2021). Pesticide users have been demonstrated to have more genetic harm (Castillo Cadena et al., 2006; Sailaja et al., 2006). Additionally, it is detrimental as it can cause miscarriages, birth defects and cancer (Sharma et al., 2009), all of which have a detrimental effect on the economies and societies of developing nations. Due to their detrimental impact on the country's social and economic structure, some pesticides have been prohibited in India. This includes the use of pesticides like, captafol, cypermethrin, aluminium phosphide, diazinon, dazomet, dichlorodiphenyltrichloroethane (DDT), fenthion, fenitrothion, methyl bromide, methoxyethyl mercuric chloride (MEMC), monocrotophos and methyl parathion (www.cibre.nic.in).

2. *Trichoderma* as a compelling microbial contender

Trichoderma is a group of filamentous fungi that reproduce asexual. They can be observed in all soils, as well as humus layers in forests (Wardle et al., 1993) and other natural places with or made up of organic matter (Domsch et al., 1980; Papavizas, 1985). Diffusible and non-diffusible chemicals produced by *Trichoderma* spp., are fungistatic in nature (Dennis and Webster, 1971a, b; Moss et al., 1975; Bruce et al., 1984; Corley et al., 1994; Horvarth et al., 1995). It seems to indicate that successful antagonism may depend on a combination of these kinds of behavior (Ghisalberti and Sivasithamparam, 1991). They are found on the root surfaces of several plants and on decaying bark, especially when the latter has been infected by other fungus (Caldwell, 1958). *Trichoderma* spp., release a chemically diverse range of secondary metabolites, some of which exhibit broad spectrum antibacterial activities *in vitro.* The 'inverted plate technique' established by Dennis and Webster (1971a, b) can be used to assess the effect of *Trichoderma* isolates' generation of Volatile Organic compounds (VOCs). It has been observed that volatile metabolites play a crucial role in pathogen control (Scarselletti and Faull, 1994; Lopes et al., 2012; Patil and Lunge, 2012; Keswani et al., 2014). The poisoned food technique was used to assess the influence of non volatile compounds/metabolites produced by *Trichoderma* spp., on pathogen growth (Nene and Thapliyal, 1993).

Trichoderma is a secondary invader, fast-growing fungus, a source of cell wall disintegrating enzymes and instigator of antibiotics (volatile and non-volatile) (Vinale et al., 2008; Zaid et al., 2022). It illustrates the antagonistic potential of fungal biological control agents (Patil and Lunge, 2012; Lopes et al., 2012; Pakdaman et al., 2013; Ziedan et al., 2021; Rajput et al., 2013), as well as the competitive interaction by the pathogen at the center and pathogen at periphery techniques (Rajput et al., 2013; Asalmol and Awasthi, 1990). The antagonistic potential of *Trichoderma* can be screened using further modified methodologies, such as the slide contact methodology, direct assay method and modified bilayer poison agar method (Rahman et al., 2009). The disparities between the two measures multiplied by 100 were regarded as the percentages of mycelial growth weight inhibition using the modified Skidmore and Dickinson (1976) technique, i.e., Percentages of Inhibition of mycelial Growth Weight (PIWG). Different methods, like the confronting test (also known as the dual culture test), which was first developed by Weindling (1932), the discrepancies between the two measurements multiplied by 100 were measured as the PIWG, can be used to calculate the diffusible metabolites. *Trichoderma* isolates from a wide variety of environments have been tested for their potential to inhibit the growth of pathogens during the course of the past seven decades (Monte, 2001).

3. Voyage and chronology of the *Trichoderma* species

In 1865, the scientists Tulasne and Tulasne (1860) published their discoveries regarding the sexual stage, teleomorph of *T. viride* Pers as *Hypocrea rufa*. Till 1969, *Trichoderma* had only *T. viride* (Bisby, 1939). Rifai, in 1969 identified nine collective species': *T. viride*, *T. hamatum* (Bonord.) Bainier, *T. harzianum* Rifai, *T. koningii* (Oudem.) Duché and *T. piluliferum* J. Webster, *T. polysporum* (Link) Rifai, *T. longibrachiatum* Rifai, *T. pseudokoningii* Rifai and *T. aureoviride* Rifai. Early in the 1990s (Bissett, 1991a; Bissett, 1991b; and Bissett, 1991c) found that the genus *Trichoderma* had five sections and 27 biological species. The use of molecular techniques helped identify species more precisely, and from the late 1990s to 2002, 47 new species of *Trichoderma* were discovered (Kullnig-Gradinger et al., 2002). The key processes by which *Trichoderma* genus interacts with phytopathogenic fungi include hyperparasitism, competition and antibiosis. This adversary is capable of colonizing a number of ecological niches (Herrera-Estrella and Chet, 2004; Harman, 2006; Vinale et al., 2008). Based on phylogenetic analysis, the International Sub-commission on *Hypocrea/Trichoderma* recognized 104 species until 2005 (Druzhinina et al., 2006). There are 252 species, one variant and one form as of the year 2015. Additionally, two other names for the species were proposed: *T. patellotropicum* Samuels (syn. *Hypocrea patella* f. *tropica* Yoshim. Doi) and *T. neocrassum* Samuel (syn. *Hypocrea crassa* P. Chaverri and Samuels) (Bissett et al., 2015). Persoon (1794) coined the term "*Trichoderma*" for the first time. It is crucial to concentrate on a few key elements of it before discussing its practical uses in agriculture (Topolovec-Pintarić, 2019).

4. Biocontrol mechanisms of *Trichoderma* spp.

4.1 Mycoparasitism/Hyperparasitism

Trichoderma genomes are rich in chitinases and glucanases, that play important roles in mycoparasitism and biocontrol. Endo-β-1,4-glucanases, exo-β-1,4-glucanases and β-glucosidases, which are ceratin cellulase enzymes generated by antagonists, also play a crucial role in hyperparasitism (Enari and Niku-Paavola, 1987). Endochitinase, chitobiosidase and β-N-acetylglucosaminidase are the chitinolytic enzymes formed by the pathogenic fungus *T. atroviride*, *P. Karst*, *T. harzianum* and *T. asperellum* Samuels, Lieckf. and Nirenberg, respectively, are crucial for the breakdown of the cell walls of other plants (Haran et al., 1996). These enzymes breakdown the chitin and glucan polysaccharides of pathogen cell walls. The enzymes such as, β-1,3- and β-1,6-glucanases govern *Trichoderma*'s hyperparasitic capabilities against *Phytophthora* sp. and *Pythium* spp. (Lorito et al., 1994; Thrane et al., 1997). Direct conflict with mycopathogens, production of cell-wall-degrading enzymes with lytic activity (Lorito et al., 1996; Lorito et al., 1998) and subsequent penetration and killing are essential for this mechanism of defense (Ayers and Adams, 1981; Chet et al., 1998; Woo and Lorito, 2007). In the first stage, known as chemotrophy, the target fungus releases a chemical signal that attracts an antagonist fungi (Steyaert et al., 2003). In the second, "specific recognition," the antagonist fungus identifies the pathogen's cell surface. In the third, "coiling," *Trichoderma* mycelium surrounds the host; and in the fourth, "intimate hyphal engagement and touch," *Trichoderma* hyphae just grow along the hyphae of the host. The final stage involves the release of enzymes that digest the host cell wall (β-chitinase and proteinases) (Chet et al., 1998). Cellulases, the first extracellular enzymes produced by *Trichoderma* spp., were shown to be very efficient (Mandels, 1975). A mutant strain of *T. harzianum* displayed more chitinase, β-1,3-glucanase and β-1,6-glucanase activity than the wild type (Rey et al., 2001). *Trichoderma* endo- and exoproteases are responsible for the secretion of proteolytic enzymes for the prevention of *Rhizoctonia solani*, *Botrytis cinerea* and *Fusarium culmorum* (Viterbo et al., 2004; Sharon et al., 2001).

Trichoderma directly contacts a pathogen during hyperparasitism, ultimately leading to the death of the pathogen's cells (Vinale et al., 2008). The ability of *Trichoderma* species to suppress an infection is greatly influenced by the secretion

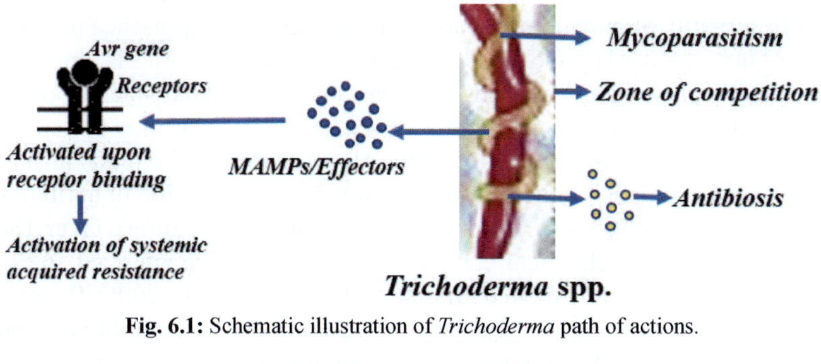

Fig. 6.1: Schematic illustration of *Trichoderma* path of actions.

of lytic enzymes (Viterbo et al., 2002; Mukherjee, 2011). *Trichoderma* species, specifically *T. atroviride* and *T. virens* comprise the furtherest genes for chitinolytic enzymes. These two mycoparasitic *Trichoderma* strains possess a larger arsenal of genes for secondary metabolism than *T. reesei*. In comparison to *T. atroviride, T. reesei* and *T. virens* the genome has 28 non-ribosomal peptide synthetases (containing the gene for gliotoxin, a potentially antifungal compound). Some secondary metabolism gene clusters are shared by all three *Trichoderma* spp., whereas others are characteristic to certain species (Mukherjee, 2011). Mycolytic enzymes, mainly chitinases and glucanases, are secreted to partially degrade the host mycelium's cell wall, allowing for its penetration (Viterbo et al., 2002). Several enzymes, including N-acetylglucosaminidases (Lorito et al., 1994), chitinases-chitin-degrading enzymes (Sahai and Manocha, 1993), chitobiosidases (Harman et al., 1993), glucanases (Thrane et al., 2001), endochitinases (Carsolio et al., 1994; Garcia et al., 1994; Lorito et al., 1998) and proteases (Antal et al., 2000; Delgado-Jarana et al., 2002), revealed the potentiality of *Trichoderma* spp., in managing the plant pathogens. Synergistic activity of lytic enzymes and antibiotics is an additional essential component that can enhance the ability of *Trichoderma* spp., to control diseases of plants (Steyaert et al., 2003). The transduction mechanisms of *T. atroviride* during mycoparasitism have led to the separation of vital aspects of the cAMP and MAP kinase signaling molecules, such as α-subunits of G proteins (G-a), which regulate extracellular enzyme, antibiotic synthesis and coiling around host hyphae (McIntyre et al., 2004). *Trichoderma* produces several volatile metabolites, such as 6-n-pentyl-2H- -pyran-2-one (6-PAP), for plant protection (Jelen et al., 2014; Salwan et al., 2019). *Trichoderma* produce Cell Wall Degrading Enzymes (CWDEs) for this purpose, including cellulase, xylanase, pectinase, lipase, glucanase, amylase, protease, arabinose and lytic enzymes (Strakowska et al., 2014; Druzhinina et al., 2018).

4.2 Antibiosis

Antibiosis is a distinctive mode of antagonism interaction between *Trichoderma* fungi and allied plant pathogenic fungi. The basis for this phenomenon is the production of SMs that have a deadly or inhibiting effect on a parasitic fungus. Non-ribosomal peptides produced by *Trichoderma* spp., include siderophores and polythiodioxopiperazines (ETPs). Small peptides of non-ribosomal origins known as peptaibols, a family of peptaibiotics, are characterized by the presence of significant amounts of non-standard amino acids. The generation of antibiotics is thought to be a species-specific phenomenon that has an impact on the target pathogen by preventing it from growing, producing primary metabolites, nutrients absorbtion and sporulating (Ghisalberti et al., 1990; Howell et al., 1993; Howell, 1998). The biocontrol activity of *Trichoderma* strains is frequently associated with the formation of antibiotic SMs (Ghisalberti et al., 1990; Worasatit et al., 1994; Vinale et al., 2006). Over 180 SMs from several groups of chemical compounds have been characterized so far from the genus *Trichoderma* (Gams and Bisset, 1998; Reino et al., 2008; Bansal et al., 2021). These substances can be separated into peptaibols, water-soluble substances and volatile antibiotics. A key component in the biocontrol of *R. solani*, *Fusarium oxysporum* and *Botrytis cinerea*, is the volatile antimicrobial antibiotic (6-PAP), which is formed by *T. harzianum*, *T. viride* and *T. koningii*. Peptaibols are polypeptide antibiotics with approximately 500–2200 Da, high in non-proteinogenic amino acids, especially α–aminoisobutyric acid. They also have C–end amino alcohols and N–acetylated ends as their distinguishing features. According to research by Reino et al. (2008), *Trichoderma* spp., create a wide range of SMs (natural products), including non-polar low molecular mass chemicals such as, terpenoids, pyrones, polyketides and steroids. It has been established that secondary metabolite production occurs. SMs are organic substances that support the producing organism's survival and fundamental processes like symbiosis, competition and differentiation (Shwab and Keller, 2008; Dini et al., 2021).

4.3 Competition

Trichoderma spp., participate in a form of contact with plant pathogens known as "competition," which is another method of interaction that contributes to the prevention of plant diseases (Mukhopadhyay and Kumar, 2020). This process occurs for a number of reasons, including the consumption of nutrients, the occupation of ecological position and the establishment of infection sites on plant roots. Certain strains of *Trichoderma* are responsible for the production of siderophores, which are chemicals that chelate iron. The antagonist, by character of its siderophores, removes iron from the environment in which it is found, causing a nutrient shortfall. The growth of harmful fungus, such as *Botrytis cinerea*, is thereby inhibited. *Trichoderma* provides an acidic environment, which limits the development and growth of infective fungi, and vigorously colonizes the plant's host root system due to the improved hydrophobins activity. This is due to the fact that *Trichoderma* produces hydrophobins (Benitez et al., 2004; Nawaz et al., 2021).

5. Proposed biopesticide claims of *Trichoderma* metabolites

SMs are a assembly of chemically diverse natural composites that usually have a molecular weight of less than 3000 daltons. They may help the organism that makes them survive in ways like competition with other symbiosis, micro/macroorganisms and metal transport (Demain and Fang, 2000). *Trichoderma* SMs may function as auxin-like chemicals, which typically exhibit maximal action at low concentrations (10^{-5} and 10^{-6} M) and an inhibitory impact at higher concentrations (Brenner, 1981; Cleland, 1972). Ghisalberti and Sivasithamparam (1991) classified chemical compounds into three groups, and these substances fall under them. Class I antibiotics group include 6-PAP and the vast majority of isocyanide derivatives; class II antibiotics, include water-soluble compounds like koningic acid or heptelidic acid; and iii) peptaibols, which are N-acetylated sequential oligopeptides of 12–22 amino acids rich in -aminoisobutyric acid (Le Doan et al., 1986; Rebuffat et al., 1989). Cutler et al. (1989, 1986) and Parker et al. (1997, 1995a, b) illustrated the isolation, characterization and biological properties of *T. koningii*-derived SMs (koninginins A-C, E, G and 1–5). Four carotane-based metabolites, named trichocaranes A through D, were produced by *T. virens*. At concentrations of 10 to 4 molar and 10 to 3

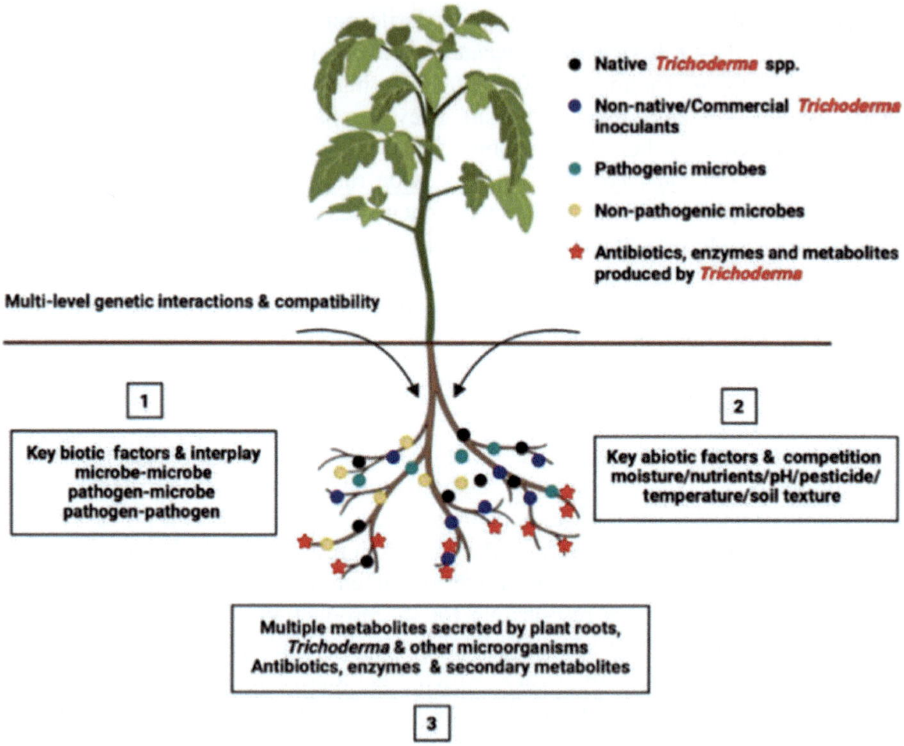

Fig. 6.2: *Trichoderma* bioinoculants' survival, establishment, and function in soil rhizosphere: Loam ecological challenges and determinants (Source: Shahriar et al., 2022).

molar, these compounds significantly inhibited etiolated wheat coleoptile formation (Macias et al., 2000).

Trichoderma harzianum was isolated in a pure culture by Cutler and Jacyno (1991), fermented on solid medium and the crude extract was examined for natural biologically active compounds. The SMs (–) harzianopyridone was isolated using bioassay guidance. This metabolite has abilities to control plant growth (Cutler and Jacyno, 1991). Cyclonerodiol, a sesquiterpene isolated from *T. harzianum* and *T. koningii*, regulates growth and development of the plant (Cutler et al., 1991; Ghisalberti and Rowland, 1993). *T. virens* generates viridiol, a metabolite comparable to the antibiotic viridin, on substrates with high C/N ratios, which inhibits plant growth (Howell and Stipanovic, 1994). Trichosetin was created by the co-culture of *T. harzianum* and *Catharathus roseus* calli. However, this substance was not formed in either the C. roseus or the *T. harzianum* callus individual cultures and appears to be created by the fungus (Marfori et al., 2002). Aflatoxin biosynthesis enzyme 5'-hydroxyaverantin dehydrogenase is inhibited by this chemical (Sakuno et al., 2000; Wipf and Kerekes, 2003). *T. harzianum* has been found to have several strains of *T. harzianolide* and its derivatives, such as deydro-harzianolide and T39 butenolide (Almassi et al., 1991; Ordentlich et al., 1992; Vinale et al., 2006). Antibiotics are natural compounds that suppress or kill microbes (Chiang et al.,

2009). Recently, harzianic acid, an additional nitrogen heterocyclic compound with growth-promoting characteristics, was identified from the culture filtrate of *T. harzianum* obtained from composted pine bark (Vinale et al., 2009). Harzianic acid affects canola seedling growth dose-dependently (Vinale et al., 2009). *Trichoderma* strains can enhance plant biomass production, stimulate growth of lateral root via an auxin-independent process and/or produce Indole-3-Acetic Acid (IAA) or auxin mimics (Hoyos-Carvajal et al., 2009).

Several *Trichoderma* species had been the focus of reports on the production of antibiotics. The growth of fungi is restricted at low concentrations by the sesquiterpenoid metabolites of trichodermin (Godtfredsen and Vangedal, 1965; Dennis and Webster, 1971b; Fedorinchik et al., 1975) and viridin (Weindling and Emerson, 1936; Sivasithamparam and Ghisalberti, 1998), an antibiotic agent. Ergokonin (Kumeda et al., 1995), viridin (Grove et al., 1996; Brian and McGowan, 1945) and viridian fungin A, B, and C have all been associated with pathogen biocontrol (Harris et al., 1985). Trichorzianines and a number of closely related peptaibols are produced by *Trichoderma* spp. These antibiotics change the mobility of liposome membranes in the absence of an applied voltage and establish voltage-gated ion channels in lipid membranes (Molle et al., 1987; Ghisalberti and Sivasithamparam, 1991; El-Hajji et al., 1989; Doan et al., 1986). Harzianolide has been shown to have antifungal properties against a number of plant diseases, as well as plant-growth promoting properties (Vinale et al., 2008). Cerinolactone was identified along with three major butenolides (18, 19 and 20) in *Trichoderma* cerinum culture filtrates. Compound 21 inhibited tomato seedlings development 3 d after the treatment and displayed antifungal activity against various plant diseases (Vinale et al., 2012). Tissues of root and leaf of different plants treated with live *Trichoderma* microbial preparations have already shown that proteins involved in the plant's defense response are changed. This can lead to a coherent or incoherent quantifiable tendency of a single protein, depending on the organism being studied, the fungal formulation and when it was applied (Manganiello et al., 2018; Marra et al., 2006; Segarra et al., 2007; Shoresh and Harman, 2008; Perazzolli et al., 2016; Nogueira-Lopez et al., 2018; De Palma et al., 2019; Mulatu et al., 2022). *Trichoderma* inhibited the development and survival of *Ralstonia* spp., a pathogenic Gram-negative bacterium, in tomato plants, which was linked to the release of different chemicals such as lysosime, viridiofungin and trichokonin (Yan and Khan, 2021).

5.1 Peptaibols

Trichoderma spp., produce SMs called peptaibols, which are made up of unique amino acids. A number of acetylated N- and C-terminal amino alcohols, like isovaline (Iva) and α-amino isobutyric acid (Aib), are found in these linear hydrophobic peptides. Peptaibols, the unique amino acids play a major role in the biomolecules' intermediate production. Alamethicin, one of the peptaibols, was initially isolated from *T. viride*, yet it was later shown that alamethicin is made up of at least 12 distinct chemicals (Meyer and Reusser, 1967; Brewer et al., 1987). *Trichoderma* was found to be primarily responsible for peptaibol production. According to the length of the peptide chains, peptaibols are divided into three subclasses: long-sequence peptaibols, which have 18–20 residues, short-sequence peptaibols, which have 11–16 residues and lipopeptaibols, which have 7–11 residues and have their N-terminal amino acids acylated by a short lipid chain (Auvin-Guette et al., 1992). Harzianic Acid (HA), a new siderophore poduced by *T. harzianum*, has been found to have plant-growth promoting activity (Vinale et al., 2009b; Vinale et al., 2013).

Peptaibols cause *Phytophthora cactorum* lysis and *Rhizoctonia solani* chemical leaking (El-Hajji et al., 1989). The most significant and well-researched of the volatile antifungal substances produced by *Trichoderma* strains is an polyketide, 6-PAP with a very pleasant, coconut like odor. Numerous *Trichoderma* strains' cultures have been found to contain 6-PAP and other -pyrone analogs, which have been connected to the control of plant growth and the induction of plant defense mechanisms (Collins and Halim, 1972; Claydon et al., 1987; Simon et al., 1988; Bonnarme et al., 1997; Reithner et al., 2005, 2007; Vinale et al., 2008a; El-Hassan and Buchennauer, 2009; Urbaniak et al., 2021). *T. atroviride* strain 11 was found to contain the volatile metabolites, 5,5-dimethyl-2H-pyran2-one, 2-methoxy-1,3-dioxolane, 2-n-heptyl-8-hydroxy-2H-1-benzopyran-5-one, and methyl acetate (Keszler et al., 2000). Solid-phase microextraction (SPME) has been reported to yield volatile metabolites such as 2-pentanone, -p-xylene, -pinene, 2-heptanol, ethyl decanoate, ethyl octanoate, methyl benzoate, -farnesene and -curcumene (Jeleń et al., 2014). Similar substances generated by *Trichoderma* species include 3-ethyl-5-methylphenol, 1-pentanol, 2-pentanone, 2-hexanone, geranyl acetone, methyl benzoate, cyclohept-3-en-1-one, α-pinene and β-pinene (Korpi et al., 2009; Müller et al., 2013).

5.2 Polyketides

Trichoderma spp., create polyketides that are harmful to plant pathogens like, *Gaeumannomyces gaminis* var. *tritici*, *Rhizoctonia solani*, *Phytophthora cinnamomi*, *Fusarium oxysporum*, *Bipolaris sorokiniana* and *Pythium middletonii*. The actinomycetes present in the same environment as *Aspergillus nidulans* activate the transcription of genes involved in the fungus' polyketide production (Schroeckh et al., 2009; Zhu et al., 2021). For beneficial uses, *Trichoderma* spp., are the most common producers of polyketide synthases (PKSs) in agriculture. Trichoharzins, trichodimerols, *Trichoderma*tides A, B, C and D, koninginins A, B, D, E and G and koninginins L and M are examples of PKS (Almassi et al., 1991; Ghisalberti, 1993; Kobayashi et al., 1993; Lang et al., 2015; Nuansri et al., 2021; Vicente et al., 2022).

5.3 Terpenoids

Trichoderma spp., exhibited a diverse spectrum of terpenoids with powerful biological activities. *T. harzianum* and *T. koningii* both contain a sesquiterpene, cyclonerodiol that promotes growth of the plant (Cutler et al., 1991; Ghisalberti and Rowland, 1993; Salwan et al., 2019; Amirzakariya and Shakeri, 2022). *Trichoderma* spp., produced a number of triterpenes and sterols, including ergosterol, lanostadiol, pyrocalciferol and others (Kamal et al., 1971). Harziandione, a diterpene obtained from *T. harzianum*, lacked antifungal action; nevertheless, a comparable molecule with antifungal potential was discovered from *T. viride*, and the topologies were similar to harziandione (Sivasithamparam and Ghisalberti, 1998). *Trichoderma* spp. does not include monoterpene chemicals. Due to their volatile nature, it is possible that monoterpenes are only created in very small quantities (Sivasithamparam and Ghisalberti, 1998). *T. virens* produced a steroid called viridiol, which is comparable to viridin (formerly known as *Gliocladium virens*). It has been shown to hinder the growth of plants (Howell and Stipanovic, 1994). *T. hamatum* also produces viridiol, which decreases the formation of aflatoxin in plant pathogens (Sakuno et al., 2000; Aloj et al., 2009). *T. virens* has been shown to have anticancer-active trichodermic acids A and B (Yamaguchi et al., 2010). Harzianone and trichoacorenol were discovered in the *Trichoderma* spp., in the *Xylocarpus granatum*, i.e., mangrove plant, respectively (Zhang et al., 2014).

5.4 Pyrones

The first volatile antifungal chemical identified from *T. viride* was 6-PAP (Collins and Halim, 1972). *Trichoderma* cultures smell like coconuts due to the 6-PAP; these metabolites also significantly reduced *Rhizoctonia solani* fungal growth (Dennis and Webster 1971a, b). Some other research supports, the polyketide synthesis of 6-PAP, inspite of Serrano-Carreon et al. (1993) who attempted to understand the metabolic pathways for 6-PAP. They proposed linolenic acid as the origin of 6-PAP (Sivasithamparam and Ghisalberti, 1998). In *T. viride* and *T. koningii*, a dehydroderivative both with a coconut-scented compound was reported by Moss et al., in 1975 that is similar to 6-PAP. Hill et al. (1995) used massoilactone and decanolactone to present *Trichoderma* as a biocontrol agent. Viridepyronone's shows MIC against *Sclerotium rolfsii* with 196 g/ml (Evidente et al., 2003). In *T. atroviride*, the G-protein Tga1 was found to be the major regulator of 6-PAP biosynthesis, whereas in *T. harzianum*, a similar transcription factor thctf1 was found to be a regulator of 6PP. The mutant lines lacking the thctf1 gene did not show the two metabolites, 6-[(1'R,2'S)-dihydroxypentyl] and 6-[(1'R,2'S)-dihydroxypentyl], -2H-pyran-2-one, 6-PAP and -2'-propyloxiran-1-yl)-2Hpyran-2-one 6-((1'S,2'R)-pyran-2-one (Daoubi et al., 2009; Rubio et al., 2009). Cerinolactone, a *Trichoderma* cerinum-derived hydroxyl lactone (Vinale et al., 2012). *Trichoderma*erin is an antifungal diterpene lactone from *T. asperellum* (Chantrapromma et al., 2014; Stracquadanio et al., 2020).

5.5 Isocyano Metabolites

The unpleasant odor molecules with a distinctive five-membered ring are recognized as isocyano metabolites. In 1956, the first naturally occurring isocyano metabolite was xanthocillin, which was discovered in *Penicillium notatum* (Scheuer, 1992). After 10 yr, dermadin was isolated from *Trichoderma* spp. To combat rumen bacteria, *T. hamatum* was used by Liss et al. (1985) to isolate the antibiotic metabolite 3-(3-isocyanocyclopent-2-enylidene) propionic acid. First discovered in *T. koningii*, isonitrin C (trichoviridin) inhibits the formation of melanin, according to Tamura et al., (1975). *Trichoderma* spp., also produced isonitrins A, B, C and D (Fujiwara et al., 1982). Homothallin I (Pratt et al., 1972) and Homothallin II, which inhibit Phytophthora spp., were generated by *Trichoderma* koningii (Edenborough and Herbert, 1988). Isocyano metabolites limit melanin biosynthesis in mammals by inhibiting tyrosinase (Brewer et al., 1982), stimulate oospores of the A2 mating type of *Phytophthora* spp. (Reeves and Jackson, 1972; Pratt et al., 1972; Brasier, 1975) and induce bacteriostasis (Brewer et al., 1982; Ni et al., 2019; Contreras-Cornejo et al., 2020).

5.6 Diketopiperazines

Weindling and Emerson (1936) first described it as a metabolic byproduct of *T. lignorum* (Tode), Brian (1944) later identified it as the gliotoxin from *T. viride*. Gliotoxin was the first diketopiperazines (DKPs) that was isolated from *Trichoderma*. Gliotoxin and gliovirin are two examples of members of the toxin family known as epipolythiodioxopiperazine. Both examples demonstrated the presence of disulfide bridges across cyclic dipeptides. One of these two DKPs, gliotoxin, is substantially more effective than gliovirin in inhibiting the growth of *Rhizoctonia solani* (Howell et al., 1993; Mukherjee et al., 2018; Jayalakshmi et al., 2021). Due to its unusual structure and potent biological activity, gliotoxin has attracted the interest of biologists. They hypothesized that L-phenylalanine and L-serine formed the gliotoxin via a cyclic dipeptide. Sivasithamparam and Ghisalberti (1998) mapped out the biochemical process for gliotoxin production. It was found that the enzyme called dioxopiperazine synthase (gliP) was engaged in the biosynthesis process. This enzyme is responsible for producing the distinctive diketopiperazine ring (Dagenais and Keller, 2009; Shi et al., 2018).

5.7 Harzianic acid

Trichoderma harzianum, which was isolated from decaying pine bark in western Australia, demonstrated suppression of the phytopathogen *Pythium irregulare*. The fungus principally produced harzianic acid, a completely characterized tetramic acid derivative (Sawa et al., 1994). Biosynthetic tetramic acids are thought to be formed via the synthesis of an activated acyl and amino acid group (Sodeoka et al., 2001). Harzianic acid was obtained from liquid cultures of a fungal strain derived from a soil sample in Amagi, Japan, in association to N-demethyl analog tetramic acids or pyrrolidinediones (Kawada et al., 2004). According to Reino et al. (2008), *Trichoderma* species create about a 100 different SMs (natural products), including pyrones, terpenoids, steroids and polyketides, all of which have low molecular masses. Bioactive chemical harzianic acid has been identified as a potential alternative to live antagonists because of its antifungal and plant growth-promoting properties. Canola and tomato seedlings treated with harzianic acid and isoharzianic acid both exhibited antifungal action and grew more shoots and roots (Vinale et al., 2009a, 2014). In plant nutrient regulation, microbial siderophores are iron-chelating agents that aid in iron solubilization (Hider and Kong, 2010). Harzianic acid is thought to have growth-promoting effects on plants because of its strong siderophoric characteristics (Vinale et al., 2013). The uptake of harzianic acid by plants and an increase in seedling

Table 6.1: An overview of *Trichoderma* spp. and its active metabolites.

Class	*Trichoderma* spp.	Secondary metabolites	References
Azapilones	*T. harzianum* T22	T22 azaphilone (12)	Vinale et al. (2006)
Butenolides	*T. harzianum*	Harzianolide (13-A) T39, butenolide (13-B), and dehydro-harzianolide (13-C)	Claydon et al. (1991); Almassi et al. (1991); Ordentlich et al. (1992); Vinale et al. (2006)
	T. cerinum	Harzianolide, Dehydroharzianolide, T39butenolide, Cerinolactone	Vinale et al. (2012)
Diketopiperazines	*T. virens*	Gliotoxin (19), Gliovirin (20)	Howell (1991)
Heterocyclic compounds	*T. harzianum*	Harzianopyridone (10), Harzianic acid (11)	Dickinson et al. (1989); Vinale et al. (2006); Vinale et al. (2009a)
Isocyano metabolites	*T. viride*, *T. koningii*, *T. hamatum*	Dermadin (14)	Pyke and Dietz (1966); Meyer (1966); Coats et al. (1971)
	T. viride, *T. koningii*,	isonitrin B (17), Isonitrile trichoviridin (15), isonitrin A (16)	Tamura et al. (1975)
	T. koningii	Homothallin I, homothallin II (18)	Pratt et al. (1972); Edenborough and Herbert (1988)
	T. harzianum	Homothallin II (18)	Faull et al. (1994)
Koninginins	*T. harzianum*, *T. aureoviride*, *T. koningii*	Koninginins A (3), B (4), D (5), E (6) and G (7)	Dunlop et al. (1989); Almassi et al. (1991); Ghisalberti et al. (1993)
Lactones	*T. cerinum*	Cerinolactone	Vinale et al. (2012)
Pyrones	*T. viride*, *T. harzianum*, *T. atroviride*, *T. koningii*	6-pentyl-2H-pyran-2-one (6-pentyl-α-pyrone) (1), cytosporone S (2)	Worasatit et al. (1994); Scarselletti et al. (1994); El-Hassan et al. (2009)
Peptaibols	*T. atroviride*	Atroviridin A Atroviridin B Atroviridin C Neoatroviridins A-D	Oh et al. (2000); Blaszczyk et al. (2014)
	T. viride	Trichotoxin A 50 I Trichodecenin I Suzukacillin Alamethicin F50	Bruckner and Przybylski (1984); Rebufatt et al. (1992); Fujita et al. (1994); Goulard et al. (1995),
		Richotoxins A and B, Trichorovins, Trichodecenins, Trichocellins Trichokinidins	Blaszczyk et al. (2014)

Table 6.1 contd. ...

...*Table 6.1 contd.*

Class	*Trichoderma* spp.	Secondary metabolites	References
	T. citrinoviride	Trichotetronine, Dihydrotrichodimerol, Dihydrobislongiquinolide,	Balde et al. (2010)
	T. harzianum	Trichokindin IIIa Trichorzin HA I Harzianin HB I Paracelsin (23) Trichotoxins (21), Trichorzins	Iida et al. (1994); Goulard et al. (1995); Augeven-Bour et al. (1997); Daniel and Filho (2007)
	T. koningii	Trichokonin Ib Trikoningin KB I	Auvin-Guette et al. (1993); Huang et al. (1995)
	T. saturnisporum	Paracelsin E Saturnisporin SA I Saturnisporin SA II Saturnisporin SA III Saturnisporin SA IV	Ritieni et al. (1995); Rebuffat et al. (1993); Goulard et al. (1995)
	T. stromaticum	Trichostromaticins A, B, C, D and E	Degenkolb et al. (2006)
	T. strigosum	Tricholongin B I, II and III, Lipostrigocin A1–A6, B1–B9,	Degenkolb et al. (2006)
	T. longibrachiatum	Trichobranchin A-I Tricholongins BI and BII, and longibrachins	Mohamed-Benkada et al. (2006); Rebuffat et al. (1991); Blaszczyk et al. (2014)
	T. polysporum	Trichosporin TS-B-1a-1 Polysporin A Polysporin B	Iida et al. (1993); New et al. (1996); Sharman et al. (1996)
	T. pseudokoningii	Harzianin HK-VI Pseudokinin KL-VI	Rebuffat et al (1996); Rebuffat et al. (2000)
	T. reesei	Paracelsin A Paracelsin B	Bruckner and Przybylski (1984)
Viridins	*T. koningii,* *T. virens,* *T. viride*	Viridin (8)	Brian and McGowan (1945); Reino et al. (2008)
	T. hamatum, *T. viride,*	Viridiol (9)	Moffatt et al. (1969); Howell and Stipanovic (1994)
Other metabolites	*T. viride,* *T. virens,* *T. polysporum,* *T. koningii,* *T. longibrachiatum* *T. viride,* *T. harzianum,* *T. harzianum 25*	6-nonylene alcohol, 3-methyl-heptadecanol, 2-methyl-heptadecanol ketotriol, Massoilactone, Massoilactone, 2-Phenylethanol, tyrosol, sorbicillin, Anthraquinones	Score and Palfreyman (1994); Dickinson et al. (1995); Tarus et al. (2003); Dubey et al. (2011)

Table 6.2: Outline of marketable products based on *Trichoderma* spp.

***Trichoderma* spp.**	**Product name**	**Country**	**Pathogens controlled**
T. polysporum, *T. harzianum* (ATCC 20, 476)	Binab T	BINAB Bio-Innovation AB, Helsingborg, Sweden Henry Doubleday Research Association, United Kingdom	*Chondrostereum, Didymella, Verticillium, Heterobasidion, Fusarium, Botrytis, Pythium, Rhizoctoni, Phytophthora* spp., and Tree wound pathogens
Trichoderma harzianum T22	Plant Shield Root shield,	BioWorks, Inc., USA	Soil-borne pathogens, *Fusarium, Pythium, Rhizoctonia, Botrytis, Cylindrocladiu, Thielaviopsis* spp.
Trichoderma harzianum T39	Trichodex	Bio works, Victor, NY, USA	*Botrytis cinerea*
Trichoderma spp., *T. koningii*, *T. harzianum*	Promot PlusWP; Pro-mot PlusDD	Tan Quy, Vietnam	Root rot and Wilt diseases
Trichoderma spp.	Antagon	DeCeusterMeststoffen N.V. (DCM), Belgium	Damping-off diseases
T. harzianum	Trichostar	Green Tech, Agroproducts, Rajaji Road, Coimbatore	*Macrophomina* spp.
T. viride	Antagon TV	Green Tech, Agroproducts, Rajaji Road Coimbatore	*Macrophomina* spp.
Trichoderma spp.	Monitor	Agricultural and Biotech Pvt. Ltd. Gujarat Department of Plant Pathology, MPKV, Rahuri	-
T. virens	Gliostar	GBPUAT, Pantnagar	*Fusarium, Rhizoctonia, Sclerotium, Pythium* spp.
T. viride/ *T. harzianum*	Bioderma	Biotech International Ltd. India	-
T. viride	Ecofit	Hoechst Schering Afgro Evo Ltd, India	-
T. viride	Bio Fit	Ajay Biotech (India) Ltd., India	*Fusarium, Pythium, Rhizoctonia, Sclerotium* spp. *Botrytis* spp. and other root rots pathogens.
T. viride	Biocon	Tocklai Experimental Station Tea Research Association, Jorhat (Assam), I	-
T. viride	Trichoguard	Anu Biotech Int. Ltd. Faridabad	-

growth even in iron-deficient environments have both been documented (Vinale et al., 2013; Xie et al., 2021). ETPs and siderophores are examples of non-ribosomal peptides produced by *Trichoderma* spp. (Shokrollahi et al., 2021).

6. Significance and mechanism of *Trichoderma* secondary metabolites

SMs with antibacterial characteristics are secreted by *Trichoderma* (Ghisalberti and Sivasithamparam, 1991; Mathivanan et al., 2008; Sivasithamparam and Ghisalberti, 1998; Vicente et al., 2022). Both volatile (e.g., alcohols, ethylene, hydrogen cyanide, ketones and aldehydes upto C4 chain length) and nonvolatile (peptides) SMs are produced from *Trichoderma* spp. (Keszler et al., 2000; Khan et al., 2020). *Trichoderma* spp., appear to be an inexhaustible supply of antibiotics, including viridin and gliotoxin from acetaldehyde (Dennis and Webster, 1971a, b), alpha-pyrones (Keszler et al., 2000), polyketides, terpenes, piperacines, isocyanide derivatives and peptaibols complex families (Sivasithamparam and Ghisalberti, 1998). These compounds, in concert with CWDEs, inhibit numerous fungal plant diseases (Schirmbock et al., 1994; Lorito et al., 1996). In medicine, antibiotic biosynthesis-related genes, such as peptaibols (Wiest et al., 2002) and polyketides (Sherman, 2002), has not been studied.

Trichoderma spp., produce the elicitor Sm1/Epl1. This is the small cysteine-rich hydrophobin-like protein of the ceratoplatanin class (Djonovic et al., 2006). *Trichoderma* species are among the most potent agents for biological control (Pandey et al., 2018). According Verma et al. (2007), up to 60% of the biofungicides registered for use in modern agriculture are *Trichoderma*-based formulations (Verma et al., 2007). In India, there are 250 field-use products, however the proportion of biofungicides is low. It is dominated by manmade substances (Singh et al., 2009). *Trichoderma* spp., can provide hormonal signals that promote root colonization. This results in the secretion of auxins that stimulate root growth, which, due to the increased surface area, may encourage greater colonization (Contreras-Cornejo et al., 2009). Gene knockout was used to study the role of ACC deaminase in regulating root development of canola with *T. asperellum* (Viterbo et al., 2010). According to extensive research, *Trichoderma* cells secrete cysteine-rich hydrophobin proteins to facilitate adhesion. Two proteins, TasHyd1 (*T. asperellum*) and Qid74 (*T. harzianum*), have been identified as helping in root attachment (Samolski et al., 2012). Biofertilizers and biopesticides are essential alternatives for sustaining production potential with minimal environmental harm (Govind et al., 2016). *T. virens* is able to manufacture auxins such as indole-3-ethanol (IEt), Indole-3- Acetic Acid (IAA) and indole-3-acetaldehyde (IAAld). These contribute to the growth and development of plants (Contreras-Cornejo et al., 2009). Harzianolide, produced by *Trichoderma* spp., promotes vigorous early plant development by lengthening the root system (Cai et al., 2013; Manganiello et al., 2018). *Trichoderma* has a beneficial influence on plant development. In particular, they release koninginins, trichocaranes A–D, harzianopyridone, 6-PAP, harzianolide, harzianic acid and cyclonerodiol. They exert their influence based on a dose dependent manner (Vinale et al., 2014). This method is being used for the biocontrol of wilt diseases (Kumar and Khurana, 2016), that prevents seed-borne fungus in *Vigna radiata*, i.e., mung bean seeds (Kumar and Khurana, 2018) and is able to resist wilt disease caused by *Fusarium* spp., in bananas (Kumar et al., 2021; Kumar and Khurana, 2019).

Gliovirin and gliotoxin are the Q and P group *Trichoderma* secondary metabolites. *T. virens* strains of P group are active against *P. ultimum* but not against *R. solani*, while the Q group strains are more active against *R. solani* (Howell, 2000; Zaid et al., 2022). The *T. virens* veA ortholog vel1 regulates gliotoxin production, biocontrol function and numerous secondary metabolism-related genes (Mukherjee et al., 2004). The antibiotic action was shown by Trichokonins VI, a type of peptaibol produced from *T. pseudokoningii* SMF2. This was established by the extensive production of programmed apoptotic cell death in fungal infections (Shi et al., 2012). Koninginin D also inhibited the development of soil-borne plant diseases such as *Pythium middletonii*, *Rhizoctonia solani*, *Fusarium oxysporum*, *Bipolaris sorokiniana* and *Phytophthora cinnamomi* (Dunlop et al., 1989; Verma et al., 2020; Sridharan et al., 2021). Viridins obtained from various species of *Trichoderma* (*T. viride*, *T. koningii* and *T. virens*) suppress the germination of spores of *Botrytis allii*, *Aspergillus niger*, *Penicillium expansum*, *Fusarium caeruleum*, *Colletotrichum lini* and *Stachybotrys atra* (Singh et al., 2005). Harzianic acid, from the strain of *T. harzianum*, illustrates antibiotic activity against *Sclerotinia sclerotiorum*, *Pythium irregulare* and *Rhizoctonia solani* under *in vitro* conditions (Vinale et al., 2009). Trichodiene synthase (tri5) was identified and found to be overexpressed in the transformant of *T. brevicompactum* Tb41tri5, and also observed the increased production of trichodermin and as a response, the antifungal activity against *Aspergillus fumigates* and *Fusarium* spp. was augmented (Tijerino et al., 2011; Taylor et al., 2022). *Trichoderma* genus includes a collection of filamentous fungus that are prolific in nature. In order to suppress numerous pathogenic fungus, these species function as BCAs (Biological Control Agents). The potency of these organisms is the consequence of a combination of mechanisms, which in turns stimulates the production of SMs (Nandini et al., 2021). They produce peptaibols, an antibiotic belonging to the peptide family. These possess nonproteinogenic amino acid residues like, Aib (-aminoisobutyrate) as well as C-terminal amino alcohols and N-terminal modifications. The *T. asperellum* strains produced five trichotoxins (T5D2, T5E, T5F, T5G and 1717) and two antibiosis-related asperelines (A and E) (Brito et al., 2014; Alfaro-Vargas et al., 2022).

7. Multifaceted profile of *Trichoderma* with its pathogens

It has been demonstrated that *T. virens* is involved in the expression of hydrophobins to enable host adherence and mycoparasitism (Zeilinger et al., 2005). Through the activation of G-protein cascades, the binding of ligands to these receptors induces downstream signaling processes. Certain species, such as *T. atroviride*, generate, 6-PAP, a volatile metabolite that plays a role in *Trichoderma* fungal interrelations (El-Hasan et al., 2008; Vinale et al., 2009a; Singh et al., 2022). *Trichoderma* spp., has MAPKK, MAPKKK and MAPK signaling cascades. This contributes to mycoparasitism and biocontrol mechanism (Kumar et al., 2010). The synthesis of CWDEs and antibiotics provides them with the ability to eliminate infections. *Trichoderma* produces glucan and synthases to repair damage to its own cell wall caused by a pathogen during *Trichoderma*-pathogen interactions. Most likely genes encoding hydrolytic enzymes like chitinases and glucanases, as well as secondary metabolic genes like Non-Ribosomal Peptide Synthetases (NRPSs), restrict pathogenic development (Kubicek et al., 2011). Through a single NRPS Tex2 of *T. virens*, genetic data has been reported for the assembling of 11 and 14 modules peptaibols (Mukherjee et al., 2011). Gpr1 is a seven-transmembrane G protein-coupled receptor that senses fungal prey in the immediate area (Omann et al., 2012). Peptaibiotics have been shown to have significant antibacterial effects (such as the ability to produce voltagegated membranes receptors like peptaibol trichokonin VI in *T. pseudokoningii*) and to be involved in the induction of Fusarium oxysporum programmed cell death (Shi et al., 2012).

Trichoderma genes encoding proteases, particularly those belonging to the subtilisin-like serine protease group and oligopeptide carriers are stimulated before and during contact between diverse species of *Trichoderma* and the host (Suarez et al., 2007; Seidl et al., 2009; Morais et al., 2022). Additionally, there is biochemical proof that Gα is involved in *Trichoderma* hyphae coiling around the host hyphae. Benitez et al. (2004) discovered that the presence of G-protein activators (mastoparan and fluoroaluminate) increased the coiling of antagonist hyphae around nylon strands. The conserved G proteins signaling cascade consists of three subunits, Gα, Gβ and Gγ and is responsible for further signal transduction. Mutants of *T. atroviride* that lack the Gα subunit are completely incapable of mycoparasis, have diminished chitinolytic activity and are unable to produce the antifungal chemical 6-PAP (Rocha-Ramirez et al., 2002; Reithner et al., 2005; Druzhinina et al., 2011). In addition, during mycoparasitism in *Trichoderma*, Mitogen-Activated Protein Kinases (MAPKs), particularly pathogenesis MAPK (Tmk1and TmkA), are implicated in signal transduction pathways (Schmoll, 2008; Druzhinina et al., 2011). Moreover, it has been proposed that the pathogenic host sensing mechanism involves class IV, receptors for oligopeptides and compounds released from phytopathogen cell walls by the action of protease enzymes are G-protein coupled receptors (GPCRs) (Druzhinina et al., 2011). Analysis of the *Trichoderma* genomes that have been sequenced (e.g., *T. reesei*, *T. virens*, *T. atroviride*, *T. asperellum*, *T. gamsii*, *T. harzianum*, etc.,) has allowed the discovery of a few genes linked to the production of SMs (Mukherjee et al., 2012). Even though *Trichoderma* genomes are replete with PKS-encoding genes, only a small number of studies have focused on the genetics and polyketide synthesis of these fungi. The PKS-coding genes of *T. reesei*, *T. virens* and *T. atroviride* phylogenomic research showed that most of the polyketide synthases belong to either the lovastatin/citrinin or fumonisin clades and are found in all three species as orthologs (Baker et al., 2012). In addition, the PKS gene known as pks4 that is present in *T. reesei* is the one that accounts for green pigmentation, the integrity of the conidial cell wall and the ability to act as an antagonist towards other fungi (Atanasova et al., 2013). Furthermore during interactions with the phytopathogens *R. solani* and *F. oxysporum*, two PKS-encoding genes from *T. harzianum*, known as pksT-1 and pksT-2, showed variable levels of expression (Yao et al., 2016).

T. harzianum strain T22 can solubilize metal oxides through chelation and reduction (Altomare et al., 1999). Harzianic acid is a tetramic acid derivative isolated from *T. harzianum* culture filtrate (M10). This chemical promotes plant development (Vinale et al., 2009). The increased expression of defense-related genes in maize, such as pal1 and aos, in response to interaction with *T. virens* is related to the SMs that are synthesized by the beneficial fungus due to the activity of the PKS/NPRS activity of the Tex13 gene (Mukherjee et al., 2012). *T. virens* and *T. atroviride*, two mycoparasitic species, have an increased number of genes associated to secondary metabolism when compared to degrading enzyme manufacturers (such as, *T. reesei*) (Mukherjee et al., 2012). Lehner et al. (2013) devised a method to identify *Trichoderma* extracellular siderophores. Aspinolide C from *T. arundinaceum* stimulated PR1b1 and PR-P2 genes implicated in Salicylic Acid signaling (SA). Aspinolide B (an aspinolide C derivative) inhibited SA-related genes considerably, whereas the effect on jasmonic acid-related genes was not consistent (Malmierca et al., 2015). Using TvTex10 mutants, an enzyme is up-regulated in iron-deficient conditions that is involved in the generation of intracellular siderophores, researchers have established the interactions of *Trichoderma*–plant. Genotypes disrupting tex10 gene production resulted in reduced production and enhanced ability to colonize maize roots. The lower concentration of *Cochliobolus heterostrophus* induces systemic resistance in maize also has been demonstrated (Mukherjee et al., 2018).

Microbe-associated molecular patterns (MAMPs) such as those seen in *Trichoderma* spp., trigger signaling cascades, which in turn activate phytohormone networks to regulate pathogen assault and environmental stress, and also growth signaling pathways (Hermosa et al., 2013; Wang et al., 2022). They include genes for enzymes including PKSs, oxidoreductases, methyl transferases and NRPS, as well as genes for transporters and transcription factors that are involved in biosynthesis (Zeilinger et al., 2016). For instance, the biosynthesis of gliotoxin gene clusters in *T. virens* and *T. reesei* contain eight varied genes that code for different enzymes. These genes are as follows: Glutathione-S-Transferase (GST), non-ribosomal peptide synthetase, dipeptidase, cytochrome P450 (P450), N-Methyl Transferase (NMT), O-methyl transfer (Zeilinger et al., 2016). In many cases, the generation of fungal SMs involves unusual metabolic processes that culminate in the formation of various natural chemicals from a small number of precursors that are the products of primary metabolism (such as mevalonate, acetyl-CoA and amino acids) (Zeilinger et al., 2016). Researchers have discovered that *T. harzianum* is capable of producing harzianins, trichorzins (HA, MA and PA), trichokindins and trichotoxin (Siddiquee, 2017). *T. atroviride*, on the other hand, is responsible for the release of peptaibols such as, neoatroviridins A–D, and atroviridins A–C. whereas, *T. viride* manages the production of trichotoxins A and B, trichorovins, trichodecenins and trichocellins (Siddiquee, 2017; Sood et al., 2020).

8. Socio-economic value for biopesticides

Pesticides are compounds or mixtures of substances used to prevent, eradicate, repel or mitigate pests. Pesticides are the final input in an agricultural operation and are used to prevent crop spoiling from pests such as insects, fungi, weeds and other organisms, hence enhancing agricultural output (https://www.imarcgroup.com/indian-pesticides-market). Pesticides have increased significantly over the past few years, driven by the need to advance total farming invention and also to provide suitable food obtainability aimed at the country's constantly rising populace. Pests and diseases in India, on an average, account for 20 to 25% of overall food grain production (https://www.imarcgroup.com/indian-pesticides-market). The global market for biopesticides increased from US $1,213 million in 2010 to US $3,222 million in 2017 at an annual rate of 15.8% between 2012 and 2017. In 2011, bioinsecticides accounted for 46% of this industry and biofungicides were worth US $600.5 million before reaching US $1,477 million in 2017. The pesticides market of India was appraised in 2018 at INR 197 billion and is anticipated to reach by 2024 INR 316 billion, expanding at a CAGR of 8.1% from 2019 to 2024 (https://www.imarcgroup.com/indian-pesticides-market). On evaluating the pattern of consumption of pesticide in Union Territories (UTs) and 29 Indian states from 2000 to 2013, a constructive progress inclination was found in UTs/17 states. An optimistic increase was greatest in Andaman and Nicobar Islands, Jammu and Kashmir and Tripura, but Andhra Pradesh, Uttar Pradesh, Punjab, Maharashtra and Haryana reported for 70% of overall pesticide usage (Devi et al., 2017; Satapute et al., 2019). The annual growth rate between 2012 and 2017 was 16.1%. Given the current market need for free products derived from pesticide waste, enormous agrochemical corporations are purchasing biocontrol companies and creating novel biotechnological technologies (Hernández-Castillo et al., 2020).

9. Conclusion

Plant pathogenic bacteria are largely suppressed and the rate of plant growth is regulated by *Trichoderma* spp. Root rot disease, damping-off, wilt and other common plant diseases can be effectively managed by *Trichoderma* spp. In recent studies, many phytopathogens, including nematodes, bacteria and fungi, can be inhibited by *Trichoderma* either through direct interactions (such as, hyperparasitism, competing for nutrition and space or antibiosis) or indirectly through improved plant growth. There are numerous ways in which *Trichoderma* can benefit plants, including promoting plant growth, increasing mineral solubilization, producing secondary metabolites, stimulating plant defense and producing siderophores. *Trichoderma* species boost plant growth and prevent the spread of plant diseases. In addition to destroying other fungi enzymatically, *Trichoderma* spp., produce anti-microbial chemicals that kill pathogenic fungus and outcompete them for nutrients and space, which makes them good biopesticides. The use of products containing *Trichoderma* has piqued the interest of researchers, looking into the possibility of further advantages derived from *Trichoderma* spp. As a result, one is able to present the achievements of *Trichoderma* spp., in relation to phytopathogens and plant growth, attributed to the work done by researchers around the world. In addition to this, the production of their SMs in an agroecosystem is described in this chapter. These startling findings present a significant opportunity for the agricultural sector to enhance its sustainability performance through the implementation of more self-sustaining farming practices.

Abbrevations

CWDEs	cell wall degrading enzymes
DDT	dichlorodiphenyltrichloroethane
DKPs	diketopiperazines
ETPs	polythiodioxopiperazines
GPCRs	G-protein coupled receptors
GST	glutathione-S-transferase
IAA	indole-3-acetic acid
IAAld	indole-3-acetaldehyde
IEt	indole-3-ethanol
MAMPs	Microbe-associated molecular patterns
MAPKs	mitogen-activated protein kinases
MEMC	methoxyethyl mercuric chloride
NMT	N-methyl transferase
NRPS	non-ribosomal protein synthases
NRPSs	non-ribosomal peptide synthetases
6-PAP	6-n-pentyl-2H-pyran-2-one
PKSs	polyketide synthases
SMs	secondary metabolites
SPME	Solid-phase microextraction
UTs	Union Territories
VOCs	volatile organic compounds

References

Aaseth, J., Alexander, J., Alehagen, U., Tinkov, A., Skalny, A., Larsson, A., Crisponi, G. and Nurchi, V.M. (2021). The aging kidney—as influenced by heavy metal exposure and selenium supplementation. Biomolecules, 11(8): 1078.

Alfaro-Vargas, P., Bastos-Salas, A., Muñoz-Arrieta, R., Pereira-Reyes, R., Redondo-Solano, M., Fernández, J., Mora-Villalobos, A. and López-Gómez, J.P. (2022). Peptaibol production and characterization from *Trichoderma asperellum* and their action as Biofungicide. bioRxiv. In Press.

Almassi, F., Ghisalberti, E.L., Narbey, M.J. and Sivasithamparam, K. (1991). New antibiotics from strains of *Trichoderma harzianum*. Journal of Natural Products, 54(2): 396–402.

Aloj, V., Vinale, F., Woo, S., Marra, R., Ruocco, M., Lanzuise, S., Ritieni, A., Campanile, G., Scala, F., Cavallo, P. and Lorito, M. (2009). Use of a *Trichoderma* spp. enzyme mixture to increase feed digestibility and degrade mycotoxins. Journal of Plant Pathology, 91(4).

Altomare, C., Norvell, W.A., Björkman, T.H.O.M.A.S. and Harman, G. (1999). Solubilization of phosphates and micronutrients by the plant-growth-promoting and biocontrol fungus *Trichoderma harzianum* Rifai 1295-22. Applied and Environmental Microbiology, 65(7): 2926–2933.

Amirzakariya, B.Z. and Shakeri, A. (2022). Bioactive terpenoids derived from plant endophytic fungi: An updated review (2011–2020). Phytochemistry, 113130.

Anonymous. (2009). Turkish The Benefist of Biotechnology Comperdium (accessed: http:// www.census. gov/ipc/,www/idp/worldpopinfo. html).

Antal, Z., Manczinger, L., Szakacs, G., Tengerdy, R.P. and Ferenczy, L. (2000). Colony growth, in vitro antagonism and secretion of extracellular enzymes in cold-tolerant strains of *Trichoderma* species. Mycological Research, 104(5): 545–549.

Asalmol, M.N. and Awasthi, J. (1990). Role of temperature and pH in antagonism of *Aspergillus niger* and *Trichoderma viride* against *Fusarium solani*. *In*: Proceedings of the All India Phytopathological Society (West Zone). MPAU, Pune, pp 11–13.

Asalmol, M.N., Sen, B. and Awasthi, J. (1990). Role of temperature and pH in antagonism of *Aspergillus niger* and *Trichoderma* viride against *Fusarium solani*. Proceedings Indian Phytopathological Society, 11–13.

Atanasova, L., Knox, B.P., Kubicek, C.P., Druzhinina, I.S. and Baker, S.E. (2013). The polyketide synthase gene pks4 of *Trichoderma reesei* provides pigmentation and stress resistance. Eukaryotic Cell, 12(11): 1499–1508.

Augeven-Bour, I., Rebuffat, S., Auvin, C., Goulard, C., Prigent, Y. and Bodo, B. (1997). Harzianin HB I, an 11-residue peptaibol from *Trichoderma harzianum*: Isolation, sequence, solution synthesis and membrane activity. Journal of the Chemical Society, Perkin Transactions, 1(10): 1587–1594.

Auvin-Guette, C., Rebuffat, S., Prigent, Y. and Bodo, B. (1992). Trichogin A IV, an 11-residue lipopeptaibol from *Trichoderma longibrachiatum*. Journal of the American Chemical Society, 114(6): 2170–2174.

Auvin-Guette, C., Rebuffat, S., Vuidepot, I., Massias, M. and Bodo, B. (1993). Structural elucidation of trikoningins KA and KB, peptaibols from *Trichoderma koningii*. Journal of the Chemical Society, Perkin Transactions, 1(2): 249–255.

Ayers, W.A. and Adams, P.B. (1981). Mycoparasitism and its application to biological control of plant diseases. Biological Control in Crop Production, 91–103.

Baker, R. (1988). *Trichoderma* spp. as plant-growth stimulants. Critical Reviews in Biotechnology, 7(2): 97–106.

Baker, S.E., Perrone, G., Richardson, N.M., Gallo, A. and Kubicek, C.P. (2012). Phylogenomic analysis of polyketide synthase-encoding genes in *Trichoderma*. Microbiology, 158(1): 147–154.

Balde, E.S., Andolfi, A., Bruyere, C., Cimmino, A., Lamoral-Theys, D., Vurro, M., Damme, M.V., Altomare, C., Mathieu, V., Kiss, R. and Evidente, A. (2010). Investigations of fungal secondary metabolites with potential anticancer activity. Journal of Natural Products, 73(5): 969–971.

Bansal, R., Pachauri, S., Gururajaiah, D., Sherkhane, P.D., Khan, Z., Gupta, S., Banerjee, K., Kumar, A. and Mukherjee, P.K. (2021). Dual role of a dedicated GAPDH in the biosynthesis of volatile and non-volatile metabolites-novel insights into the regulation of secondary metabolism in *Trichoderma virens*. Microbiological Research, 253: 126862.

Benítez, T., Rincón, A.M., Limón, M.C. and Codon, A.C. (2004). Biocontrol mechanisms of *Trichoderma* strains. International Microbiology, 7(4): 249–260.

Bisby, G.R. (1939). *Trichoderma* viride Pers. ex Fries, and notes on Hypocrea. Transactions of the British Mycological Society, 23(2): 149–168.

Bissett, J. (1991a). A revision of the genus *Trichoderma*. II. Infrageneric classification. Canadian Journal of Botany, 69(11): 2357–2372.

Bissett, J. (1991b). A revision of the genus *Trichoderma*. III. Section Pachybasium. Canadian Journal of Botany, 69(11): 2373–2417.

Bissett, J. (1991c). A revision of the genus *Trichoderma*. IV. Additional notes on section Longibrachiatum. Canadian Journal of Botany, 69(11): 2418–2420.

Bissett, J., Gams, W., Jaklitsch, W. and Samuels, G.J. (2015). Accepted *Trichoderma* names in the year 2015. IMA Fungus, 6(2): 263–295.

Blaszczyk, L.M.S.K.S., Siwulski, M., Sobieralski, K., Lisiecka, J. and Jedryczka, M. (2014). *Trichoderma* spp.—application and prospects for use in organic farming and industry. Journal of Plant Protection Research, 54(4).

Bodo, B., Rebuffat, S., El Hajji, M. and Davoust, D. (1985). Structure of trichorzianine A IIIc, an antifungal peptide from *Trichoderma harzianum*. Journal of the American Chemical Society, 107(21): 6011–6017.

Bonnarme, P., Djian, A., Latrasse, A., Féron, G., Ginies, C., Durand, A. and Le Quéré, J.L. (1997). Production of 6-pentyl-α-pyrone by *Trichoderma* spp. from vegetable oils. Journal of Biotechnology, 56(2): 143–150.

Brasier, C.M. (1975). Stimulation of sex organ formation in phytophthora by antagonistic species of *Trichoderma*: I. the effect *in vitro*. New Phytologist, 74(2): 183–194.

Brenner, M.L. (1981). Modern methods for plant growth substance analysis. Annual Review of Plant Physiology, 32(1): 511–538.

Brewer, D., Feicht, A., Taylor, A., Keeping, J.W., Taha, A.A. and Thaller, V. (1982). Ovine ill-thrift in Nova Scotia. 9. Production of experimental quantities of isocyanide metabolites of *Trichoderma hamatum*. Canadian Journal of Microbiology, 28(11): 1252–1260.

Brewer, D., Mason, F.G. and Taylor, A. (1987). The production of alamethicins by *Trichoderma* spp. Canadian Journal of Microbiology, 33(7): 619–625.

Brian, P.W. (1944). Production of gliotoxin by *Trichoderma viride*. Nature, 154(3917): 667–668.

Brian, P.W. and McGowan, J.G. (1945). Viridin: A highly fungistatic substance produced by *Trichoderma viride*. Nature, 156(3953): 144–145.

Brito, J.P., Ramada, M.H., de Magalhães, M.T., Silva, L.P. and Ulhoa, C.J. (2014). Peptaibols from *Trichoderma asperellum* TR356 strain isolated from Brazilian soil. *SpringerPlus*, 3(1): 1–10.

Bruce, A., Austin, W.J. and King, B. (1984). Control of growth of Lentinus lepideus by volatiles from *Trichoderma*. Transactions of the British Mycological Society, 82(3): 423–428.

Brückner, H. and Graf, H. (1983). Paracelsin, a peptide antibiotic containing α-aminoisobutyric acid, isolated from *Trichoderma ressei* Simmons Part A. Experientia, 39(5): 528–530.

Brückner, H. and Przybylski, M. (1984). Isolation and structural characterization of polypeptide antibiotics of the peptaibol class by high-performance liquid chromatography with field desorption and fast atom bombardment mass spectrometry. Journal of Chromatography A, 296: 263–275.

Cai, F., Yu, G., Wang, P., Wei, Z., Fu, L., Shen, Q. and Chen, W. (2013). Harzianolide, a novel plant growth regulator and systemic resistance elicitor from *Trichoderma harzianum*. Plant Physiology and Biochemistry, 73: 106–113.

Caldwell, R. (1958). Fate of spores of *Trichoderma viride* Pers. ex Fr. introduced into soil. Nature, 181(4616): 1144–1145.

Carsolio, C., Gutiérrez, A., Jiménez, B., Van Montagu, M. and Herrera-Estrella, A. (1994). Characterization of ech-42, a *Trichoderma harzianum* endochitinase gene expressed during mycoparasitism. Proceedings of the National Academy of Sciences, 91(23): 10903–10907.

Castillo-Cadena, J., Tenorio-Vieyra, L.E., Quintana-Carabia, A.I., García-Fabila, M.M., Juan, E. and Madrigal-Bujaidar, E. (2006). Determination of DNA damage in floriculturists exposed to mixtures of pesticides. Journal of Biomedicine and Biotechnology, 2006.

Chantrapromma, S., Jeerapong, C., Phupong, W., Quah, C.K. and Fun, H.K. (2014). Trichodermaerin: A diterpene lactone from *Trichoderma asperellum*. Acta Crystallographica Section E: Structure Reports Online, 70(4): o408–o409.

Chen, M., Liu, Q., Gao, S.S., Young, A.E., Jacobsen, S.E. and Tang, Y. (2019). Genome mining and biosynthesis of a polyketide from a biofertilizer fungus that can facilitate reductive iron assimilation in plant. Proceedings of the National Academy of Sciences, 116(12): 5499–5504.

Chet, I. (1987). *Trichoderma*: Application, mode of action, and potential as biocontrol agent of soilborne plant pathogenic fungi. Innovative Approaches to Plant Disease Control, 137–160.

Chet, I., Benhamou, N. and Haran, S. (1998). *Trichoderma* and Gliocladium. In Mycoparasitism and Lytic Enzymes (pp. 153–172). Taylor and Francis London.

Chiang, Y.M., Lee, K.H., Sanchez, J.F., Keller, N.P. and Wang, C.C. (2009). Unlocking fungal cryptic natural products. Natural Product Communications, 4(11): 1934578X0900401113.

Claydon, N., Allan, M., Hanson, J.R. and Avent, A.G. (1987). Antifungal alkyl pyrones of *Trichoderma harzianum*. Transactions of the British Mycological Society, 88(4): 503–513.

Claydon, N., Hanson, J.R., Truneh, A. and Avent, A.G. (1991). Harzianolide, a butenolide metabolite from cultures of *Trichoderma harzianum*. Phytochemistry, 30(11): 3802–3803.

Cleland, R. (1972). The dosage-response curve for auxin-induced cell elongation: A reevaluation. Planta, 104(1): 1–9.

Coats, J.H., Meyer, C.E. and Pyke, T.R. (1971). U.S. Patent No. 3,627,882. Washington, DC: U.S. Patent and Trademark Office.

Collins, R.P. and Halim, A.F. (1972). Characterization of the major aroma constituent of the fungus *Trichoderma viride*. Journal of Agricultural and Food Chemistry, 20(2): 437–438.

Contreras-Cornejo, H.A., Macías-Rodríguez, L., Cortés-Penagos, C. and López-Bucio, J. (2009). *Trichoderma virens*, a plant beneficial fungus, enhances biomass production and promotes lateral root growth through an auxin-dependent mechanism in Arabidopsis. Plant Physiology, 149(3): 1579–1592.

Contreras-Cornejo, H.A., Macías-Rodríguez, L., del-Val, E. and Larsen, J. (2020). Interactions of *Trichoderma* with plants, insects, and plant pathogen microorganisms: Chemical and molecular bases. Co-evolution of Secondary Metabolites, 263–290.

Copping, L.G. and Menn, J.J. (2000). Biopesticides: A review of their action, applications and efficacy. Pest Management Science: Formerly Pesticide Science, 56(8): 651–676.

Corley, D.G., Miller-Wideman, M. and Durley, R.C. (1994). Isolation and structure of harzianum A: A new trichothecene from *Trichoderma harzianum*. Journal of Natural Products, 57(3): 422–425.

Cutler, H.G. and Jacyno, J.M. (1991). Biological activity of (–)-harziano-pyridone isolated from *Trichoderma harzianum*. Agricultural and Biological Chemistry, 55(10): 2629–2631.

Cutler, H.G., Cox, R.H., Crumley, F.G. and Cole, P.D. (1986). 6-Pentyl-α-pyrone from *Trichoderma harzianum*: Its plant growth inhibitory and antimicrobial properties. Agricultural and Biological Chemistry, 50(11): 2943–2945.

Cutler, H.G., Himmelsbach, D.S., Arrendale, R.F., Cole, P.D. and Cox, R.H. (1989). Koninginin A: A novel plant growth regulator from *Trichoderma koningii*. Agricultural and Biological Chemistry, 53(10): 2605–2611.

Cutler, H.G., Jacyno, J.M., Phillips, R.S., VonTersch, R.L., Cole, P.D. and Montemurro, N. (1991). Cyclonerodiol from a novel source, *Trichoderma koningii*: Plant growth regulatory activity. Agricultural and Biological Chemistry, 55(1): 243–244.

Dagenais, T.R. and Keller, N.P. (2009). Pathogenesis of Aspergillus fumigatus in invasive aspergillosis. Clinical Microbiology Reviews, 22(3): 447–465.

Daniel, J. and Filho, E.R. (2007). Peptaibols of *Trichoderma*. Natural Product Reports, 24: 1128–1141.

Daniel, J.F.D.S. and Rodrigues Filho, E. (2007). Peptaibols of *Trichoderma*. Natural Product Reports, 24(5): 1128–1141.

Daoubi, M., Pinedo-Rivilla, C., Rubio, M.B., Hermosa, R., Monte, E., Aleu, J. and Collado, I.G. (2009). Hemisynthesis and absolute configuration of novel 6-pentyl-2H-pyran-2-one derivatives from *Trichoderma* spp. Tetrahedron, 65(25): 4834–4840.

De Palma, M., Salzano, M., Villano, C., Aversano, R., Lorito, M., Ruocco, M., Docimo, T., Piccinelli, A.L., D'Agostino, N. and Tucci, M. (2019). Transcriptome reprogramming, epigenetic modifications and alternative splicing orchestrate the tomato root response to the beneficial fungus *Trichoderma harzianum*. Horticulture Research, 6.

Degenkolb, T., Gräfenhan, T., Nirenberg, H.I., Gams, W. and Brückner, H. (2006). *Trichoderma* brevicompactum complex: Rich source of novel and recurrent plant-protective polypeptide antibiotics (peptaibiotics). Journal of Agricultural and Food Chemistry, 54(19): 7047–7061.

Deguine, J.P., Aubertot, J.N., Flor, R.J., Lescourret, F., Wyckhuys, K.A. and Ratnadass, A. (2021). Integrated pest management: good intentions, hard realities. A review. Agronomy for Sustainable Development, 41(3): 1–35.

Delgado-Jarana, J., Rincon, A.M. and Benítez, T. (2002). Aspartyl protease from *Trichoderma harzianum* CECT 2413: Cloning and characterization the GenBank/EMBL/DDBJ accession number for the sequence reported in this paper is AJ276388. Microbiology, 148(5): 1305–1315.

Demain, A.L. and Fang, A. (2000). The natural functions of secondary metabolites. History of Modern Biotechnology I: 1–39.

Dennis, C. and Webster, J. (1971a). Antagonistic properties of species-groups of *Trichoderma*: I. Production of non-volatile antibiotics. Transactions of the British Mycological Society, 57(1): 25–IN3.

Dennis, C. and Webster, J. (1971b). Antagonistic properties of species-groups of *Trichoderma*: II. Production of volatile antibiotics. Transactions of the British Mycological Society, 57(1): 41–IN4.

Devi, P.I., Thomas, J. and Raju, R.K. (2017). Pesticide consumption in India: A spatiotemporal analysis. Agricultural Economics Research Review, 30(1): 163–172.

Dickinson, J.M., Hanson, J.R., Hitchcock, P.B. and Claydon, N. (1989). Structure and biosynthesis of harzianopyridone, an antifungal metabolite of *Trichoderma harzianum*. Journal of the Chemical Society, Perkin Transactions, 1(11): 1885–1887.

Dickinson, J.M., Hanson, J.R. and Truneh, A. (1995). Metabolites of some biological control agents. Pesticide Science, 44(4): 389–393.

Dini, I., Marra, R., Cavallo, P., Pironti, A., Sepe, I., Troisi, J., Scala, G., Lombari, P. and Vinale, F. (2021). *Trichoderma* strains and metabolites selectively increase the production of volatile organic compounds (VOCs) in Olive trees. Metabolites, 11(4): 213.

Djonović, S., Pozo, M.J., Dangott, L.J., Howell, C.R. and Kenerley, C.M. (2006). Sm1, a proteinaceous elicitor secreted by the biocontrol fungus *Trichoderma virens* induces plant defense responses and systemic resistance. Molecular Plant-microbe Interactions, 19(8): 838–853.

Doan, L.T., El-Hajii, M. and Rebuffat, S. (1986). Fluorescein studies on the interaction of trichorzianine A IIIc with model membranes. Biochimica et Biophysica Acta, 858: 1–5.

Domsch, K.H., Gams, W. and Anderson, T.H. (1980). Compendium of Soil Fungi. Volume 1. Academic Press (London) Ltd.

Druzhinina, I.S., Chenthamara, K., Zhang, J., Atanasova, L., Yang, D., Miao, Y., Rahimi, M.J., Grujic, M., Cai, F., Pourmehdi, S. and Salim, K.A. (2018). Massive lateral transfer of genes encoding plant cell wall-degrading enzymes to the mycoparasitic fungus *Trichoderma* from its plant-associated hosts. PLoS Genetics, 14(4): e1007322.

Druzhinina, I.S., Kopchinskiy, A.G. and Kubicek, C.P. (2006). The first 100 *Trichoderma* species characterized by molecular data. Mycoscience, 47(2): 55–64.

Druzhinina, I.S., Seidl-Seiboth, V., Herrera-Estrella, A., Horwitz, B.A., Kenerley, C.M., Monte, E., Mukherjee, P.K., Zeilinger, S., Grigoriev, I.V. and Kubicek, C.P. (2011). *Trichoderma*: The genomics of opportunistic success. Nature Reviews Microbiology, 9(10): 749–759.

Dubey, S.C., Aradhika, T., Dureja, P. and Grover, A. (2011). Characterization of secondary metabolites and enzymes produced by *Trichoderma* species and their efficacy against plant pathogenic fungi. Indian Journal of Agricultural Sciences, 81(5): 455–461.

Dunlop, R.W., Simon, A., Sivasithamparam, K. and Ghisalberti, E.L. (1989). An antibiotic from *Trichoderma koningii* active against soilborne plant pathogens. Journal of Natural Products, 52(1): 67–74.

Duval, D., Cosette, P., Rebuffat, S., Duclohier, H., Bodo, B. and Molle, G. (1998). Alamethicin-like behaviour of new 18-residue peptaibols, trichorzins PA. Role of the C-terminal amino-alcohol in the ion channel forming activity. Biochimica et Biophysica Acta (BBA)-Biomembranes, 1369(2): 309–319.

Edenborough, M.S. and Herbert, R.B. (1988). Naturally occurring isocyanides. Natural Product Reports, 5(3): 229–245.

El-Hajji, M., Rebuffat, S., Le Doan, T., Klein, G., Satre, M. and Bodo, B. (1989). Interaction of trichorzianines A and B with model membranes and with the amoeba Dictyostelium. Biochimica et Biophysica Acta (BBA)-Biomembranes, 978(1): 97–104.

El-Hasan, A. and Buchenauer, H. (2009). Actions of 6-pentyl-alpha-pyrone in controlling seedling blight incited by Fusarium moniliforme and inducing defence responses in maize. Journal of Phytopathology, 157(11-12): 697–707.

El-Hasan, A., Walker, F. and Buchenauer, H. (2008). *Trichoderma harzianum* and its metabolite 6-pentyl-alpha-pyrone suppress fusaric acid produced by *Fusarium moniliforme*. Journal of Phytopathology, 156(2): 79–87.

Enari, T.M. and Niku-Paavola, M.L. (1987). Enzymatic hydrolysis of cellulose: Is the current theory of the mechanisms of hydrolysis valid? Critical Reviews in Biotechnology, 5(1): 67–87.

Evidente, A., Cabras, A., Maddau, L., Serra, S., Andolfi, A. and Motta, A. (2003). Viridepyronone, a new antifungal 6-substituted 2 h-pyran-2-one produced by *Trichoderma viride*. Journal of Agricultural and Food Chemistry, 51(24): 6957–6960.

Faull, J.L., Graeme-Cook, K.A. and Pilkington, B.L. (1994). Production of an isonitrile antibiotic by an UV-induced mutant of *Trichoderma harzianum*. Phytochemistry, 36(5): 1273–1276.

Fedorinchik, N.S., Tarunina, T.A., Tyutyunnikov, M.G. and Kudryatseva, K.I. (1975). Trichodermin-4, a new biological preparation for plant disease control. Plant Prot., 3: 67–72.

Fuji, K., Fujita, E., Takaishi, Y., Fujita, T., Arita, I., Komatsu, M. and Hiratsuka, N. (1978). New antibiotics, trichopolyns A and B: Isolation and biological activity. Experientia, 34(2): 237–239.

Fujita, T., Wada, S.I., Iida, A., Nishimura, T., Kanai, M. and Toyama, N. (1994). Fungal metabolites. XIII. Isolation and structural elucidation of new peptaibols, trichodecenins-I and-II, from *Trichoderma viride*. Chemical and Pharmaceutical Bulletin, 42(3): 489–494.

Fujiwara, A., Okuda, T., Masuda, S., Shiomi, Y., Miyamoto, C., Sekine, Y., Tazoe, M. and Fujiwara, M. (1982). Fermentation, isolation and characterization of isonitrile antibiotics. Agricultural and Biological Chemistry, 46(7): 1803–1809.

Gams, W. and Bisset, J. (1998). Morphology and identification of *Trichoderma*. pp. 3–34. *In*: G.E. Harman and C.P. Kubicek (eds.). *Trichoderma* and Gliocladium. London, UK: Taylor and Francis.

García, I., Lora, J.M., de la Cruz, J., Benítez, T., Llobell, A. and Pintor-Toro, J.A. (1994). Cloning and characterization of a chitinase (CHIT42) cDNA from the mycoparasitic fungus *Trichoderma harzianum*. Current Genetics, 27(1): 83–89.

Gerhardson, B. (2002). Biological substitutes for pesticides. Trends in Biotechnology, 20(8): 338–343.

Ghisalberti, E.L. (1993). Detection and isolation of bioactive natural products. pp. 15–18. *In*: S.M. Colegate and R.J. Molyneux (eds.). Bioactive Natural Products: Detection, Isolation and Structure Elucidation. CRC Press, Boca Raton.

Ghisalberti, E.L. and Rowland, C.Y. (1993). Antifungal metabolites from *Trichoderma harzianum*. Journal of Natural Products, 56(10): 1799–1804.

Ghisalberti, E.L. and Sivasithamparam, K. (1991). Antifungal antibiotics produced by *Trichoderma* spp. Soil Biology and Biochemistry, 23(11): 1011–1020.

Ghisalberti, E.L., Narbey, M.J., Dewan, M.M. and Sivasithamparam, K. (1990). Variability among strains of *Trichoderma harzianum* in their ability to reduce take-all and to produce pyrones. Plant and Soil, 121(2): 287–291.

Godtfredsen, W.O. and Vangedal, S. (1965). Trichodermin, a new sesquiterpen antibiotic. Acta Chemica Scandinavica, 19(5): 1088–1102.

Goulard, C., Hlimi, S., Rebuffat, S. and Bodo, B. (1995). Trichorzins HA and MA, antibiotic peptides from *Trichoderma harzianum* I. Fermentation, isolation and biological properties. The Journal of Antibiotics, 48(11): 1248–1253.

Govind, S.R., Jogaiah, S., Abdelrahman, M., Shetty, H.S. and Tran, L.S.P. (2016). Exogenous trehalose treatment enhances the activities of defense-related enzymes and triggers resistance against downy mildew disease of pearl millet. Frontiers in Plant Science, 7: 1593.

Grove, J.F., McCloskey, P. and Moffatt, J.S. (1966). Viridin. Part V. Structure. Journal of the Chemical Society C: Organic, 743–747.

Gutiérrez, S., Casquero, P.A., Porteous-Álvarez, A.J., Mayo-Prieto, S., Carro-Huerga, G., Rodríguez-González, Á., Álvarez-García, S., del Ser, S., Lorenzana, A., Campelo, M.P. and Maldonado-González, M.M. (2021, March). *Trichoderma species* isolated from hop soils in the Órbigo valley, León, Spain. In V International Humulus Symposium 1328: 63–66.

Hahn, S., Zhong, X.Y. and Holzgreve, W. (2002). Single cell PCR in laser capture microscopy. Methods Enzymology, 356: 295–301.

Hajji, M.E., Rebuffat, S., Lecommandeur, D. and Bodo, B. (1987). Isolation and sequence determination of trichorzianines A antifungal peptides from *Trichoderma harzianum*. International Journal of Peptide and Protein Research, 29(2): 207–215.

Hamilton, D., Ambrus, A., Dieterle, R., Felsot, A., Harris, C., Petersen, B., Racke, K., Wong, S.S., Gonzalez, R., Tanaka, K. and Earl, M. (2004). Pesticide residues in food—acute dietary exposure. Pest Management Science: Formerly Pesticide Science, 60(4): 311–339.

Haran, S., Schickler, H., Oppenheim, A. and Chet, I. (1996). Differential expression of *Trichoderma harzianum* chitinases during mycoparasitism. Phytopathology, 86(9): 980–985.

Harman, G.E. (2000). Myths and dogmas of biocontrol changes in perceptions derived from research on *Trichoderma* harzinum T-22. Plant Disease, 84(4): 377–393.

Harman, G.E. (2006). Overview of mechanisms and uses of *Trichoderma* spp. Phytopathology, 96(2): 190–194.

Harman, G.E., Hayes, C.K., Lorito, M., Broadway, R.M., Di Pietro, A., Peterbauer, C. and Tronsmo, A. (1993). Chitinolytic enzymes of *Trichoderma harzianum*: Purification of chitobiosidase and endochitinase. Phytopathology, 83(3): 313–318.

Harris, D., Pacovsky, R.S. and Paul, E.A. (1985). Carbon economy of soybean–Rhizobium–Glomus associations. New Phytologist, 101(3): 427–440.

Henríquez-Urrutia, M., Spanner, R., Olivares-Yánez, C., Seguel-Avello, A., Pérez-Lara, R., Guillén-Alonso, H., Winkler, R., Herrera-Estrella, A., Canessa, P. and Larrondo, L.F. (2022). Circadian oscillations in *Trichoderma atroviride* and the role of core clock components in secondary metabolism, development, and mycoparasitism against the phytopathogen Botrytis cinerea. Elife, 11: e71358.

Hermosa, R., Rubio, M.B., Cardoza, R.E., Nicolás, C., Monte, E. and Gutiérrez, S. (2013). The contribution of *Trichoderma* to balancing the costs of plant growth and defense. Int. Microbiol., 16(2): 69–80.

Hernández-Castillo, F.D., Castillo-Reyes, F., Tucuch-Pérez, M.A. and Arredondo-Valdes, R. (2020). Biological efficacy of *Trichoderma* spp. and Bacillus spp. in the management of plant diseases. In Organic Agriculture. IntechOpen.

Herrera-Estrella, A. and Chet, I. (2004). The biological control agent *Trichoderma* – From fundamentals to applications. pp. 147–156. *In*: D.K. Arora and M. Dekker (eds.). Fungal Biotechnology in Agricultural, Food and Environmental Applications. 21st ed. New York, USA: CRC Press.

Hider, R.C. and Kong, X. (2010). Chemistry and biology of siderophores. Natural Product Reports, 27(5): 637–657.

Hill, R.A., Cutler, H.G. and Parker, S.R. (1995). *Trichoderma* and metabolites as control agents for microbial plant diseases. PCT Int. Appl., 9520879(10).

Hjeljord, L. and Tronsomo, A. (1988) *Trichoderma* and Gliocladium in biological control: An overview. pp. 13–151. *In*: G.E. Harman and C.P. Kubicek (eds.). *Trichoderma* and Gliocladium. Taylor and Francis, London.

Horvath, E.M., Bürgel, J.L. and Messner, K. (1995). The production of soluble antifungal metabolites by the biocontrol fungus *Trichoderma harzianum* in connection with the formation of conidiospores. Material und Organismen (Germany).

Howell, C.R. (1991). Biological control of Pythium damping-off of cotton with seed-coating preparations of *Gliocladium virens*. Phytopathology, 81(7): 738–741.

Howell, C.R. (1998). The role of antibiosis in biocontrol. pp. 173–183. *In*: G.E. Harman and C.P. Kubicek (eds.). *Trichoderma* and Gliocladium, Vol 2, Enzymes, Biological Control and Commercial Application. Taylor and Francis Ltd., London.

Howell, C.R. (2003). Mechanisms employed by *Trichoderma* species in the biological control of plant diseases: The history and evolution of current concepts. Plant Disease, 87(1): 4–10.

Howell, C.R. and Stipanovic, R.D. (1994). Effect of sterol biosynthesis inhibitors on phytotoxin (viridiol) production by *Gliocladium virens* in culture. Phytopathology 84: 969–972.

Howell, C.R., Hanson, L.E., Stipanovic, R.D. and Puckhaber, L.S. (2000). Induction of terpenoid synthesis in cotton roots and control of *Rhizoctonia solani* by seed treatment with *Trichoderma virens*. Phytopathology, 90(3): 248–252.

Howell, C.R., Stipanovic, R.D. and Lumsden, R.D. (1993). Antibiotic production by strains of *Gliocladium virens* and its relation to the biocontrol of cotton seedling diseases. Biocontrol Science and Technology, 3(4): 435–441.

Hoyos-Carvajal, L., Orduz, S. and Bissett, J. (2009). Growth stimulation in bean (*Phaseolus vulgaris* L.) by *Trichoderma*. Biological Control, 51(3): 409–416.

Huang, Q., Tezuka, Y., Hatanaka, Y., Kikuchi, T., Nishi, A. and Tubaki, K. (1995). Studies on metabolites of mycoparasitic fungi. IV. Minor peptaibols of *Trichoderma koningii*. Chemical and Pharmaceutical Bulletin, 43(10): 1663–1667.

Iida, A.S., Shingu, U.T., Okuda, M. et al. (1993) Fungal metabolites. Part 6. Nuclear magnetic study of antibiotic peptides, trichosporin Bs, from *Trichoderma polysporum*. Journal of the Chemical Society, Perkin Transactions, 1: 367–373.

Iida, A., Sanekata, M., Fujita, T., Tanaka, H., Enoki, A., Fuse, G., Kanai, M., Rudewicz, P.J. and Tachikawa, E. (1994). Fungal metabolites. XVI. Structures of new peptaibols, trichokindins I–VII, from the fungus *Trichoderma harzianum*. Chemical and Pharmaceutical Bulletin, 42(5): 1070–1075.

Iida, J., Iida, A., Takahashi, Y., Takaishi, Y., Nagaoka, Y. and Fujita, T. (1993). Fungal metabolites. Part 5. Rapid structure elucidation of antibiotic peptides, minor components of trichosporin Bs from *Trichoderma polysporum*. Application of linked-scan and continuous-flow fast-atom bombardment mass spectrometry. Journal of the Chemical Society, Perkin Transactions, 1(3): 357–365.

Irmscher, G. and Jung, G. (1977). Die hämolytischen Eigenschaften der membranmodifizierenden Peptidantibiotika Alamethicin, Suzukacillin und Trichotoxin. European Journal of Biochemistry, 80(1): 165–174.

Jayalakshmi, R., Oviya, R., Premalatha, K., Mehetre, S.T., Paramasivam, M., Kannan, R., Theradimani, M., Pallavi, M.S., Mukherjee, P.K. and Ramamoorthy, V. (2021). Production, stability and degradation of *Trichoderma* gliotoxin in growth medium, irrigation water and agricultural soil. Scientific Reports, 11(1): 1–14.

Jeleń, H., Błaszczyk, L., Chełkowski, J., Rogowicz, K. and Strakowska, J. (2014). Formation of 6-n-pentyl-2H-pyran-2-one (6-PAP) and other volatiles by different *Trichoderma* species. Mycological Progress, 13(3): 589–600.

Jørs, E., Morant, R.C., Aguilar, G.C., Huici, O., Lander, F., Bælum, J. and Konradsen, F. (2006). Occupational pesticide intoxications among farmers in Bolivia: A cross-sectional study. Environmental Health, 5(1): 1–9.

Jung, G., König, W.A., Leibfritz, D., Ooka, T., Janko, K. and Boheim, G. (1976). Structural and membrane modifying properties of suzukacillin, a peptide antibiotic related to alamethicin: Part A. Sequence and conformation. Biochimica et Biophysica Acta (BBA)-Biomembranes, 433(1): 164–181.

Kamal, A., Akhtar, R. and Qureshi, A.A. (1971). Biochemistry of microorganisms. 2, 5-Dimethoxybenzoquinone, tartronic acid, itaconic acid, succinic acid, pyrocalciferol, epifriedelinol, lanosta-7, 9 (11), 24-triene-3, 6, 21-diol, trichodermene A, methyl 2, 4, 6-octatriene and cordycepic acid, *Trichoderma* metabolites. Pak. Sci. Ind. Res., 14: 71–78.

Kawada, M., Yoshimoto, Y., Kumagai, H., Someno, T., Momose, I., Kawamura, N., Isshiki, K. and Ikeda, D. (2004). PP2A inhibitors, harzianic acid and related compounds produced by fungus strain F-1531. The Journal of Antibiotics, 57(3): 235–237.

Keswani, C., Mishra, S., Sarma, B.K., Singh, S.P. and Singh, H.B. (2014). Unraveling the efficient applications of secondary metabolites of various *Trichoderma* spp. Applied Microbiology and Biotechnology, 98(2): 533–544.

Keszler, Á., Forgács, E., Kótai, L., Vizcaíno III, J.A., Monte, E. and García-Acha, I. (2000). Separation and identification of volatile components in the fermentation broth of *Trichoderma atroviride* by solid-phase extraction and gas chromatography—mass spectrometry. Journal of Chromatographic Science, 38(10): 421–424.

Khan, R.A.A., Najeeb, S., Hussain, S., Xie, B. and Li, Y. (2020). Bioactive secondary metabolites from *Trichoderma* spp. against phytopathogenic fungi. Microorganisms, 8(6): 817.

Kleinkauf, H. and Rindfleisch, H. (1975). Non-ribosomal biosynthesis of the cyclic octadecapeptide alamethicin. Acta Microbiologica Academiae Scientiarum Hungaricae, 22(4): 411–418.

Kobayashi, M., Uehara, H., Matsunami, K., Aoki, S. and Kitagawa, I. (1993). Trichoharzin, a new polyketide produced by the imperfect fungus *Trichoderma harzianum* separated from the marine sponge Micale cecilia. Tetrahedron Letters, 34(49): 7925–7928.

Kono, Y., Knoche, H.W. and Daly, J.M. (1981). Structure of Host–specific toxin. Toxins in Plant Diseases, 221–257.

Korpi, A., Järnberg, J. and Pasanen, A.L. (2009). Microbial volatile organic compounds. Critical Reviews in Toxicology, 39(2): 139–193.

Kubicek, C.P., Herrera-Estrella, A., Seidl-Seiboth, V., Martinez, D.A., Druzhinina, I.S., Thon, M., Zeilinger, S., Casas-Flores, S., Horwitz, B.A., Mukherjee, P.K. and Mukherjee, M. (2011). Comparative genome sequence analysis underscores mycoparasitism as the ancestral life style of *Trichoderma*. Genome Biology, 12(4): 1–15.

Kullnig-Gradinger, C.M., Szakacs, G. and Kubicek, C.P. (2002). Phylogeny and evolution of the genus *Trichoderma*: A multigene approach. Mycological Research, 106(7): 757–767.

Kumar, A., Scher, K., Mukherjee, M., Pardovitz-Kedmi, E., Sible, G.V., Singh, U.S., Kale, S.P., Mukherjee, P.K. and Horwitz, B.A. (2010). Overlapping and distinct functions of two *Trichoderma* virens MAP kinases in cell-wall integrity, antagonistic properties and repression of conidiation. Biochemical and Biophysical Research Communications, 398(4): 765–770.

Kumar, N. and Khurana, S.M.P. (2019). Pathogen and management of fungal wilt of banana through biocontrol agents. *In*: Plant Microbe Interface (pp. 177–194). Springer, Cham.

Kumar, N. and Khurana, S.M. (2018.) Biocontrol of seed borne Fungi of Mung Bean, *Vigna radiata* (L.) Wiczek. Nat Con. on Diversity and Utilization of Tropical Plants (NCDUTP 22–23 Feb, 2018). ABSTRACT NO-88.

Kumar, N. and Khurana, S.P. (2016). Biomanagement of wilting of a valuable timber and medicinal plant of Shisham (Dalbergia sissoo Roxb.)—A review. International Journal of Current Microbiology and Applied Sciences, 5(1): 32–54.

Kumar, N. and Khurana, S.P. (2021). *Trichoderma*-plant-pathogen interactions for benefit of agriculture and environment. In Biocontrol Agents and Secondary Metabolites (pp. 41–63). Woodhead Publishing.

Kumar, N. and Paul Khurana, S.M. (2019). Pathogen and management of fungal wilt of banana through biocontrol agents. In Plant Microbe Interface (pp. 177–194). Springer, Cham.

Kumeda, Y., Asao, T., Iida, A., Wada, S., Futami, S. and Fujita, T. (1995). Effects of Ergokonin A Produced by *Trichoderma viride* on Growth and Morphological Development of Fungi. *In*: Proceedings-osaka prefectural institute of public health edition of food sanitation. Osaka Prefectural Institute of Public Health, 26: 95–95.

Lang, B.Y., Li, J., Zhou, X.X., Chen, Y.H., Yang, Y.H., Li, X.N., Zeng, Y. and Zhao, P.J. (2015). Koninginins L and M, two polyketides from *Trichoderma koningii* 8662. Phytochemistry Letters, 11: 1–4.

Le Doan, T., El Hajji, M., Rebuffat, S., Rajesvari, M.R. and Bodo, B. (1986). Fluorescence studies of the interaction of trichorzianine A IIIc with model membranes. Biochimica et Biophysica Acta (BBA)-Biomembranes, 858(1): 1–5.

Leclerc, G., Rebuffat, S., Goulard, C. and Bodo, B. (1998). Directed biosynthesis of peptaibol antibiotics in two *Trichoderma* strains I. Fermentation and isolation. The Journal of Antibiotics, 51(2): 170–177.

Lehner, S.M., Atanasova, L., Neumann, N.K., Krska, R., Lemmens, M., Druzhinina, I.S. and Schuhmacher, R. (2013). Isotope-assisted screening for iron-containing metabolites reveals a high degree of diversity among known and unknown siderophores produced by *Trichoderma* spp. Applied and Environmental Microbiology, 79(1): 18–31.

Liss, S.N., Brewer, D., Taylor, A. and Jones, G.A. (1985). Antibiotic activity of an isocyanide metabolite of *Trichoderma hamatum* against rumen bacteria. Canadian Journal of Microbiology, 31(9): 767–772.

Lopes, F.A.C., Steindorff, A.S., Geraldine, A.M., Brandão, R.S., Monteiro, V.N., Júnior, M.L., Coelho, A.S.G., Ulhoa, C.J. and Silva, R.N. (2012). Biochemical and metabolic profiles of *Trichoderma* strains isolated from common bean crops in the Brazilian Cerrado, and potential antagonism against Sclerotinia sclerotiorum. Fungal Biology, 116(7): 815–824.

Lorito, M., Farkas, V., Rebuffat, S., Bodo, B. and Kubicek, C.P. (1996). Cell wall synthesis is a major target of mycoparasitic antagonism by *Trichoderma harzianum*. Journal of Bacteriology, 178(21): 6382–6385.

Lorito, M., Hayes, C.K., Di Pietro, A., Woo, S.L. and Harman, G.E. (1994). Purification, characterization, and synergistic activity of a glucan 1,3-beta-glucosidase and an N-acetyl-beta-glucosaminidase from *Trichoderma harzianum*. Phytopathology (USA), 84(4): 398–405.

Lorito, M., Woo, S.L., Fernandez, I.G., Colucci, G., Harman, G.E., Pintor-Toro, J.A., Filippone, E., Muccifora, S., Lawrence, C.B., Zoina, A. and Tuzun, S. (1998). Genes from mycoparasitic fungi as a source for improving plant resistance to fungal pathogens. Proceedings of the National Academy of Sciences, 95(14): 7860–7865.

Lynch, J.M. (1990). Fungi as antagonists. New directions in biological control. Alternatives for suppressing agricultural pests and diseases. Proceedings of a UCLA colloquim held at Frisco, Colorado, January 20–27, 1989: 243–253.

Macías, F.A., Varela, R.M., Simonet, A.M., Cutler, H.G., Cutler, S.J., Eden, M.A. and Hill, R.A. (2000). Bioactive Carotanes from *Trichoderma virens*. Journal of Natural Products, 63(9): 1197–1200.

Mahato, D. (2021). *Trichoderma*'s contribution in environmentally friendly plant disease management. Biotica Research Today, 3(7): 591–594.

Malmierca, M.G., McCormick, S.P., Cardoza, R.E., Monte, E., Alexander, N.J. and Gutiérrez, S. (2015). Trichodiene production in a *Trichoderma harzianum* erg1-silenced strain provides evidence of the importance of the sterol biosynthetic pathway in inducing plant defense-related gene expression. Molecular Plant-Microbe Interactions, 28(11): 1181–1197.

Mandels, M. (1975). Microbial sources of cellulase. In Biotechnology and Bioengineering Symposium, 5: 81–105.

Manganiello, G., Sacco, A., Ercolano, M.R., Vinale, F., Lanzuise, S., Pascale, A., Napolitano, M., Lombardi, N., Lorito, M. and Woo, S.L. (2018). Modulation of tomato response to *Rhizoctonia solani* by *Trichoderma harzianum* and its secondary metabolite harzianic acid. Frontiers in Microbiology, 9: 1966.

Marfori, E.C., Kajiyama, S.I., Fukusaki, E.I. and Kobayashi, A. (2002). Trichosetin, a novel tetramic acid antibiotic produced in dual culture of *Trichoderma harzianum* and *Catharanthus roseus* callus. Zeitschrift für Naturforschung C, 57(5-6): 465–470.

Marfori, E.C., Kajiyama, S.I., Fukusaki, E.I. and Kobayashi, A. (2003). Phytotoxicity of the tetramic acid metabolite trichosetin. Phytochemistry, 62(5): 715–721.

Maroni, M., Fanetti, A.C. and Metruccio, F. (2006). Risk assessment and management of occupational exposure to pesticides in agriculture. La Medicina del lavoro, 97(2): 430–437.

Marra, R., Ambrosino, P., Carbone, V., Vinale, F., Woo, S.L., Ruocco, M., Ciliento, R., Lanzuise, S., Ferraioli, S., Soriente, I. and Gigante, S. (2006). Study of the three-way interaction between *Trichoderma atroviride*, plant and fungal pathogens by using a proteomic approach. Current Genetics, 50(5): 307–321.

Mathivanan, N., Prabavathy, V.R. and Vijayanandraj, V.R. (2008). The effect of fungal secondary metabolites on bacterial and fungal pathogens. Secondary Metabolites in Soil Ecology, 129–140.

McCauley, C.D., Drath, W.H., Palus, C.J., O'Connor, P.M. and Baker, B.A. (2006). The use of constructive-developmental theory to advance the understanding of leadership. The Leadership Quarterly, 17(6): 634–653.

Mcintyre, M., Nielsen, J., Arnau, J., Brink, H., Hansen, K. and Madrid, S. (2004). Proceedings of the 7th European Conference on Fungal Genetics. Copenhagen.

Metwally, R.A., Abdelhameed, R.E., Soliman, S.A. and Al-Badwy, A.H. (2022). Potential use of beneficial fungal microorganisms and C-phycocyanin extract for enhancing seed germination, seedling growth and biochemical traits of Solanum lycopersicum L. BMC Microbiology, 22(1): 1–17.

Meyer, C.E. (1966). U-21,963, a new antibiotic: II. Isolation and characterization. Applied Microbiology, 14(4): 511–512.

Meyer, C.E. and Reusser, F. (1967). A polypeptide antibacterial agent isolated from *Trichoderma viride*. Experientia, 23(2): 85–86.

Moffatt, J.S., Bu'Lock, J.D. and Yuen, T.H. (1969). Viridiol, a steroid-like product from *Trichoderma viride*. Journal of the Chemical Society D: Chemical Communications, (14): 839a–839a.

Mohamed-Benkada, M., Montagu, M., Biard, J.F., Mondeguer, F., Verite, P., Dalgalarrondo, M., Bissett, J. and Pouchus, Y.F. (2006). New short peptaibols from a marine *Trichoderma* strain. Rapid Communications in Mass Spectrometry: An International Journal Devoted to the Rapid Dissemination of Up-to-the-Minute Research in Mass Spectrometry, 20(8): 1176–1180.

Molle, G., Duclohier, H. and Spach, G. (1987). Voltage-dependent and multi-state ionic channels induced by trichorzianines, anti-fungal peptides related to alamethicin. FEBS Letters, 224(1): 208–212.

Monte, E. (2001). Understanding *Trichoderma*: Between biotechnology and microbial ecology. International Microbiology, 4(1): 1–4.

Morais, E.M., Silva, A.A.R., Sousa, F.W.A.D., Azevedo, I.M.B.D., Silva, H.F., Santos, A.M.G., Beserra Júnior, J.E.A., Carvalho, C.P.D., Eberlin, M.N., Porcari, A.M. and Araújo, F.D.D.S. (2022). Endophytic *Trichoderma* strains isolated from forest species of the Cerrado-Caatinga ecotone are potential biocontrol agents against crop pathogenic fungi. Plos One, 17(4): e0265824.

Moss, M.O., Jackson, R.M. and Rogers, D. (1975). The characterization of 6-(pent-1-enyl)-α-pyrone from *Trichoderma viride*. Phytochemistry, 14(12): 2706–2708.

Mukherjee, P.K. (2011). Genomics of biological control-whole genome sequencing of two mycoparasitic *Trichoderma* spp. Current Science, 101(3).

Mukherjee, P.K., Buensanteai, N., Moran-Diez, M.E., Druzhinina, I.S. and Kenerley, C.M. (2012). Functional analysis of non-ribosomal peptide synthetases (NRPSs) in *Trichoderma virens* reveals a polyketide synthase (PKS)/NRPS hybrid enzyme involved in the induced systemic resistance response in maize. Microbiology, 158(1): 155–165.

Mukherjee, P.K., Horwitz, B.A. and Kenerley, C.M. (2012). Secondary metabolism in *Trichoderma*–A genomic perspective. Microbiology, 158(1): 35–45.

Mukherjee, P.K., Hurley, J.F., Taylor, J.T., Puckhaber, L., Lehner, S., Druzhinina, I., Schumacher, R. and Kenerley, C.M. (2018). Ferricrocin, the intracellular siderophore of *Trichoderma virens*, is involved in growth, conidiation, gliotoxin biosynthesis and induction of systemic resistance in maize. Biochemical and Biophysical Research Communications, 505(2): 606–611.

Mukherjee, P.K., Latha, J., Hadar, R. and Horwitz, B.A. (2004). Role of two G-protein alpha subunits, TgaA and TgaB, in the antagonism of plant pathogens by *Trichoderma viren*s. Applied and Environmental Microbiology, 70(1): 542–549.

Mukherjee, P.K., Wiest, A., Ruiz, N., Keightley, A., Moran-Diez, M.E., McCluskey, K., Pouchus, Y.F. and Kenerley, C.M. (2011). Two classes of new peptaibols are synthesized by a single non-ribosomal peptide synthetase of *Trichoderma virens*. Journal of Biological Chemistry, 286(6): 4544–4554.

Mukhopadhyay, R. and Kumar, D. (2020). *Trichoderma*: a beneficial antifungal agent and insights into its mechanism of biocontrol potential. Egyptian Journal of Biological Pest Control, 30(1): 1–8.

Mulatu, A., Megersa, N., Abena, T., Kanagarajan, S., Liu, Q., Tenkegna, T.A. and Vetukuri, R.R. (2022). Biodiversity of the genus *Trichoderma* in the rhizosphere of coffee (*Coffea arabica*) plants in Ethiopia and their potential use in biocontrol of coffee wilt disease. Crops, 2(2): 120–141.

Müller, A., Faubert, P., Hagen, M., Zu Castell, W., Polle, A., Schnitzler, J.P. and Rosenkranz, M. (2013). Volatile profiles of fungi–chemotyping of species and ecological functions. Fungal Genetics and Biology, 54: 25–33.

Nandini, B., Geetha, N., Prakash, H.S. and Hariparsad, P. (2021). Natural uptake of anti-oomycetes *Trichoderma* produced secondary metabolites from pearl millet seedlings—A new mechanism of biological control of downy mildew disease. Biological Control, 156: 104550.

Nawaz, S., Subhani, M.N., Chattha, M.B., Saleem, Y., Abidi, S.H., Shahzad, K., Saeed, S., Syed, Q., Irfan, M. and Ambreen, A. (2021). Fungal isolates of genus *Trichoderma* induce wilt resistance to pea caused by *Fusarium oxysporum* f. sp. *pisi* through competitive inhibition. Revista Mexicana De Ingeniería Química, 20(3): Bio2475–Bio2475.

Nene, Y.L. and Thapliyal, P.N. (1993). Evaluation of fungicides *In*: Fungicides in plant disease control. International Science Publisher, New York, 531.

New, A.P., Eckers, C., Haskins, N.J., Neville, W.A., Elson, S., Hueso-Rodríguez, J.A. and Rivera-Sagredo, A. (1996). Structures of polysporins AD, four new peptaibols isolated from *Trichoderma polysporum*. Tetrahedron Letters, 37(17): 3039–3042.

Ni, M., Wu, Q., Wang, G.S., Liu, Q.Q., Yu, M.X. and Tang, J. (2019). Analysis of metabolic changes in *Trichoderma asperellum* TJ01 at different fermentation time-points by LC-QQQ-MS. Journal of Environmental Science and Health, Part B, 54(1): 20–26.

Nogueira-Lopez, G., Greenwood, D.R., Middleditch, M., Winefield, C., Eaton, C., Steyaert, J.M. and Mendoza-Mendoza, A. (2018). The apoplastic secretome of *Trichoderma virens* during interaction with maize roots shows an inhibition of plant defence and scavenging oxidative stress secreted proteins. Frontiers in Plant Science, 9: 409.

Nuansri, S., Rukachaisirikul, V., Rungwirain, N., Kaewin, S., Yimnual, C., Phongpaichit, S., Preedanon, S., Sakayaroj, J. and Muanprasat, C. (2021). α-Pyrone and decalin derivatives from the marine-derived fungus *Trichoderma harzianum* PSU-MF79. Natural Product Research, 36(21): 5462–5469.

Oh, S.U., Lee, S.J., Kim, J.H. and Yoo, I.D. (2000). Structural elucidation of new antibiotic peptides, atroviridins A, B and C from *Trichoderma atroviride*. Tetrahedron Letters, 41(1): 61–64.

Omann, M.R., Lehner, S., Rodríguez, C.E., Brunner, K. and Zeilinger, S. (2012). The seven-transmembrane receptor Gpr1 governs processes relevant for the antagonistic interaction of *Trichoderma atroviride* with its host. Microbiology, 158(Pt 1): 107.

Ordentlich, A., Wiesman, Z., Gottlieb, H.E., Cojocaru, M. and Chet, I. (1992). Inhibitory furanone produced by the biocontrol agent *Trichoderma harzianum*. Phytochemistry, 31(2): 485–486.

Pakdaman, N., Ghaderian, S.M., Ghasemi, R. and Asemaneh, T. (2013). Effects of calcium/magnesium quotients and nickel in the growth medium on growth and nickel accumulation in Pistacia atlantica. Journal of Plant Nutrition, 36(11): 1708–1718.

Pandey, A.K., Burlakoti, R.R., Kenyon, L. and Nair, R.M. (2018). Perspectives and challenges for sustainable management of fungal diseases of mung bean [*Vigna radiata* (L.) R. Wilczek var. radiata]: A review. Frontiers in Environmental Science, 6: 53.

Pandey, N., Adhikhari, M. and Bhantana, B. (2019). *Trichoderma* and its prospects in agriculture of Nepal: An overview. International Journal of Applied Sciences and Biotechnology, 7(3): 309–316.

Pandey, R.C., Cook Jr, J.C. and Rinehart Jr, K.L. (1977). High resolution and field desorption mass spectrometry studies and revised structures of alamethicins I and II. Journal of the American Chemical Society, 99(26): 8469–8483.

Papavizas, G.C. (1985). *Trichoderma* and Gliocladium: Biology, ecology, and potential for biocontrol. Annual Review of Phytopathology, 23(1): 23–54.

Parker, S.R., Cutler, H.G. and Schrelner, P.R. (1995a). Koninginin C: A biologically active natural product from *Trichoderma koningii*. Bioscience, Biotechnology, and Biochemistry, 59(6): 1126–1127.

Parker, S.R., Cutler, H.G. and Schreiner, P.R. (1995b). Koninginin E: Isolation of a biologically active natural product from *Trichoderma koningii*. Bioscience, Biotechnology, and Biochemistry, 59(9): 1747–1749.

Parker, S.R., Cutler, H.G., Jacyno, J.M. and Hill, R.A. (1997). Biological activity of 6-pentyl-2 H-pyran-2-one and its analogs. Journal of Agricultural and Food Chemistry, 45(7): 2774–2776.

Patil, A.S. and Lunge, A.G. (2012). Strain improvement of *Trichoderma harzianum* by UV mutagenesis for enhancing its biocontrol potential against aflotoxigenic Aspergillus species. Experiment, 4(2): 228–242.

Patil, A.S., Patil, S.R. and Paikrao, H.M. (2016). *Trichoderma* secondary metabolites: Their biochemistry and possible role in disease management. In Microbial-mediated Induced Systemic Resistance in Plants (pp. 69–102). Springer, Singapore.

Perazzolli, M., Palmieri, M.C., Matafora, V., Bachi, A. and Pertot, I. (2016). Phosphoproteomic analysis of induced resistance reveals activation of signal transduction processes by beneficial and pathogenic interaction in grapevine. Journal of Plant Physiology, 195: 59–72.

Persoon, C.H. (1794). Disposita methodical fungorum. Romers Neues Mag. Bot., 1: 81–128.

Pittendrigh, B.R. and Gaffney, P.J. (2001). Pesticide resistance: can we make it a renewable resource? Journal of Theoretical Biology, 211(4): 365–375.

Polizzi, V., Adams, A., Malysheva, S.V., De Saeger, S., Van Peteghem, C., Moretti, A., Picco, A.M. and De Kimpe, N. (2012). Identification of volatile markers for indoor fungal growth and chemotaxonomic classification of *Aspergillus* species. Fungal Biology, 116(9): 941–953.

Pratt, B.H., Sedgley, J.H., Heather, W.A. and Shepherd, C.J. (1972). Oospore production in *Phytophthora cinnamomi* in the presence of *Trichoderma koningii*. Australian Journal of Biological Sciences, 25(4): 861–864.

Pyke, T.R. and Dietz, A. (1966). U-21,963, a New Antibiotic: I. Discovery and Biological Activity. Applied Microbiology, 14(4): 506–510.

Rahman, M.A., Begum, M.F. and Alam, M.F. (2009). Screening of *Trichoderma* isolates as a biological control agent against Ceratocystis paradoxa causing pineapple disease of sugarcane. Mycobiology, 37(4): 277–285.

Rajput, R.B., Solanky, K.U., Prajapati, V.P., Pawar, D.M. and Suradkar, S.R. (2013). Effect of fungal and bacterial bioagents against *Alternaria alternata* (fr.) Keissler *in vitro* condition. The Bioscan, 8(2): 627–629.

Rebuffat, S., Conraux, L., Massias, M., Auvin-Guette, C.A.T.H.E.R.I.N.E. and Bodo, B. (1993). Sequence and solution conformation of the 20-residue peptaibols, saturnisporins SA II and SA IV. International Journal of Peptide and Protein Research, 41(1): 74–84.

Rebuffat, S., Duclohier, H., Auvin-Guette, C., Molle, G., Spach, G. and Bodo, B. (1992). Membrane-modifying properties of the pore-forming peptaibols saturnisporin SA IV and harzianin HA V. FEMS Microbiology Immunology, 5(1-3): 151–160.

Rebuffat, S., EL-Hajji, M.O.H.A.M.E.D., Hennig, P., Davoust, D. and Bodo, B. (1989). Isolation, sequence, and conformation of seven trichorzianines from *Trichoderma harzianum*. International Journal of Peptide and Protein Research, 34(3): 200–210.

Rebuffat, S., Goulard, C. and Bodo, B. (1995). Antibiotic peptides from *Trichoderma harzianum*: Harzianins HC, proline-rich 14-residue peptaibols. Journal of the Chemical Society, Perkin Transactions, 1(14): 1849–1855.

Rebuffat, S., Goulard, C., Hlimi, S. and Bodo, B. (2000). Two unprecedented natural Aib-peptides with the (Xaa-Yaa-Aib-Pro) motif and an unusual C-terminus: structures, membrane-modifying and antibacterial properties of pseudokonins KL III and KL VI from the fungus *Trichoderma pseudokoningii*. Journal of Peptide Science: An Official Publication of the European Peptide Society, 6(10): 519–533.

Rebuffat, S., Hlimi, S., Prigent, Y., Goulard, C. and Bodo, B. (1996). Isolation and structural elucidation of the 11-residue peptaibol antibiotic, harzianin HK VI. Journal of the Chemical Society, Perkin Transactions, 1(16): 2021–2027.

Rebuffat, S., Prigent, Y., Auvin-Guette, C. and Bodo, B. (1991). Tricholongins BI and BII, 19-residue peptaibols from *Trichoderma longibrachiatum*: Solution structure from two-dimensional NMR spectroscopy. European Journal of Biochemistry, 201(3): 661–674.

Reeves, R.J. and Jackson, R.M. (1972). Induction of Phytophthora cinnamomi oospores in soil by *Trichoderma viride*. Transactions of the British Mycological Society, 59(1): 156–159.

Reino, J.L., Guerrero, R.F., Hernández-Galán, R. and Collado, I.G. (2008). Secondary metabolites from species of the biocontrol agent *Trichoderma*. Phytochemistry Reviews, 7(1): 89–123.

Reithner, B., Brunner, K., Schuhmacher, R., Peissl, I., Seidl, V., Krska, R. and Zeilinger, S. (2005). The G protein α subunit Tga1 of *Trichoderma atroviride* is involved in chitinase formation and differential production of antifungal metabolites. Fungal Genetics and Biology, 42(9): 749–760.

Reithner, B., Schuhmacher, R., Stoppacher, N., Pucher, M., Brunner, K. and Zeilinger, S. (2007). Signaling via the *Trichoderma atroviride* mitogen-activated protein kinase Tmk1 differentially affects mycoparasitism and plant protection. Fungal Genetics and Biology, 44(11): 1123–1133.

Rey, M., Delgado-Jarana, J. and Benitez, T. (2001). Improved antifungal activity of a mutant of *Trichoderma harzianum* CECT 2413 which produces more extracellular proteins. Applied Microbiology and Biotechnology, 55(5): 604–608.

Ritieni, A., Fogliano, V., Nanno, D., Randazzo, G., Altomare, C., Perrone, G., Bottalico, A., Maddau, L. and Marras, F. (1995). Paracelsin E, a new peptaibol from *Trichoderma saturnisporum*. Journal of Natural Products, 58(11): 1745–1748.

Rocha-Ramírez, V., Omero, C., Chet, I., Horwitz, B.A. and Herrera-Estrella, A. (2002). *Trichoderma atroviride* G-protein α-subunit gene tga1 is involved in mycoparasitic coiling and conidiation. Eukaryotic Cell, 1(4): 594–605.

Rubio, M.B., Hermosa, R., Reino, J.L., Collado, I.G. and Monte, E. (2009). Thctf1 transcription factor of *Trichoderma harzianum* is involved in 6-pentyl-2H-pyran-2-one production and antifungal activity. Fungal Genetics and Biology, 46(1): 17–27.

Sahai, A.S. and Manocha, M.S. (1993). Chitinases of fungi and plants: Their involvement in morphogenesis and host—parasite interaction. FEMS Microbiology Reviews, 11(4): 317–338.

Sailaja, N., Chandrasekhar, M., Rekhadevi, P.V., Mahboob, M., Rahman, M.F., Vuyyuri, S.B., Danadevi, K., Hussain, S.A. and Grover, P. (2006). Genotoxic evaluation of workers employed in pesticide production. Mutation Research/Genetic Toxicology and Environmental Mutagenesis, 609(1): 74–80.

Sakuno, E., Yabe, K., Hamasaki, T. and Nakajima, H. (2000). A new inhibitor of 5 '-hydroxyaverantin dehydrogenase, an enzyme involved in aflatoxin biosynthesis, from *Trichoderma hamatum*. Journal of Natural Products, 63(12): 1677–1678.

Salwan, R., Rialch, N. and Sharma, V. (2019). Bioactive volatile metabolites of *Trichoderma*: An overview. Secondary Metabolites of Plant Growth Promoting Rhizomicroorganisms, 87–111.

Samolski, I., Rincon, A.M., Pinzón, L.M., Viterbo, A. and Monte, E. (2012). The qid74 gene from *Trichoderma harzianum* has a role in root architecture and plant biofertilization. Microbiology, 158(1): 129–138.

Samuels, G.J. (1996). *Trichoderma*: A review of biology and systematics of the genus. Mycological Research, 100(8): 923–935.

Satapute, P., Kamble, M.V., Adhikari, S.S. and Jogaiah, S. (2019). Influence of triazole pesticides on tillage soil microbial populations and metabolic changes. Science of the Total Environment, 651: 2334–2344.

Sawa, R., Mori, Y., Iinuma, H., Naganawa, H., Hamada, M., Yoshida, S., Furutani, H., Kajimura, Y., Fuwa, T. and Takeuchi, T. (1994). Harzianic acid, a new antimicrobial antibiotic from a fungus. The Journal of Antibiotics, 47(6): 731–732.

Scarselletti, R. and Faull, J.L. (1994). *In vitro* activity of 6-pentyl-α-pyrone, a metabolite of *Trichoderma harzianum*, in the inhibition of *Rhizoctonia solani* and *Fusarium oxysporum* f. sp. *lycopersici*. Mycological Research, 98(10): 1207–1209.

Scheuer, P.J. (1992). Isocyanides and cyanides as natural products. Accounts of Chemical Research, 25(10): 433–439.

Schirmböck, M., Lorito, M., Wang, Y.L., Hayes, C.K., Arisan-Atac, I., Scala, F., Harman, G.E. and Kubicek, C.P. (1994). Parallel formation and synergism of hydrolytic enzymes and peptaibol antibiotics, molecular mechanisms involved in the antagonistic action of *Trichoderma harzianum* against phytopathogenic fungi. Applied and Environmental Microbiology, 60(12): 4364–4370.

Schmoll, M. (2008). The information highways of a biotechnological workhorse–signal transduction in *Hypocrea jecorina*. BMC Genomics, 9(1): 1–25.

Schroeckh, V., Scherlach, K., Nützmann, H.W., Shelest, E., Schmidt-Heck, W., Schuemann, J., Martin, K., Hertweck, C. and Brakhage, A.A. (2009). Intimate bacterial–fungal interaction triggers biosynthesis of archetypal polyketides in *Aspergillus nidulans*. Proceedings of the National Academy of Sciences, 106(34): 14558–14563.

Score, A.J. and Palfreyman, J.W. (1994). Biological control of the dry rot fungus *Serpula lacrymans* by *Trichoderma* species: The effects of complex and synthetic media on interaction and hyphal extension rates. International Biodeterioration & Biodegradation, 33(2): 115–128.

Segarra, G., Casanova, E., Bellido, D., Odena, M.A., Oliveira, E. and Trillas, I. (2007). Proteome, salicylic acid, and jasmonic acid changes in cucumber plants inoculated with *Trichoderma* asperellum strain T34. Proteomics, 7(21): 3943–3952.

Seidl, V., Song, L., Lindquist, E., Gruber, S., Koptchinskiy, A., Zeilinger, S., Schmoll, M., Martínez, P., Sun, J., Grigoriev, I. and Herrera-Estrella, A. (2009). Transcriptomic response of the mycoparasitic fungus *Trichoderma atroviride t*o the presence of a fungal prey. BMC Genomics, 10(1): 1–13.

Serrano-Carreon, L., Hathout, Y., Bensoussan, M. and Belin, J.M. (1993). Metabolism of linoleic acid or mevalonate and 6-pentyl-α-pyrone biosynthesis by *Trichoderma* species. Applied and Environmental Microbiology, 59(9): 2945–2950.

Shahriar, S.A., Islam, M.N., Chun, C.N.W., Kaur, P., Rahim, M.A., Islam, M.M., Uddain, J. and Siddiquee, S. (2022). Microbial metabolomics interaction and ecological challenges of *Trichoderma* species as biocontrol inoculant in crop rhizosphere. Agronomy, 12: 900.

Sharma, K., Mishra, A.K. and Misra, R.S. (2009). Morphological, biochemical and molecular characterization of *Trichoderma harzianum* isolates for their efficacy as biocontrol agents. Journal of Phytopathology, 157(1): 51–56.

Sharman, G.J., Try, A.C., Williams, D.H., Ainsworth, A.M., Beneyto, R., Gibson, T.M., McNICHOLAS, C., Renno, D.V., Robinson, N., Wood, K.A. and Wrigley, S.K. (1996). Structural elucidation of XR586, a peptaibol-like antibiotic from *Acremonium persicinum*. Biochemical Journal, 320(3): 723–728.

Sharon, E., Bar-Eyal, M., Chet, I., Herrera-Estrella, A., Kleifeld, O. and Spiegel, Y. (2001). Biological control of the root-knot nematode *Meloidogyne javanica* by *Trichoderma harzianum*. Phytopathology, 91(7): 687–693.

Sherman, D.H. (2002). New enzymes for "warheads". Nature Biotechnology, 20(10): 984–985.

Shi, M., Chen, L., Wang, X.W., Zhang, T., Zhao, P.B., Song, X.Y., Sun, C.Y., Chen, X.L., Zhou, B.C. and Zhang, Y.Z. (2012). Antimicrobial peptaibols from *Trichoderma pseudokoningii* induce programmed cell death in plant fungal pathogens. Microbiology, 158(1): 166–175.

Shi, Z.Z., Miao, F.P., Fang, S.T., Yin, X.L. and Ji, N.Y. (2018). Sulfurated diketopiperazines from an algicolous isolate of *Trichoderma virens*. Phytochemistry Letters, 27: 101–104.

Shokrollahi, N., Ho, C.L., Zainudin, N.A.I.M., Wahab, M.A.W.B.A. and Wong, M.Y. (2021). Identification of non-ribosomal peptide synthetase in Ganoderma boninense Pat. that was expressed during the interaction with oil palm. Scientific Reports, 11(1): 1–16.

Shoresh, M. and Harman, G.E. (2008). The molecular basis of shoot responses of maize seedlings to *Trichoderma harzianum* T22 inoculation of the root: a proteomic approach. Plant Physiology, 147(4): 2147–2163.

Shwab, E.K. and Keller, N.P. (2008). Regulation of secondary metabolite production in filamentous ascomycetes. Mycological Research, 112(2): 225–230.

Siddiquee, S. (2017). Fungal volatile organic compounds: Emphasis on their plant growth-promoting. In Volatiles and Food Security (pp. 313–333). Springer, Singapore.

Simon, A., Dunlop, R.W., Ghisalberti, E.L. and Sivasithamparam, K. (1988). *Trichoderma koningii* produces a pyrone compound with antibiotic properties. Soil Biology & Biochemistry, 20(2): 263–264.

Singh, G., Tiwari, A., Choudhir, G., Kumar, A., Kumar, S., Hariprasad, P. and Sharma, S. (2022). Deciphering the role of *Trichoderma* spp. bioactives in combating the wilt causing cell wall degrading enzyme polygalacturonase produced by *Fusarium oxysporum*: An *in-silico* approach. Microbial Pathogenesis, 105610.

Singh, H.B., Singh, B.N. and Singh, S.P. (2009). Biological control of plant diseases: Current status and future prospects. pp. 193–304. *In*: B.K. Sarma, S.R. Singh and J.K. Johri (eds.). New Delhi: Biotechnological Applications. New Indian Publishing Agency.

Singh, S., Dureja, P., Tanwar, R.S. and Singh, A. (2005). Production and antifungal activity of secondary metabolites of *Trichoderma virens*. Pesticide Research Journal, 17(2): 26–29.

Sivasithamparam, K. and Ghisalberti, E. (1998). Secondary metabolism in *Trichoderma* and Gliocladium. *Trichoderma* and Gliocladium. CP Kubicek, Harman, GE London, Francis & Taylor Ltd, 1: 139–191.

Skidmore, A.M. and Dickinson, C.H. (1976). Colony interactions and hyphal interference between *Septoria nodorum* and phylloplane fungi. Transactions of the British Mycological Society, 66(1): 57–64.

Sodeoka, M., Sampe, R., Kojima, S., Baba, Y., Morisaki, N. and Hashimoto, Y. (2001). Asymmetric synthesis of a 3-acyltetronic acid derivative, RK-682, and formation of its calcium salt during silica gel column chromatography. Chemical and Pharmaceutical Bulletin, 49(2): 206–212.

Sood, M., Kapoor, D., Kumar, V., Sheteiwy, M.S., Ramakrishnan, M., Landi, M., Araniti, F. and Sharma, A. (2020). *Trichoderma*: The "secrets" of a multitalented biocontrol agent. Plants, 9(6): 762.

Sridharan, A.P., Sugitha, T., Karthikeyan, G., Nakkeeran, S. and Sivakumar, U. (2021). Metabolites of *Trichoderma longibrachiatum* EF5 inhibits soil borne pathogen, Macrophomina phaseolina by triggering amino sugar metabolism. Microbial Pathogenesis, 150: 104714.

Steyaert, M., Vanaverbeke, J., Vanreusel, A., Barranguet, C., Lucas, C. and Vincx, M. (2003). The importance of fine-scale, vertical profiles in characterizing nematode community structure. Estuarine, Coastal and Shelf Science, 58(2): 353–366.

Stoessl, A.L.B.E.R.T. (1981). Structure and biogenetic relations: Fungal nonhost-specific. Toxins in Plant Disease, 109–219.

Stracquadanio, C., Quiles, J.M., Meca, G. and Cacciola, S.O. (2020). Antifungal activity of bioactive metabolites produced by *Trichoderma asperellum* and *Trichoderma atroviride* in liquid medium. Journal of Fungi, 6(4): 263.

Strakowska, J., Błaszczyk, L. and Chełkowski, J. (2014). The significance of cellulolytic enzymes produced by *Trichoderma* in opportunistic lifestyle of this fungus. Journal of Basic Microbiology, 54(S1): S2–S13.

Suárez, M.B., Vizcaíno, J.A., Llobell, A. and Monte, E. (2007). Characterization of genes encoding novel peptidases in the biocontrol fungus *Trichoderma harzianum* CECT 2413 using the TrichoEST functional genomics approach. Current Genetics, 51(5): 331–342.

Sun, F., Zhang, W., Hu, H., Li, B., Wang, Y., Zhao, Y., Li, K., Liu, M. and Li, X. (2008). Salt modulates gravity signaling pathway to regulate growth direction of primary roots in Arabidopsis. Plant Physiology, 146(1): 178–188.

Tamura, A., Kotani, H. and Naruto, S. (1975). Trichoviridin and dermadin from *Trichoderma* spp. TK-1. The Journal of Antibiotics, 28(2): 161–162.

Tarus, P.K., Lang'at-Thoruwa, C.C., Wanyonyi, A.W. and Chhabra, S.C. (2003). Bioactive metabolites from *Trichoderma harzianum* and *Trichoderma longibrachiatum*. Bulletin of the Chemical Society of Ethiopia, 17(2).

Taylor, L., Gutierrez, S., McCormick, S.P., Bakker, M.G., Proctor, R.H., Teresi, J., Kurtzman, B., Hao, G. and Vaughan, M.M. (2022). Use of the volatile trichodiene to reduce Fusarium head blight and trichothecene contamination in wheat. Microbial Biotechnology, 15(2): 513–527.

Thrane, C., Tronsmo, A. and Jensen, D.F. (1997). Endo-1, 3-β-glucanase and cellulase from *Trichoderma harzianum*: purification and partial characterization, induction of and biological activity against plant pathogenic Pythium spp. European Journal of Plant Pathology, 103(4): 331–344.

Thrane, U., Poulsen, S.B., Nirenberg, H.I. and Lieckfeldt, E. (2001). Identification of *Trichoderma* strains by image analysis of HPLC chromatograms. FEMS Microbiology Letters, 203(2): 249–255.

Tijerino, A., Cardoza, R.E., Moraga, J., Malmierca, M.G., Vicente, F., Aleu, J., Collado, I.G., Gutiérrez, S., Monte, E. and Hermosa, R. (2011). Overexpression of the trichodiene synthase gene tri5 increases trichodermin production and antimicrobial activity in *Trichoderma brevicompactum*. Fungal Genetics and Biology, 48(3): 285–296.

Topolovec-Pintarić, S. (2019). *Trichoderma*: Invisible partner for visible impact on agriculture. *Trichoderma*-The Most Widely Used Fungicide, 15.

Tulasne, L. (1865). Some fungal spheres, in connection with a memoir by M. Antoine de Bary on the Nyctalis. Ann. Sci. Nat. Bot., 13: 5–19.

Tulasne, L. and Tulasne, R. (1860). De quelques Sphéries fungicoles, àpropos d'un mémoire de M. Antoine de Barysur les Nyctalis. Ann. des. Sci. Nat. Bot., 13: 5–19.

Tyśkiewicz, R., Nowak, A., Ozimek, E. and Jaroszuk-Ściseł, J. (2022). *Trichoderma*: The current status of its application in agriculture for the biocontrol of fungal phytopathogens and stimulation of plant growth. International Journal of Molecular Sciences, 23(4): 2329.

Urbaniak, M., Waśkiewicz, A., Koczyk, G., Błaszczyk, L., Uhlig, S. and Stępień, L. (2021). Naturally-produced beauvericins and divergence of BEAS gene among *Fusarium* and *Trichoderma* species. DNA, 1(3): 4.

Verma, C., Jandaik, S., Gupta, B.K., Kashyap, N., Suryaprakash, V.S., Kashyap, S. and Kerketta, A. (2020). Microbial metabolites in plant disease management: Review on biological approach. International Journal of Chemical Studies, 8(4): 2570–2581.

Verma, M., Brar, S.K., Tyagi, R.D., Surampalli, R.N. and Valero, J.R. (2007). Antagonistic fungi, *Trichoderma* spp.: panoply of biological control. Biochemical Engineering Journal, 37(1): 1–20.

Vesonder, R.F., Tjarks, L.W., Rohwedder, W.K., Burmeister, H.R. and Laugal, J.A. (1979). Equisetin, an antibiotic from Fusarium Eq uiseti NRRL 5537, Identified as a derivative of N-methyl -2, 4-pyrollidone. The Journal of Antibiotics, 32(7): 759–761.

Vicente, I., Baroncelli, R., Hermosa, R., Monte, E., Vannacci, G. and Sarrocco, S. (2022). Role and genetic basis of specialised secondary metabolites in *Trichoderma* ecophysiology. Fungal Biology Reviews, 39: 83–99.

Vinale, F., Flematti, G., Sivasithamparam, K., Lorito, M., Marra, R., Skelton, B.W. and Ghisalberti, E.L. (2009). Harzianic acid, an antifungal and plant growth promoting metabolite from *Trichoderma harzianum*. Journal of Natural Products, 72(11): 2032–2035.

Vinale, F., Ghisalberti, E.L., Sivasithamparam, K., Marra, R., Ritieni, A., Ferracane, R., Woo, S. and Lorito, M. (2009a). Factors affecting the production of *Trichoderma harzianum* secondary metabolites during the interaction with different plant pathogens. Letters in Applied Microbiology, 48(6): 705–711.

Vinale, F., Girona, I.A., Nigro, M., Mazzei, P., Piccolo, A., Ruocco, M., Woo, S., Rosa, D.R., Herrera, C.L. and Lorito, M. (2012). Cerinolactone, a hydroxy-lactone derivative from *Trichoderma cerinum*. Journal of Natural Products, 75(1): 103–106.

Vinale, F., Marra, R., Scala, F., Ghisalberti, E.L., Lorito, M. and Sivasithamparam, K. (2006). Major secondary metabolites produced by two commercial *Trichoderma* strains active against different phytopathogens. Letters in Applied Microbiology, 43(2): 143–148.

Vinale, F., Nigro, M., Sivasithamparam, K., Flematti, G., Ghisalberti, E.L., Ruocco, M., Varlese, R., Marra, R., Lanzuise, S., Eid, A. and Woo, S.L. (2013). Harzianic acid: A novel siderophore from *Trichoderma harzianum*. FEMS Microbiology Letters, 347(2): 123–129.

Vinale, F., Sivasithamparam, K., Ghisalberti, E.L., Marra, R., Barbetti, M.J., Li, H., Woo, S.L. and Lorito, M. (2008). A novel role for *Trichoderma* secondary metabolites in the interactions with plants. Physiological and Molecular Plant Pathology, 72(1-3): 80–86.

Vinale, F., Sivasithamparam, K., Ghisalberti, E.L., Woo, S.L., Nigro, M., Marra, R., Lombardi, N., Pascale, A., Ruocco, M., Lanzuise, S. and Manganiello, G. (2014). *Trichoderma* secondary metabolites active on plants and fungal pathogens. The Open Mycology Journal, 8(1).

Viterbo, A., Harel, M. and Chet, I. (2004). Isolation of two aspartyl proteases from *Trichoderma asperellum* expressed during colonization of cucumber roots. FEMS Microbiology Letters, 238(1): 151–158.

Viterbo, A., Landau, U., Kim, S., Chernin, L. and Chet, I. (2010). Characterization of ACC deaminase from the biocontrol and plant growth-promoting agent *Trichoderma asperellum* T203. FEMS Microbiology Letters, 305(1): 42–48.

Viterbo, A., Ramot, O., Chernin, L. and Chet, I. (2002). Significance of lytic enzymes from *Trichoderma* spp. in the biocontrol of fungal plant pathogens. Antonie Van Leeuwenhoek, 81(1): 549–556.

Wang, H., Ma, S., Xia, Q., Zhao, Z., Chen, X., Shen, X., Yin, C. and Mao, Z. (2022). The interaction of the pathogen *Fusarium proliferatum* with *Trichoderma asperellum* characterized by transcriptome changes in apple rootstock roots. Physiological and Molecular Plant Pathology, 121: 101894.

Wardle, D.A., Parkinson, D. and Waller, J.E. (1993). Interspecific competitive interactions between pairs of fungal species in natural substrates. Oecologia, 94(2): 165–172.

Weindling, R. (1932). *Trichoderma lignorum* as a parasite of other soil fungi. Phytopathology, 22(8): 837–845.

Weindling, R. and Emerson, O.H. (1936). The isolation of a toxic substance from the culture filtrates of *Trichoderma*.

WHO. (1992). Cadmium. Environmental Health Criteria, vol 134, Geneva.

Wiest, A., Grzegorski, D., Xu, B.W., Goulard, C., Rebuffat, S., Ebbole, D.J., Bodo, B. and Kenerley, C. (2002). Identification of peptaibols from *Trichoderma* virens and cloning of a peptaibol synthetase. Journal of Biological Chemistry, 277(23): 20862–20868.

Wipf, P. and Kerekes, A.D. (2003). Structure reassignment of the fungal metabolite TAEMC161 as the phytotoxin viridiol. Journal of Natural Products, 66(5): 716–718.

Woo, S.L. and Lorito, M. (2007). Exploiting the interactions between fungal antagonists, pathogens and the plant for biocontrol. In Novel Biotechnologies for Biocontrol Agent Enhancement and Management. Springer, Dordrecht, pp. 107–130.

Worasatit, N., Sivasithamparam, K., Ghisalberti, E.L. and Rowland, C. (1994). Variation in pyrone production, lytic enzymes and control of *Rhizoctonia* root rot of wheat among single-spore isolates of *Trichoderma koningii*. Mycological Research, 98(12): 1357–1363.

Xie, L., Zang, X., Cheng, W., Zhang, Z., Zhou, J., Chen, M. and Tang, Y. (2021). Harzianic acid from *Trichoderma harzianum* is a natural product inhibitor of acetohydroxyacid synthase. Journal of the American Chemical Society, 143(25): 9575–9584.

Yamaguchi, Y., Manita, D., Takeuchi, T., Kuramochi, K., Kuriyama, I., Sugawara, F., Yoshida, H. and Mizushina, Y. (2010). Novel terpenoids, trichoderonic acids A and B isolated from *Trichoderma virens*, are selective inhibitors of family X DNA polymerases. Bioscience, Biotechnology, and Biochemistry, 1002261878–1002261878.

Yan, L. and Khan, R.A.A. (2021). Biological control of bacterial wilt in tomato through the metabolites produced by the biocontrol fungus, *Trichoderma harzianum*. Egyptian Journal of Biological Pest Control, 31(1): 1–9.

Yao, L., Tan, C., Song, J., Yang, Q., Yu, L. and Li, X. (2016). Isolation and expression of two polyketide synthase genes from *Trichoderma harzianum* 88 during mycoparasitism. Brazilian Journal of Microbiology, 47: 468–479.

Zaid, R., Koren, R., Kligun, E., Gupta, R., Leibman-Markus, M., Mukherjee, P.K., Kenerley, C.M., Bar, M. and Horwitz, B.A. (2022). Gliotoxin, an immunosuppressive fungal metabolite, primes plant immunity: Evidence from *Trichoderma virens*-tomato interaction. Mbio, 13(4): e00389-22.

Zeilinger, S., Gruber, S., Bansal, R. and Mukherjee, P.K. (2016). Secondary metabolism in *Trichoderma*–chemistry meets genomics. Fungal Biology Reviews, 30(2): 74–90.

Zeilinger, S., Reithner, B., Scala, V., Peissl, I., Lorito, M. and Mach, R.L. (2005). Signal transduction by Tga3, a novel G protein α subunit of *Trichoderma atroviride*. Applied and Environmental Microbiology, 71(3): 1591–1597.

Zhang, M., Li, N., Chen, R., Zou, J. and Wang, C. (2014). Two terpenoids and a polyketide from the endophytic fungus *Trichoderma* spp. Xy24 isolated from mangrove plant Xylocarpus granatum. Journal of Chinese Pharmaceutical Sciences, 23(6): 421–424.

Zhu, Y., Wang, J., Mou, P., Yan, Y., Chen, M. and Tang, Y. (2021). Genome mining of cryptic tetronate natural products from a PKS-NRPS encoding gene cluster in *Trichoderma harzianum* t-22. Organic & Biomolecular Chemistry, 19(9): 1985–1990.

Ziedan, E.S.H., Alamri, S.A., Hashem, M. and Mostafa, Y.S. (2021). Secondary invader bacteria associated with the red pest weevil infestation in date palm trees. Agronomy Journal, 113(5): 4271–4279.

Mechanisms

Chapter 7

Exploring Mechanisms of Disease Suppression using Endophytic Fungi as Biocontrol Agents

*Silju Juby, E.K. Radhakrishnan** and *K. Jayachandran*

1. Introduction

Agriculture is the backbone sector of many countries all over the world and has been practised for thousands of years. Due to the growing global population, it is expected that food consumption would increase by 60% no later than 2050. Therefore, new possibilities in the agricultural field should be explored and agriculture also needs to be sustainable. Sustainable development is a development that meets the needs of the present without compromising the ability of future generations to meet their own requirements (Gimenez et al., 2018). Several factors represent a major hazard to farms, resulting in yield decline when not monitored and controlled correctly. These factors are classified into three categories: technological, biological and environmental (Ngoune Liliane and Shelton Charles, 2020). This chapter delves into the positive impacts of biocontrol agents as a means of managing plant pathogenic fungus.

1.1 Threats and challenges to sustainable agriculture

Sustainable agricultural development aims to conserve and develop existing natural resources because future generations need higher quantities and quality of agricultural and food products (Shalaby et al., 2011). However, some stress factors affect the agriculture sector globally. These will be described in brief.

School of Biosciences Mahatma Gandhi University, Kerala, India.
* Corresponding author: radhakrishnanek@mgu.ac.in

1.1.1 Climate change

One of the most important threats faced by the agriculture industry is, climate change. Climate change is expected to have a significant impact on the water cycle, changing rainfall patterns and influencing surface and groundwater availability and quality, agricultural production and associated ecosystems. Insect infestations, shifting seasons, superstorms, drought, heat, flooding and other signs of a warming world will cause a 2 to 6% decrease in global crop yields every decade onwards, potentially wiping out millions of acres annually, according to a recent IPCC assessment. Different parts of the world are affected by climate change in different ways, but the negative effects are widespread. Due to these climate changes, farmers are subject to stringent limitations on the techniques they can employ.

1.1.2 Water scarcity

Water scarcity is another growing threat to agriculture. As a result of increased evaporation and an increase in the capacity of the atmosphere to hold water, there is an imbalance between areas that experience extreme dryness and areas that experience significant rainfall.

1.1.3 Land scarcity

Land is a resource that is quickly running out and is putting pressure on the agriculture sector. The competition for industrialization, urbanization, housing, etc., has a detrimental impact on agricultural growth as the population expands.

1.1.4 Invasive species and disease

Around the world, invasive species have posed a threat to native species and agriculture. As foreign bodies come with foreign pathogens, new diseases appear. Any external condition that adversely affects the growth, development or productivity of plants can be considered stress (Grover et al., 2010). Plants are subjected to several environmental stresses which reduce and limit their growth. Environmental stresses are categorized as abiotic and biotic (Shameer and Prasad, 2018). Abiotic stresses, such as radiation, salt, floods, droughts, extremely high or low temperatures and heavy metals, among others, result in the loss of significant crop plants across the world . Conversely, biotic stresses involve invasions by numerous pathogens such as fungi, bacteria, oomycetes, nematodes and herbivores (Atkinson and Urwin, 2012). As plants are sessile organisms, they cannot escape from these types of environmental cues. Plants respond to these challenges by a number of morphological, biochemical and molecular mechanisms, and evidence suggests that these signaling pathways interact with each other (Nejat and Mantri, 2017). Pests, parasites and pathogens have long been known to generate biotic stressors in plants. Pathogens that cause plant diseases include fungi, bacteria, nematodes and viruses (Gimenez et al., 2018). Plant disease lowers the quantity and quality of food, fiber, and biofuel crops as agriculture tries to feed the world's fast-expanding population. Farmers spend billions of dollars managing diseases, frequently without enough technical support, which has negative effects such as contamination of soil and poor disease control. Additionally, plant diseases have the power to completely destroy natural ecosystems, exacerbating the environmental issues brought on by habitat loss and bad land management.

1.1.5 Plant diseases

When the plant is not developing and functioning in the manner it is expected, it is called diseased. Plant disease is a normal part of nature that helps to keep living plants and animals in balance with one another (Singh, 2018). Some definitions for plant disease given by researchers are described. In 1858, Julius Kuhn defined abnormal changes in physiological processes that disturb the normal activity of plant organs (Wilhelm and Tietz, 1978). Ward (1901) defined plant disease as a condition in which the functions of the organism are improperly discharged (Talhinhas et al., 2017). Plant diseases are the major cause of reduced crop production in all agricultural and horticultural systems. Any factor or agent that incites disease is called a pathogen. The disease is always associated with a pathogen. In a precise sense, a pathogen does not necessarily belong to a living or animate group, they can be living or non-living (Singh, 2018). Plant diseases increase mortality, impair plant fitness, rapidly deplete populations of certain hosts or drastically alter the structure and makeup of plant communities (Gilbert, 2002). In natural systems, seed and seedling illnesses typically account for the highest rates of plant disease-related death (Lambert, 2001). In general, around 25% of the world's annual agricultural production is lost due to plant diseases (Lugtenberg, 2015).

Plant fungal infections cause the majority of diseases in agricultural and horticultural settings (Hariharan and Prasannath, 2021). Phytopathogens have developed mechanisms and methods to attack any plant (Knogge, 1996b) seeking access and finding nutrients for growth and development (Horbach et al., 2011). The plant fungal pathogen should be able to germinate on the surface of a suitable host for plant fungal infections to appear. Plant fungal pathogen spores can only germinate in favorable conditions. This includes the availability of low molecular mass nutrients, as well as an appropriate host (Osherov and May, 2001). Fungal spores can survive for several years by employing self-inhibitors to prevent germination until favorable conditions are met (Chitarra et al., 2004). When conditions are favorable, plant fungus spores create infection structures such as the appressorium and the infection peg, which allow the hyphae to penetrate the host (Schäfer, 1994). Pathogens that cause diseases in avocado, such as *Colletotrichum gloeosporioides*, can exploit their host waxes to infect their host (Podila). Although not all fungal species infect plants, plant fungal pathogens target all plant groups (Knogge, 1996a). These fungi are jointly responsible for 80% of plant diseases (El Hussein, 2014). Approximately 8000 fungal species cause almost 100,000 plant diseases (George N. Agrios, 2005). Plant fungal pathogens are classified as biotrophs, necrotrophs or hemibiotrophs based on their mode of infection. Biotrophs are organisms that reside on living tissues and infect the host without killing it. Pathogens use the appressorium to enter the host and feeding structures such as the haustoria to obtain nutrients from the surrounding cells. Biotrophs have a limited host range, such as powdery mildew and rust fungus (De Silva, 2016). Necrotrophs infect the living host and finally kill it. This is because necrotrophic fungi can only finish their life cycles on dead tissues. These infections are constantly producing hydrolytic enzymes and poisons in order to kill plant cells. Necrotrophs produce two types of toxins: host specific toxins that are specific to the plant host and allow the pathogen to cause disease on a specific host, such as *Cochliobolus carbonum*. Second, infections with broad range toxins, such as *Sclerotinia sclerotiorum*, *Alternaria brassicicola* and *Botrytis cinerea*, can infect and harm unrelated plant species (Wen, 2013). Hemi-biotrophs, such as *Colletrichium* use similar mechanisms to biotrophs to produce infections and afterwards kill their hosts as necrotrophs (Hadwiger, 2005).

2. Problems caused by plant fungal pathogens in agricultural production and food spoilage

Agricultural production can provide long-term plant products that help alleviate poverty and hunger. Agricultural productivity has been severely harmed by epidemics like late blight disease of potatoes, cereal rusts and smuts, ergot of rye and wheat, the brown spot of rice, coffee rust, the downy and powdery mildews of grape, etc. Reduced plant quality and quantity caused by plant fungal infections can potentially be a concern to human health. This could result in forced food displacement, altered country economic outlooks, increased political unpredictability and forced migration of people (Anderson et al., 2004; Gould, 2009). Due to these phytopathogens, farmers, decision-makers, researchers and consumers face huge challenges (Fletcher et al., 2006).

Destructive diseases are caused worldwide by plant fungal pathogens such as, *Magnaporthe oryzae* and *Colletotrichum* spp. *Blumeria graminis* decreases crop amount while *Fusarium graminearum* lowers crop quality. In a single field, one pathogen such as *Mycosphaerella graminicola* can mutate and infect different plant species, diminishing the ability of plants to fight off the infection. Similarly, as *Fusarium oxysporum* has over 70 formae speciales, it can be challenging to pinpoint the important pathogen. A single species can spread disease to a wide range of plant species; for instance, *F. oxysporum* and *Botrytis cinerea* each have 200 plant hosts. A single pathogen, such as *M. oryzae* can cause a significant reduction in grain output. Infections can completely shut down plants when they coexist with other pathogens, such as *F. graminearum* and other *Fusarium* species. Additionally, pathogens like *M. graminicola* have a longer than 7-d symptomless colonization period. Due to, it is challenging to know if the plant is diseased. Some pathogens such as *Ustilago maydis* can finish their lifecycle in as little as 2 wk, making them extremely harmful. It becomes challenging to even produce other crops in crop rotation when pathogens like *Puccinia* spp., repeatedly cause crop failures (Vollmeister et al.). Some pathogens can infect new plant species, affect several plant species or induce disease in related plant species (Burdon and Thrall, 2009). Numerous weak infections can cause destruction and start epidemics in closely related species across continents. It is obvious that fungi pose a wide range of issues and have the ability to have a significant impact on plant growth.

3. Fungi

Only 70,000 of the estimated 1.5 million species of fungi that exist on Earth are known and have been investigated. About 70% of plant diseases are caused by fungal pathogens (Yang et al., 2017). More than 10,000 species of known fungi can cause disease in plants. Pathogenic fungi belong to different lifestyles such as necrotrophic, hemibiotrophic, biotrophic or obligate biotrophic (Yang et al., 2017). While non-obligate parasites either require a living host plant for part of their life cycles, but can complete those cycles on dead organic matter or can grow and multiply on both dead organic matter and living plants (necrotrophs), obligate parasites (biotrophs) can only grow and multiply in constant association with their living hosts (Almeida et al., 2019).

The vast majority of fungal species can decompose large amounts of plants, especially wood and also animal tissue produced and dying each year. Fungal pathogens are known to use well-conserved proteins for their infection processes (Zhang et al., 2016). To infect, pathogenic fungi can develop specialized infection structures to penetrate host cells (Yuan et al., 2018). Different approaches that fungi has devised to colonize plant cells have a wide range of effects, from advantageous interactions to the host's death. The phyla Ascomycota and Basidiomycota constitute major fungal plant pathogens. Among Ascomycetes, plant pathogens are classified as Dothideomycetes (*Cladosporium* spp.), Sordariomycetes (*Magnaporthe* spp.) or Leotiomycetes (*Botrytis* spp.). Whereas among Basidiomycetes, the rusts (Pucciniomycetes) and the smuts (distributed throughout the subphylum of Ustilaginomycotina) are the two major plant disease groups.) (Gunther Doehlemann et al., 2017). Penicillin is an antibiotic produced by Penicillium a pathogenic fungus that is beneficial to both humans and animals. Trichoderma and Aspergillus are two other fungi that can biologically manage soil-borne plant pathogenic fungi (Agrios, 2009).

3.1 Impacts of fungal pathogens on different crops

In the field of fungal plant diseases, the first scientific contribution was attributed to Theophrastus (BC 370-288). In his botanical investigations, he revealed how rust diseases manifest on various host plants. There has not been much scientific advancement in our understanding of plant diseases for approximately 2000 yr since they have consistently been perceived as God's wrath (Gunther Doehlemann et al., 2017). A fungal pathogen called *Claviceps purpurea* has had a significant impact on agriculture and civilization throughout human history. It is the cause of rye, barley, oat and wheat ergot. LSD is one of the several alkaloids found in ergot sclerotia, which can replace kernels in plant heads and contaminate harvested grains and their flour (Doehlemann et al., 2017). Agriculture has always been seriously threatened by smuts and rust. It was discovered in the 17th century that *Berberis vulgaris* plants growing close to wheat fields were associated with *Puccinia graminis*-caused wheat stem rust. Barberry plants were destroyed to limit rust infection in wheat plants because the *P. graminis* sexual life cycle requires them as an intermediary host that is infected by monokaryotic basidiospores to create spermatia (Doehlemann et al., 2017).

Plant fungal pathogens can cause post-harvest diseases and food deterioration at any point in the processing chain, including harvesting, handling, storage, packaging and transportation (Hadwiger, 2005). Fresh fruit and vegetable rot in post-harvest processes is primarily caused by fungi (Gatto et al., 2011) Most post-harvest diseases are caused by more than 100 fungus species (Tripathi and Dubey, 2004) and postharvest illnesses can reduce crop production by 10 to 30% (Hadwiger, 2005). In tropical and underdeveloped nations there can be a 50% loss of essential commodities in some areas (Tripathi and Dubey, 2004).

Due to microbial spoilage, certain plant fungal diseases generate considerable losses during storage. Botrytis cinerea (fruits - raspberries, strawberries, grapes, kiwi fruit, pears, peaches, plums and cherries; vegetables - carrots, lettuce, peas and beans), *B. allii* (onions and related crops such as garlic), *Penicillium italicum*, *Penicillium digitatum* (green rot of citrus), *Penicillium expansum* (blue rot of apples and pears), *Penicillium glabrum* (onion) and *Penicillium funiculosum* (onion) are some of the examples of fungi (Moss, 2008). *Fusarium*, *Geotrichum* and *Aspergillus* are a few fungal species that cause fruit and vegetable deterioration, leading to considerable financial losses and unpleasant features in plant-based products (Hadwiger, 2005). Fruits that have been preserved can be destroyed by colletotrichum infections (Dean et al., 2012). Fungal infections also have significant social and economic impacts. Until the 1870s, Sri Lanka was one of the greatest coffee producers in the world. But it changed dramatically after the arrival of the coffee rust, *Hemileia vastatrix*. The fungus destroyed the coffee plantations in this region and production dropped by about 95% from 1870 to 1885. Even today, coffee rust is still a significant threat to coffee production (Avelino et al., 2015).

3.2 Problems caused by mycotoxins

Low molecular weight substances called mycotoxins are generated by molds. These secondary fungal metabolites have no impact on the formation or growth of the fungus and are poisonous to vertebrates in very low quantities (Hussein and Brasel, 2001). Mycotoxins are present in a wide variety of foods and are significant because they can result in both human and animal diseases like cancer and dermatitis (Awuchi et al., 2021). Consumption of tainted plant foods or animal feed containing mycotoxins can also result in a number of metabolic issues such as liver function reduction, interference with protein synthesis or other conditions such as, skin sensitivity, necrosis or severe immunodeficiency (Sweeney and Dobson, 1998). Besides having detrimental impacts on human and animal health, toxins such as aflatoxins, ochratoxins, trichothecenes, zearalenone, fumonisins, tremorgenic toxins and ergot alkaloids also have a negative economic and agricultural impact (Hussein and Brasel, 2001).

Therefore, plant fungal infections have a largely detrimental role in the efforts to feed the world's ever-increasing population. Plant fungal infections and the toxins they create seriously endanger agricultural production by reducing the quality and quantity of plant commodities, which causes financial losses both before and after harvesting. Various management and control methods are utilized to try and reduce the harm that fungi can create in agricultural settings.

3.3 *Some fungal pathogenic species in agriculture*

3.3.1 Cladosporium fulvum *(*Passalora fulva*)*

It is a nonobligate biotrophic fungal pathogen causing leaf mold in tomato plants (Joosten and de Wit, 1999; Thomma et al., 2005). A severe disease that devastated the tomato industry was leaf mold. This mold has been controlled, nevertheless, by introducing various *cf* resistance genes into commercial cultivars. Today, it serves as a useful model for research on plant-fungus interactions. The infection process affects plants in the following manner. Conidia on the beneath tomato leaves germinate to signal the beginning of the infection. The runner hypha merely colonizes the extracellular space of tomato leaves after entering the host leaf through the open stomata without differentiating an appressorium. The production of many conidiophores by *C. fulvum* on the 12th d after infection blocks the stomata and causes minor chlorosis or necrosis on the leaf (Thomma et al., 2005).

3.3.2 Botrytis *and* Sclerotinia

Examples of necrotrophic plant pathogens with a wide host range include *Botrytis cinerea* and *Sclerotinia sclerotiorum*. According to phylogeny, *B. cinerea* is a member of the family Sclerotiniaceae and the genus Botrytis, and its sexual form *Botryotinia whetzel* (Staats et al., 2005). There are 22 species in the genus *Botrytis*, and they can infect both monocotyledonous and dicotyledonous plants with diseases. All other species are specialists with a limited range of hosts, except *B. cinerea* and *B. fabae*. Species from the genera Vicia, Lens, Psium and Phaseolus can be infected by *B. fabae* (Walker et al., 2011). Gray mold is the ailment that *B. cinerea* most frequently causes. This disease is pervasive and causes significant agricultural loss throughout the world. Senescent or mature tissues of dicotyledonous hosts are where it causes the most damage. Early in the development of the crop, the fungus enters these tissues and remains dormant until conditions are right and the host physiology changes. Therefore, when crops that appear to be in good health are harvested and stored, significant damage is done. Before harvest, *B. cinerea* also causes significant loss in horticultural crops grown in fields and greenhouses, with a focus on fruits, vegetables and ornamental flowers. Typically, *B. cinerea* spreads infection through its conidia, which use nonmelanized appressoria to pierce the cuticle (Gunther Doehlemann et al., 2017).

3.3.3 Magnaporthe oryzae

The causal agent of rice blast disease and are a hemibiotrophic ascomycete. Rice blast is a destructive disease affecting planted rice crops. This disease can cause an annual output loss of up to 30% of the overall rice crop (Talbot, 2003; Valent and Chumley, 1991). *M. oryzae* can also infect other grains like wheat (Cruz et al., 2012). Due to its economic importance and ease of genetic and molecular modification, *M. oryzae* has become a model organism for studying plant-microbe interactions (Ebbole, 2007). *M. oryzae* conidia generate melanized appressorium (a specific infection structure) for successful colonization in rice leaves, allowing them to pierce the cell membrane by massive turgor pressure (Dagdas et al., 2012; de Jong et al., 1997). The initial hyphae spread through plant cells after penetrating the cell via the penetration peg and developing into bulbous invasive hyphae (Kankanala et al., 2007). *M. oryzae* also forms a biotrophic interfacial complex structure with an invasive hyphae. During the biotrophic stage, it is important for the accumulation and secretion of effectors that aid to avoid or reduce plant defensive responses (Khang et al., 2010). *M. oryzae* initial hyphae infiltrate adjacent cells via plasmodesmata and switch to the necrotrophic stage, resulting in necrotrophic lesions on the surface of rice leaves (de Jong et al., 1997; Ebbole, 2007).

3.3.4 Zymoseptoria tritici *(*Mycosphaerella graminicola*)*

It is a hemibiotrophic or necrotrophic fungal pathogen causing Septoria Tritici Blotch (STB) of wheat (*Triticum aestivum*) (Duncan and Howard, 2000; Kema, 1996). *Z. tritici* is a dothidiomycete with sexual and asexual phases found in all STB-infested locations (Quaedvlieg et al., 2011). STB is one of the most economically significant wheat foliar diseases, accounting for up to 30 to 50% of yearly production losses (Fraaije et al., 2005; Suffert et al., 2011). The fungus has developed resistance to numerous fungicides, making control of this disease challenging (Cools and Fraaije, 2008).

3.3.5 Ustilago maydis

It is a member of the smut fungus family, which includes over 1500 species that parasitize a wide variety of monocotyledonous and dicotyledonous plants. Due to its ease of reverse genetics, simple axenic development and small haploid genome with little redundancy, *U. maydis* is an ideal tool for molecular fungal genetics (Kamper et al., 2006). It has a biphasic lifecycle that includes a non-pathogenic, saprophytic phase as well as an infectious phase. *U. maydis* infects primordia from all aerial sections of the plant and forms a biotrophic relationship with them, allowing the fungus to propagate both inter- and intracellularly. The vast growth of fungal hyphae culminates in the production of plant tumors, which are the most visible signs of *U. maydis* caused maize smut disease (Doehlemann et al., 2008). In recent years, studies have been conducted to identify and characterize *U. maydis* virulence factors, particularly secreted effector proteins that operate in the host tissue to support fungal virulence and reprogram host tissue, hence increasing tumor growth (Lo Presti et al., 2015).

3.4 Fungal pathogenesis and its lifestyle

Fungal spores can adhere to a number of surfaces, including host and non-host plants. To reach their hosts, the spores are distributed by wind, water or insect vectors. The strong connection keeps the spores from being washed away from the surface before they can penetrate. For firm adhesion, they secrete an adhesive extracellular matrix (Kumari et al., 2021). The infection cycle begins with spore germination and germ tube production, both of which necessitate complete cell reprogramming and the establishment of specialized regulatory networks (Ebbole, 2007). Plant pathogenic fungi are classified into distinct groups based on their pathogenic lifestyle and how they feed on the host.

3.4.1 Biotrophs

Obligate biotrophs have developed to fit the lifespan of their hosts and rely on them to complete their life cycle They are the agents that cause powdery mildew and rust diseases (Schulze-Lefert and Panstruga, 2003). Obligate biotrophs have a limited ability to use common substrates as energy sources, therefore they must rely entirely on their host for energy (Schulze-Lefert and Panstruga, 2003; Wernegreen, 2005). According to genomic investigations, this is attributed to the loss of genes in important metabolic pathways such as sulfur and nitrogen absorption (Duplessis et al., 2011; Kemen and Jones, 2012; Links et al., 2011; Spanu et al., 2010). However, there is no strong evidence that the loss of biosynthetic and metabolic ability is the limiting factor in obligatory parasites thriving outside of their host (Both et al., 2005; Spanu, 2006). Another theory regarding obligatory biotrophs is that the regulation of metabolic gene expression is dependent on signals and signal gradients found exclusively in host plants (Both et al., 2005). Nonobligate biotrophs are more taxonomically diverse and can be found in a greater number of genera. They can survive and thrive in the absence of a host (Tudzynski and Scheffer, 2004).

3.4.2 Necrotrophs

Unlike biotrophic pathogens, necrotrophic pathogens feed on dead tissue. True necrotrophic pathogens attack and kill healthy plants, whereas secondary necrotrophic-like organisms are saprophytic but may occasionally infect previously compromised plants. The infection kills the tissue before colonization in necrotrophy (Oliver and Solomon, 2010).

3.4.3 Hemibiotrophs

Hemibiotrophic organisms are those that live in both biotrophic and necrotrophic environments. Hemibiotrophs are species that have a varied length of a biotrophic phase before transitioning to a necrotrophic phase, according to most experts (O'Connell et al., 2012; Perfect et al., 1998; Yi and Valent, 2013). There is an initial true biotrophic phase in this case, which is mediated by unique biotrophic organs. Pathogens produce effectors at this stage to decrease plant defense. At the end of this stage, the pathogen goes through a huge developmental transformation that allows it to shift from a biotrophic to a necrotrophic state (Yi and Valent, 2013).

All fungal pathogens can be classified as either killers or nonkillers. Nonkiller pathogens strive to keep their host alive, while killer pathogens turn living tissue into dead organic substances (Sharon and Shlezinger, 2013). As a result, cell-death regulating mechanisms are believed to be crucial in mediating pathogenic fungal-host infections (Eichmann et al., 2004; Imani et al., 2006).

3.5 Virulence factors of plant pathogenic fungi

Fungal infections use a variety of virulence factors to establish disease on host plants, including effector proteins, secondary metabolites and even tiny RNA molecules. They aid in host penetration, defense prevention and suppression and nutrition uptake. Besides these, morphological transitions are essential for host colonization and the completion of their life cycles (Lo Presti et al., 2015).

4. Control of fungal pathogens

Plant diseases pose a significant risk to global food security. Among plant infections, fungal pathogens cause the most damaging plant diseases and incur significant losses in agricultural systems (Alfen, 2001). Every plant serves as a host for different fungi, and each fungal pathogen can target a variety of plants (Li et al., 2020). Crop diseases must be adequately controlled to maintain an acceptable level of both qualitative and quantitative yields. Many people will be hungry and suffer from some of the dreadful diseases and outbreaks that occur from time to time if plant defenses are not controlled. Fungicides are widely used by farmers to manage fungal diseases due to their quick and effective action. However, its continued use causes less soil contamination, global warming and health issues in the living system (Saba, 2012).

4.1 Induction of resistance against fungi

Utilizing plant defense molecules in agricultural production to promote resistance to invasive fungal infections is the foundation of natural resistance (Nega, 2014). When different crops are affected by diseases, salicylic acid and its analogs are used to create systemic acquired resistance. According to studies, 30 g of benzo(1,2,3) thiadiazole-7-carbothioic acid S-methyl ester (BTH) can shield wheat crops from *Puccinia recondita* and Septoria species for a full growing season. Jasmonic acid and its derivatives can increase crop resistance and the synthesis of chemical that are beneficial to health (Wasternak, 2914). Methyl jasmonates prevent degradation of “Marsh Seedless” grapefruit by *P. digitatum* and post-harvest infections of strawberries by *B. cinerea* in agricultural settings (Tripathi and Dubey, 2004). Other natural substances that promote resistance against fungi include chitosan, -aminobutyric acid, glucosinolates, propolis, fusapyrone, deoxyfusapyrone, ethephon, microbial products and plant extracts. These products are used in agricultural settings all over the world to improve quality and yields (Thakur and Sohal, 2013; Tripathi and Dubey, 2004).

4.2 Biological control

Scientists are now looking for o alternatives to synthetic chemicals that are biological to address the issues with those substances (Ambrose and Belanger, 2012; Nega, 2014). Chemicals generated from microbial, botanical and animal sources are classified as biological pesticides. At present, there are hundreds of products using more than 245 registered biopesticide-active chemicals in the USA. These make up about 20% of all pesticide-active chemicals that are registered in the nation (Yoon et al., 2013). Based on the fact that some bacteria are antagonistic to the fungi that cause fungal diseases in plants, they may be used in their treatment. The discoveries of investigations using microbial antagonists against plant fungal infections in the laboratory and the field are encouraging. There is an urgent need for some eco-friendly and cost-effective approaches to resolve these issues. An effective alternative method to the use of chemical pesticides is biocontrol (Cook et al., 2007) and microbial biocontrol agents (MBCAs). Farmers are paying more attention to using organic products in the manufacture of organic food, which is largely backed by environmentalists and some customers.

The term, biological control means, the use of special microorganisms or their products that interfere with the growth of plant pathogens and pests (He et al., 2021). Generally, biocontrol agents decrease the negative potential of pathogens either directly via their antagonistic activities or indirectly through their modifying effects on plant physiology and anatomy. The key advantages of using MBCAs for plants include the establishment of an antagonistic microbial community in the rhizosphere, suppression of pathogens, overall improvement of plant health, growth promotion, increased nutrient availability and uptake and enhanced host resistance to both biotic and abiotic stresses (Shoresh and Harman, 2008; Vinale et al., 2008). The improved nutritional status of plants can also potentially lead to the alleviation of the adverse effects caused by certain pathogens (Dordas, 2008). For example, higher nitrogen content in plants can decrease their susceptibility to facultative parasites such as *Alternaria solani* (Blachinski et al., 1996) and enhanced levels of manganese can reduce the severity of diseases such as powdery mildew, downy mildew and several other diseases (Brennan, 1992; Dordas, 2008). Since MBCAs have a complex mode of action, it is difficult for a pest to fight with MBCAs. MBCAs include viruses, bacteria, nematodes and fungi. But fungal biocontrol agents are the most important among them, because of their properties such as easy delivery, improving formulation, a vast number of known pathogenic strains, easy engineering techniques and over-expression of its endogenous proteins or exogenous toxins (Butt et al., 2001; St Leger et al., 1996).

5. Endophytic fungi as biocontrol agnets

In this chapter, fungal biocontrol agents are called beneficial fungi. Beneficial fungi colonize the habitats which are exposed to pathogens such as the rhizosphere, phyllosphere and plant organs. The successful establishment, activity and proliferation of beneficial fungi depend on their adaptation and performance within physically, biologically and spatially complex systems, where a broad range of trophic and non-trophic interspecific interactions occur (Knudsen and Dandurand, 2014). When compared with synthetic fungicides, beneficial fungi have fewer non-target and environmental effects, higher efficacy against fungicide-resistant pathogens and a lower risk of resistance development in the target pathogens (Brimner and Boland, 2003). Beneficial fungi with biocontrol properties include Trichoderma, Arbuscular Mycorrhizas (AM), ectomycorrhizas (ECM), root endophytes, yeasts, *Fusarium oxysporum* (non-pathogenic strains), *Cryphonectria parasitica* (hypovirulent strains) and *Muscodor albus* (Fan et al., 2020).

5.1 Mechanism of endophytic biocontrol agents

A thorough understanding of the general biology of the three-way interaction between the endophyte, host plant and pathogen is essential to optimize the selection

Table 7.1: Examples of some beneficial fungi.

Biocontrol agent	**Target pathogen**	**Host plant**
Trichoderma asperellum	*Fusarium oxysporum* f. sp. *lycopersici*	Tomato (Patel and Saraf, 2017)
T. koningii, T. viride, T. harzianum	*Sclerotium rolfsii*	Groundnut (Hirpara et al., 2017)
T. harzianum	*Alternaria brassicae* and *A. brassicicola*	Mustard (Jagana et al., 2013)
T. harzianum	*Meloidogyne javanica*	Tomato (Sahebani and Hadavi, 2008)
T. harzianum	*Botrytis cinerea*	Bean
T. hamatum	*Phytophthora*	Rhododendron (Hoitink et al., 2006)
AM: *Glomus mosseae*	*Ralstonia solanacearum*	Tobacco (Yuan et al., 2016)
AM: *G. intraradices*	*Nacobbus aberrans*	Tomato (Maron et al., 2011)
AM: *Glomus fasciculatum*	*Alternaria alternata*	Tomato (Nair et al., 2015)
ECM: *Thelephora terrestris* and *Pisolithus tinctorius*	*Phytophthora cinnamomi*	Shortleaf pine (Colpaert, 1999)
ECM: *Laccaria bicolor*	*F. moniliforme*	Scots pine (Estefania et al., 2017)
ECM: *Suillus luteus*	*F. oxysporum*	Stone pine Scots pine (Estefania et al., 2017)
Endophyte: *Piriformospora indica*	*R. solani*	Rice (Nassimi and Taheri, 2017)
Endophyte: *Cryptosporiopsis quercina*	*Pyricularia oryzae*	Rice (Li et al., 2000)
Endophyte: *Penicillium* sp.	*Pseudomonas syringae* pv. tomato	Arabidopsis (Hossain et al., 2008)
Yeast: *Cryptococcus laurentii* and *Sporobolomyces roseus*	*B. cinerea*	Apple (Filonow et al., 1996)
Yeast: *Pichia anomala*	*B. cinerea*	Apple (Jijakli and Lepoivre, 1998)
Avirulent/Hypovirulent strains: *Cryphonectria parasitica*	*Cryphonectria parasitica*	Chestnut (Lee et al., 2006)
Verticillium nigrescens	*V. dahliae*	Cotton (Lee et al., 2006)

of endophytic biocontrol agents. Above all, the underlying physiological mechanism should be understood. Beneficial fungi decrease plant diseases directly or indirectly through the following biocontrol mechanisms: (a) competing with pathogens for space and nutrients (b) mycoparasitism (c) antibiosis (d) Mycovirus Mediated Cross-Protection (MMCP) (e) developed systemic resistance.

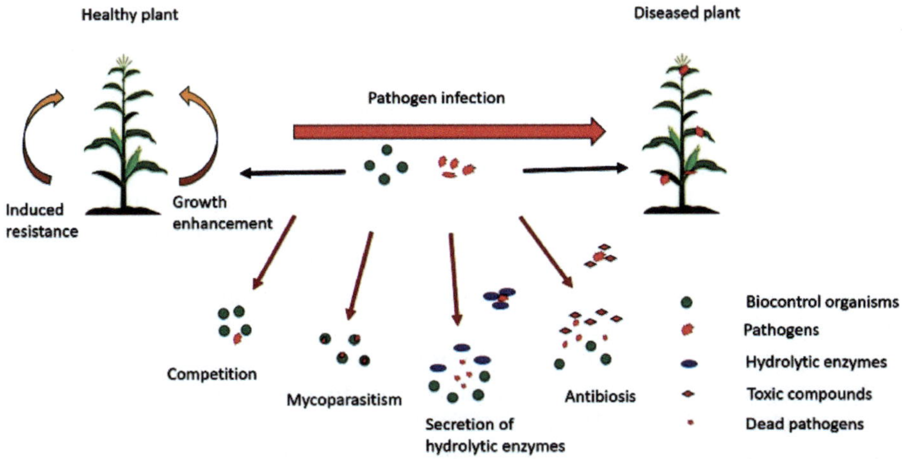

Fig. 7.1: Shows different aspects of interactions between endophytes and plants (Ghorbanpour et al., 2018).

5.1.1 Competing with pathogens for space and nutrients

Competitive exclusion is an important factor in determining the composition of the plant microbiome and prevents the colonization of the host by pathogens (Zabalgogeazcoa, 2008). Fungal endophytes can colonize different plant tissues intercellularly or intracellularly (Boyle et al., 2001). Beneficial fungus should have good plant colonization techniques and maintain a high population density inside the habitat overlap to outcompete invading pathogens for niche and resources. *Trichoderma* species have well-developed and highly diverse strategies for colonizing a wide range of niches; as a result, they are widespread across ecosystems and can be found in nearly all soil types (crop fields, forests, marshes and deserts), all climatic zones (including temperate and tropical regions, Antarctica and tundra), and unusual niches such as air, lakes, marine bivalves, shellfish and termites (Montero-Barrientos et al., 2011; Mukherjee et al., 2018).

Competitive exclusion is a biological control mechanism that is more likely to be used in conjunction with other mechanisms than on its own. Since the control effect is only local, the host plant section that the pathogen might target would need to be systemically colonized. Accordingly, clubroot symptoms were inversely linked with the presence of the dark-septate endophyte Heteroconium chaetospira in oilseed rape roots as detected by qPCR (Wison et al., 2014). But when the pathogen inoculum increased, the control effect shrank, highlighting the limitations of competition as a defense mechanism in the face of intense disease load. (2003) discovered that localized reduction of *Phytophthora* spp., disease symptoms on foliage after foliar application of a mixture of frequently isolated endophytes from cacao tree leaves to endophyte-free seedlings suggests competition as a possible mechanism. *In vitro* confrontation assays, however, it also revealed metabolite synthesis by the chosen strains, indicating that competition may not be the only mechanism at work. A promising method to study competitive exclusion involves microscopy studies, assessment of endophyte biomass *in planta* in response to disease severity, and microscopy of the broad colonization. However, (Mengistu, 2020) hypothesized that competitive exclusion is probably not crucial for endophytes acting as BCAs because the pathogen's predicted entrance point requires highly compact colonization.

5.1.2 Mycoparasitism

Mycoparasitism refers to the parasitism of one fungus by another. In a parasitic manner, the parasitic fungus obtains at least some of its nutrition from the host fungus (Jeffries, 1995). The route of entry of mycoparasite is physical penetration into the host hyphae by the development of specialized organs such as haustoria and production of various enzymes or secondary metabolites leading to the breakdown of fungal structures followed by nutrient/metabolite intake from the host fungus (Daguerre et al., 2014). Since the nutrient transfer from one fungus to another is difficult to confirm in plants, it is a challenge to prove that mycoparasitism occurs *in planta* (Jeffries, 1995). The presence of two fungi nearby does not necessarily indicate mycoparasitism. They are referred to as fungicolous if only a tight interaction between the organisms is observed. Depending on whether the parasite creates substances intended to release nutrients from the prey at a distance or is in direct physical touch, mycoparasitism can be both indirect and direct (Jeffries, 1995). In either case, the parasite produces substances such as antibiotics, toxins and Cell Wall Degrading Enzymes (CWDEs) for the release and acquisition of nutrients (Kim and Vujanovic, 2016).

Since it is simpler to spot parasitic indications in a Petri dish than on a plant, studies to confirm mycoparasitic behaviour are typically undertaken *in vitro*. It is often feasible to observe direct interactions between couples using a number of microscopy techniques. The mycoparasite may come into direct contact with the prey's hyphae, enter them, coil around them and then disrupt or deplete them. A work by Donayre and Dalisay (2016), which examined the parasitism of Geotrichum sp. (isolate EF-ds104-16), on Thanatephorus cucumeris, provides an example of mycoparasitism by endophytic fungus. However, the studies were carried out *in vitro*. The mycoparasite was isolated as an endophyte from the grass Echinochloa glabrescens. Similarly (Cao et al., 2009), used microscopy to test three fungal isolates obtained from Phragmites australis as endophytes against eight soil-borne pathogens and discovered penetration, hyphal coiling around pathogen hyphae and cytoplasm destruction. Additionally, it was shown that the extracellular CWDEs chitinases and -1,3-glucanases were produced. Even though the mycoparasitic fungi were initially isolated as endophytes, they were only evaluated *in vitro* in the presence of a plant pathogen, which means that the plant was not present to obstruct the relationship as was the fact with the cases above. It is difficult to confirm mycoparasitism for an endophyte that is developing endophytically. It would be necessary for the parasite to get close enough to its target inside the plant for parasitism to start for it to occur. Therefore, as noted by (Hyde et al., 2019) as well, it is likely that mycoparasitism, as a control principle, has little significance for endophytes.

5.1.3 Antibiosis

Biocontrol agents create a number of antimicrobial compounds that suppress or diminish phytopathogen growth and/or proliferation (Askary, 2015). This phenomenon is known as antibiosis. Endophytic fungi are rich sources of specialized secondary metabolites and other antimicrobial compounds (Thines et al., 2004). Identification and development of fungal metabolites or metabolite-producing organisms for plant disease control are of commercial importance (Schulz et al., 2002; Suryanarayanan, 2013). Alkaloids, flavonoids, peptides, phenols, quinones, steroids, terpenoids, polyketides and volatile organic compounds are examples of natural products with antimicrobial activity, with terpenoids and polyketides being the most purified anti-microbial specialized metabolites (Mousa and Raizada, 2013; Yu et al., 2010).

Different microbial species can cause the host and/or endophyte to produce metabolites, which can inhibit the growth of potentially harmful organisms (Kusari et al., 2013). In some instances, the host and the endophyte share components of a particular pathway and jointly contribute to the creation of metabolites or one partner stimulates the metabolism of the other or metabolizes its products (Aly et al., 2013; Kusari et al., 2013; Ludwig-Muller, 2015). The finding that numerous fungal endophytes were involved in the Pacific yew tree *Taxus brevifolia*'s manufacture of the anticancer medicine taxol is a well-known example. Although many of the endophyte strains were thought to be independent producers (Heinig et al., 2013), came to a different conclusion. The ability of endophytes to produce specialized metabolites and the pathogen-inhibiting properties of endophyte cultures or crude extracts have both been the subject of substantial research (Daguerre et al., 2017; Suryanarayanan, 2013). The effects of metabolites are frequently evaluated *in vitro* settings. While this indicates that an organism may have the ability to create compounds that can inhibit the pathogen, it does not suggest that pathogen inhibition will occur in plants (Köhl et al., 2011; Laur et al., 2018). While the plant's disease may be controlled by direct toxicity or by generating resistance when *in vitro* produced metabolites are applied directly to the plant (Mathivanan et al., 2008; Sinha and Trivedi, 1969), these

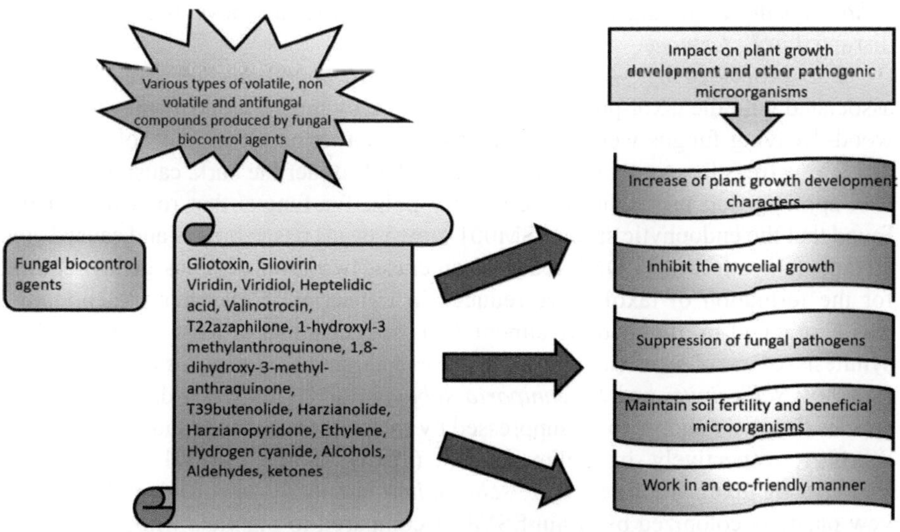

Fig. 7.2: Different biocontrol agents produced by endophytic fungi and their actions on plant growth development and other pathogenic microorganisms.

substances might not even be produced *in planta* by the endophyte. This is a result of significant variations in nutrient availability, environmental factors and interactions with other species. For example, in the case of *planta*, there are intricate interactions between the BCA (biocontrol agents), plant, innate microbiota and disease. Some bioactive substances made by the BCA's biosynthetic machinery are only expressed in connection with plants or potentially in tripartite interactions with pathogens. It is optimal to confirm that an in planta-generated antimicrobial chemical is in contact with the pathogen to show that it is involved in biocontrol induced by an endophyte. Given that the endophyte is confined to one area of the plant and that the amount of metabolite generated may be very little, it could be challenging to quantify. Furthermore, metabolites can be translocated in the plant from the endophyte to the site of pathogen infection, whereas volatile organic compounds produced by a BCA can diffuse to the site of the pathogen infection, as reported in some studies (Gabriel et al., 2018). It should still be determined whether there is enough of the active ingredient at the pathogen site to control the disease or if there are additional mechanisms at work (Lugtenberg et al., 2016).

In order to manage *Sclerotinia homoeocarpa* in *Festuca rubra* ssp. *rubra*., Tian et al. (2017) studied the role of *planta*-released antifungal protein of *Epichlo festucae*. *F. rubra* ssp., *rubra*'s above-ground regions are systemically colonized by *E. festucae*, a characteristic that has not been demonstrated for other fescues (Tian et al., 2017). According to Ambrose and Belanger (2012), a quantitative transcriptome analysis of *F. rubra* ssp. *rubra* infected with *E. festucae* revealed the secreted protein Efe-AfpA, which made up 6% of the fungal transcriptome and shared similarities with antifungal proteins secreted by *Penicillium* sp., and *Aspergillus* spp. Additionally, recombinant Efe-afpA gene product produced in *Pichia pastoris* and partly purified Efe-AfpA isolated from apoplastic fluid of endophyte-infected red fescue both shown *in vitro* antifungal efficacy against *S. homoeocarpa*. It was demonstrated that EfeAfpA specifically targeted the cell membrane of *S. homoeocarpa* utilizing a viability/vital dye. The gene has not yet been successfully disrupted in *E. festucae*.

The processes by which the endophyte *Paraconiothyrium* strain SSM001 associated with the taxol-producing yew tree (*Taxus* spp.) warded against hazardous wood-decaying fungus were examined by Soliman et al., 2015 and Soliman et al., 2013. Yew trees frequently hyperbranch from buds under the bark, causing continual peel splitting and providing an easy entry point for fungus that rot wood. It was found that the endophytic strain SSM001 grew toward tissue breaks and caused cuts. Taxadiene synthase and DXP reductoisomerase, two crucial Taxus genes required for the formation of taxol, were reduced in transcript levels after the endophyte was eliminated by fungicide treatment. This suggests that at least a portion of the synthesis of taxol was carried out by the endophyte. *Heterobasidion annosum*, *Phaeolus schweinitzi* and *Perenniporia subacida* are three wood-decaying fungus species that were shown to be suppressed by treatment with taxol and the endophyte SSM001, respectively, by Soliman et al. (2015). Strain SSM001 itself was not inhibited by taxol. Additionally, *P. schweinitzii* biomass was considerably lower in yew plantlets colonized by strain SSM001 compared to plantlets whose endophyte colonization was earlier decreased by fungicide injection. An *in vitro* investigation revealed that *P. schweinitzii*, chitin and an elicitor from *P. schweinitzii* all contributed to the production of taxol by the strain SSM001. It would be interesting to determine which endophyte genes contribute to the manufacture of taxol and validate that their deletion results in the loss of the ability to control fungi.

5.1.4 Mycovirus-mediated cross-protection

In mycovirus-mediated cross-protection, first the plants are pre-inoculated with a hypovirulent/avirulent strain of the fungal pathogen, and then the isolate reduces the virulence of the pathogenic isolate via transmission of a hypovirulence-associated mycovirus (Kyrychenko et al., 2018). The biocontrol mechanism is considered an independent process in this case because it is carried out by anastomosis of a hypovirulence-associated mycovirus (mostly dsRNA) from an avirulent isolate to a virulent isolate. That is, rather than affecting the plant defense system, the biocontrol fungus directly impacts the pathogen (Ghabrial and Suzuki, 2009).

5.1.5 Induced systemic resistance

Induced resistance to diseases is the process of active resistance that is dependent on the physical or chemical barriers of the host plant and is activated by biotic or abiotic agents (Kloepper et al., 1992). The inducing substance leads the host to produce a translocatable signal, which causes the host to react defensively to a future pathogen attack on plants. Through the plant signal, the influencing substance causes specific metabolic changes, protein synthesis and differential gene expression. Reduced disease levels are the result of these changes in plant metabolism, which alter the plant's suitability as a host. Both live entities and chemical substances can act as inducing agents, and the host's response might be local or systemic.

In plants, several defense pathways exist that can prevent or alleviate the adverse impact of pathogens, which may be activated on exposure to pests, pathogens, beneficial microbes or certain chemical substances. These defense pathways involve the evolution of MAMPs (microbial-associated molecular patterns), PAMPs (pathogen-activated molecular patterns) or DAMPs (Damage-Associated Molecular Patterns) (Boller and Felix, 2009). Compounds of microbial origin are frequently recognized by plant receptors and trigger MAMPs or PAMPs pathways (Nrnberger and Kemmerling, 2008). The basic components of fungal cell walls, chitin and -glucans are recognized by plant receptors and both have been demonstrated to cause MAMPs pathway (Lyon, 2014). Usually reported inducing agents include endophyte-secreted proteins and peptides. The host may recognize some secreted enzymes during infection and colonization, such as xylanases, cellulases and chitinases, and this recognition could trigger defense responses (Belien et al., 2006; Druzhinina et al., 2011; Rotblat et al., 2002). During both pathogenic and endophytic colonization processes, fungal effectors, that are frequently tiny cysteine-rich proteins, are produced to increase compatibility with the host plant by modifying defense responses and physiology (Bent and Mackey, 2007; Hacquard et al., 2016; Nogueira-Lopez et al., 2018). When recognized by the host, these tiny proteins can also trigger reactions in plants (Djonovic et al., 2007; Sáenz-Mata et al., 2014). Additionally, there is proof that certain fungi chemicals, created to combat rival organisms, can also cause resistance (Luo et al., 2010). These instances demonstrate the number of endophyte-derived chemicals that may trigger plant defense responses.

Microorganisms confer two types of systemic resistance to host plants named Induced Systemic Resistance (ISR) and systemic acquired resistance (SAR) (Birkenbihl et al., 2017). SAR is salicylic acid-dependent, whereas ISR is a salicylic acid-independent pathway. The hormones Jasmonic Acid (JA) and ethylene (ET) are important in ISR (Pieterse et al., 2012). The plant may accurately modify the defense response by cross-communication between the SA, JA and ET signals. *Trichoderma asperellum*, *Penicillium* sp., and the endophyte *Serendipita indica*, which are associated with plants, have all been discovered to have JA/ET-dependent systemic resistance (Hossain et al., 2008; Shoresh et al., 2005; Stein et al., 2008). On the other hand, in other pathosystems, *S. indica* produced resistance without involving the JA/ET route (Waller et al., 2005) and *T. asperellum* associated with plants activated resistance in an SA-reliant manner (Yoshioka et al., 2012). This suggests that the functions of hormones and their potential interactions are complicated, and the introduction of a microbe to a plant is likely to alter the entire profile of hormones rather than just their levels. Additionally, depending on the host and the strain of the inducing substance, the function of hormones may change. Usually, pathogens trigger the SAR pathway and beneficial microbes often set off ISR, however, exceptions are there (Nie et al., 2017). Beneficial fungi can trigger the defense signaling pathways via MAMPs before infection by pathogens and confer ISR to the host plants, which is mainly regulated by plant hormones (Romera et al., 2019). During initial interactions, local suppression of the root immune responses has been proposed as the common feature of ISR-eliciting beneficial microorganisms that can facilitate their accommodation within roots (Pieterse et al., 2014).

5.2 Selected classes of beneficial fungi

5.2.1 *Trichoderma* spp.

Trichoderma species can be found in practically all soil types, including crop fields, forests, marshes and deserts, as well as all climatic zones, including temperate and tropical regions, Antarctica and tundra, as well as uncommon niches including air, lakes, marine bivalves, shellfishes and termites (Gal-Hemed et al., 2011; Montero-Barrientos et al., 2011; Rajesh et al., 2016). They are common because they have well-developed and very diversified tactics for colonizing such different habitats. Several bacterial and fungal biocontrol agents are already accessible, with *Trichoderma* spp., being one of the most adaptable, having long been used to manage plant pathogenic fungi. An earlier study found that treating sunflower and mugbean seeds with conidial suspensions of *Trichoderma* spp., reduced seed rot, damping-off and root rot (Yaqub and Shahzad, 2008). *Trichoderma* spp., is abundant in agricultural soils as well as other substrates such as decomposing wood (Samuels, 1996). They are typically anaerobic, facultative and cosmopolitan fungi. They belong to the subdivision Deuteromycetes and most of the strains are adapted to an asexual lifecycle (Harman et al., 2004). Besides its role as a biocontrol agent, *Trichoderma* spp., can stimulate the colonization in rhizospheres, stimulate plant growth and root growth and enhance plant defense mechanisms (Harman et al., 2004; Vinale et al., 2008).

Table 7.2: Different kinds of fungal biocontrol agents and the biocontrol mechanisms involved have shown in the table.

Agents/ Mechanisms	**Competing with Pathogens for Space and Nutrients**	**Mycoparasitism**	**Antibiosis**	**Mycovirus-Mediated Cross Protection**	**Induced Systemic Resistance**
Trichoderma	+	+	+	–	+
Arbuscular Mycorrhizas	+	–	–	–	+
Ectomycorrhizas	+	–	+	–	NIL
Endophytes	+	–	+	–	+
Yeasts	+	+	+	–	+
Avirulent/ hypovirulent strains of *Cryphonectria parasitica*	–	–	–	+	–

5.2.2 Arbuscular mycorrhizal fungi

Mycorrhizae are symbiotic connections between plant roots and fungi that are distinguished by the two-way transfer of vital nutrients from the plant to fungus and inorganic nutrients from the fungus to plant. The majority of land plants have a symbiotic relationship with fungi. Only seven varieties of mycorrhizae have been identified based on form and anatomy: ectomycorrhiza, endomycorrhiza or arbuscular mycorrhiza, ericoid mycorrhiza, arbutoid mycorrhiza, monotropoid mycorrhiza, ect-endomycorrhiza and orchidaceous mycorrhiza. Among these, arbuscular mycorrhiza is an important biocontrol agent. They are obligate root symbionts that are thought to colonize more than 80% of all land plant species. They promote plant development in a number of ways, most notably by enhancing nutrient absorption in exchange for photosynthetic carbon from the host (Smith et al., 2009). Alleviation of plant stress caused by different biotic and abiotic stress is another way (Gianinazzi et al., 2010; Vos et al., 2011). Thirdly, Arbuscular Mycorrhizal Fungi (AMF) show biocontrol effects against a wide range of plant pathogens, most of which are soil-borne fungal pathogens (Fritz et al., 2006; Harrier and Watson, 2004; Whipps, 2004). AMF can suppress both biotrophic and necrotrophic pathogens either directly or indirectly (Horn et al., 2014).

5.2.3 Entomopathogenic fungi

Entomopathogenic fungi (EPF) can infect both insects and plants. They can control arthropods, insects and harmful plant fungi. Entomopathogenic fungi were the first organisms to be utilized as biocontrol agents against pests (Khan et al., 2012). Causative organisms to insects are approximately 90 genera with over 700 species (Khachatourians and Qazi, 2008). The majority of EPF species are classified as Ascomycota or Zygomycota. As common natural adversaries of aphids and other agricultural pests, entomopathogenic fungi such as *Verticillium lecanii*, *Beauveria bassiana* and *Metarhizium anisopliae* have been extensively studied (Mouatcho et al., 2011; Roberts and St. Leger, 2004; Wang and Xu, 2012). When a pathogen infects a plant, *Beauveria bassiana* colonizes endophytically and generates systemic resistance. They can also help to minimize diseases caused by soil-borne pathogens like *Pythium*, *Fusarium* and *Rhizoctonia* (Ownley et al., 2008). Beauveria bassiana filtrate was found to contain low molecular weight compounds such as cyclic peptides and Cyclosporins A and C with insecticidal properties such as beauvericin, enniatins, bassianolide (Roberts and Humber, 1981), Oosporein (a red-colored dibenzoquinone with antibiotic activity against gram-positive bacteria) and immunosuppressive cyclic peptides (Dannon et al., 2020).

5.3 Antimicrobial secondary metabolites produced by endophytic fungi

Extraneous microbial compounds can be recognized as alien by plants. The plant works by rapidly inducing defense responses. The activity of pathogenic and endophytic fungi in inducing plant responses differs (Wani et al., 2015; Xu et al., 2015). Many endophytic fungi can create secondary metabolites with promising antagonistic activity against horticultural and agricultural plant diseases and pests. Epicoccum nigrum, an endophytic fungus found in sugarcane, produces secondary metabolites such as epicorazines A-B (Baute et al., 1978), epirodines A-B (Ben, 2016), flavipin (Bissett, 1987), epicoccines A-D,epipiridones and epicocarines (KemamiWangunandHertweck,2007).FlavipinandepicorazinesA-Bhavebeenlinked to *E. nigrum* biocontrol action (Madrigal and Melgarejo, 1995). *Pestalotiopsis microspore* produces ambuic acid, which is an antifungal and anti-oomycete substance. It is effective against *Fusarium* spp., and *Pythium ultimum* (Strobel et al., 2002). *Colletotrichum* sp., isolated from *Artemisia annua*, produce Colletonoic acid (Bills et al., 2002), which has good anti-bacterial, anti-fungal and anti-algal activities (Bills et al., 2002). Colletotric acid is another metabolite generated by *Colletotrichum gloeosporioides*, an *A. mongolica* endophytic fungus. It is antibacterial against bacteria as well as the fungus *Helminthosporium sativum* (Zou et al., 2000). Cordycepsidone A is a plant pathogenic fungus that was created by the endophytic fungus Cordyceps dipterigena. It shows significant antifungal action against *Gibberella fujikuroi* (Varughese et al., 2012). Pestacin and isopestacin are antimicrobial anti-oxidants produced by *Pestalotiopsis icrospore*, an endophyte of *Terminalia morobensis* (Harper et al., 2003;). Cryptocandin is a lipopeptide that inhibits the growth of plant-pathogenic fungi such as *Sclerotinia sclerotiorum* and *Botrytis cinerea* (Strobel et al., 1999) isolated from *Cryptosporiopsis quercina*, a fungus commonly associated with hardwood species in Europe.

6. Conclusion and future perspectives

The study of fungal endophytes with biocontrol capabilities is currently a subject that requires substantial research. In the last few decades, advanced molecular tools for detailed analyses of the tripartite interaction between endophyte, plant and pathogen have advanced substantially. Techniques such as genomics, transcriptomics, proteomics and metabolomics can be used to elucidate the pathways that affect different species and their capacity for metabolite production (Kaul et al., 2016; Laur et al., 2018). RNA-seq is currently emerging as a technique for understanding the tripartite connections between all three organisms. Light microscopy, scanning electron microscopy and transmission electron microscopy studies to see multiple phenotypes at various time points during the interaction are critical to supporting this evidence (Laur et al., 2018). Other techniques, such as gene expression investigations, enzyme assays and the use of knockout mutants or RNA interference, provide a deep understanding of the molecular basis of endophytic antagonists. Mass spectrometry-based molecular 3-D cartography has enormous potential as a method for detecting the distribution of individual metabolites with microorganisms *in situ* and, when combined with molecular genetic research, for elucidating the role of certain metabolites (Floros et al., 2017).

It can be concluded that investigating the nature of plant-endophyte interaction and the conditions for successful establishment is critical. It will aid in the identification of fungal roles and chemicals needed for disease reduction, as well as plant responses generated by the interaction. This results in the development of indicators for the efficiency of a specific biocontrol agent, as well as testing the impacts of plant genotype, intrinsic microbial population and the environment, as well as offering a structured approach for the discovery of novel endophytes with desired features.

Acknowledgement

The authors acknowledge DST-PURSE PII programme, the state of Kerala plan fund project and Jaivam Project for the support they provided.

References

Agrios, G.N. (2005). Plant pathology. from https://search.ebscohost.com/login.aspx?direct=true&scope=site&db=nlebk&db=nlabk&AN=189453.

Agrios, G.N. (2009). Plant Pathogens and Disease: General Introduction, 613–646. doi: 10.1016/b978-012373944-5.00344-8.

Alfen, N.K.V. (2001). Fungal Pathogens of Plants. doi: 10.1038/npg.els.0000362.

Almeida, F., Rodrigues, M.L. and Coelho, C. (2019). The still underestimated problem of fungal diseases worldwide. Frontiers in Microbiology, 10. doi: 10.3389/fmicb.2019.00214.

Aly, A.H., Debbab, A. and Proksch, P. (2013). Fungal endophytes—secret producers of bioactive plant metabolites. Pharmazie, 68(7): 499–505.

Ambrose, K.V. and Belanger, F.C. (2012). SOLiD-SAGE of endophyte-infected red fescue reveals numerous effects on host transcriptome and an abundance of highly expressed fungal secreted proteins. PLoS One, 7(12): e53214. doi: 10.1371/journal.pone.0053214.

Anderson, P.K., Cunningham, A.A., Patel, N.G., Morales, F.J., Epstein, P.R. and Daszak, P. (2004). Emerging infectious diseases of plants: Pathogen pollution, climate change and agrotechnology drivers. Trends Ecol. Evol., 19(10): 535–544. doi: 10.1016/j.tree.2004.07.021.

Arnold, A.E., Mejía, L.C., Kyllo, D., Rojas, E.I., Maynard, Z., Robbins, N. and Herre, E.A. (2003). Fungal endophytes limit pathogen damage in a tropical tree. Proceedings of the National Academy of Sciences, 100(26): 15649–15654. doi: 10.1073/pnas.2533483100.

Askary, T.H. (2015). Nematophagous fungi as biocontrol agents of phytonematodes, 81–125. doi: 10.1079/9781780643755.0081.

Atkinson, N.J. and Urwin, P.E. (2012). The interaction of plant biotic and abiotic stresses: From genes to the field. Journal of Experimental Botany, 63(10): 3523–3543. doi: 10.1093/jxb/ers100.

Avelino, J., Cristancho, M., Georgiou, S., Imbach, P., Aguilar, L., Bornemann, G., Peter Läderach, Francisco Anzueto, Allan J. Hruska and Morales, C. (2015). The coffee rust crises in Colombia and Central America (2008–2013): Impacts, plausible causes and proposed solutions. Food Security, 7(2): 303–321. doi: 10.1007/s12571-015-0446-9.

Awuchi, C.G., Ondari, E.N., Ogbonna, C.U., Upadhyay, A.K., Baran, K.R., Okpala, C.O., Korzeniowska, M.F. and Guiné, R.P. (2021). Mycotoxins affecting animals, foods, humans, and plants: Types, occurrence, toxicities, action mechanisms, prevention, and detoxification strategies—A revisit. Foods, 10(6). https://doi.org/10.3390/foods10061279.

Baute, M.A., Deffieux, G., Baute, R. and Neveu, A. (1978). New antibiotics from the fungus Epicoccum nigrum. I. Fermentation, isolation and antibacterial properties. J. Antibiot. (Tokyo), 31(11): 1099–1101. doi: 10.7164/antibiotics.31.1099.

Belien, T., Van Campenhout, S., Robben, J. and Volckaert, G. (2006). Microbial endoxylanases: Effective weapons to breach the plant cell-wall barrier or, rather, triggers of plant defense systems? Mol. Plant Microbe Interact., 19(10): 1072–1081. doi: 10.1094/MPMI-19-1072.

Ben, J.J. Lugtenberg and others, (2016). Fungal endophytes for sustainable crop production. FEMS Microbiology Ecology, 92(12): fiw194. https://doi.org/10.1093/femsec/fiw194.

Bent, A.F. and Mackey, D. (2007). Elicitors, effectors, and R genes: The new paradigm and a lifetime supply of questions. Annu. Rev. Phytopathol., 45: 399–436. doi: 10.1146/annurev.phyto.45.062806.094427.

Bills, G.F., Dombrowski, A.W., Peldez, F., Polishook, J.D., An, Z., Watling, R. and Robinson, C.H. (2002). Recent and future discoveries of pharmacologically active metabolites from tropical fungi.

Birkenbihl, R.P., Liu, S. and Somssich, I.E. (2017). Transcriptional events defining plant immune responses. Curr. Opin. Plant Biol., 38: 1–9. doi: 10.1016/j.pbi.2017.04.004.

Bissett, J. (1987). Fungi associated with urea-formaldehyde foam insulation in Canada. Mycopathologia, 99(1): 47–56. doi: 10.1007/BF00436681

Blachinski, D., Shtienberg, D., Dinoor, A., Kafkafi, U., Sujkowski, L.S., Zitter, T.A. and Fry, W.E. (1996). Influence of foliar application of nitrogen and potassium onalternaria diseases in potato, tomato and cotton. Phytoparasitica, 24(4): 281–292. doi: 10.1007/bf02981411.

Boller, T. and Felix, G. (2009). A renaissance of elicitors: Perception of microbe-associated molecular patterns and danger signals by pattern-recognition receptors. Annu. Rev. Plant Biol., 60: 379–406. doi: 10.1146/annurev.arplant.57.032905.105346.

Both, M., Csukai, M., Stumpf, M.P. and Spanu, P.D. (2005). Gene expression profiles of Blumeria graminis indicate dynamic changes to primary metabolism during development of an obligate biotrophic pathogen. Plant Cell, 17(7): 2107–2122. doi: 10.1105/tpc.105.032631.

Boyle, C., Götz, M., Dammann-Tugend, U. and Schulz, B.J. (2001). Endophyte-host interactions III. Local vs. systemic colonization. Symbiosis, 31: 259–281.

Brennan, R.F. (1992). The role of manganese and nitrogen nutrition in the susceptibility of wheat plants to take-all in Western Australia. Fertilizer Research, 31(1): 35–41. doi: 10.1007/bf01064225.

Brimner, T.A. and Boland, G.J. (2003). A review of the non-target effects of fungi used to biologically control plant diseases. Agriculture, Ecosystems & Environment, 100(1): 3–16. doi: 10.1016/s0167-8809(03)00200-7.

Burdon, J.J. and Thrall, P.H. (2009). Coevolution of plants and their pathogens in natural habitats. Science, 324(5928): 755–756. doi: 10.1126/science.1171663.

Butt, T.M., Jackson, C.W. and Magan, N. (2001). Introduction - fungal biological control agents: Progress, problems and potential. 1–8. doi: 10.1079/9780851993560.0001.

Cao Ronghua, Xiaoguang Liu, Kexiang Gao, Kurt Mendgen, Zhensheng Kang, Jianfeng Gao, Yang Dai and Xue Wang. (2009). Mycoparasitism of endophytic fungi isolated from reed on soilborne phytopathogenic fungi and production of cell wall-degrading enzymes *in vitro*. Curr. Microbiol., 59(6): 584–592. doi: 10.1007/s00284-009-9477-9.

Card, S., Johnson, L., Teasdale, S. and Caradus, J. (2016). Deciphering endophyte behaviour: The link between endophyte biology and efficacious biological control agents. FEMS Microbiol. Ecol., 92(8). doi: 10.1093/femsec/fiw114.

Card, S., Johnson, L., Teasdale, S., Caradus, J. and Muyzer, G. (2016). Deciphering endophyte behaviour: the link between endophyte biology and efficacious biological control agents. FEMS Microbiology Ecology, 92(8): fiw114. doi: 10.1093/femsec/fiw114.

Chitarra, G.S., Abee, T., Rombouts, F.M., Posthumus, M.A. and Dijksterhuis, J. (2004). Germination of penicillium paneum Conidia is regulated by 1-octen-3-ol, a volatile self-inhibitor. Appl. Environ. Microbiol., 70(5): 2823–2829. doi: 10.1128/AEM.70.5.2823-2829.2004.

Colpaert, J.V. (1999). Thelephora, 325–345. doi: 10.1007/978-3-662-06827-4_14.

Cook, S.M., Khan, Z.R. and Pickett, J.A. (2007). The use of push-pull strategies in integrated pest management. Annual Review of Entomology, 52(1): 375–400. doi: 10.1146/annurev.ento.52.110405.091407.

Cools, H.J. and Fraaije, B.A. (2008). Are azole fungicides losing ground against Septoria wheat disease? Resistance mechanisms in Mycosphaerella graminicola. Pest Management Science, 64(7): 681–684. doi: 10.1002/ps.1568.

Cruz, C.D., Bockus, W.W., Stack, J.P., Tang, X., Valent, B., Pedley, K.F. and Peterson, G.L. (2012). Preliminary assessment of resistance among U.S. wheat cultivars to the triticum pathotype of Magnaporthe oryzae. Plant Dis., 96(10): 1501–1505. doi: 10.1094/PDIS-11-11-0944-RE.

Dagdas, Y.F., Yoshino, K., Dagdas, G., Ryder, L.S., Bielska, E., Steinberg, G. and Talbot, N.J. (2012). Septin-mediated plant cell invasion by the rice blast fungus, Magnaporthe oryzae. Science, 336(6088): 1590–1595. doi: 10.1126/science.1222934.

Daguerre, Y., Edel-Hermann, V. and Steinberg, C. (2017). Fungal genes and metabolites associated with the biocontrol of soil-borne plant pathogenic fungi, 33–104. doi: 10.1007/978-3-319-25001-4_27.

Daguerre, Y., Siegel, K., Edel-Hermann, V. and Steinberg, C. (2014). Fungal proteins and genes associated with biocontrol mechanisms of soil-borne pathogens: A review. Fungal Biology Reviews, 28(4): 97–125. doi: 10.1016/j.fbr.2014.11.001.

Dannon H. Fabrice, Dannon A. Elie, Douro-Kpindou O. Kobi, Zinsou A. Valerien, Houndete A. Thomas, Toffa-Mehinto Joëlle, Elegbede I.A.T. Maurille, Olou B. Dénis and Tamò Manuele. (2020). Toward the efficient use of Beauveria bassiana in integrated cotton insect pest management. Journal of Cotton Research, 3(1). doi: 10.1186/s42397-020-00061-5.

Devi, R. et al. 2020. Beneficial fungal communities from different habitats and their roles in plant growth promotion and soil health. Microbial Biosystems, 5(1): 21–47.

de Jong, J.C., McCormack, B.J., Smirnoff, N. and Talbot, N.J. (1997). Glycerol generates turgor in rice blast. Nature, 389(6648): 244–244. doi: 10.1038/38418.

De Silva, N.I. (2016). Mycosphere Essays 9: Defining biotrophs and hemibiotrophs. Mycosphere, 7(5): 545–559. doi: 10.5943/mycosphere/7/5/2.

Dean Ralph, Jan A.L. Van Kan, Zacharias A. Pretorius, Kim E. Hammond-Kosack, Antonio Di Pietro, Pietro D. Spanu, Jason J. Rudd, Marty Dickman, Regine Kahmann, Jeff Ellis and Gary D. Foster. (2012). The Top 10 fungal pathogens in molecular plant pathology. Mol. Plant Pathol., 13(4): 414–430. doi: 10.1111/j.1364-3703.2011.00783.x.

Djonovic, S., Vargas, W.A., Kolomiets, M.V., Horndeski, M., Wiest, A. and Kenerley, C.M. (2007). A proteinaceous elicitor Sm1 from the beneficial fungus Trichoderma virens is required for induced systemic resistance in maize. Plant Physiol., 145(3): 875–889. doi: 10.1104/pp.107.103689.

Doehlemann, G., Okmen, B., Zhu, W. and Sharon, A. (2017). Plant pathogenic Fungi. Microbiol. Spectr., 5(1). doi: 10.1128/microbiolspec.FUNK-0023-2016.

Doehlemann, G., Ökmen, B., Zhu, W., Sharon, A., Heitman, J. and Howlett, B.J. (2017). Plant pathogenic Fungi. Microbiology Spectrum, 5(1). doi: 10.1128/microbiolspec.FUNK-0023-2016.

Doehlemann, G., Wahl, R., Vranes, M., de Vries, R.P., Kamper, J. and Kahmann, R. (2008). Establishment of compatibility in the Ustilago maydis/maize pathosystem. J. Plant Physiol., 165(1): 29–40. doi: 10.1016/j.jplph.2007.05.016.

Donayre, D.K.M. and Dalisay, T.U. (2016). Identities, characteristics, and assemblages of dematiaceous-endophytic fungi isolated from tissues of barnyard grass weed. Phil. J. Sci., 145: 153–164.

Dordas, C. (2008). Role of nutrients in controlling plant diseases in sustainable agriculture. A review. Agronomy for Sustainable Development, 28(1): 33–46. doi: 10.1051/agro:2007051.

Druzhinina, I.S., Verena Seidl-Seiboth, Alfredo Herrera-Estrella, Benjamin A. Horwitz, Charles M. Kenerley, Enrique Monte, Prasun K. Mukherjee, Susanne Zeilinger, Igor V. Grigoriev and Christian P. Kubicek. (2011). Trichoderma: The genomics of opportunistic success. Nat. Rev. Microbiol., 9(10): 749–759. doi: 10.1038/nrmicro2637.

Duncan, K.E. and Howard, R.J. (2000). Cytological analysis of wheat infection by the leaf blotch pathogen Mycosphaerella graminicola. Mycological Research, 104(9): 1074–1082. doi: 10.1017/s0953756299002294.

Duplessis, S., Cuomo, C.A., Lin, Y.C., Aerts, A., Tisserant, E., Veneault-Fourrey, C. and Martin, F. (2011). Obligate biotrophy features unraveled by the genomic analysis of rust fungi. Proc. Natl. Acad. Sci. USA, 108(22): 9166–9171. doi: 10.1073/pnas.1019315108.

Ebbole, D.J. (2007). Magnaporthe as a model for understanding host-pathogen interactions. Annu. Rev. Phytopathol., 45: 437–456. doi: 10.1146/annurev.phyto.45.062806.094346.

Eichmann, R., Schultheiss, H., Kogel, K.H. and Huckelhoven, R. (2004). The barley apoptosis suppressor homologue BAX inhibitor-1 compromises nonhost penetration resistance of barley to the inappropriate pathogen Blumeria graminis f. sp. tritici. Mol. Plant Microbe Interact., 17(5): 484–490. doi: 10.1094/MPMI.2004.17.5.484.

El Hussein, A. (2014). Isolation and identification of Streptomyces rochei strain active against phytopathogenic Fungi. British Microbiology Research Journal, 4(10): 1057–1068. doi: 10.9734/bmrj/2014/11074.

Estefania, M., Jaime, O., Juan, A.P., Valentín, P. and Julio, J.D. (2017). Influence of Suillus luteus on Fusarium damping-off in pine seedlings. African Journal of Biotechnology, 16(6): 268–273. doi: 10.5897/ajb11.1164.

Fan, X.L., Zhang, Y., Li, X. and Fu, Q.L. (2020). Mechanisms underlying the protective effects of mesenchymal stem cell-based therapy. Cell Mol. Life Sci., 77(14): 2771–2794. doi: 10.1007/s00018-020-03454-6.

Filonow, A.B., Vishniac, H.S., Anderson, J.A. and Janisiewicz, W.J. (1996). Biological control of Botrytis cinereain apple by yeasts from various habitats and their putative mechanisms of antagonism. Biological Control, 7(2): 212–220. doi: 10.1006/bcon.1996.0086.

Fletcher, J., Bender, C., Budowle, B., Cobb, W.T., Gold, S.E., Ishimaru, C.A., Luster, D., Melcher, U., Murch, R., Scherm, H., Seem, R.C., Sherwood, J.L., Sobral, B.W. and Tolin, S.A. (2006). Plant pathogen forensics: Capabilities, needs, and recommendations. Microbiol. Mol. Biol. Rev., 70(2): 450–471. doi: 10.1128/MMBR.00022-05.

Floros, D.J., Petras, D., Kapono, C.A., Melnik, A.V., Ling, T.J., Knight, R. and Dorrestein, P.C. (2017). Mass spectrometry based molecular 3D-cartography of plant metabolites. Front. Plant Sci., 8: 429. doi: 10.3389/fpls.2017.00429.

Fontana, D.C., de Paula, S., Torres, A.G., de Souza, V.H.M., Pascholati, S.F., Schmidt, D. and Dourado Neto, D. (2021). Endophytic Fungi: Biological control and induced resistance to phytopathogens and abiotic stresses. Pathogens, 10(5): 570. doi: 10.3390/pathogens10050570. PMID: 34066672; PMCID: PMC8151296.

Fraaije, B.A., Cools, H.J., Fountaine, J., Lovell, D.J., Motteram, J., West, J.S. and Lucas, J.A. (2005). Role of ascospores in further spread of QoI-resistant Cytochrome b Alleles (G143A) in field populations of Mycosphaerella graminicola. Phytopathology®, 95(8): 933–941. doi: 10.1094/phyto-95-0933.

Fritz, M., Jakobsen, I., Lyngkjaer, M.F., Thordal-Christensen, H. and Pons-Kuhnemann, J. (2006). Arbuscular mycorrhiza reduces susceptibility of tomato to Alternaria solani. Mycorrhiza, 16(6): 413–419. doi: 10.1007/s00572-006-0051-z.

Gabriel, K.T., Joseph Sexton, D. and Cornelison, C.T. (2018). Biomimicry of volatile-based microbial control for managing emerging fungal pathogens. J. Appl. Microbiol., 124(5): 1024–1031. doi: 10.1111/jam.13667.

Gal-Hemed, I., Atanasova, L., Komon-Zelazowska, M., Druzhinina, I.S., Viterbo, A. and Yarden, O. (2011). Marine isolates of Trichoderma spp. as potential halotolerant agents of biological control for arid-zone agriculture. Applied and Environmental Microbiology, 77(15): 5100–5109. doi: 10.1128/aem.00541-11.

Gatto, M.A., Ippolito, A., Linsalata, V., Cascarano, N.A., Nigro, F., Vanadia, S. and Di Venere, D. (2011). Activity of extracts from wild edible herbs against postharvest fungal diseases of fruit and vegetables. Postharvest Biology and Technology, 61(1): 72–82. doi: 10.1016/j.postharvbio.2011.02.005.

Ghabrial, S.A. and Suzuki, N. (2009). Viruses of plant pathogenic Fungi. Annu. Rev. Phytopathol., 47(1): 353–384. doi: 10.1146/annurev-phyto-080508-081932.

Ghorbanpour, M., Omidvari, M., Abbaszadeh-Dahaji, P., Omidvar, R. and Kariman, K. (2018). Mechanisms underlying the protective effects of beneficial fungi against plant diseases. Biological Control, 117: 147–157. doi: 10.1016/j.biocontrol.2017.11.006.

Gianinazzi, S., Gollotte, A., Binet, M.-N., van Tuinen, D., Redecker, D. and Wipf, D. (2010). Agroecology: The key role of arbuscular mycorrhizas in ecosystem services. Mycorrhiza, 20(8): 519–530. doi: 10.1007/s00572-010-0333-3.

Gilbert, G.S. (2002). Evolutionary ecology of plant diseases in natural ecosystems. Annual Review of Phytopathology, 40(1): 13–43. doi: 10.1146/annurev.phyto.40.021202.110417.

Gimenez, E., Salinas, M. and Manzano-Agugliaro, F. (2018). Worldwide research on plant defense against biotic stresses as improvement for sustainable agriculture. Sustainability, 10(2): 391. doi: 10.3390/su10020391.

Gould, A.B. (2009). Fungi: Plant Pathogenic, 457–477. doi: 10.1016/b978-012373944-5.00347-3.

Grover, M., Ali, S.Z., Sandhya, V., Rasul, A. and Venkateswarlu, B. (2010). Role of microorganisms in adaptation of agriculture crops to abiotic stresses. World Journal of Microbiology and Biotechnology, 27(5): 1231–1240. doi: 10.1007/s11274-010-0572-7.

Hacquard Stéphane, Barbara Kracher, Kei Hiruma, Philipp C. Münch, Ruben Garrido-Oter, Michael R. Thon, Aaron Weimann, Ulrike Damm, Jean-Félix Dallery, Matthieu Hainaut, Bernard Henrissat Olivier Lespinet, Soledad Sacristán, Emiel Ver Loren van Themaat, Eric Kemen, Alice C. McHardy, Paul Schulze-Lefert and Richard J. O'Connell. (2016). Survival trade-offs in plant roots during colonization by closely related beneficial and pathogenic fungi. Nat. Commun., 7: 11362. doi: 10.1038/ncomms11362.

Hadwiger, L.A. (2005). Plant Science, 168(6): 1643. doi: 10.1016/j.plantsci.2005.02.019.

Hariharan, G. and Prasannath, K. (2021). Recent advances in molecular diagnostics of fungal plant pathogens: A mini review. Frontiers in Cellular and Infection Microbiology, 10. doi: 10.3389/fcimb.2020.600234.

Harman, G.E. (2011). Trichoderma—not just for biocontrol anymore. Phytoparasitica, 39(2): 103–108. doi: 10.1007/s12600-011-0151-y.

Harman, G.E., Howell, C.R., Viterbo, A., Chet, I. and Lorito, M. (2004). Trichoderma species—Opportunistic, avirulent plant symbionts. Nat. Rev. Microbiol., 2(1): 43–56. doi: 10.1038/nrmicro797.

Harper, J.K., Atta M. Arif, Eugene J. Ford, Gary A. Strobel, John A. Porco, David P. Tomer, Kim L. Oneill, Elizabeth M. Heider and David M. Grant. (2003). Pestacin: A 1,3-dihydro isobenzofuran from Pestalotiopsis microspora possessing antioxidant and antimycotic activities. Tetrahedron, 59(14): 2471–2476. doi: 10.1016/s0040-4020(03)00255-2.

Harrier, L.A. and Watson, C.A. (2004). The potential role of arbuscular mycorrhizal (AM) fungi in the bioprotection of plants against soil-borne pathogens in organic and/or other sustainable farming systems. Pest Management Science, 60(2): 149–157. doi: 10.1002/ps.820.

He, D.C., He, M.H., Amalin, D.M., Liu, W., Alvindia, D.G. and Zhan, J. (2021). Biological control of plant diseases: An evolutionary and eco-economic consideration. Pathogens, 10(10): 1311. doi: 10.3390/pathogens10101311. PMID: 34684260; PMCID: PMC8541133.

Heinig, U., Scholz, S. and Jennewein, S. (2013). Getting to the bottom of Taxol biosynthesis by fungi. Fungal Diversity, 60(1): 161–170. doi: 10.1007/s13225-013-0228-7.

Hirpara, D.G., Gajera, H.P., Hirpara, H.Z. and Golakiya, B.A. (2017). Antipathy of <b><i>Trichoderma</i></b> against <b><i>Sclerotium rolfsii </i></b>Sacc.: Evaluation of cell wall-degrading enzymatic activities and molecular diversity analysis of antagonists. Journal of Molecular Microbiology and Biotechnology, 27(1): 22–28. doi: 10.1159/000452997.

Hoitink, H.A.J., Madden, L.V. and Dorrance, A.E. (2006). Systemic resistance induced by Trichoderma spp.: Interactions between the host, the pathogen, the biocontrol agent, and soil organic matter quality. Phytopathology®, 96(2): 186–189. doi: 10.1094/phyto-96-0186.

Horbach, R., Navarro-Quesada, A.R., Knogge, W. and Deising, H.B. (2011). When and how to kill a plant cell: Infection strategies of plant pathogenic fungi. J. Plant Physiol., 168(1): 51–62. doi: 10.1016/j.jplph.2010.06.014.

Horn, S., Caruso, T., Verbruggen, E., Rillig, M.C. and Hempel, S. (2014). Arbuscular mycorrhizal fungal communities are phylogenetically clustered at small scales. ISME J., 8(11): 2231–2242. doi: 10.1038/ismej.2014.72.

Hossain, M.M., Sultana, F., Kubota, M. and Hyakumachi, M. (2008). Differential inducible defense mechanisms against bacterial speck pathogen in Arabidopsis thaliana by plant-growth-promoting-fungus Penicillium sp. GP16-2 and its cell free filtrate. Plant and Soil, 304(1-2): 227–239. doi: 10.1007/s11104-008-9542-3.

Hussein, H.S. and Brasel, J.M. (2001). Toxicity, metabolism, and impact of mycotoxins on humans and animals. Toxicology, 167(2): 101–134. doi: 10.1016/s0300-483x(01)00471-1.

Hyde, K.D., Xu, J., Rapior, S. et al. (2019). The amazing potential of fungi: 50 ways we can exploit fungi industrially. Fungal Diversity, 97: 1–136. https://doi.org/10.1007/s13225-019-00430-9.

Imani, J., Baltruschat, H., Stein, E., Jia, G., Vogelsberg, J., Kogel, K.H. and Huckelhoven, R. (2006). Expression of barley BAX Inhibitor-1 in carrots confers resistance to Botrytis cinerea. Mol. Plant Pathol., 7(4): 279–284. doi: 10.1111/j.1364-3703.2006.00339.x.

Jagana, M., Zacharia, S., Lal, A.A. and Basayya. (2013). Management of Alternaria blight in Mustard. Annals of Plant Protection Sciences, 21: 441–442.

Jeffries, P. (1995). Biology and ecology of mycoparasitism. Canadian Journal of Botany, 73(S1): 1284–1290. doi: 10.1139/b95-389.

Jijakli, M.H. and Lepoivre, P. (1998). Characterization of an Exo-β-1,3-Glucanase produced by Pichia anomala strain K, antagonist of Botrytis cinereaon apples. Phytopathology®, 88(4): 335–343. doi: 10.1094/phyto.1998.88.4.335.

Joosten, M. and de Wit, P. (1999). The tomato-cladosporium fulvum interaction: A versatile experimental system to study plant-pathogen interactions. Annu. Rev. Phytopathol., 37: 335–367. doi: 10.1146/annurev.phyto.37.1.335.

Kämper Jörg, Regine Kahmann, Michael Bölker, Li-Jun Ma, Thomas Brefort, Barry J. Saville, Flora Banuett, James W. Kronstad, Scott E. Gold, Olaf Müller, Michael H. Perlin, Han A.B. Wösten, Ronald de Vries, José Ruiz-Herrera, Cristina G. Reynaga-Peña, Karen Snetselaar, Michael McCann, José Pérez-Martín, Michael Feldbrügge, Christoph W. Basse, Gero Steinberg, Jose I. Ibeas, William Holloman, Plinio Guzman and Bruce W. Birren. (2006). Insights from the genome of the biotrophic fungal plant pathogen Ustilago maydis. Nature, 444(7115): 97–101. doi: 10.1038/nature05248.

Kankanala, P., Czymmek, K. and Valent, B. (2007). Roles for rice membrane dynamics and plasmodesmata during biotrophic invasion by the blast fungus. Plant Cell, 19(2): 706–724. doi: 10.1105/tpc.106.046300.

Kaul, S., Sharma, T. and Dhar, K.M. (2016). "Omics" tools for better understanding the plant–endophyte interactions. Frontiers in Plant Science, 7. doi: 10.3389/fpls.2016.00955.

Kema, G. (1996). Histology of the pathogenesis of mycosphaerella graminicolain wheat. Phytopathology, 86(7): 777. doi: 10.1094/Phyto-86-777.

Kemami Wangun, H.V. and Hertweck, C. (2007). Epicoccarines A, B and epipyridone: Tetramic acids and pyridone alkaloids from an Epicoccum sp. associated with the tree fungus Pholiota squarrosa. Organic & Biomolecular Chemistry, 5(11): 1702. doi: 10.1039/b702378b.

Kemen, E. and Jones, J.D. (2012). Obligate biotroph parasitism: Can we link genomes to lifestyles? Trends Plant Sci., 17(8): 448–457. doi: 10.1016/j.tplants.2012.04.005.

Khachatourians, G.G. and Qazi, S.S. (2008). Entomopathogenic Fungi: Biochemistry and molecular biology, 33–61. doi: 10.1007/978-3-540-79307-6_3.

Khan, S., Guo, L., Maimaiti, Y., Mijit, M. and Qiu, D. (2012). Entomopathogenic Fungi as microbial biocontrol agent. Molecular Plant Breeding. doi: 10.5376/mpb.2012.03.0007.

Khang, C.H., Romain Berruyer, Martha C. Giraldo, Prasanna Kankanala, Sook-Young Park, Kirk Czymmek, Seogchan Kang and Barbara Valent (2010). Translocation of Magnaporthe oryzae effectors into rice cells and their subsequent cell-to-cell movement. Plant Cell, 22(4): 1388–1403. doi: 10.1105/tpc.109.069666.

Kim, S.H. and Vujanovic, V. (2016). Relationship between mycoparasites lifestyles and biocontrol behaviors against Fusarium spp. and mycotoxins production. Appl. Microbiol. Biotechnol., 100(12): 5257–5272. doi: 10.1007/s00253-016-7539-z.

Kloepper, J.W., Tuzun, S. and Kuć, J.A. (1992). Proposed definitions related to induced disease resistance. Biocontrol Science and Technology, 2(4): 349–351. doi: 10.1080/09583159209355251.

Knogge, W. (1996a). Fungal infection of plants. Plant Cell, 8(10): 1711–1722. doi: 10.1105/tpc.8.10.1711.

Knogge, W. (1996b). Fungal infection of plants. The Plant Cell, 1711–1722. doi: 10.1105/tpc.8.10.1711.

Knudsen, G.R. and Dandurand, L.-M.C. (2014). Ecological complexity and the success of fungal biological control agents. Advances in Agriculture, 2014: 1–11. doi: 10.1155/2014/542703.

Knudsen, I.M.B., Hockenhull, J., Funck Jensen, D., Gerhardson, B., Hökeberg, M., Tahvonen, R., Teperi, E., Sundheim, L. and Henriksen, B. (1997). European Journal of Plant Pathology, 103(9): 775–784. doi: 10.1023/a:1008662313042.

Köhl, J., Postma, J., Nicot, P., Ruocco, M. and Blum, B. (2011). Stepwise screening of microorganisms for commercial use in biological control of plant-pathogenic fungi and bacteria. Biological Control, 57(1): 1–12. doi: 10.1016/j.biocontrol.2010.12.004.

Kumari, A., Tripathi, A.H., Gautam, P., Gahtori, R., Pande, A., Singh, Y., Madan, T. and Upadhyay, S.K. (2021). Adhesins in the virulence of opportunistic fungal pathogens of human. Mycology, 12(4): 296–324. doi: 10.1080/21501203.2021.1934176. PMID: 34900383; PMCID: PMC8654403.

Kumar, N. and Khurana, S.M.P. (2021). Trichoderma-plant-pathogen interactions for benefit of agriculture and environment, 41–63. doi: 10.1016/b978-0-12-822919-4.00003-x.

Kumar, V., Basu, M.S. and Rajendran, T.P. (2008). Mycotoxin research and mycoflora in some commercially important agricultural commodities. Crop Protection, 27(6): 891–905. doi: 10.1016/j.cropro.2007.12.011.

Kusari, S., Hertweck, C. and Spiteller, M. (2012). Chemical ecology of endophytic fungi: Origins of secondary metabolites. Chemistry & Biology, 19(7): 792–798. doi: 10.1016/j.chembiol.2012.06.004.

Kusari, S., Pandey, S.P. and Spiteller, M. (2013). Untapped mutualistic paradigms linking host plant and endophytic fungal production of similar bioactive secondary metabolites. Phytochemistry, 91: 81–87. doi: 10.1016/j.phytochem.2012.07.021.

Kyrychenko, A.N., Tsyganenko, K.S. and Olishevska, S.V. (2018). Hypovirulence of mycoviruses as a tool for biotechnological control of phytopathogenic Fungi. Cytology and Genetics, 52(5): 374–384. doi: 10.3103/s0095452718050043.

Lambert, J.E. (2001). International Journal of Primatology, 22(2): 189–201. doi: 10.1023/a:1005667313906.

Laur, J., Ramakrishnan, G.B., Labbe, C., Lefebvre, F., Spanu, P.D. and Belanger, R.R. (2018). Effectors involved in fungal-fungal interaction lead to a rare phenomenon of hyperbiotrophy in the tritrophic system biocontrol agent-powdery mildew-plant. New Phytol., 217(2): 713–725. doi: 10.1111/nph.14851.

Lee, S.H., Moon, B.J. and Lee, J.K. (2006). Characteristics of hypovirulent strains of chestnut blight fungus,cryphonectria parasitica, isolated in Korea. Mycobiology, 34(2): 61. doi: 10.4489/myco.2006.34.2.061.

Li, J., Cornelissen, B. and Rep, M. (2020). Host-specificity factors in plant pathogenic fungi. Fungal Genetics and Biology, 144: 103447.

Li, J., Strobel, G., Harper, J., Lobkovsky, E. and Clardy, J. (2000). Cryptocin, a potent tetramic acid antimycotic from the endophytic fungus Cryptosporiopsis cf. quercina. Organic Letters, 2: 767–770.

Libera Lo Presti, Daniel Lanver, Gabriel Schweizer, Shigeyuki Tanaka, Liang Liang, Marie Tollot, Alga Zuccaro, Stefanie Reissmann and Regine Kahmann (2015). Fungal effectors and plant susceptibility. Annu. Rev. Plant Biol., 66: 513–545. doi: 10.1146/annurev-arplant-043014-114623.

Links, M.G., Eric Holub, Rays H.Y. Jiang, Andrew G. Sharpe, Dwayne Hegedus, Elena Beynon, Dean Sillito, Wayne E. Clarke, Shihomi Uzuhashi and Mohammad H. Borhan. (2011). *De novo* sequence assembly of Albugo candida reveals a small genome relative to other biotrophic oomycetes. BMC Genomics, 12: 503. doi: 10.1186/1471-2164-12-503.

Ludwig-Muller, J. (2015). Plants and endophytes: equal partners in secondary metabolite production? Biotechnol. Lett., 37(7): 1325–1334. doi: 10.1007/s10529-015-1814-4.

Lugtenberg, B. (2015). doi: 10.1007/978-3-319-08575-3.

Lugtenberg, B.J., Caradus, J.R. and Johnson, L.J. (2016). Fungal endophytes for sustainable crop production. FEMS Microbiol. Ecol., 92(12). doi: 10.1093/femsec/fiw194.

Lugtenberg, B.J.J., Caradus, J.R., Johnson, L.J. and Muyzer, G. (2016). Fungal endophytes for sustainable crop production. FEMS Microbiology Ecology, 92(12): fiw194. doi: 10.1093/femsec/fiw194.

Luo, Y., Dan-Dan Zhang, Xiao-Wei Dong, Pei-Bao Zhao, Lei-Lei Chen, Xiao-Yan Song, Xing-Jun Wang, Xiu-Lan Chen, Mei Shi and Yu-Zhong Zhang. (2010). Antimicrobial peptaibols induce defense responses and systemic resistance in tobacco against tobacco mosaic virus. FEMS Microbiol. Lett., 313(2): 120–126. doi: 10.1111/j.1574-6968.2010.02135.x.

Lyon, G.D. (2014). Agents That Can Elicit Induced Resistance, 11–40. doi: 10.1002/9781118371848.ch2.

Madrigal, C. and Melgarejo, P. (1995). Morphological effects of Epiccocum nigrum and its antibiotic flavipin on Monilinia laxa. Canadian Journal of Botany, 73(3): 425–431. doi: 10.1139/b95-043.

Maron, J.L., Marler, M., Klironomos, J.N. and Cleveland, C.C. (2011). Soil fungal pathogens and the relationship between plant diversity and productivity. Ecology Letters, 14(1): 36–41. doi: 10.1111/j.1461-0248.2010.01547.x.

Mathivanan, N., Prabavathy, V.R. and Vijayanandraj, V.R. (2008). The effect of fungal secondary metabolites on bacterial and fungal pathogens, 14: 129–140. doi: 10.1007/978-3-540-74543-3_7.

Mengistu, A.A. 2020. Endophytes: Colonization, behaviour, and their role in defense mechanism. Int. J. Microbiol., 2020: 6927219. doi: 10.1155/2020/6927219. PMID: 32802073; PMCID: PMC7414354.

Montero-Barrientos, M., Hermosa, R., Cardoza, R.E., Gutiérrez, S. and Monte, E. (2011). Functional analysis of the Trichoderma harzianum nox1 gene, encoding an NADPH oxidase, relates production of reactive oxygen species to specific biocontrol activity against pythium ultimum. Applied and Environmental Microbiology, 77(9): 3009–3016. doi: 10.1128/aem.02486-10.

Moss, M.O. (2008). Fungi, quality and safety issues in fresh fruits and vegetables. Journal of Applied Microbiology, 104(5): 1239–1243. doi: 10.1111/j.1365-2672.2007.03705.x.

Mouatcho, J.C., Koekemoer, L.L., Coetzee, M. and Brooke, B.D. (2011). The effect of entomopathogenic fungus infection on female fecundity of the major malaria vector, anopheles funestus. African Entomology, 19(3): 725–729. doi: 10.4001/003.019.0311.

Mousa, W.K. and Raizada, M.N. (2013). The diversity of anti-microbial secondary metabolites produced by fungal endophytes: An interdisciplinary perspective. Front. Microbiol., 4: 65. doi: 10.3389/fmicb.2013.00065.

Mukherjee, S., Joardar, N., Sengupta, S. and Sinha Babu, S.P. (2018). Gut microbes as future therapeutics in treating inflammatory and infectious diseases: Lessons from recent findings. J. Nutr. Biochem., 61: 111–128. doi: 10.1016/j.jnutbio.2018.07.010.

Nair, A., Kolet, S.P., Thulasiram, H.V., Bhargava, S. and Bisseling, T. (2015). Systemic jasmonic acid modulation in mycorrhizal tomato plants and its role in induced resistance against Alternaria alternata. Plant Biology, 17(3): 625–631. doi: 10.1111/plb.12277.

Nassimi, Z. and Taheri, P. (2017). Endophytic fungus Piriformospora indica induced systemic resistance against rice sheath blight via affecting hydrogen peroxide and antioxidants. Biocontrol Science and Technology, 27(2): 252–267. doi: 10.1080/09583157.2016.1277690.

Nega, A. (2014). Review on concepts in biological control of plant pathogens. Journal of Biology, Agriculture and Healthcare, 4: 33–54.

Nejat, N. and Mantri, N. (2017). Plant immune system: Crosstalk between responses to biotic and abiotic stresses the missing link in understanding plant defence. Current Issues in Molecular Biology, 1–16. doi: 10.21775/cimb.023.001.

Ngoune Liliane, T. and Shelton Charles, M. (2020). Factors Affecting Yield of Crops. doi: 10.5772/intechopen.90672.

Nie, P., Li, X., Wang, S., Guo, J., Zhao, H. and Niu, D. (2017). Induced systemic resistance against Botrytis cinerea by Bacillus cereus AR156 through a JA/ET- and NPR1-dependent signaling pathway and activates PAMP-triggered immunity in arabidopsis. Front. Plant Sci., 8: 238. doi: 10.3389/fpls.2017.00238.

Nogueira-Lopez, G., Greenwood, D.R., Middleditch, M., Winefield, C., Eaton, C., Steyaert, J.M. and Mendoza-Mendoza, A. (2018). The apoplastic secretome of trichoderma virens during interaction with maize roots shows an inhibition of plant defence and scavenging oxidative stress secreted proteins. Front. Plant Sci., 9: 409. doi: 10.3389/fpls.2018.00409.

Nrnberger, T. and Kemmerling, B. (2008). Pathogen-Associated Molecular Patterns (PAMP) and PAMP-Triggered Immunity, 16–47. doi: 10.1002/9781444301441.ch2.

O'Connell, R.J., Michael R. Thon, Stéphane Hacquard, Stefan G. Amyotte, Jochen Kleemann, Maria F. Torres, Ulrike Damm, Ester A. Buiate, Lynn Epstein, Noam Alkan, Janine Altmüller, Lucia Alvarado-Balderrama, Christopher A. Bauser, Christian Becker, Bruce W. Birren, Zehua Chen, Jaeyoung Choi, Jo Anne Crouch, Jonathan P. Duvick, Mark A. Farman, Pamela Gan, David Heiman, Bernard Henrissat, Richard J. Howard and Lisa J. Vaillancourt. (2012). Lifestyle transitions in plant pathogenic Colletotrichum fungi deciphered by genome and transcriptome analyses. Nat. Genet., 44(9): 1060–1065. doi: 10.1038/ng.2372.

Oliver, R.P. and Solomon, P.S. (2010). New developments in pathogenicity and virulence of necrotrophs. Curr. Opin. Plant Biol. doi: 10.1016/j.pbi.2010.05.003.

Osherov, N. and May, G.S. (2001). The molecular mechanisms of conidial germination. FEMS Microbiol. Lett., 199(2): 153–160. doi: 10.1111/j.1574-6968.2001.tb10667.x.

Ownley, B.H., Griffin, M.R., Klingeman, W.E., Gwinn, K.D., Moulton, J.K. and Pereira, R.M. (2008). Beauveria bassiana: Endophytic colonization and plant disease control. J. Invertebr. Pathol., 98(3): 267–270. doi: 10.1016/j.jip.2008.01.010.

Patel, S. snd Saraf, M. (2017). Biocontrol efficacy of Trichoderma asperellum MSST against tomato wilting by Fusarium oxysporum f. sp. lycopersici. Archives of Phytopathology and Plant Protection, 50(5-6): 228–238. doi: 10.1080/03235408.2017.1287236.

Perfect, S.E., O'Connell, R. ., Green, E.F., Doering-Saad, C. and Green, J.R. (1998). Expression cloning of a fungal proline-rich glycoprotein specific to the biotrophic interface formed in the Colletotrichum-bean interaction. Plant J., 15(2): 273–279. doi: 10.1046/j.1365-313x.1998.00196.x.

Pieterse, C.M.J., Van der Does, D., Zamioudis, C., Leon-Reyes, A. and Van Wees, S.C.M. (2012). Hormonal modulation of plant immunity. Annual Review of Cell and Developmental Biology, 28(1): 489–521. doi: 10.1146/annurev-cellbio-092910-154055.

Pieterse, C.M., Zamioudis, C., Berendsen, R.L., Weller, D.M., Van Wees, S.C. and Bakker, P.A. (2014). Induced systemic resistance by beneficial microbes. Annu. Rev. Phytopathol., 52: 347–375. doi: 10.1146/annurev-phyto-082712-102340.

Quaedvlieg, W., Kema, G.H.J., Groenewald, J.Z., Verkley, G.J.M., Seifbarghi, S., Razavi, M., Mirzadi Gohari, A., Mehrabi, R. and Crous, P.W. (2011). <I>Zymoseptoria</I> gen. nov.: A new genus to accommodate <I>Septoria-</I>like species occurring on graminicolous hosts. Persoonia-Molecular Phylogeny and Evolution of Fungi, 26(1): 57–69. doi: 10.3767/003158511x571841.

Rajesh, R.W., Rahul, M.S. and Ambalal, N.S. (2016). Trichoderma: A significant fungus for agriculture and environment. African Journal of Agricultural Research, 11(22): 1952–1965. doi: 10.5897/ajar2015.10584.

Roberts, D.W. and Humber, R.A. (1981). Entomogenous Fungi, 201–236. doi: 10.1016/b978-0-12-179502-3.50014-5.

Roberts, D.W. and St. Leger, R.J. (2004). Metarhizium spp., Cosmopolitan insect-pathogenic Fungi. Mycological Aspects, 54: 1–70. doi: 10.1016/s0065-2164(04)54001-7.

Romera, F.J., María J. García, Carlos Lucena, Ainhoa Martínez-Medina, Miguel A. Aparicio, José Ramos, Esteban Alcántara, Macarena Angulo and Rafael Pérez-Vicente. (2019). Induced Systemic Resistance (ISR) and Fe deficiency responses in dicot plants. Front. Plant Sci., 10: 287. doi: 10.3389/fpls.2019.00287.

Rotblat, B., Enshell-Seijffers, D., Gershoni, J.M., Schuster, S. and Avni, A. (2002). Identification of an essential component of the elicitation active site of the EIX protein elicitor. Plant J., 32(6): 1049–1055. doi: 10.1046/j.1365-313x.2002.01490.x.

Saba, H. (2012). Trichoderma—A promising plant growth stimulator and biocontrol agent. Mycosphere, 3(4): 524–531. doi: 10.5943/mycosphere/3/4/14.

Sáenz-Mata, J., Salazar-Badillo, F.B. and Jiménez-Bremont, J.F. (2014). Transcriptional regulation of Arabidopsis thaliana WRKY genes under interaction with beneficial fungus Trichoderma atroviride. Acta Physiologiae Plantarum, 36(5): 1085–1093. doi: 10.1007/s11738-013-1483-7.

Sahebani, N. and Hadavi, N. (2008). Biological control of the root-knot nematode Meloidogyne javanica by Trichoderma harzianum. Soil Biology and Biochemistry, 40(8): 2016–2020. doi: 10.1016/j.soilbio.2008.03.011.

Samuels, G.J. (1996). Trichoderma: A review of biology and systematics of the genus. Mycological Research, 100(8): 923–935. doi: 10.1016/s0953-7562(96)80043-8.

Schäfer, W. (1994). Molecular mechanisms of fungal pathogenicity to plants. Annual Review of Phytopathology, 32(1): 461–477. doi: 10.1146/annurev.py.32.090194.002333.

Schulz, B., Boyle, C., Draeger, S., Römmert, A.-K. and Krohn, K. (2002). Endophytic fungi: A source of novel biologically active secondary metabolites. Mycol. Res., 106(9): 996–1004. doi: 10.1017/s0953756202006342.

Schulze-Lefert, P. and Panstruga, R. (2003). Establishment of biotrophy by parasitic fungi and reprogramming of host cells for disease resistance. Annu. Rev. Phytopathol., 41: 641–667. doi: 10.1146/annurev.phyto.41.061002.083300.

Shalaby, M.Y., Al-Zahrani, K.H., Baig, M.B., Straquadine, G.S. and Aldosari, F. (2011). Threats and challenges to sustainable agriculture and rural development in Egypt: Implications for agricultural extension. Journal of Animal and Plant Sciences, 21: 581–588.

Shameer, S. and Prasad, T.N.V.K.V. (2018). Plant growth promoting rhizobacteria for sustainable agricultural practices with special reference to biotic and abiotic stresses. Plant Growth Regulation, 84(3): 603–615. doi: 10.1007/s10725-017-0365-1.

Sharon, A. and Shlezinger, N. (2013). Fungi infecting plants and animals: Killers, non-killers, and cell death. PLoS Pathog., 9(8): e1003517. doi: 10.1371/journal.ppat.1003517.

Shoresh, M. and Harman, G.E. (2008). The relationship between increased growth and resistance induced in plants by root colonizing microbes. Plant Signal Behav., 3(9): 737–739. doi: 10.4161/psb.3.9.6605.

Shoresh, M., Yedidia, I. and Chet, I. (2005). Involvement of jasmonic acid/ethylene signaling pathway in the systemic resistance induced in cucumber by Trichoderma asperellum T203. Phytopathology, 95(1): 76–84. doi: 10.1094/PHYTO-95-0076.

Singh, D., Jackson, G., Hunter, D., Fullerton, R., Lebot, V., Taylor, M., Iosefa T., Okpul, T. and Tyson, J. (2012). Taro leaf blight—a threat to food security. Agriculture, 2(3): 182–203. doi: 10.3390/agriculture2030182.

Singh, R.S. (2018). Plant Diseases: Oxford & Ibh Publishing Company Pvt Limited.

Sinha, A.K. and Trivedi, N. (1969). Immunization of rice plants against helminthosporium infection. Nature, 223(5209): 963–964. doi: 10.1038/223963a0.

Smith, S.E., Facelli, E., Pope, S. and Andrew Smith, F. (2009). Plant performance in stressful environments: Interpreting new and established knowledge of the roles of arbuscular mycorrhizas. Plant and Soil, 326(1-2): 3–20. doi: 10.1007/s11104-009-9981-5.

Soliman, S.S., Greenwood, J.S., Bombarely, A., Mueller, L.A., Tsao, R., Mosser, D.D. and Raizada, M.N. (2015). An endophyte constructs fungicide-containing extracellular barriers for its host plant. Curr Biol., 25(19): 2570–2576. doi: 10.1016/j.cub.2015.08.027.

Soliman, S.S.M., Trobacher, C.P., Tsao, R., Greenwood, J.S. and Raizada, M.N. (2013). A fungal endophyte induces transcription of genes encoding a redundant fungicide pathway in its host plant. BMC Plant Biology, 13(1). doi: 10.1186/1471-2229-13-93.

Spanu, P.D. (2006). Why do some fungi give up their freedom and become obligate dependants on their host? New Phytol., 171(3): 447–450. doi: 10.1111/j.1469-8137.2006.01802.x.

Spanu, P.D., James C. Abbott, Joelle Amselem, Timothy A. Burgis, Darren M. Soanes, Kurt Stüber, Emiel Ver Loren van Themaat, James K.M. Brown, Sarah A. Butcher, Sarah J. Gurr, Marc-Henri Lebrun, Christopher J. Ridout, Paul Schulze-Lefert, Nicholas J. Talbot, Nahal Ahmadinejad, Christian Ametz, Geraint R. Barton, Mariam Benjdia, Przemyslaw Bidzinski, Laurence V. Bindschedler, Maike Both, Marin T. Brewer, Lance Cadle-Davidson, Molly M. Cadle-Davidson, Jerome Collemare, Rainer Cramer, Omer Frenkel, Dale Godfrey, James Harriman, Claire Hoede, Brian C. King, Sven Klages, Jochen Kleemann, Daniela Knoll, Prasanna S. Koti, Jonathan Kreplak, Francisco J. López-Ruiz, Xunli Lu, Takaki Maekawa, Siraprapa Mahanil, Cristina Micali, Michael G. Milgroom, Giovanni Montana, Sandra Noir, Richard J. O'Connell, Simone Oberhaensli, Francis Parlange, Carsten Pedersen, Hadi Quesneville, Richard Reinhardt, Matthias Rott, Soledad Sacristán, Sarah M. Schmidt, Moritz Schön, Pari Skamnioti, Hans Sommer, Amber Stephens, Hiroyuki Takahara, Hans Thordal-Christensen, Marielle Vigouroux, Ralf Wessling, Thomas Wicker and Ralph Panstruga. (2010). Genome expansion and gene loss in powdery mildew fungi reveal tradeoffs in extreme parasitism. Science, 330(6010): 1543–1546. doi: 10.1126/science.1194573.

St Leger, R., Joshi, L., Bidochka, M.J. and Roberts, D.W. (1996). Construction of an improved mycoinsecticide overexpressing a toxic protease. Proc. Natl. Acad. Sci. USA, 93(13): 6349–6354. doi: 10.1073/pnas.93.13.6349.

Staats, M., van Baarlen, P. and van Kan, J.A. (2005). Molecular phylogeny of the plant pathogenic genus Botrytis and the evolution of host specificity. Mol. Biol. Evol., 22(2): 333–346. doi: 10.1093/molbev/msi020.

Stein, E., Molitor, A., Kogel, K.-H. and Waller, F. (2008). Systemic resistance in arabidopsis conferred by the Mycorrhizal fungus Piriformospora indica requires jasmonic acid signaling and the cytoplasmic function of NPR1. Plant and Cell Physiology, 49(11): 1747–1751. doi: 10.1093/pcp/pcn147.

Strobel, G., Daisy, B. and Castillo, U. (2005). Novel Natural Products From Rainforest Endophytes, 329–351. doi: 10.1007/978-1-59259-976-9_15.

Strobel, G., Ford, E., Worapong, J., Harper, J.K., Arif, A.M., Grant, D.M., Peter C.W. Fung and Ming Wah Chau, R. (2002). Isopestacin, an isobenzofuranone from Pestalotiopsis microspora, possessing antifungal and antioxidant activities. Phytochemistry, 60(2): 179–183. doi: 10.1016/s0031-9422(02)00062-6.

Strobel, G.A., Miller, R.V., Martinez-Miller, C., Condron, M.M., Teplow, D.B. and Hess, W.M. (1999). Cryptocandin, a potent antimycotic from the endophytic fungus Cryptosporiopsis cf. quercina. Microbiology (Reading), 145(Pt 8): 1919–1926. doi: 10.1099/13500872-145-8-1919.

Suffert, F., Sache, I. and Lannou, C. (2011). Early stages of septoria tritici blotch epidemics of winter wheat: Build-up, overseasoning, and release of primary inoculum. Plant Pathology, 60(2): 166–177. doi: 10.1111/j.1365-3059.2010.02369.x.

Suryanarayanan, T.S. (2013). Endophyte research: going beyond isolation and metabolite documentation. Fungal Ecology, 6(6): 561–568. doi: 10.1016/j.funeco.2013.09.007.

Sweeney, M.J. and Dobson, A.D. (1998). Mycotoxin production by Aspergillus, Fusarium and Penicillium species. Int. J. Food Microbiol., 43(3): 141–158. doi: 10.1016/s0168-1605(98)00112-3.

Talbot, N.J. (2003). On the trail of a cereal killer: Exploring the biology of Magnaporthe grisea. Annu. Rev. Microbiol., 57: 177–202. doi: 10.1146/annurev.micro.57.030502.090957.

Talhinhas Pedro, Dora Batista, Inês Diniz, Ana Vieira, Diogo N. Silva, Andreia Loureiro, Sílvia Tavares, Ana Paula Pereira, Helena G. Azinheira, Leonor Guerra-Guimarães, Vítor Várzea and Maria do Céu Silva. (2017). The coffee leaf rust pathogen Hemileia vastatrix: One and a half centuries around the tropics. Molecular Plant Pathology, 18(8): 1039–1051. doi: 10.1111/mpp.12512.

Thakur, M. and Sohal, B.S. (2013). Role of elicitors in inducing resistance in plants against pathogen infection: A review. ISRN Biochemistry, 2013: 1–10. doi: 10.1155/2013/762412.

Thambugala, K.M., Daranagama, D.A., Phillips, A.J.L., Kannangara, S.D. and Promputtha, I. (2020). Fungi vs. Fungi in biocontrol: An Overview of fungal antagonists applied against fungal plant pathogens. Front. Cell Infect. Microbiol., 10: 604923. doi: 10.3389/fcimb.2020.604923. PMID: 33330142; PMCID: PMC7734056.

Thines, E., Anke, H. and Weber, R.W. (2004). Fungal secondary metabolites as inhibitors of infection-related morphogenesis in phytopathogenic fungi. Mycol. Res., 108(Pt 1): 14–25. doi: 10.1017/s0953756203008943.

Thomma, B.P., HP, V.A.N.E., Crous, P.W. and PJ, D.E.W. (2005). Cladosporium fulvum (syn. Passalora fulva), a highly specialized plant pathogen as a model for functional studies on plant pathogenic Mycosphaerellaceae. Mol. Plant Pathol., 6(4): 379–393. doi: 10.1111/j.1364-3703.2005.00292.x.

Tian Binnian, Jiatao Xie, Yanping Fu, Jiasen Cheng, Bo Li, Tao Chen, Ying Zhao, Zhixiao Gao, Puyun Yang, Martin J. Barbetti, Brett M. Tyler and Daohong Jiang. (2020). A cosmopolitan fungal pathogen of dicots adopts an endophytic lifestyle on cereal crops and protects them from major fungal diseases. The ISME Journal, 14(12): 3120–3135. doi: 10.1038/s41396-020-00744-6.

Tian, Z., Wang, R., Ambrose, K.V., Clarke, B.B. and Belanger, F.C. (2017). The Epichloe festucae antifungal protein has activity against the plant pathogen Sclerotinia homoeocarpa, the causal agent of dollar spot disease. Sci. Rep., 7(1): 5643. doi: 10.1038/s41598-017-06068-4.

Tripathi, P. and Dubey, N.K. (2004). Exploitation of natural products as an alternative strategy to control postharvest fungal rotting of fruit and vegetables. Postharvest Biology and Technology, 32(3): 235–245. doi: 10.1016/j.postharvbio.2003.11.005.

Tudzynski, P. and Scheffer, J. (2004). Claviceps purpurea: Molecular aspects of a unique pathogenic lifestyle. Mol. Plant Pathol., 5(5): 377–388. doi: 10.1111/j.1364-3703.2004.00237.x.

Valent, B. and Chumley, F.G. (1991). Molecular genetic analysis of the rice blast fungus, magnaporthe grisea. Annu. Rev. Phytopathol., 29: 443–467. doi: 10.1146/annurev.py.29.090191.002303.

Varughese Titto, Nivia Riosa, Sarah Higginbotham, A. Elizabeth Arnold, Phyllis D. Coley, Thomas A. Kursar, William H. Gerwick and L. Cubilla Rios. (2012). Antifungal depsidone metabolites from Cordyceps dipterigena, an endophytic fungus antagonistic to the phytopathogen Gibberella fujikuroi. Tetrahedron Letters, 53(13): 1624–1626. doi: 10.1016/j.tetlet.2012.01.076.

Vinale, F., Sivasithamparam, K., Ghisalberti, E.L., Marra, R., Woo, S.L. and Lorito, M. (2008). Trichoderma–plant–pathogen interactions. Soil Biology and Biochemistry, 40(1): 1–10. doi: 10.1016/j.soilbio.2007.07.002.

Vos, C., Geerinckx, K., Mkandawire, R., Panis, B., De Waele, D. and Elsen, A. (2011). Arbuscular mycorrhizal fungi affect both penetration and further life stage development of root-knot nematodes in tomato. Mycorrhiza, 22(2): 157–163. doi: 10.1007/s00572-011-0422-y.

Walker, A.S., Gautier, A.L., Confais, J., Martinho, D., Viaud, M., Le, P.C.P. and Fournier, E. (2011). Botrytis pseudocinerea, a new cryptic species causing gray mold in French vineyards in sympatry with Botrytis cinerea. Phytopathology, 101(12): 1433–1445. doi: 10.1094/PHYTO-04-11-0104.

Waller, F., Achatz, B., Baltruschat, H., Fodor, J., Becker, K., Fischer, M. and Kogel, K.H. (2005). The endophytic fungus Piriformospora indica reprograms barley to salt-stress tolerance, disease resistance, and higher yield. Proc. Natl. Acad. Sci. USA, 102(38): 13386–13391. doi: 10.1073/pnas.0504423102.

Wang, C. and St Leger, R.J. (2007). The Metarhizium anisopliae Perilipin Homolog MPL1 regulates lipid metabolism, appressorial Turgor Pressure, and Virulence. J. Biol. Chem., 282(29): 21110–21115. doi: 10.1074/jbc.M609592200.

Wang, Q. and Xu, L. (2012). Beauvericin, a bioactive compound produced by fungi: A short review. Molecules, 17(3): 2367–2377. doi: 10.3390/molecules17032367.

Wani, Z.A., Ashraf, N., Mohiuddin, T. and Riyaz-Ul-Hassan, S. (2015). Plant-endophyte symbiosis, an ecological perspective. Appl. Microbiol. Biotechnol., 99(7): 2955–2965. doi: 10.1007/s00253-015-6487-3.

Wasternack, C. (2014). Action of jasmonates in plant stress responses and development-applied aspects. Biotechnol. Adv., 32(1): 31–39. doi: 10.1016/j.biotechadv.2013.09.009.

Wen, L. (2013). Cell death in plant immune response to necrotrophs. Journal of Plant Biochemistry & Physiology, 1(1). doi: 10.4172/2329-9029.1000e103.

Wernegreen, J.J. (2005). For better or worse: Genomic consequences of intracellular mutualism and parasitism. Curr. Opin. Genet. Dev., 15(6): 572–583. doi: 10.1016/j.gde.2005.09.013.

Whipps, J.M. (2004). Prospects and limitations for mycorrhizas in biocontrol of root pathogens. Canadian Journal of Botany, 82(8): 1198–1227. doi: 10.1139/b04-082.

Wilhelm, S. and Tietz, H. (1978). Julius kuehn-his concept of plant pathology. Annual Review of Phytopathology, 16(1): 343–358. doi: 10.1146/annurev.py.16.090178.002015.

Wilson, R.A., Lahlali, R., McGregor, L., Song, T., Gossen, B.D., Narisawa, K. and Peng, G. (2014). Heteroconium chaetospira induces resistance to clubroot via upregulation of host genes involved in jasmonic acid, ethylene, and auxin biosynthesis. PLoS ONE, 9(4): e94144. doi: 10.1371/journal.pone.0094144.

Wilson, R.A. and Talbot, N.J. (2009). Under pressure: Investigating the biology of plant infection by Magnaporthe oryzae. Nat. Rev. Microbiol., 7(3): 185–195. doi: 10.1038/nrmicro2032.

Xu Xi-Hui, Chen Wang, Shu-Xian Li, Zhen-Zhu Su, Hui-Na Zhou, Li-Juan Mao, Xiao-Xiao Feng, Ping-Ping Liu, Xia Chen, John Hugh Snyder, Christian P. Kubicek, Chu-Long Zhang and Fu-Cheng Linb. (2015). Friend or foe: Differential responses of rice to invasion by mutualistic or pathogenic fungi revealed by RNAseq and metabolite profiling. Sci. Rep., 5: 13624. doi: 10.1038/srep13624.

Yang, J., Hsiang, T., Bhadauria, V., Chen, X.-L. and Li, G. (2017). Plant Fungal pathogenesis. BioMed Research International, 2017: 1–2. doi: 10.1155/2017/9724283.

Yaqub, F. and Shahzad, S. (2008). Effect of seed pelleting with Trichoderma spp., and Gliocladium virens on growth and colonization of roots of sunflower and mung bean by Sclerotium rolfsii. Pak. J. Bot., 40(2): 947–953.

Yi, M. and Valent, B. (2013). Communication between filamentous pathogens and plants at the biotrophic interface. Annu. Rev. Phytopathol., 51: 587–611. doi: 10.1146/annurev-phyto-081211-172916.

Yoon, M.-Y., Cha, B. and Kim, J.-C. (2013). Recent trends in studies on botanical fungicides in agriculture. The Plant Pathology Journal, 29(1): 1–9. doi: 10.5423/ppj.rw.05.2012.0072.

Yoshioka, Y., Ichikawa, H., Naznin, H.A., Kogure, A. and Hyakumachi, M. (2012). Systemic resistance induced in Arabidopsis thaliana by Trichoderma asperellum SKT-1, a microbial pesticide of seedborne diseases of rice. Pest. Manag. Sci., 68(1): 60–66. doi: 10.1002/ps.2220.

Yu, H., Zhang, L., Li, L., Zheng, C., Guo, L., Li, W., Sun, P. and Qin, L. (2010). Recent developments and future prospects of antimicrobial metabolites produced by endophytes. Microbiol. Res., 165(6): 437–449. doi: 10.1016/j.micres.2009.11.009.

Yuan, J., Wang, Z., Xing, J., Yang, Q. and Chen, X.-L. (2018). Genome-wide identification and characterization of circular RNAs in the rice blast fungus Magnaporthe oryzae. Scientific Reports, 8(1). doi: 10.1038/s41598-018-25242-w.

Yuan, S., Li, M., Fang, Z., Liu, Y., Shi, W., Pan, B., Wu, K., Shi, J., Shen, B. and Shen, Q. (2016). Biological control of tobacco bacterial wilt using Trichoderma harzianum amended bioorganic fertilizer and the arbuscular mycorrhizal fungi Glomus mosseae. Biological Control, 92: 164–171. doi: 10.1016/j.biocontrol.2015.10.013

Zabalgogeazcoa, I. (2008). Fungal endophytes and their interaction with plant pathogens. Spanish Journal of Agricultural Research, 6: 138–146.

Zhang, X., Jain, R. and Li, G. (2016). Roles of Rack1 proteins in fungal pathogenesis. BioMed Research International, 2016: 1–8. doi: 10.1155/2016/4130376.

Zou, W.X., Meng, J.C., Lu, H., Chen, G.X., Shi, G.X., Zhang, T.Y. and Tan, R.X. (2000). Metabolites of Colletotrichum gloeosporioides, an endophytic fungus in Artemisia mongolica. J. Nat. Prod., 63(11): 1529–1530. doi: 10.1021/np000204t.

Chapter 8

Biofungicides:

Antifungal Biomaterials and Mechanisms

Nasir A. Rajput, Muhammad Atiq, Ghalib A. Kachelo, Azeem Akram, Nuzhat Jamal* and *Muniza Baig*

1. Introduction

Biofungicides are derivatives of two words bio means (living organisms) and fungicide (to kill the plant pathogenic bacteria and fungi). It is a general name that is given to the micro-organisms and naturally occurring products through which one can control the pathogenic activity of bacteria and fungi (Roger and Keinath, 2010). The concept of biofungicides is based on beneficial microorganisms and plant extracts including phenolic compounds and oils, that have the potential to defeat fungal pathogens by direct interaction, competition for survival and induction of localized or systemic resistance in plants. The modern agricultural system is facing a wide range of diseases, pests, droughts, a decrease in soil fertility due to excessive use of chemicals and various environmental problems. Disease caused by fungus is one of the main constraints responsible for yield losses (Mukesh et al., 2016). Since the mid-20th century crop diseases by fungal pathogens have increased in severity and scale and now is a high risk to world food security, where the world is losing approximately 10–23% of the crops pre-harvest and 10–20% post-harvest due to disease intervention (Savary et al., 2019; Fisher et al., 2012). To overcome crop losses by fungal pathogens, a vast number of fungicides are used regularly in the world, that is why pesticides play a key role in the protection of valuable crops (Sharma et al., 2019). There is no doubt that the large use of synthetic chemicals is creating human and animal health and environmental hazards, especially fungicides used against post-harvest diseases of fruits and vegetables that are directly in contact with

Department of Plant Pathology, University of Agriculture, Faisalabad, Pakistan.
* Corresponding author: nasirrajput81@gmail.com; nasir.ahmed@uaf.edu.pk

humans, causing severe health threats. The U.S Environmental Protection Agency banned the usage of different fungicides, i.e., Benomyl, Thiophanate, Captan for some or all post-harvest uses due to scientific concern about the presence of chemical residues in food supply (Droby et al., 1992). Therefore, it needs time to produce new and applicable approaches toward controlling fungal diseases which pose less or no loss to human and animal heath and the environment.

Biocontrol agents that have potential against pathogenic fungi are an attractive substitute for harmful agrochemicals, these may be used as controlling agents or in a combination with other materials for integrated disease management (Das and Abdulhameed, 2020). Different kinds of biofungicides are being used, including microorganisms, their secondary metabolites, plant extracts and their essential oils and phenolic compounds. Due to antimicrobial properties, numerous plant extracts are used as biocontrol agents against bacterial and fungal plant diseases instead of chemical/synthetic fungicides (Sernaite et al., 2011). Biofungicides are prepared from populations of microorganisms which are isolated from the natural environment to inhibit the activity of the phytopathogen. These biological agents can enhance and induce resistance, compete with microorganisms for nutrients and employ hyperparasitism against different fungal and bacterial pathogens.

2. Biofungicidal potential of fungi

Different types of biofungicides are used in agricultural production containing bio-fungus such as (*Trichoderma* spp., Belgium), Biotrek (*T. harzianum*, USA), Binab-T (*Trichoderma harzianum* and *T. polysporum*, Sweden) and some bacterial preparations Serenada, Rhizo-plus (*B. subtilis*, Germany), Kodiak (*Bacillus subtilis*, USA), Planriz (*Pseudomonas fluorescens*, Belarus and Russia), Phytolavin (*Streptomyces griseus*, Russia) and Baktofit, Phytosporin (*B. subtilis*, Russia) (Dyakunchak and Kolomiyets, 2006). Various studies are discussed here where the use of fungus against pathogenic fungi was significantly evaluated (Table 8.1).

A unique and highly effective biofungicide, *Trichoderma harzianum* CH_1 has been introduced due to its potential for limiting pathogens like Phytophthora spp., Fusarium spp., and Rhizoctonia spp., on ginger and pepper crops (Das et al., 2021). *Trichoderma harzianum* T39 was used in place of synthetic fungicides to control gray mold on cucumber, it showed comparably the same results as synthetic chemicals. An active enzyme "Proteases" is secreted by *T. harzianum* T39 that is found as a suppressive factor against *B. Cinerea* (Elad, 2000; Zimand et al., 1996). A study by Sudantha and Suwardji (2021) described different formulations of Trichoderma spp., such as liquid, tablet and powder biofungicides against Fusarium wilt of shallots, where results declared 100% disease depression by application of all these formulations. Panama wilt of banana is a serious threat caused by *Fusarium oxysporum*, it has different strains that are pathogenic to plants. Four strain of *F. oxysporum* were reported to cause Panama wilt. The use of suppressing microorganisms is an efficient management strategy against *F. oxysporum* (Rakh et al., 2011). *T. harzianum* has the potential to control plant diseases effectively by

Table 8.1: Microbes used as biofungicides against plant diseases.

No.	Biofungicides/ specie/strain	Diseases control	Trade name	Target pathogen	References
1.	*Streptomyces griseochromogenes*	Rice blast	Blasticidin S	*Magnaporthe oryzae*	(Takeuchi et al., 1957)
2.	Subtilis QST713	Club root of canola Plasmodiophora brassicae	Serenade	*Plasmodiophora brassicae*	(Lahlali et al., 2011)
3.	Validamycin A	Rice sheath blight	Streptomyces hygroscopicus	*Rhizoctonia solani*	(Iwasa et al., 1970)
4.	*Pseudomonas* sp.		Pyrrolnitrin	*Botrytis cinerea, Magnaporthe grisea*	(Arima et al., 1964)
5.	*Fusarium semitectum*	Gray mold of strawberry	Fusapyrone and deoxyfusapyrone	*Botrytis cinerea*	(Altomare et al., 2004)
6.	*Acremonium strictum*	Barley powdery mildew	Verlamelin	*Blumeria graminis* f. sp. *hordei*	(Kim et al. 2002)
7.	*Bacillus subtilis* strain VKM-B-2604D + *Bacillus subtilis* strain VKM-B-2605D	Fusarium root rot	Vitaplan	*Fusarium* spp.	(Rezvyakova et al., 2021)
8.	*Bacillus subtillis* GB03	Powdery mildew, Botrytis, Leaf spots, Pythium, Phytophthora, Rhizocotonia	Companion Liquid	*Botrytis* spp. *Rhizoctonia* spp. *Pythium* spp.	(Khakimov et al., 2020)
9.	*Bacillus pumilus* QST 2808	Powdery mildew, Cercospora and Rust	Ballad Plus Biofungicide	Cercospora spp.	
10.	*Pantoea agglomerans* strain E325	Fireblight disease	Bloomtime Biological FD3	*Erwinia amylovora*	
11.	*Clonostachys rosea* (*Gliocladium catenulatum*)	Damping off. Wilt diseases and Root rot	Primastop	*Fusarium* spp. *Pythium* spp.	(Thambugala et al., 2020)
12.	*Candida oleophila* + *Paraphaeosphaeria minitans*	Post-harvest diseases	Aspire Contans WG KONI	*Sclerotinia sclerotium* Sclerotium minor	
13.	*Trichoderma virens* (*Glicladium virens*)	Soil Borne Pathogens	Soligard	*Rhizoctonia* spp. *Pythium* spp.	

decreasing disease severity. It was found to be a highly effective biological control agent against banana wilt (Khan et al., 2017). *Ulocladium* spp., was an antagonistic fungus found to be effective against *B. Cinerea*. Research conducted on *Ulocladium atrum* showed that it effectively controlled 90% sporulation of *B. Cinerea* on Lily leaves (Köhl et al., 1995).

As in the case study of Fusarim Head Blight (FHB) which is the prominent disease of wheat (*Triticum aestivum* L.) caused by the fungus *Gibberella zea*, and was once reported and spread throughout the world (McMullen et al., 1997, 2012; Gilbert and Tekauz, 2000). The result of this disease was contaminated harvested grains with mycotoxic substances deoxynivalenol (DON) and also a reduction in yield and quality of the grain, later these toxins were found responsible for health concerns in livestock (Desjardins, 2006; Del Ponte et al., 2012; Placinta et al., 1999; Miller, 1995). Controlled environmental conditions have been used to control (FHB) with fungicides to maintain grain yield and improve quality in many regions where wheat is produced (Tekauz et al., 2004; Jones, 2000; Matthies and Buchenauer, 2000). By using different biofungicides this disease was controlled, various studies discussed here as the foliar application of ACM941 strain was used as an alternative fungicide that effectively controls the Fusarium head blight severity (Xue et al., 2009). The treatment of crops remaining with fungal strains inhibited the perithecial structure of *G. zeae* and helped to diminish the primary inoculum of the disease (Xue et al., 2009). The strain of Trichoderma (*T. asperellum*) was identified as a significant biocontrol agent against Fusarium head blight (Kolombet et al., 2008; Kolombet et al., 2005). Recently the new strain of *Trichoderma asperellum* (GJS 03-35) has also been recognized (pers.).

Coumarins: The three strains of yeast *Metschnikowia fructicola* strain AL27, *Metschnikowia pulcherrima* strain GS9 and *Metschnikowia pulcherrima* MACH1, showed efficient antagonistic effect toward *Penicillium expansum* and patulin. To improve the potential range of application as well as to identify the efficacy of biocontrol agents (BCAs) of all three yeast strains were evaluated on four different varieties of apple, i.e., Granny Smith, Golden Delicious, Royal Gala and Red Chief. AL27 showed the most effective antagonistic ability against blue mold rot and reduced the accumulation of patulin as compared to MACH1 and GS9 (Spadaro et al., 2002). The highly reactive lactone is a non-saturated Patulin produced by *P. exanpsum* that may cause acute and chronic toxicity, including mutagenic, teratogenic and carcinogenic effects (McCallum et al., 2002; Hasan, 2000; Beretta et al., 2000). It also disturbs the male reproductive system through interaction with hormone production. The mycotoxin causes disturbance of kidney functions, oxidative damage and weakness of the immune system (Fuchs et al., 2008; Selmanoglu and Kockaya, 2004). The Internal Transcribed Spacer 1(ITS1), 5.85 ribosomal RNA gene and internal transcribed spacer 2 (ITS2) were used to identify the resistant effect of yeast strain *Metschnikowia fructicola* AL27 (White et al., 1990). It showed an antagonistic effect against post-harvest rots development in grapes and strawberries.

Golden Delicious is one of the varieties of apples. It was harvested in organic orchards located in piedmont and was used to extract the strain of the yeast named *M. pulcherrima* (MACHH1), which was found effective against *B. cinerea, A. alternata* as well as *P. expansum*. Furthermore, the strain showed promising results against alternaria rot and gray mold, but it expressed lower competence against blue mold rot disease (Saravanakumar et al., 2008). It was further reported that through the competition mechanism, the strain MACH1 compete for nutrients and finally released hydrolases enzymes specially chitinase (Saravanakumar et al., 2009).

Rhizoctonia solani is a fungal pathogen responsible for causing damping-off of different crops (Vinodhini and Narayanan, 2008; Lida et al., 1983; Abbasi et al., 2004). Pythium spp., was shown to often be the cause of damping-off in cucumber (Gubler and Davis, 1996; Deadman et al., 2002; Stanghellini and Phillips, 1975) where Trichomix (*Trichoderma harzianum*) was used *in vivo* as well *in vitro* to suppress the soil-borne phytopathogen *Pythium aphanidermatun* (Obire and Anyanwu, 2009).

3. Different endophytic fungus used as Biofungicide

Cryptosporiopsis cf. quercina is an endophytic fungus of *Tripterygium wilfordii* containing antifungal compound crytocandin which has antagonistic properties against diseases caused by fungus *Sclerotinia sclerotium* and *Botrytis cinerea* (Strobel et al., 1999). Furthermore, the endophytic fungus of *Phleum pratense* has been used to isolate sesquiterpenes, chokols A-G (19–25) which is a fungitoxin compound against the pathogen *Cladosporium* spp., causing leaf spot diseases (Koshino et al., 1989). The fumitremorgin-C is a mycotoxin secreted by *Aspergillus fumigatus* and found as an antifungal agent against many fungal pathogens, that is extracted from *Cynodon dactylon* plant (Cole et al., 1977). In addition to that *Acremonium zeae* is an endophytic fungus of host plant maize that contain pyrocidine and some compound that is antagonistic to microorganisms (Wicklow et al., 2005).

Artimisia annua is the host plant of *Collectotrichum* spp. It is an endophytic fungus. Moreover, it exhibits antagonistic effects against diseases caused by fungus, i.e., *Phytophthora capsici, Rhizoctonia cerealis, Helminthosporim sativum* and *Gaeumannomyces graminis* Var. *tritici* (Lu et al., 2000). The endophytes of *Calliandra calothyrsus, Leucaena diversifolia* and *Sebania sesban* are essential bio agents for diseases like gray leaf spot of maize and wilt of bananas caused by *Cercospora zeae-maydis* and *Fusarium oxysporum* f. spp. *cubense* (Omuketi, 2020).

Phomopsis sp., is an endophytic fungus mixed to crushed wheat, containing antimicrobial compounds (Horn et al., 1995). In addition to that one of the isolates of *Phomopsis* sp., was extracted from *Gossypium hirsutum* and produced antifungal compounds epoxycytochalasin H (4), Cytochalasins N (5) as well Cytochalasins and was found efficient against *Bipolaris maydi, Sclerotinia sclerotiorum, Rhizoctonia cerealis, Fusarium oxysporum, Gaeumannomyces graminis* Var. *tritici Bipolaris sorokiniana* and *Botrytis cinerea* (Fu et al., 2011).

4. Fungicidal metabolites: An emerging tool to fight against phytopathogens

The phylum Basidiomycete contain fungicidal metabolites strobilurin A and Strobilurus tenacellus (Anke et al., 1977) as well as oudemansin A, Oudemansiella mucida (Anke et al., 1997). The two compounds fusapyrone and deoxyfusapyrone (3-substituted -4-hydroxyp-6-alkyl-2-pyrones) extracted from *Fusarium semitectum*, have a significant effect against fungal diseases caused by *Botrytus cinerea*, *Aspergillus parasiticus* and *Penicillium* spp. (Altomare et al., 2004). The compound streptimidone contains a strong antagonistic effect against *Phytophthora capsici* that caused phytophthora blight in pepper (Kim et al., 1999). It was isolated from *Micromonospora coerulea*, about 500 mg/I quantity, which was equal to the efficacy of the inorganic fungicide metalaxyl. In addition to that, the culture of *Lechevalieria aerocolonigenes* was used to isolate the butanoic acid thiobutacin, 4-(-2-aminophenyl)-4-oxo-2-methylthiobutanoic acid, that exhibited antioomycete as well as antifungal activities against *P. capsici* and *B. cinerea* (Lee et al., 2005). *Colletotrichum gloeosporioides* produced gloeosporone that was a self–an inhibitor, which restrain the growth of conidia of *F. oxysporum* and *Colletotrichum* spp. (Lax et al., 1985). The chitin synthase I and II of *Saccharomyces cereisiae* Meyen were used in cell-free screening. This method was used to find Phellinsin A, an antifungal phenolic compound that controlled diseases caused by pathogen *M. grisea*, *R. soloni* and *C. lagenarium* (Hwang et al., 2000).

Many compounds that can inhibit the disease development caused by ethylene production, aminooxyvinylglycine benzyl adenine (AVG), cobalt ion, aminoxyacetic acid (AOA) and 2,5-norbornadiene (2,5-NBD) by hindering the activity of ethylene production, controlled the *B. cinerea* infection as well as increased the length of survival of cut flowers (Yu and Yang, 1979; Elad, 1993; Baker et al., 1977, 1982).

Choi and Huber (2008) extracted the fungus *Simplicillium lamellicola* BCP from *B. cinerea mycelia*, it produced the compound verlamelin which was the broad range and antagonistic to diseases caused by a fungus (Kim et al., 2002). *S. lamellicola* BCP can considerably reduce the prevalence of gray mold diseases in many crops living in growth chambers, tunnels and fields, i.e., cucumber, strawberry and tomato (Choi and Huber, 2008).

5. Fungal species used as Biofungicide according to their mode of actions

The species of fungus phylum Ascomycota *Chaetomium globosum* is a regular established interaction with soil and substrates that contain cellulose (Domsch and Gams, 1972). The fungus was found to be a biocontrol agent of different seed-borne as well as soil-borne phytopathogens (Tiwari et al., 2004). The *C. globosum* act as an antifungal agent against disease-causing agents of plants based on their three different mechanisms of action, competition, myco-parasitism and antibiosis. It is reported that the *C. globosum* strain can produce the substance chaetomin in culture, that correlated with the activity of pathogens like *Pythium ultimum* causing damping–off of sugar beet (Di Pietro et al., 1993).

The strain of *C. globosum* F0142 was extracted from barnyard grass, it was found to be a potential antagonistic against *Magnaporthe grisea* causing rice blast and *Puccinia recondita* on wheat leaf rust, furthermore, it showed a lower antagonistic effect against *Phytophthora infestans* causing late blight of tomato (Park et al., 2005). Yeast was classified as a biofungicide based on anti-fungal patterns, according to their mode of action. Thirty extracts of yeasts showed two modes of action. It was found that the yeast species used in the wine production industry, showed biocontrol activity against many fungi causing different diseases in plants by inhibiting the mycelial growth through metabolites (Nally et al., 2015), laminarinases, competition, inhibiting sporulation and lowering the length of the germinal tube.

Pycnidia is the fruiting body of fungi, the genus Ampleomyces is associated with pycnidia and can parasitize the fungus. It was the first fungus used as an antagonistic against diseases caused by fungal phytopathogens. Ampleomyces showed mycoparasitism mechanism by killing the pathogen cells causing powdery mildew by rapid degradation of the cytoplasm (Hashioka and Nakai, 1980). The mycoparasite can inhibit both the reproduction procedures (sexual and asexual).

According to Agrios (2005), Trichoderma genus showed good results as an antagonistic against many phytopathogens by following a mechanism of competition for a food source, by secreted lytic enzymes, secretion of toxic metabolites and direct attack on the host (parasitism). For the first time root rot diseases of citrus caused by *Armillaria mellea*, and initially Trichoderma was used against it. Furthermore, it has been also effective against soil-borne pathogens that caused diseases in plants like Rhizoctonia, Sclerotium, Pythium, Fusarium, Sclerotinia and Gaeumannomyces.

6. Bacterial species used as Biofungicide against phytopathogens

Along with antagonistic fungi, there are various bacteria that are also used as biological control agents against numerous fungal pathogens. There are different studies described here for a better understanding of the bacterial role in fungal disease management (Table 8.1). *Bacillus subtilis* BD0310 a biocontrol agent against *Colletotrichum theae-sinensis*, was extracted from tea trees of phylloplane at a tea plantation in Korea. Using living organisms to control phytopathogens is found to be a beneficial method (Borowicz, 2001; Atkinson et al., 1994). *Bacillus* spp., *Rhizobacteria* (PGPR) and *Pseudomonas* spp., are regularly used as antagonists against diseases caused by a fungus (Chen et al., 2000). These BCAs control most of the soil-borne pathogens and are also known to increase the growth and production of the crop (Schmiedeknecht et al., 1998; Ryder et al., 1999; Krebs et al., 1998). PGPR can enhance the effectivity of BCAs when mixed (Duffy et al. 1996). There are several strains of *Bacillus subtilis* and *Pseudomonas flourescens* that have been reported to have significant potential against plant diseases caused by *Pythium*, *Rhizoctonia*, *Gaeumannomyces*, *Sclerotinia*, *Fusarium* and others (Bacon et al., 2001; Schmiedeknecht et al., 1998; Zhang et al., 1996).

Q125B, Q112B, Q004B and Q110B are different strains extracted from *Pseudomonas*, having the ability to suppress the pathogen of citrus (*Penicillium digitatum*). Many scientists have found that most species of the genus Pseudomonas are efficient antagonistic agents against *P. digitatum* (Kaur et al., 2007; Gao et al., 2018; Qessaoui et al., 2020). Besides, Pseudomonas, the specie of *Bacillus subtilis* QST-713 produced about 30 different strains that are lipopeptide based. It was found that lipopeptide plays an important role to increase bacterial colonization on plant tissues and also increases an antagonistic effect against pathogenic microbes (Ongea and Jacques, 2007). Extensive biological active compounds (antibiotics) were also produced by Bacillus strains, which were suppressive against plant pathogens (Stein, 2005; Emmert and Handelsman, 1999). *B. licheniformis* N1 is a strain of Bacillus that was used to produce N1E by using specific formulations and preparing bacterial culture through fermentation. It successfully managed gray mold diseases in tomatoes (Lee et al., 2006) and strawberry plants (Kim et al., 2007).

Lysobacter genus (Christensen and Cook, 1978), a gram-negative bacteria showed gliding motility. It has been extensively studied that it contains the potential to suppress phytopathogens and also boost plant growth. *Lysobacteria enzymogenes* strain C3, which was earlier determined as *Stenotrophomonas maltophilia* (Giesler and Yuen, 1998), was considered a successful biocontrol agent against phytopathogens in vertical farms as well as in fields (Zhang and Yuen, 1999; Giesler and Yuen, 1998; Zhang et al., 2001; Zhang and Yuen, 2000; Li et al., 2008; Jochum et al., 2006). Likewise, *L. enzymogenes* OH11, the most important strain that was studied in China, *L. enzymogenes* 3.1T8 and *L. enzymogenes* C3 produced and released a large number of extracellular hydrolases and antibiotics (Qian et al., 2009). By using broth culture of *L. enzymogenes* OH11 was cultured in 109 cells MI1, it potentially suppressed the rice blast, tomato bacterial wilt and canola white mold diseases (Qian et al., 2009). This strain was used in China as a biocontrol agent against many vegetable pathogens.

In recent times, a biofungicide identified as “Serenade”, was prepared from the commercial formulation of the bacterium *Bacillus subtilis* (Ehrenberg) Cohn, an efficient product for the control of *Monilinia vaccinii-corymbosi* (Reade) Honey, a fungus that caused mummy berry disease of cherry (Scherm et al., 2004). Bacillus spp., produced many peptides in the class of bacteriocins that are used against different microorganisms (Klaenhammer, 1993). Moreover, biosurfactants that are surface active are also used against microorganisms, including lipopeptides and glycopeptides that are cyclic nonribosomally formed (Mukherjee et al., 2006; Rodrigues et al., 2006). *Bacillus subtilis* was found to be a successful control of many phytopathogens in an eco-friendly manner. They are used to control a variety of phytopathogens of pre-harvest and post-harvest nature (Sharga and Lyon, 1998; Collins et al., 2003; Rodger, 1989).

Damping-off disease of many horticultural crops is caused by myco pathogen *Sclerotium rolfsii* Sacc and *Rhizoctonia solani* Kuhn with continuous loss of yield under vertical farming and in the field (Berta et al., 2005; De Curtis et al., 2001; Punja 1985). Biocontrol and integrated controls are the most advanced methods to manage phytopathogens by using only suppressing agents or adding chemical fungicides or natural adjuvants (Lima and Cicco, 2006; Paulitz and Belanger, 2001). Many bacteria that can act as biocontrol agents were earlier extracted as well as selected from organic compost and amendments, forestry peats in addition to agricultural soil which contain biocontrol agents (Vitullo et al., 2008; Lima and Cicco, 2006). In some cases, some interesting results are found by applying some bacteria in the root zone portion activated the systematic acquired resistance, moreover lowering the effect of some phytopathogens, like crown, root and foliar diseases by inducing the mechanisms of induced systemic resistance (ISR) and Rhizobacteria–mediated Induced Systemic Resistance (RISR) (Pieteres et al., 2002; Pieteres and Loon, 1999; Van der Ent et al., 2008; Jones and Dangal, 2006).

The most important disease in oil palm farms is - basal stem rot caused by the pathogen *Ganoderma boninense*. Older palm trees are more affected by the pathogen Ganoderma. About 80% of oil palm trees were attacked by this pathogen in Indonesian farms, causing a reduction in palm oil yield per unit area (Sustano, 2002). Different methods have been recommended for controlling basal-stem rot disease compromising cultural, chemical as well as biocontrol agents extracted from the root portion (Rhizosphere) of trees. Where using biocontrol agents to manage pathogen *Ganoderma boninense* decreased the use of chemical-based fungicides and established resistance against phytopathogens (Harjono and Widyastuti, 2001). Presently, using biological organisms including endophytic fungus and bacteria for the management of plant diseases has received a lot of interest in research. The species of bacillus form a colony with the plant roots thereby suppressing the development of *Ganoderma boninense* by producing antimicrobic substances and competing for food and habitat also causing resistance in the host plant (Bacon and Hinton, 2007).

7. Phyto extracts as Biofungicides and their fungicidal activities

Plant extracts with diverse fungicidal activities have played a role in the enhancement of structural strength of products and increase the resistance of produce by strengthening the membrane (Table 8.2). Edible films and coatings are associated with improving the characteristics of fresh produce (Vukić et al., 2017). Biofungicides have a different mode of action. Several plant extracts have been used against fungi and bacteria to prevent disease occurrence. The anti-fungal activity of plant extracts is mainly the preferred use in the control of post-harvest deterioration of fresh commodities. Fruit coating with the oil of sunflower and alginate has led to an increase in the strength of fresh pineapple pieces (Azarakash et al., 2012). Alginate-based coatings give strength and freshness to slices of watermelon (Sipahi et al., 2013). When the fruit's outer covering is strengthened, the chance of the produce to external damage is ultimately reduced, which in turn reduces the occurrence of fungal infections.

Plant extracts-based fungicides have been achieved as a sustainable alternative to chemicals against pathogens by inhibition mechanism (Polo et al., 2021). Plant extracts from four plant materials namely basil, marjoram, peppermint and spearmint were taken to access the activity of extracts on the mycelial growth inhibition of

Table 8.2: Plant extracts used as biofungicides against plant diseases.

No.	Plant extracts	Composition/ Compounds	Targeted pathogens	Growth inhibition (%)	References
1.	*Moringa oleifera* (Moringa)	Cytokinins, auxins, zeatin	*Fusarium verticilliodies*	94.57	(Foidl et al., 2001; Nagar et al., 2006; Mohamed et al., 2020)
2.	*Origanum. majorana* L. (Marjoram)	2 -methylbutanoic acid methyl ester, 3-Thujene	*Trichoderma harzianum, Aspergillus niger*	87.77	(Salem et al., 2019)
3.	*Allium cepa* (Onion)	Flavanols	*Cercospora arachidicola, Cercosporidium personatum*	68	(Khalil and Javeed, 2019)
4.	*Allium ampeloprasum* (Wild leek)	Cinnamic acid, Steroidal saponins	*Alternaria alternata, Mortierella ramanniana*	100	(Sadeghi et al., 2013; Sata et al., 1998)
5.	*Allium sativum* (Garlic)	Allicin, ajoene	*Alternaria alternate, Colletotrichum graminicola*	100	(De Falco et al., 2018; Liaqat et al., 2019)
6.	*Azadirachta indica* (Neem)	Azadirachitin	*Fusarium solani, Colletotrichum capsici*	47.65, 42.90	(Rajput et al., 2011; Abarna et al., 2019)
7.	*Aloe vera*	Anthraquinones, glycosides, aloine	*Rhizoctonia solani, Colletotrichum coccodes, Penicillium gladioli*	23.08	(Saks and Barkai-Golan, 1995; Hamman et al., 2008; Bruneton, 1993)
8.	*Capsicum annum*	Capsaicin	*Botrytis cineria*	80	(Jiménez Reyes et al., 2019; Silva and Fernandes Júnior, 2010)
9.	*Glycyrrhiza uralensis*	Glycyrrhizin	*Rhizoctonia solani, Pythium aphanidermatum*	87.3, 90.3	(Rajput et al., 2018)
10.	*Mentha pulegium* L	pulégone Menthol	*Alternaria alternata, Botrytis cinereal, Penicillium expansum*	100	(Hmiri et al., 2011)
11.	*Thymus danensis, Thymus carmanicus*	le thymol α-terpinene, carvacrol p-cymène	*Rhizopus stolonifer Botrytis cinereal*	100	(Nabigol and Morshedi, 2011)
12	*Citrus limon, Citrus aurantifolia*	limonène γ-terpinène	*Botrytis cinereal*	84	(Mbili et al., 2015)
13	*Warionia saharae*	β-eudesmol nerolidole linalol	Alternaria sp., *Penicillium expansum, Rhizopus stolonifer*	85.51	(Zinini et al., 2014)

some pathogenic fungi. The plant materials were homogenized into a paste by blending and extracted by the hot or cold extraction method according (Wokocha and Okereke, 2005) to evaluate the inhibition of mycelial growth, the mycelial extension of pathogenic fungi was placed on one disk of active cultures in five petri plates containing the media and leaf extract. The mycelial growth inhibition occurred to a certain percentage due to the activity of plant extracts. The inhibitory activity of extracts was rated by using scales designed by (Sangoyouni, 2004). The results of this experiment showed that with increasing concentration of the plant extracts, the mycelial growth of the disease-causing fungi decreased among which basil and marjoram were more effective as compared to the activity of peppermint and spearmint, which were the least effective (Baraka et al., 2011).

The activity of *Ocimum basilicum*, *Azadiracht indica*, *Eucalyptus chamadulonsis*, *Datura stramonium*, *Nerium oleander* and *Allium sativum*, contributed to the reduction in the linear growth of *Alternaria solani*. When the concentrations of the extracts were increased, further reduction in the growth of fungus was observed (Nashwa and Elyousr, 2013). Many other authors have reported effective plant products against the species of Alternaria (Latha et al., 2009; Goussous et al., 2010). Likewise, the extracts from the bulb of *Allium sativum*, *Aegle marmelos* and extract of *Catharanthus roseus* are effective in the inhibition of spore germination and mycelial growth of *A. solani* stated by Vijayan (1989). The mechanism of action of inhibitory activity has revealed that the plant products may directly act on the pathogens (Amadioha, 2000) or lead to the induction of host plant systemic resistance, which ultimately show the inhibition of disease development (Kagale et al., 2004).

Medicinal plants have been reported as antifungal compounds. Among 103 Chinese medicinal plant extracts used against Rhizoctonia solani and Pythium aphanidermatum, 12 crude methanolic extracts expressed durable anti-microbial activities against both pathogens (Rajput et al., 2018). The species of genus Allium are very good against antifungal activities. The presence of compounds like organosulfur, allicin and ajoene in allium species infer inhibitory action toward disease-causing fungi (Lanzotti et al., 2013; Sharanappa et al., 2013). The anti-fungal activity of *A. ampeloprasum* against the pathogens of the coffee tree showed that its extracts can cause a full reduction in mycelial growth and is also efficient at reducing spore germination (Silva et al., 2014). The mechanism of action of *A. ampeloprasum* is due to the inhibition of enzymes containing sulfhydryl and also causes the inhibition of lipid bio synthesis, which leads to damage of the cell wall, the same process was conducted in the case of *Candida albicans* (Hughes and Lawson, 1991; Adetumbi et al., 1986).

A study for the management of a well known destructive plant pathogen *Alternaria solani* revealed that out of 29 different plant extracts, nine were efficient at antifungal activity, which was around 31%. While *Cynara scolymus*, *Lippia alba*, *Salvia Sclarea officinalis* inhibit the growth of *A. solani* up to 98%, which is comparable to the effect of fungicides used commercially (Dellavalle et al., 2011). It was shown that when 5% of *A. stavium* was used to reduce the mycelial growth of *Alternaria alternata* by 100%, the rate of reduction was highest among all other extracts used. The complete inhibition of sporulation of the pathogen by using aqueous and methanolic extracts of *A. sativum* was found effective (Tata et al., 2018). When one considers the extract of *Allium stavium*, it is found that it contains biologically active components. Such components are rich in sulfur compounds, i.e., allicin, saponins, flavonoids, ajoene, allylmethyltrisulfide and diallyldisulfide. These components actively retard the growth of bacteria and fungi (De Falco et al., 2018; Fufa, 2019; Liaqat et al., 2019).

The mechanism of action of various plant extracts is the inhibition of the mycelial growth of the targeted fungus. *Allium cepa* contain secondary metabolites such as flavanols, which play an important role in the reduction of plant pathogenic microorganisms (Franco and AR et al., 2018). When the *Allium cepa* is used with the combination of *A. sativum* and *Z. officinale*, a 15% dose of these biofungicides proved to be effective in the inhibition of mycelial growth of *Fusarium oxysporum*, when incubated for 3 days, due to the activity of components like citral, verbenyl ethyl ether, geranic acids and artemisiole present in the phytoextract (Gholamnezhad, 2019).

Similarly, when 20% of neem extract was used against *C. capsici*, the pathogen was effectively controlled (Abarna et al., 2019), this finding is similar to those of Nikhil and Sahu (2014) who showed, the extract of Ocimum, ginger, garlic, turmeric and onion extract was fruitful in the reduction of growth and sporulation of *C. falcatum*. Masum et al. (2009) reported from their discoveries that, neem extract is better at controlling seed-borne diseases, of fungi like *C. lunata* and *B. sorghicola*. Based on the seed health test, the neem extract can be used for the control of *C. lunata* and *B. sorghicola* infection in sorghum. When the neem extract was applied to the plants of groundnut, it increased the growth of roots and plant length, but the germination rate of groundnut seeds was reduced (Hasan et al., 2014). Moreover, the neem extract has given positive results in the protection of sesame plants against Alternaria leaf spot disease by inducing changes in the metabolism of sesame . In addition, neem extract played a role in boosting the natural defense response of plants, while it is not directly involved in the inhibition of spores of pathogen.

Guleria and Kumar (2006) reported that the level of bioactive compounds like PAL, PO and phenols increases as a result of the application of the neem extract. A programmed response period was observed, in which the activity of PAL, PO and phenols was increased after 48 hr of treatment. After 48 hr the level of these bioactive constituents decreased which is due to the mechanism that regulates the sensitivity of cells in response to the extract of neem. In short, neem extract works as a biofungicide in sesame by increasing the activity of PAL, leading to the synthesis of salicylic acid, which plays a role in resistance induction (Guleria and Kumar, 2006). From South America, the antifungal activity of neem extract was reported against *Crinipellis perniciosa* and phytophthora which causes Witches broom and Pot Not of cocoa (Ramos et al., 2007). Furthermore, the antifungal activity of seed extracts and neem leaves which were extracted by ethanol, hexane and petroleum ether were tested *in vitro* against *Fusarium oxysporum*, *R. solani*, *A. solani* and *S. sclerotiorum*, showed that these extracts lead to the reduction in the growth of tested fungi at a certain level. For the detection of azadirachtin, nimonol and epoxyazadiradione which are components of neem extracts, High-Performance Liquid Chromatography

(HPLC) was used. As a result of detection, it was concluded that the leaves and seeds extracts were truly effective against the tested fungi, among which *F. oxysporum* and *R. solani* were most sensitive (Moslem and EI-Kholie, 2009).

Different medicinal properties of cinnamon extracts and essential oils have been identified (Absalan et al., 2012; Soliman et al., 2015). Besides, plant pathogens, human pathogens, like fungi have been effectively controlled by cinnamon extracts (Ahmed et al., 2019). The frequent use of clove has been very common in traditional medicines (Cock et al., 2018). Several amounts of evidence has been provided in the support of its antimicrobial and anti-fungal activities (Daniel et al., 2015; Hamini et al., 2014; Gurjar et al., 2012). Eugenol is the essential component of clove oil (Yazdani et al., 2005; Xie et al., 2015). In addition, potent application of clove nano-emulsion in the control of *F. oxysoprum* f.sp. *lycopersici* was studied (Sharma et al., 2018).

Moreover, the biofungicidal action of clove and cinnamon was accessed by measuring the colony growth of *B. cinerea* being reinoculated. Cinnamon extract gave higher fungicidal activity as compared to clove. The extract from cinnamon has a Minimal Fungicidal Concentration (MFC) of 600 µL/L, showing that no obvious growth of the pathogen was observed after 48 hr of inoculation. On the other hand, the MFC of clove extracts was 1400 µL/L. Moreover, the application of the plant extract for biofungicidal activity also played a preventive role in the strawberry plant as it is safe for the fruits, soil and water contamination with chemical residues (Sernaite et al., 2020).

Moringa extract is also widely used against numerous fungal pathogens, as it contains antifungal substances. The aqueous and organic extract of moringa seeds showed that the seeds have both polar and non-polar components. Gharibzahedi et al. (2013) reported that the seeds of *M. peregrina* consist of oleic acid, 9.3% palmitic acid, 3.5% stearic acid and 2.6% behenic acid. The fatty acid components have antifungal properties, and alter the permeability of the membrane (Liu et al., 2008; Pohl et al., 2011). Oleic acid and palmitic acid can thus be useful for antifungal activity. Isothiocyanates which are components of Iranian moringa seeds have a volatile compound (Afsharypour et al., 2010) which effectively acts against the microbes stated by (Romeo et al., 2018).

Besides, the seed extract of moringa can also stop the growth of *M. perniciosa* and *A. bisporus*. The control of wet bubble disease in mushrooms is possible by using the moringa extract. The future prospective of using moringa extract in mushroom cultivation in place of fungicides would be safe and effective (Shokouhi and Seifi, 2020). Moringa extract is on the front line in control of soil-borne fungi such as Rhizoctonia, Pythium and Fusarium (Moyo et al., 2012). Dwivedi (2012) reported that leaves, bark and seeds extract 75% v/v of *Moringa oleifera* have shown a remarkable reduction in the mycelial growth of *Fusarium solani* and *Fusarium oxysporum* f. sp *lycopersici*.

The leaves of moringa contain zeatin, cytokinin as well as other compounds like ascorbates, phenolic and minerals such as Ca, K and Fe that have a role in the growth of the plant. In addition, *M. oleifera* also provides the combination of zeatin, quercetin, b-sitsterol, caffeoylquinic acid and kaempferol hence, exhibit action against fungi and bacteria (Anjorin et al., 2010). In most cases, seed treatment is done with *M. oliefera* extract against soil-borne fungi (Talreja, 2010; Yasmeen et al., 2011, Foidl et al., 2001). When the crude leaf extract of *M. oliefera* was used at a concentration of 100 mg/ml, it largely controlled the Penicillium fungus species with a 33 mm zone inhibition. Least zone inhibition (20 mm) of *A. flavus* was observed with the application of n-hexane extract of moringa (Oniha et al., 2021).

Saks and Barkai-Golan (1995) showed the activity of *aloe vera* against the phytopathogenic fungi *Fusarium oxysporum*, *Alternaria solani* and *Colletotrichum coccodes*. The pathogens of ornamental plants like *F. oxysporum* f. sp *gladioli*, *Heterosporium pruneti* and *Penicillium gladioli* are suppressed by the use of aloe vera extract (Rosca et al., 2007). The chemical composition of *A. vera* transmit it with fungicidal properties (Casian et al., 2007). Tannins, glycosides, anthraquinone and other enzymes are part of aloe vera's yellowish exudates (Hamman, 2008).

8. Plant essential oils as Biofugicides

Aromatic compounds which are extracted from vegetables have complex compositions and high volatility and are essential oils (Burt, 2004). They are known worldwide for antimicrobial and antifungal activity and are extensively used in agriculture, food, pharmaceutical and cosmetics (Burt, 2004; Bakkali et al., 2008). Essential oils are good as an antifungal activity. A number of studies had shown that EO can inhibit the growth of different types of fungi such as Aspergillus, Mucor and Penicillium in food products (Cosentino et al., 2003: Carson et al., 2011) (Table 8.3).

The use of Tea Tree Oil (TTO) against *Botrytis cinerea* has been investigated in depth. It was observed that the oil caused cell wall destruction by changing the composition of fatty acids in the cell membrane of *B. cineria*. As a result of the membrane destruction, the membrane became more permeable and leakage of cellular content occurred leading to the death of fungus (Shao et al., 2013). The exact mechanism of action of plant essential oils is not very clear (Tajkarimi et al., 2010). When the mechanism of action of essential oils is studied, they act on the cytoplasmic cell membrane of microorganisms. As the EO is hydrophobic, which is why they are grouped in the cell membrane of the pathogens. The disruption of the cell membrane cause an increase in cell permeability, that leads to leakage of cell constituents (Diao et al., 2014).

Similarly, the antifungal activity of Turmeric (*Curcuma longa* L) essential oil is also reported against phytopathogens., the ergosterol production is inhibited in the plasma of *Aspergillus flavus*, due to the activity of EO. Moreover, it also reduced the function of ATPase in mitochondria (Hu et al., 2017). Mostly, plant extracts worked by interfering with the normal activity of mitochondria and plasma of the fungus. The pattern of activity may change concerning the interaction between fungi and an extract. A study of Askarne et al. (2012), showed that apart from essential oil, its extract from the bark has a higher MIC than leaves. The essential oil of *T. vulgaris* extracted from aerial parts especially act as a good biofungicide due to the presence of thymol, which is considered the active substance of EO (Ismaili et al., 2014).

Table 8.3: Essential oils as biofungicides against fungal plant diseases.

No.	Essential oils	Chemical composition	usage	Reference
1.	*Cuminum cyminum*	α-pinene, α-thujene, Cuminic acid	*Fusarium oxysporum*	(Ustuner et al., 2018; Sun et al., 2017)
2.	*Mentha longifolia*	Sabinene, limonene	*Fusarium oxysporum*	(Ustuner et al., 2018)
3.	*Azadirachta indica*	Oleic acid, Linoleic acid, Palmitic acid	*Rhizoctonia solani, Fusarium solani*	(Rangiah and Gowda, 2019; Sturrock et al., 2015; Rajput et al., 2011)
4.	*Ocimum basilicum*	Methylchavicol, Eugenol, linalol	*Botrytis fabae*	(Bozin et al., 2006; Oxenham et al., 2005)
5.	*Thymus Vulgaris*	Carvacrol, estrgole, linalool	*Alternaria Niger, Curvularia lunata*	(Salem et al., 2019; Mansour et al., 2015)
6.	*Melaleuca alternifolia*	Terpinene 4-ol	*Fusarium fijiensis*	(Muchembled et al., 2018; Reuveni et al., 2020)
7.	*Seasum Indicum*	Palmitic acid, Stearic acid, Oleic acid	*Botrytis cinerea, Fusarium oxysporum*	(Fernandez and Laurentin, 2016; Bawazir and Shantaram, 2018)
8.	*Syzygium aromaticum*	Eugenol	*Rhizoctonia solani, Botrytis cineria*	(Wang et al., 2010; Olea et al., 2019)
9.	*Cinnamomum cassia* L	Cinnamaldehyde and trans-cinnamaldehyde	*Fusarium oxysoprum* f. sp. *fragariae*	(Moghadam et al., 2019; Park et al., 2017)
10.	*Origanum majorana* L	Phenolic thymol, carvacrol	*Botrytis oryzae, Curvularia lunata*	(Mohamed et al., 2020)
11.	*Origanum vulgare*	α-thujene, α-pinene, sabinene	*Penicillium allii*	(Cibanal et al., 2021)
12.	*Lavender angustifolia*	α-terpinene, Lavandulol, β trans-oscimane	*Fusarium solani, Fusarium oxysporum* f. sp *melonis*	(Dhaoudi et al., 2018)

As a preventive measure, before the onset of disease neem oil is used. Initially, it coats the surface of the leaf, in this way the germination of fungal spores is suspended. Mildews, rusts, scabs, leaf spots, blight and rot diseases are effectively controlled by the application of neem oil. Likewise, the inhibition of aflatoxins production due to the application of neem oil and the growth of *Aspergillus parasiticus* is also decreased (Ghorbanian et al., 2008; Allameh et al., 2002). Besides, the biofungicidal application of neem oil, neem seed oil is also used to produce insecticides against different crops.

Marjoram essential oil has been studied for its activity against fungi, bacteria, viruses and as well as insects (EI-Sabrout et al., 2019; Tripathy et al., 2017). Its EO contains compounds like terpinene-4-ol, cis-sabinene hydrate, α- and γ-terpinene, α-phyllandrene and carvacrol (EI-Sabrout et al., 2019; Salem et al., 2019; Rowayshed et al., 2014). The essential oil of majorana inhibits the growth of fungal pathogens which cause the rotting of grains and seedling abnormalities in rice. Before sowing the rice seeds, rice grains are treated with EO of O. Majorana helps in the inhibition of rotting and also enhances the germination of grains, which ultimately leads to good crop yield (Khan et al., 2000; Neergrad, 1970). EOs play their role as an antifungal agent and reduce the formation of biofilms. Mycotoxins production and cellular communication are also interrupted (Moghaddam et al., 2016). When the methanol extract of *O. Majorana* was used against *Aspergillus niger*, *Fusarium solani* and *Bacillus subtilis*, methanol extract was more effective against *Aspergillus niger*. Farooqi and Sreeramu (2004) reported that the leaves of marjoram possess antimicrobial activity against pathogens like *Proteus vulgaris*, *Alternaria fumigatus* and *Bacillus anthracis*. The methanol extract of Majorana due to its effective microbial activity has replaced commercial microbicides. It may also prove to be an effective herbal protectant, as herbal microbicides are non-toxic, ecofriendly and provide protection against pathogenic bacteria and fungi.

In addition to majorana essential oil, oregano essential oil is also known for its antifungal properties. It can reduce the growth of fungi like *Aspergillus*, *Fusarium* and *Penicillium* species (Pereira et al., 2006; Botre et al., 2010). Its main components are carvacrol, thymol, cymene and terpinene (Lambert et al., 2001; Rhayour et al., 2011). The presence of bioactive compounds in *Origanum vulgare* has been confirmed through chromatographic analysis. The antifungal susceptibility assays showed that oregano oil in combination with Propolis Extract (PE), worked effectively against the mycelial growth and conidial germination of *P. allii* (Cibanal et al., 2021).

Moreover, cinnamaldehyde and trans cinnamaldehyde are reported to be part of cinnamon Essential Oil (EO) (Moghadam et al., 2019). It has been found that both oils were efficient at the antifungal activities against *Rhizoctonia solani* and *Fusarium Oxysoprum* (Xie et al., 2017). Additionally, cinnamon oil exhibited moderate antifungal activity against *F. oxysoprum* f. sp. *fragariae* (Park et al., 2017). The essential oil of clove and cinnamon is highly effective in the control of gray mold disease in strawberries by disturbing cell walls and cell membranes.

Additionally, Tea Tree Oil (TTO) acts on pathogenic fungi and other microbes due to its ability to disturb the permeability of cellular membranes (Carson et al., 2006; Carson et al., 2002; Cox et al., 2001). *M. alternifolia* components are active at disturbing cell integrity, inhibiting respiration, ion transport chain and also lead to an increase in membrane permeability (Carson et al., 2006; Carson et al., 2002; Cox et al., 2001). The formulation of tea tree oil Timorex gold can disturb the permeability of the membranes of plant pathogenic fungi so it can be effectively used as a fungicide (Shao et al., 2013). It acts against *M. fijiensis* at the four and five stages by disturbing the cell membrane and cell wall in the intracellular space of the banana leaf mesophyll. This is the main reason for the reduction of Sigatoka disease development treated with Timorex Gold and the mesophyll tissues have few numbers of *M. fijiensis* hyphae as compared to plant leaves treated with synthetic fungicides. Therefore, the use of Timorex Gold is highly recommended for the control of Sigatoka disease in bananas at the commercial level (Reuveni et al., 2020; Martillo and Reuveni, 2009).

Tea tree oil act differently from sterol-inhibiting fungicides. It works by interrupting the cell membrane and the cell wall. Cell wall disruption affects the growth and morphology of the fungal cell, ultimately causing the death and lysis of the pathogen (Adams, 2007). Timorex Gold, a tea tree oil-based product has been efficiently used against the powdery mildew of cucumber. The lesion formation on the cucumber was controlled with the application. The mycelia, conidia and conidiophore of the pathogen were shrunk and acquired irregular shapes (Reuveni et al., 2020).

On the other hand, sesame leave extracts had been used against fungus and bacteria, i.e., *Candida albicans* and *Streptococcus pneumoniae* (Shittu et al., 2006; Ahmed et al., 2009). Stem, root and leaves extract of sesame reduce the growth of *M. phaseolina* and *F. oxysporum* (Laurentin, 2007). The lignan which is known as sesamol takes part in the slowdown of entry of pathogens mostly decreasing the charcoal rot in soyabean caused by *Phaseolina* (Brooker et al., 2000) and also plays a vital role in the inhibition of growth of *Mucor circinelloides* (Wynn et al., 1997).

After sesame oil, cuminic acid (p- isoprophylbenzoic acid) which is a part of the benzoic acid chemical group and is the major component of *C. cyminum*, has been isolated from the seeds (Wang et al., 2016). The presence of cuminic acid helped in the control of phytopathogens such as *Sclerotina sclerotiorum*, *Phytophthora capsici*, *Rizoctonia cerealis* and *Fusarium oxysporum.* When the effect of cuminic acid was tested on the FON (*Fusarium Oxysporum* f. sp. *Niveum*) its growth was inhibited. The antifungal activity of cuminic acid is significant in PDA plates as compared to cinnamic acid (Wu et al., 2018), gallic acid and sinapic acid (Wu et al., 2009). In the same way, cuminic acid harms the morphology of FNO. After 7 d of incubation, the color of mycelia becomes lighter with lower mycelial growth as compared to the control one (Sun et al., 2017).

Moreover, *Cinamon cyminum* EO is used against *B. cinerea*, *R. stolonifer* and *A. niger* when the oil is mixed with PDA in an amount of 750 µl/L (Hadian et al., 2008). In addition, the EO of *C. cyminum* in combination with *M. logifolia* is used to control *V. dahlia* kleb by arresting the growth (Üstüner et al., 2018). The mode of action of Cumin Essential Oil (CEO) is by disturbing the membrane integrity of pathogens shown in an electron micrograph. Thus, cummin essential oil is preferred to control natural agents to avoid health hazards and economic losses (Behbahani et al., 2019). It has been investigated that, the essential oil of cumin is widely used against fungi in foods (Kedla et al., 2013).

9. Bioactive compounds in plants and their key role in the fungicidal activity

Secondary metabolites have a crucial role in plant disease management by protecting plants from pathogens and predators. They also play a role in the defense against stress caused by abiotic factors. Moreover, they participate in the interaction of plants with other organisms (Schafer and Wink, 2009) (Table 8.4).

The bioactive compounds of different plant extracts play a key role in fungicidal activity. Biochemicals include alkaloids, tannins, lectins, terpenoids and many other chemicals. These chemicals are toxic to the pathogen by depleting the presence of ions. Some chemicals may destroy the plasma membrane and configuration of DNA, in this way fungi lose the integrity of the membrane by granule formation in the cytoplasm and leakage of cellular content. These chemicals are potent at the

Table 8.4: Secondary metabolites used as biofungicides.

No.	Compounds	Mode of action	Source plant	Reference
1.	Alkaloids	Cause leakage of cellular content	Moringa	(Cowan, 1999; Ferreira et al., 2008)
2.	Tannins	Cause deprivation of substrate by binding with the protein enzyme	*Punica granatum*	(Qnais et al., 2007; Cowan, 1999)
3.	Terpenoids	Membrane disruption of the fungal pathogen	Mentha	(Cowan, 1999)
4.	Flavanoids	Fungal enzymes inactivation	Garlic, Oregano	(Fufa, 2019; Liaqat et al., 2019; De Falco et al., 2018; Cowan, 1999)
5.	Coumarins/ Cumaric acid	Destroy the DNA of the fungal pathogen	*Origanum vulgare, Allium stavium*	(Koldas et al., 2015; Huzaifa et al., 2014; Cowan, 1999)
6.	Saponins	Disrupt membrane integrity by forming pores	*Medicago sativa, Medicago hybrida*	(Ito et al., 2005; Jarecka et al., 2008; Cowan, 1999)
7.	Phenolic acid/ Phenolic compounds	Adhesion binding and inactivation of fungal enzymes	Neem, Basil, Moringa	(Gradinariu et al., 2013; Cowan, 1999)

inhibition of fungal ATPases, which cause the dissolution of a chaperone of fungi. In this way, the activity of bioactive chemicals plays a role in the occurrence of disease by causing the death of the pathogen (Cowan, 1999). In medicinal plants, there are several bioactive compounds, but very few are explored (Nastic et al., 2018). The specific parts or whole medicinal plants are a rich source of bioactive compounds also known as phytochemicals (Nastic et al., 2018; Ibrahim and Kebede, 2020). These compounds have a role in medicines, cosmetics, food and supplements (Ibrahim and Kebede, 2020; Li et al., 2020).

Medicinal plants and aromatic compounds contain phytochemicals which include flavonoids, saponins, polyphenols and alkaloids in leaves, bark, flowers, seeds or branches. These bioactive compounds are used as antioxidants and antimicrobes and also have pharmaceutical and biopesticide potential (Al-Huqail et al., 2019). The phytochemical analysis of moringa extract showed the presence of alkaloids, saponins, lavonoids and tanins in all extracts except the root where saponins and tannins were absent (Abdulkadir et al., 2015). Pathogens destroy the cell wall of the plant through the initiation of cutinases and laccases pathways. In this way, by inhibiting the synthesis of enzymes bioactive compounds protect plants from fungal damage (Goetz et al., 1999; Bostock et al., 1999). There is a diversity in the composition of plant extracts even if they belong to the same species giving different responses in the aspects of microorganism inhibition (Sales et al., 2016). When the occurrence of bioactive compounds is considered, the concentration of each bioactive compound depends on the environmental conditions and pathosystem (Balakumar et al., 2011; Gahukar, 2012; Gillitzer et al., 2012). Secondary metabolites are divided into three categories called terpenes, phenolics and nitrogen and sulfur-containing compounds (Alkaloids) (Mazid et al., 2011).

9.1 Phenolic compounds

Phenolic compounds are the main constituents of most medicinal plants (Manousi et al., 2019). These are formed when more than one hydroxyl group are glycosylated or methylated in two ways by shikimic acid and malonic acid (Mahajan et al., 2020). Phenol compounds have an important role in maintaining plant health and determining the quality of fruits and vegetables (Guerreiro et al., 2015). These are associated with the reduction of decay severity of fruits and post-harvest disease management (Shen et al., 2013). These phenolic compounds have a major role in the defense mechanism of plants against pathogens by increasing antioxidant activity (Tzortzakis et al., 2010).

Phenolic compounds include coumarins, lignans, stilbenes, flavanoids, tannins and lignins (Agostini-Costa et al., 2012). These are responsible for different pigments in plants and help in growth and development. They also serve plants by strengthening the defense mechanism against pathogens, parasites and predators (Baidez et al., 2007). Phenolic compounds are mainly produced in plants in response to environmental stress, physiological conditions, pathogens and insects attack as well as UV exposure (Crozier et al., 2006; Chung et al., 2003; Kennedy and Wightman, 2011; Diaz Napal et al., 2010).

Plant tissues like leaves, flowers, stems and roots are rich in flavonoids and phenolic acids, that participate against biotic and abiotic stress such as insect attack, injury and UV radiation (Khoddami et al., 2013; Cheynier et al., 2013). Due to the presence of phenolic compounds, the plant can easily adapt itself to environmental fluctuations through different physiological processes. Acids of phenols like hydroxycinnamic acid derivatives act as precursor molecules for stilbenes, lignans, flavonoids, anthocyanins and chalcones (Alam et al., 2016).

The presence of these compounds in tissues as a conjugate of esters of carboxylic acids or sterols or insolublly bound, attach with structural components of the plant cell wall. They are rarely found in free form as monomers or dimers (El-Seedhi et al., 2012). Phenolic compounds are mostly used against fungi. Phenolics work by disturbing the cell membrane of the pathogen and also depolarizes the potential of the mitochondrial membrane in a concentration-dependent manner (Tian et al., 2012; Wu et al., 2008). Bostock et al. (1999) stated that in the epidermis and cell layers of plants, phenolic compounds are found in the form of chlorogenic and caffeic acid, having an important role in the inhibition of oxidase formation of the pathogen such as *Monilia fruticola* causing brown rot.

The analysis of plant extracts of *A. indica* and *A. sativum* showed the presence of flavanoids, saponins, terpenoids, steroids, tannins and coumarins. These bio compounds are also present in *Ocimum sanctum* except for cardiac glycosides. Moreover, coumarin-like compounds are responsible for antifungal activity mostly in active plant extracts. Many other studies have confirmed the role of coumarin derivatives in antimicrobial activity (Smyth et al., 2009).

Rhizophora mangle (CEMV6) showed antifungal activity due to the presence of secondary metabolites like tannins; these bioactive compounds are produced by herbal medicines rich in polyphenols (vegetable tannins) (Sales et al., 2015). As it is known, essential oils have antifungal properties, which is due to the presence of terpenoids and phenolic compounds that actively participate in the structural and functional damage of the cell and also disturb the membrane permeability of the cell (Prakash et al., 2015; Grata, 2016; Kalagatur et al., 2015).

Plants that belong to the *Lamiaceae* family have generated great attention among scientists, there are two types of phenols; carvacrol and thymol are present in such plants (Pinto et al., 2006). These substances have been described to inhibit the growth of fungi and bacteria (Sokovic et al., 2009, Kedia et al., 2014). Carvacrol and thymol change the integrity of cell membrane, inorganic ionic balance and pH of homeostasis in microorganisms (Lambert et al., 2001). In origanum spp., carvacrol is present up to 50 to 86% depending on the species of the plant (Kulisic et al., 2004). The role of phenolic compounds in the control of fungal disease is more dynamic as compared to insect control.

Whenever resistance is produced , there could be a deposition of wound plugs from phenylpropanoids like papillae, that have the ability to hinder fungal entry (Aist, 1976, 1983; Sahasbi and Shishiyama, 1986; Aist and Gold, 1987). Moreover, the structural barrier formed by phenolic compounds is the cross-linking of polysaccharides of the cell wall by the formation of ferulic acid dimers which are catalyzed by peroxidase dimers. This cross-linking is strong enough to bear the cell wall degrading enzymes of fungi (Fry, 1986). When resistance in pea is examined against powdery mildew (*Erysiphe polygonii*) it is due to the concentration of ortho-dihydroxyphenols in the leaves (Kalia and Sharma, 1988).

A number of plants produce secondary metabolites called saponins which are glycosidic, these compounds have shown significant effects against fungi, nematodes, bacteria and viruses (Tava and Avato, 2006). Jarecka et al. (2008) described the effect of saponins from *Medicago arabica*, *M. hybrida* and *M. sativa* on the growth of *F. oxysporum* f. sp *tulipae in vitro* and *in vivo*, where the findings showed that at a concentration of 0.01, 0.05, 0.1%, saponins from the root part of *M. hybrida* and *M. sativa* showed potent activity against *F. oxysporum* f. sp *tulipae*. In short, it is proved that saponins from medicago can be used as a fungicide. The reason why saponins are harmful and toxic to fungi is the formation of a complex with membrane sterols which result in pore formation (Price et al., 1987; Fenwick, 1992). Similarly, a plant extract containing saponin in the form of tomatine creates a complex with the sterol-free hydroxyl group of the fungal membrane. Thus, forming pores and reducing the integrity of the membrane of fungi (Ito et al., 2005).

In some cases, saponins become ineffective against fungi. Fungi are biotrophic for example, a pathogen of tomato *Curvularia fulvum*, that mainly restricts its growth in the intercellular spaces of tomato leaves through the release of a-tomatine. The fungi attack saponin-containing plants but also manage to adopt resistance against saponins *in vitro*. It is clear from the given data that saponin resistance is helpful in the initiation of infection (Osbourn, 1996), as the a-tomatine membranolytic action is pH dependent (Roddick and Drysdale, 1884). *Alternaria solani* which also infects tomatoes overcome the effect of a-tomatine by decreasing the pH at the point of infection until the level of saponin becomes ineffective (Schonbeck and Schlosser, 1976).

Overall, flavonoids are found as signaling molecules during plants-microbe interaction, also attracting pollinators and protecting cells from UV light exposure (Falcone et al., 2012). The interaction between plant and microbes is aided by the secretion of flavanoids in soil (Hassan et al., 2012; Sugiyama and Yazaki, 2014). For example, the roots of soybean secrete isoflavones for the formation of symbiosis with rhizobia (Sugiyama et al., 2016). The defense against root infecting pathogens and MtABCG10, G-type transporter from *M. truncatula* is ensured by the production of flavonoid by modulating the level of isoflavone in roots (Banasiak et al., 2013). More than 4,000 groups of compounds represent flavonoids, which are phenolic compounds but hydroxylated and formed in plants in the response to infection by a pathogen. Isoflavone is a subgroup of flavonoids, several studies revealed the activity of isoflavones against fungi (Fukui et al., 1973). Luteone is an isoflavone isolated from the fruits of *Lupinus luteus* (*Leguminosae*) by using methanol extraction. The conidia of *Cochliobolus mivabeanlls* (*Helminthosporium oryzae*), were inhibited by the activity of luteone (Kumar and Pandey, 2013; Dixon et al., 1983). Moreover, a compound prominent in the exudates of alfalfa roots which is deficient in Fe, that can dissolve ferric phosphate easily. The release of 2-(3-5-dihydroxyphenyl)-5, 6-dihydroxy benzofuran from roots of alfalfa, acts as phytoalexin against the pathogens like *F. oxysporum* f. sp *Phaseoli* (Masaoka et al., 1993).

Terpenes are a group of compounds which are not soluble in water. They are odoriferous constituents of plants due to the presence and release of volatile molecules which have 10, 15 and 20 carbon atoms (Naboulsi et al., 2018). Terpenes are also actively used against pathogens. The mode of action of terpenes is by disturbing the membrane of the pathogen and is also involved in fungal mitochondrial destruction by inhibiting electron transport and inhibition of ATPase (Walker et al., 2012; Tian et al., 2012). Harris and Dennis (1976, 1977) showed that the bioactive chemical type of terpenoid leads to the development of cytoplasmic granulation and disturbance of the membrane and leakage of cellular content in the zoospores of *Phytopthora infestans, P. porri* and *P. cactorum*.

A plant extract containing steroidal glycol alkaloids interferes with the membrane of fungi (Keukens et al., 1995). At very low concentrations, some alkaloids affect plant fungal pathogens. Allosecurinine a form of the alkaloid is the main component of *Phyllantus amarus*, the plant extract suppresses the germination of fungal spores of Heterosporium species, *Curvularia lunata* and *Colletotrichum musae* (Pezit and Pout, 1990). The presence of alkaloids in the plant body is toxic to prokaryotic as well as eukaryotic cells. For instance, berberine, a benzylisoquinoline alkaloid is formed in different plant species which are responsible for the inhibition of DNA and protein formation in bacteria and also protects against bacteria.

These compounds are harmful to plant cells which do not produce them, but cells producing such metabolites are insensitive to them, this indicates their role in the self-tolerance mechanism. When the transportation of alkaloids is observed, berberine and nicotine are translocated from the source of production to the organ where it accumulates (sink organ). In short, intracellular, extracellular and inter-organ transportation of alkaloids is of great importance (Shitan, 2016).

10. Biofungicides and their mechanisms of action

10.1 Nature

In the last few years, as an alternative to synthetic fungicides, pure biological control measures against various plant diseases are being paid greater attention. Around 40 types of biological products are produced globally (Golosova, 2003). The most advanced trend to keep crops safe from pathogenic diseases is to induce resistance among hosts against phytopathogens and unfavorable environmental elements through these bio products (Fig. 8.1).

Biofungicide is the collective name for all the products developed from microorganisms, their vital products which largely affect plant disease (Roger and Keinath, 2010). Their organic nature and low concentration of active compounds have made these products ecologically pure. With less toxic effects these preparations have a wide range of anti-pathogenic properties and are also favorable for resistance enhancement of plants.

Biofungicides are organic and isolated from the soil and plant extracts. Trichoderma is used as a biocontrol agent to inhibit diseases caused by many pathogens including Botrytis spp., *Cladosporium*, *Sclerotinia sclerotiorum*, *Fusarium*

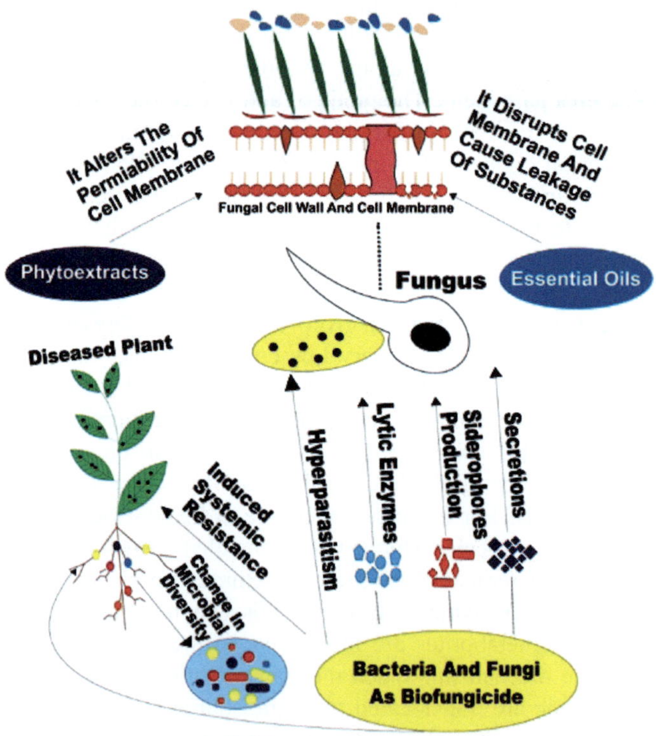

Fig. 8.1: Biofungicides work by using natural mechanisms to control or eliminate fungal pathogens. Some modes of action of biofungicides includings competition, antibiosis, hydrolysis, and parasitism. Biofungicides have multiple modes of action that work in synergy to control fungal infections and provide sustainable plant protection.

oxysporum and *Collectotrichum acutatum* (Freeman et al., 2004; Patel and Saraf, 2017; Elad, 2000). There are different mechanisms of action required by microbes (biological control agents) to protect crops from disease impact. If microbes are not directly in contact with the pathogen, they may cause enhanced resistance or induced resistance against infections by a pathogen in plant tissues (Conrath et al., 2015). Competition for nutrients and space is another indirect contact with pathogens (Spadaro and Droby, 2016). Additionally, MBCAs can interact with the pathogen directly through antibiosis or hyperparasitism. The mycelium, spores and dormant structures of fungal pathogens as well as the cells of bacterial pathogens are invaded and killed by hyperparasitism (Ghorbanpour et al., 2018). Additionally, siderophores produced by bacteria that compete for iron, antibiotics such as DAPG, pyocyanin, biosurfactants and Volatile Organic Compounds (VOCs) like 2R, 3R-butanediol made by *B. subtilis* GB03 (130) and a C13 volatile released by *Paenibacillus polymyxa* can act as elicitors inducing ISR (Pieterse et al., 2014). Plant extracts are used as biofungicides to inhibit plant diseases, some examples of biofungicides based on extracts are *Allium sativum*, *Allium cepa*, *Zingiber officinale* and oil (Polo et al., 2021). Shisham seedlings infected with *Fusarium solani* were treated with 15% neem oil which decreased the infection by 85% (Rajput et al., 2011). Steroidal glycol alkaloids in plant extracts worked by interfering with the integrity of the fungal membrane to kill plant fungal infections (Keukens et al., 1992). According to Davidson (1997), tannins' toxic action mechanism targets fungal pathogens directly on the cell membrane by depleting their metal deposits. The mechanism of action of cucurbitacin I bind with the steroid components of the fungal membrane (Zhao et al., 2010). Similarly, the plant extracts that include saponin in the form of tomatin create a complex with the free hydroxyl group in the sterol of the fungal membrane, leading to the creation of pores and the consequent loss of integrity of the fungal membrane (Ito et al., 2005). Essential oils are thought to work against the cytoplasmic cell membranes of fungi (Diao et al., 2014). Essential oils are considered to accumulate in cell membranes due to their hydrophobic nature (Sikkema et al., 1995). Research has shown that the mechanisms by which the essential oils (EOs) work include accumulation of EOs in the cell, effect on cell permeability, disturbance of major organelle membranes, variation of the general morphology, leakage of cell organelles which ultimately leads to the death of the cell (Bajpai et al., 2013).

11. Conclusions

Agriculture is facing a wide range of plant diseases causing severe yield losses to crops, around 10% of world food production is lost due to plant diseases. The major part of these losses is comprised of fungal pathogens, due to their wide host range and diversity. The management of fungal plant diseases is dependent on synthetic chemical fungicides, which create serious threats to environmental and human/animal health due to their residual effects. These residues in a food chain are cancerous to humans and animals when taken in high doses and also harmful to the ecosystem of the beneficial microbes. Thus researchers are frequently searching for different biofungicides (biological fungicides) to combat fungal plant diseases and to save environmental and human/animal health as it is the only way to minimize the application of chemical fungicides against fungal diseases. By studying the mechanism of action of these biofungicides, it is concluded that finding the natural sources for the management of plant diseases is a current need for sustainable agriculture.

References

Abarna, T., Raheem, A.A., Livingstone, R.A., Abichandra, K., Abikannan, A. and Raj, T.S. (2019). Bio efficacy of *Bacillus subtilis* and Plant extracts against *Colletotrichum capsici* (Syd.) Butler and Bisby under *in vitro* condition. International Journal of Applied Sciences and Computations, 6(4): 2110–2123.

Abbasi, P.A., Conn, K.L. and Lazarovits, G. (2004). Suppression of Rhizoctonia and Pythium damping-off of radish and cucumber seedlings by addition of fish emulsion to peat mix or soil. Canadian Journal of Plant Pathology, 26(2): 177–87.

Abdulkadir, I.S., Nasir, I.A., Sofowora, A., Yahaya, F., Ahmad, A.A. and Hassan, I.A. (2015). Phytochemical screening and antimicrobial activities of ethanolic extracts of *Moringa oleifera* Lam on isolates of some pathogens. Journal of Applied Pharmacy, 7(4): 2–7.

Absalan, A., Mohiti-Ardakani, J., Hadinedoushan, H. and Khalili, M.A. (2012). Hydro-alcoholic cinnamon extract, enhances glucose transporter isotype-4 translocation from intracellular compartments into the cytoplasmic membrane of C2C12 myotubes. Indian Journal of Clinical Biochemistry, 27(4): 351–356.

Adams, R.P. (2007). Identification of essential oil components by gas chromatography/mass spectrometry. Carol Stream: Allured Publishing Corporation, 456: 544–545.

Adetumbi, M., Javor, G.T. and Lau, B.H. (1986). *Allium sativum* (garlic) inhibits lipid synthesis by *Candida albicans*. Antimicrobial Agents and Chemotherapy, 30(3): 499–501.

Adrian, M., Jeandet, P., Veneay, J.I, Western, L.A. and Besis, R. (1977). Biological activity of re; a stilbene compound from grape vines against *Botrytis cinerea*, the causal agent of grey mould. Journal of Chemical Ecology, 23: 1689–1702.

Afsharypuor, S., Asghari, G., Mohagheghzadeh, A. and Dehshahri, S. (2010). Volatile constituents of the seed kernel and leaf of *Moringa peregrina* (Forssk.) Fiori, Agricolt. Cultivated in Chabahar (Iran). Iranian Journal of Pharmaceutical Sciences, 6(2): 141–144.

Aggarwal, R., Tewari, A.K., Srivastava, K.D. and Singh, D.V. (2004). Role of antibiosis in the biological control of spot blotch (*Cochliobolus sativus*) of wheat by *Chaetomium globosum*. Mycopathologia, 157(4): 369–377.

Agostini-Costa, T.D.S., Vieira, R.F., Bizzo, H.R., Silveira, D. and Gimenes, M.A. (2012). Secondary metabolites, chromatography and its applications. InTech, London, 131–132.

Agrios, G.N. (2005). Plant Pathology (5th edition). Elsevier Academic Press, NewYork, 633.

Ahmed, J., Altun, E., Aydogdu, M.O., Gunduz, O., Kerai, L., Ren, G. and Edirisinghe, M. (2019). Anti-fungal bandages containing cinnamon extract. International Wound Journal, 16(3): 730–736.

Ahmed, T., Shittu, L.A.J., Bankole, M.A., Shittu, R.K., Adesanya, O.A., Bankole, M.N. and Ashiru, O.A. (2009). Comparative studies of the crude extracts of sesame against some common pathogenic microorganisms. Scientific Research and Essays, 4(6): 584–589.

Aist, J.R. (1976). Papillae and related wound plugs of plant cells. Annual Review of Phytopathology, 14(1): 145–163.

Aist, J.R. (1983). Structural responses as resistance mechanisms. The Dynamics of Host Defence, 33–70.

Aist, J.R. and Gold, R.E. (1987). Prevention of fungal ingress: The role of papillae and calcium. pp. 47–58. *In*: S. Nishimura (ed.). Molecular Determinants of Plant Diseases. Tokyo/Btrlin: Japan Set. Soc. Press/Springer-Verlag.

Alam, M.A., Subhan, N., Hossain, H., Hossain, M., Reza, H.M., Rahman, M.M. and Ullah, M.O. (2016). Hydroxycinnamic acid derivatives: A potential class of natural compounds for the management of lipid metabolism and obesity. Nutrition & Metabolism, 13(1): 1–13.

Al-Huqail, A.A., Behiry, S.I., Salem, M.Z.M., Ali, H.M., Siddiqui, M.H. and Salem, A.Z.M. (2019). Antifungal, antibacterial, and antioxidant activities of *Acacia saligna* (Labill.) H. L. Wendl. flower extract: HPLC analysis of phenolic and flavonoid compounds. Molecules, 24: (4).

Allameh, A., Abyaneh, M.R., Shams, M.R., Rezaee, M.B. and Jaimand, K. (2002). Effects of neem leaf extract on production of aflatoxins and activities of fatty acid synthetase, isocitrate dehydrogenase and glutathione S-transferase in *Aspergillus parasiticus*. Mycopathologia, 154(2): 79–84.

Altomare, C., Pengue, R., Favilla, M., Evidente, A. and Visconti, A. (2004). Structure—activity relationships of derivatives of fusapyrone, an antifungal metabolite of *Fusarium semitectum*. Journal of Agricultural and Food Chemistry, 52(10): 2997–3001.

Amadioha, A.C. (2000). Controlling rice blast *in vitro* and *in vivo* with extracts of *Azadirachta indica*. Crop Protection, 19(5): 287–290.

Anjorin, T.S., Ikokoh, P. and Okolo, S. (2010). Mineral composition of *Moringa oleifera* leaves, pods and seeds from two regions in Abuja, Nigeria. International Journal of Agriculture and Biology, 12(3): 431–434.

Anke, T., Hecht, H.J., Schramm, G. and Steglich, W. (1979.) Antibiotics from basidiomycetes. IX. Oudemansin, an antifungal antibiotic from *Oudemansiella mucida* (Schrader ex Fr.) *Hoehnel (Agaricales)*. Journal of Antibiotic, 32: 1112–1117.

Anke, T., Oberwinkler, F., Steglich, W. and Schramm, G. (1977). The strobilurins-new antifungal antibiotics from the basidiomycete *Strobilurus tenacellus* (Pers. ex Fr.) Sing. The Journal of Antibiotics, 30(10): 806–810.

Arima, K., Imanaka, H., Kousaka, M., Fukuta, A. and Tamura, G. (1964). Pyrrolnitrin, a new antibiotic substance, produced by Pseudomonas. Agricultural and Biological Chemistry, 28(8): 575–576.

Askarne, L., Talibi, I., Boubaker, H., Boudyach, E.H., Msanda, F., Saadi, B., ... and Aoumar, A.A.B. (2012). *In vitro* and *in vivo* antifungal activity of several Moroccan plants against *Penicillium italicum*, the causal agent of citrus blue mold. Crop Protection, 40: 53–58.

Atkinson, D., Berta, G. and Hooker, J.E. (1994). Impact of mycorrhizal colonisation on root architecture, root longevity and the formation of growth regulators. In Impact of Arbuscular Mycorrhizas on Sustainable Agriculture and Natural Ecosystems (pp. 89–99). Birkhäuser, Basel.

Azarakhsh, N., Osman, A., Ghazali, H.M., Tan, C.P. and Mohd Adzahan, N. (2012). Optimization of alginate and gellan-based edible coating formulations for fresh-cut pineapples. International Food Research Journal, 19(1).

Bacon, C.W. and Hinton, D.M. (2007). Bacterial endophytes: The endophytic niche, its occupants, and its utility. In Plant-associated Bacteria (pp. 155–194). Springer, Dordrecht.

Bacon, C.W., Yates, I.E., Hinton, D.M. and Meredith, F. (2001). Biological control of Fusarium moniliforme in maize. Environmental Health Perspectives, 109: 325–332.

Báidez, A.G., Gómez, P., Del Río, J.A. and Ortuño, A. (2007). Dysfunctionality of the xylem in *Olea europaea* L. plants associated with the infection process by *Verticillium dahliae* Kleb. Role of phenolic compounds in plant defense mechanism. Journal of Agricultural and Food Chemistry, 55(9): 3373–3377.

Bajpai, V.K., Sharma, A. and Baek, K.H. (2013). Antibacterial mode of action of Cudrania tricuspidata fruit essential oil, affecting membrane permeability and surface characteristics of food-borne pathogens. Food Control, 32(2): 582–590.

Baker, J.E., Anderson, J.D., Adams, D.O., Apelbaum, A. and Lieberman, M. (1982). Biosynthesis of ethylene from methionine in aminoethoxyvinylglycine-resistant avocado tissue. Plant Physiology, 69(1): 93–97.

Baker, J.E., Wang, C.Y., Lieberman, M. and Hardenburg, R. (1977). Delay of senescence in carnations by a rhizo-biotoxine analog and sodium benzoate. Horticultural Science, 12: 38–39.

Bakkali, F., Averbeck, S., Averbeck, D. and Idaomar, M. (2008). Biological effects of essential oils—A review. Food and Chemical Toxicology, 46(2): 446–475.

Balakumar, S., Rajan, S., Thirunalasundari, T. and Jeeva, S. (2011). Antifungal activity of *Aegle marmelos* (L.) Correa (*Rutaceae*) leaf extract on dermatophytes. Asian Pacific Journal of Tropical Biomedicine, 1(4): 309–312.

Banasiak, J., Biała, W., Staszków, A., Swarcewicz, B., Kępczyńska, E., Figlerowicz, M. and Jasiński, M. (2013). A Medicago truncatula ABC transporter belonging to subfamily G modulates the level of isoflavonoids. Journal of Experimental Botany, 64(4): 1005–1015.

Baraka, M.A., Radwan, F.M., Shaban, W.I. and Arafat, K.H. (2011). Efficiency of some plant extracts, natural oils, biofungicides and fungicides against root rot disease of date palm. Journal of Biology Chemistry and Environmental Sciences, 6(2): 405–429.

Bawazir, A.M.A. and Shantaram, M. (2018). Effect of yemeni sesame oil against some pathogenic bacteria and fungi. Int. J. Pharm. Sci. Rev. Res., 9(6): 2507–12.

Behbahani, B.A., Noshad, M. and Falah, F. (2019). Cumin essential oil: Phytochemical analysis, antimicrobial activity and investigation of its mechanism of action through scanning electron microscopy. Microbial Pathogenesis, 136: 103716.

Beretta, B., Gaiaschi, A., Galli, C.L. and Restani, P. (2000). Patulin in apple-based foods: Occurrence and safety evaluation. Food Additives & Contaminants, 17(5): 399–406.

Berta, G., Sampo, S., Gamalero, E., Massa, N. and Lemanceau, P. (2005). Suppression of Rhizoctonia root-rot of tomato by *Glomus mossae* BEG12 and *Pseudomonas fluorescens* A6RI is associated with their effect on the pathogen growth and on the root morphogenesis. European Journal of Plant Pathology, 111(3): 279–288.

Bolgar, M., Cunningham, J., Cooper, R., Kozloski, R., Hubball, J., Miller, D.P., ... and Fairless, B. (1995). Physical, spectral and chromatographic properties of all 209 individual PCB congeners. Chemosphere, 31(2): 2687–2705.

Borowicz, V.A. (2001). Do arbuscular mycorrhizal fungi alter plant–pathogen relations? Ecology, 82(11): 3057–3068.

Bostock, R.M., Wilcox, S.M., Wang, G. and Adaskaveg, J.E. (1999). Suppression of *Monilinia fructicola* cutinase production by peach fruit surface phenolic acids. Physiological and Molecular Plant Pathology, 54(1-2): 37–50.

Botre, D.A., Soares, N.D.F.F., Espitia, P.J.P., Sousa, S.D. and Renhe, I.R.T. (2010). Avaliação de filme incorporado com óleo essencial de orégano para conservação de pizza pronta. Revista Ceres, 57: 283–291.

Bozin, B., Mimica-Dukic, N., Simin, N. and Anackov, G. (2006). Characterization of the volatile composition of essential oils of some Lamiaceae spices and the antimicrobial and antioxidant activities of the entire oils. Journal of Agricultural and Food Chemistry, 54(5): 1822–1828.

Brooker, N.L., Long, J.H. and Stephan, S.M. (2000). Field assessment of plant derivative compounds for managing fungal soybean diseases.

Bruneton, J. (1993). Pharmacognosie: Phytochimie plantes médicinales (No. 581.634 B7).

Burt, S. (2004). Essential oils: Their antibacterial properties and potential applications in foods—A review. International Journal of Food Microbiology, 94(3): 223–253.

Carson, C.F. and Hammer, K.A. (2011). Chemistry and bioactivity of essential oils. Lipids Essent Oils Antimicrob Agents, 25: 203–38.

Carson, C.F., Hammer, K.A. and Riley, T.V. (2006). *Melaleuca alternifolia* (tea tree) oil: A review of antimicrobial and other medicinal properties. Clinical Microbiology Reviews, 19(1): 50–62.

Carson, C.F., Mee, B.J. and Riley, T.V. (2002). Mechanism of action of *Melaleuca alternifolia* (tea tree) oil on *Staphylococcus aureus* determined by time-kill, lysis, leakage, and salt tolerance assays and electron microscopy. Antimicrobial Agents and Chemotherapy, 46(6): 1914–1920.

Chen, C., Belanger, R.R., Benhamou, N. and Paulitz, T.C. (2000). Defense enzymes induced in cucumber roots by treatment with plant growth-promoting rhizobacteria (PGPR) and *Pythium aphanidermatum*. Physiological and Molecular Plant Pathology, 56(1): 13–23.

Cheynier, V., Comte, G., Davies, K.M., Lattanzio, V. and Martens, S. (2013). Plant phenolics: Recent advances on their biosynthesis, genetics, and ecophysiology. Plant Physiology and Biochemistry, 72: 1–20.

Choi, S.T. and Huber, D.J. (2008). Influence of aqueous 1-methylcyclopropene concentration, immersion duration, and solution longevity on the postharvest ripening of breaker-turning tomato (*Solanum lycopersicum* L.) fruit. Postharvest Biology and Technology, 49(1): 147–154.

Christensen, P. and Cook, F.D. (1978). Lysobacter, a new genus of nonfruiting, gliding bacteria with a high base ratio. International Journal of Systematic and Evolutionary Microbiology, 28(3): 367–393.

Chung, I.M., Park, M.R., Chun, J.C. and Yun, S.J. (2003). Resveratrol accumulation and resveratrol synthase gene expression in response to abiotic stresses and hormones in peanut plants. Plant Science, 164(1): 103–109.

Cibanal, I.L., Fernandez, L.A., Murray, A.P., Pellegrini, C.N. and Gallez, L.M. (2021). Propolis extract and oregano essential oil as biofungicides for garlic seed cloves: *in vitro* assays and synergistic interaction against *Penicillium allii*. Journal of Applied Microbiology, 131(4): 1909–1918.

Cock, I.E. and Cheesman, M.A.T.T.H.E.W. (2018). Plants of the genus *Syzygium* (*Myrtaceae*): A review on ethnobotany, medicinal properties and phytochemistry. Bioactive Compounds of Medicinal Plants: Properties and Potential for Human Health, 35–84.

Cole, R.J., Kirksey, J.W., Dorner, J.W., Wilson, D.N., Johnson, J.C., Johnson, J.A.N., Bedell, D.M., Springer, J.P., Chexal, K.K., Clardy, J.C. and Cox, R.H. (1977). Mycotoxins produced by Aspergillus fumigatus species isolated from molded silage. Journal of Agriculture Food and Chemistry, 25: 826–830.

Collins, D.P., Jacobsen, B.J.and Maxwell, B. (2003). Spatial and temporal population dynamics of a phyllosphere colonizing *Bacillus subtilis* biological control agent of sugar beet cercospora leaf spot. Biological Control, 26(3): 224–232.

Conrath, U., Beckers, G.J., Langenbach, C.J. and Jaskiewicz, M.R. (2015). Priming for enhanced defense. Annual Review of Phytopathology, 53(1): 97–119.

Cosentino, S., Barra, A., Pisano, B., Cabizza, M., Pirisi, F.M. and Palmas, F. (2003). Composition and antimicrobial properties of *Sardinian juniperus* essential oils against foodborne pathogens and spoilage microorganisms. Journal of Food Protection, 66(7): 1288–1291.

Cowan, M.M. (1999). Plant products as antimicrobial agents. Clinical Microbiology Reviews, 12(4): 564–582.

Cox, S.D., Mann, C.M., Markham, J.L., Gustafson, J.E., Warmington, J.R. and Wyllie, S.G. (2001). Determining the antimicrobial actions of tea tree oil. Molecules, 6(2): 87–91.

Crozier, A., Jaganath, I.B. and Clifford, M.N. (2006). Phenols, polyphenols and tannins: An overview. Plant Secondary Metabolites: Occurrence, Structure and Role in the Human Diet, 1: 1–25.

Daniel, C.K., Lennox, C.L. and Vries, F.A. (2015). *In vivo* application of garlic extracts in combination with clove oil to prevent postharvest decay caused by *Botrytis cinerea*, *Penicillium expansum* and *Neofabraea alba* on apples. Postharvest Biology and Technology, 99: 88–92.

Das, M.M. and Abdulhameed, S. (2020). Agro-processing Residues for the production of fungal bio-control agents. In Valorisation of Agro-industrial Residues: Non-Biological Approaches. Springer, Cham., 2: 107–126.

Das, M.M., Aguilar, C.N., Haridas, M. and Sabu, A. (2021). Production of bio-fungicide, *Trichoderma harzianum* CH1 under solid-state fermentation using coffee husk. Bioresource Technology Reports, 15: 100708.

Davidson, P.M. (1997). Chemical Perspectives and natural anti-nucrobial compound. pp. 520–556. *In*: M.P. Doyle, L.R. Beuchat and T.J. Montville (eds). Food Microbiology, Fundamentals and Frontiers. Washington D.C. ASM Press.

De Curtis, F., Spina, A.M. and Lima, G. (2001). Gravi attacchi di *Sclerotium rolfsii* su pomodoro in pieno campo, 70–71.

de Falco, B., Bonanomi, G. and Lanzotti, V. (2018). Dithiosulfinates and sulfoxides with antifungal activity from bulbs of Allium sativum L. var. Voghiera. Natural Product Communications, 13(9): 1934578X1801300916.

de Rezende Ramos, A., Falcao, L.L., Barbosa, G.S., Marcellino, L.H. and Gander, E.S. (2007). Neem (*Azadirachta indica* a. Juss) components: Candidates for the control of *Crinipellis perniciosa* and Phytophthora ssp. Microbiological Research, 162(3): 238–243.

Deadman, M.L., Al-Saadi, A.M., Al-Mahmuli, I., Al-Maqbali, Y.M., Al-Subhi, R., Al-Kiyoomi, K., ... and Thacker, J.R.M. (2002). Management of *Pythium aphanidermatum* in greenhouse cucumber production in the Sultanate of Oman. In The BCPC Conference: Pests and diseases, Volumes 1 and 2. Proceedings of an international conference held at the Brighton Hilton Metropole Hotel, Brighton, UK, 18–21 November 2002 (171–176). British Crop Protection Council.

Dedej, S., Delaplane, K.S. and Scherm, H. (2004). Effectiveness of honey bees in delivering the biocontrol agent *Bacillus subtilis* to blueberry flowers to suppress mummy berry disease. Biological Control, 31(3): 422–427.

Del Ponte, E.M., Garda-Buffon, J. and Badiale-Furlong, E. (2012). Deoxynivalenol and nivalenol in commercial wheat grain related to Fusarium head blight epidemics in southern Brazil. Food Chemistry, 132(2): 1087–1091.

Dellavalle, P.D., Cabrera, A., Alem, D., Larrañaga, P., Ferreira, F. and Dalla Rizza, M. (2011). Antifungal activity of medicinal plant extracts against phytopathogenic fungus Alternaria spp. Chilean Journal of Agricultural Research, 71(2): 231.

Desjardins, A.E. (2006). Fusarium Mycotoxins: Chemistry, Genetics, and Biology. American Phytopathological Society (APS Press).

Dhaouadi, S., Rouissi, W., Mougou-Hamdane, A., Hannachi, I. and Nasraoui, B. (2018). Antifungal activity of essential oils of *Origanum majorana* and *Lavender angustifolia* against Fusarium wilt and root rot disease of melon plants. Tunisian Journal of Plant Protection, 13: 39–55.

Di Pietro, A., Lorito, M., Hayes, C.K., Broadway, R.M. and Harman, G.E. (1993). Endochitinase from *Gliocladium virens*: Isolation, characterization, and synergistic antifungal activity in combination with gliotoxin. Phytopathology.

Diao, W.R., Hu, Q.P., Zhang, H. and Xu, J.G. (2014). Chemical composition, antibacterial activity and mechanism of action of essential oil from seeds of fennel (*Foeniculum vulgare* Mill.). Food Control., 35(1): 109–116.

Diaz Napal, G.N., Defagó, M.T., Valladares, G.R. and Palacios, S.M. (2010). Response of *Epilachna paenulata* to two flavonoids, pinocembrin and quercetin, in a comparative study. Journal of Chemical Ecology, 36(8): 898–904.

Dill-Macky, R. and Jones, R.K. (2000). The effect of previous crop residues and tillage on Fusarium head blight of wheat. Plant Disease, 84(1): 71–76.

Dixon, R.A., Dey, P.M. and Lamb, C.J. (1983). Phytoalexins: enzymology and molecular biology. Advances in Enzymology and Related Areas of Molecular Biology, 55(1): 69.

Domsch, K.H. and Gams, W. (1972). Fungi in agricultural soils. Fungi in agricultural soils.

Donald, T., Shoshannah, R.O.T.H., Deyrup, S.T. and Gloer, J.B. (2005). A protective endophyte of maize: *Acremonium zeae* antibiotics inhibitory to *Aspergillus flavus* and *Fusarium verticillioides*. Mycological Research, 109(5): 610–618.

Droby, S., Chalutz, E., Wilson, C.L. and Wisniewski, M.E. (1992). Biological control of postharvest diseases: A promising alternative to the use of synthetic fungicides. Phytoparasitica, 20(1): S149–S153.

Duffy, B.K., Simon, A. and Weller, D.M. (1996). Combination of Trichoderma koningii with fluorescent pseudomonads for control of talk-all on wheat. Phytopathology (USA).

Dwivedi, S.K. (2012). Effectiveness of extract of some medicinal plants against soil-borne fusaria causing diseases on *Lycopersicon esculantum* and *Solanum melongena* plants. International Journal of Pharma and Bio Sciences, 3(4).

Dyakunchak, S.A. and Koroleva, S.V. (2006). Cabbage fusarium and its control measures. Agro XXI agro-industrial portal. Newspaper Plant Protection, 2: 19–21.

Elad, Y. (1993). Regulators of ethylene biosynthesis or activity as a tool for reducing susceptibility of host plant tissues to infection by *Botrytis cinerea*. Netherlands Journal of Plant Pathology, 99: 105–113.

Elad, Y. (2000). Biological control of foliar pathogens by means of *Trichoderma harzianum* and potential modes of action. Crop Protection, 19(8-10): 709–714.

Elad, Y. (2000). *Trichoderma harzianum* T39 preparation for biocontrol of plant diseases-control of *Botrytis cinerea*, *Sclerotinia sclerotiorum* and *Cladosporium fulvum*. Biocontrol Science and Technology, 10(4): 499–507.

Elad, Y., Barak, R., Chet, I. and Henis, Y. (1983). Ultrastructural studies of the interaction between Trichoderma spp. and plant pathogenic fungi. Journal of Phytopathology, 107(2): 168–175.

El-Sabrout, A.M., Salem, M.Z., Bin-Jumah, M. and Allam, A.A. (2019). Toxicological activity of some plant essential oils against *Tribolium castaneum* and *Culex pipiens* larvae. Processes, 7(12): 933.

El-Seedi, H.R., El-Said, A.M., Khalifa, S.A., Goransson, U., Bohlin, L., Borg-Karlson, A.K. and Verpoorte, R. (2012). Biosynthesis, natural sources, dietary intake, pharmacokinetic properties, and biological activities of hydroxycinnamic acids. Journal of Agricultural and Food Chemistry, 60(44): 10877–10895.

Emmert, E.A. and Handelsman, J. (1999). Biocontrol of plant disease: A (Gram-) positive perspective. FEMS Microbiology Letters, 171(1): 1–9.

Engelmeier, D. and Hadacek, F. (2006). Antifungal natural products: assays and applications. Advances in Phytomedicine, 3: 423–467.

Falcone Ferreyra, M.L., Rius, S.P. and Casati, P. (2012). Flavonoids: Biosynthesis, biological functions, and biotechnological applications. Frontiers in Plant Science, 3: 222.

Farooqi, A.A. and Sreeramu, B.S. (2004). Cultivation of medicinal and aromatic crops. Universities Press, Indian Plant Pathology, 465–470.

Fenwick, G.R. (1992). Saponins. Toxic Substances in Crop Plants, 285–327.

Fernández, P. and Laurentin, H. (2016). Efecto de extractos etanólicos de ajonjolí (*Sesamum indicum* L.) sobre *Fusarium oxysporum* f. sp. *Sesami*. Acta Agronómica, 65(1): 104–108.

Ferreira, P.M.P., Farias, D.F., Oliveira, J.T.D.A. and Carvalho, A.D.F.U. (2008). *Moringa oleifera*: bioactive compounds and nutritional potential. Revista de Nutrição, 21: 431–437.

Fisher, M.C., Henk, D., Briggs, C.J., Brownstein, J.S., Madoff, L.C., McCraw, S.L. and Gurr, S.J. (2012). Emerging fungal threats to animal, plant and ecosystem health. Nature, 484(7393): 186–194.

Foidl, N., Makkar, H.P.S. and Becker, K. (2001). The potential of *Moringa oleifera* for agricultural and industrial uses. What Development Potential for Moringa Products, 20.

Franco, D.M. and AR, L.L.G. (2018). Use of alcoholic extract of onion peel (*Allium cepa* L.) to control *Fusarium moniliforme*. Journal of Analytical & Pharmaceutical Research, 7(4): 491–492.

Freeman, S., Minz, D., Kolesnik, I., Barbul, O., Zveibil, A., Maymon, M., ... and Elad, Y. (2004). Trichoderma biocontrol of *Colletotrichum acutatum* and *Botrytis cinerea* and survival in strawberry. European Journal of Plant Pathology, 110(4): 361–370.

Fry, S.C. (1986). Cross-linking of matrix polymers in the growing cell walls of angiosperms. Annual Review of Plant Physiology, 37(1): 165–186.

Fu, J., Zhou, Y., Li, H.F., Ye, Y.H. and Guo, J.H. (2011). Antifungal metabolites from Phomopsis sp. By254, an endophytic fungus in *Gossypium hirsutum*. African Journal Microbiology Research, 5(10): 1231–1236.

Fuchs, S., Sontag, G., Stidl, R., Ehrlich, V., Kundi, M. and Knasmüller, S. (2008). Detoxification of patulin and ochratoxin A, two abundant mycotoxins, by lactic acid bacteria. Food and Chemical Toxicology, 46(4): 1398–1407.

Fufa, B.K. (2019). Anti-bacterial and anti-fungal properties of garlic extract (*Allium sativum*): A review. Microbiology Research Journal International, 28: 1–5.

Fukui, H., Egawa, H., Koshimizu, K. and Mitsui, T. (1973). A new isoflavone with antifungal activity from immature fruits of *Lupinus luteus*. Agricultural and Biological Chemistry, 37(2): 417–421.

Gahukar, R.T. (2012). Evaluation of plant-derived products against pests and diseases of medicinal plants: A review. Crop Protection, 42: 202–209.

Gao, P., Qin, J., Li, D. and Zhou, S. (2018). Inhibitory effect and possible mechanism of a Pseudomonas strain QBA5 against gray mold on tomato leaves and fruits caused by *Botrytis cinerea*. PLoS One, 13(1): e0190932.

Gharibzahedi, S.M.T., Ansarifard, I., Hasanabadi, Y.S., Ghahderijani, M. and Yousefi, R. (2013). Physicochemical properties of *Moringa peregrina* seed and its oil. Quality Assurance and Safety of Crops & Foods, 5(4): 303–309.

Gholamnezhad, J. (2019). Effect of plant extracts on activity of some defense enzymes of apple fruit in interaction with *Botrytis cinerea*. Journal of Integrative Agriculture, 18(1): 115–123.

Ghorbanian, M., Razzaghi-Abyaneh, M., Allameh, A., Shams-Ghahfarokhi, M. and Qorbani, M. (2008). Study on the effect of neem (*Azadirachta indica* A. juss) leaf extract on the growth of *Aspergillus parasiticus* and production of aflatoxin by it at different incubation times. Mycoses, 51(1): 35–39.

Ghorbanpour, M., Omidvari, M., Abbaszadeh-Dahaji, P., Omidvar, R. and Kariman, K. (2018). Mechanisms underlying the protective effects of beneficial fungi against plant diseases. Biological Control, 117: 147–157.

Giesler, L.J. and Yuen, G.Y. (1998). Evaluation of *Stenotrophomonas maltophilia* strain C3 for biocontrol of brown patch disease. Crop Protection, 17(6): 509–513.

Gilbert, J. and Tekauz, A. (2000). Recent developments in research on Fusarium head blight of wheat in Canada. Canadian Journal of Plant Pathology, 22(1): 1–8.

Gillitzer, P., Martin, A.C., Kantar, M., Kauppi, K., Dhalberg, S. and Lis, D. 2012. Optimization of screening of native and naturalized plants. From Minnesota for antimicrobial activity. Journal of Medicinal Plants Research, 6(6): 938–49.

Goetz, G., Fkyerat, A., Métais, N., Kunz, M., Tabacchi, R., Pezet, R. and Pont, V. (1999). Resistance factors to grey mould in grape berries: Identification of some phenolics inhibitors of *Botrytis cinerea* stilbene oxidase. Phytochemistry, 52(5): 759–767.

Gold, R.E. (1987). Prevention of fungal ingress: The role of papillae and calcium. Molecular determinants of plant diseases/edited by Syoyo Nishimura, Carroll P. Vance, and Noriyuki Doke, 47–48.

Golosova, M.A. (2003). Microbiological Plant Protection: Tutorial for Students of Specialty. Moscow: MSUF.

Goussous, S.J., Abu el-Samen, F.M. and Tahhan, R.A. (2010). Antifungal activity of several medicinal plants extracts against the early blight pathogen (*Alternaria solani*). Archives of Phytopathology and Plant Protection, 43(17): 1745–1757.

Grădinariu, V., Cioancă, O., Gille, E., Aprotosoaie, A.C., Hriţcu, L. and Hăncianu, M. (2013). The chemical profile of basil biovarieties and its implication on the biological activity. Farmacia, 61(4): 632–639.

Grata, K. (2016). Sensitivity of *Fusarium solani* isolated from asparagus on essential oils. Ecological Chemistry and Engineering. A, 23(4): 453–464.

Gubler, W.D., Davis, R.M., Zitter, T.A., Hopkins, D.L. and Thomas, T.C. (1996). Compendium of cucurbit diseases. Pythium and phytophthora damping off and root rot. St. Paul, MN, USA., APS Press.

Guerreiro, A.C., Gago, C.M., Faleiro, M.L., Miguel, M.G. and Antunes, M.D. (2015). The effect of alginate-based edible coatings enriched with essential oils constituents on *Arbutus unedo* L. fresh fruit storage. Postharvest Biology and Technology, 100: 226–233.

Guleria, S. and Kumar, A. (2006). *Azadirachta indica* leaf extract induces resistance in sesame against Alternaria leaf spot disease. Journal of Cell and Molecular Biology, 5(2): 81–86.

Gurjar, M.S., Ali, S., Akhtar, M. and Singh, K.S. (2012). Efficacy of plant extracts in plant disease management.

Hadian, J., Ghasemnezhad, M., Ranjbar, H., Frazane, M. and Ghorbanpour, M. (2008). Antifungal potency of some essential oils in control of postharvest decay of strawberry caused by *Botrytis cinerea*, *Rhizopus stolonifer* and *Aspergillus niger*. Journal of Essential Oil Bearing Plants, 11(5): 553–562.

Hamini, K.N., Hamdane, F., Boutoutaou, R., Kihal, M. and Henni, J.E. (2014). Antifungal activity of clove (*Syzygium aromaticum* L.) essential oil against phytopathogenic fungi of tomato (*Solanum lycopersicum* L) in Algeria. Journal of Experimental Biology and Agricultural Sciences, 2(5): 447–454.

Hamman, J.H. (2008). Composition and applications of Aloe vera leaf gel. Molecules, 13(8): 1599–1616.

Harjonoand, S. and Widyastuti, M. (2001). Antifungal activity of purified endochitinase produced by biocontrol agent *Trichoderma reseei a*gainsts *Ganoderma philippii*. Pakistan Journal of Biological Sciences, 4(10): 1232–1234.

Harris, J.E. and Dennis, C. (1976). Antifungal activity of post-infectional metabolites from potato tubers. Physiological Plant Pathology, 9(2): 155–165.

Hasan, H.A.H. (2000). Patulin and aflatoxin in brown rot lesion of apple fruits and their regulation. World Journal of Microbiology and Biotechnology, 16(7): 607–612.

Hasan, M.M., Islam, M.R., Hossain, I. and Shirin, K. (2014). Biological control of leaf spot of groundnut. Journal of Bioscience and Agriculture Research, 1(2): 66–78.

Hashioka, Y. and Nakai, Y. (1980). Ultrastructure of pycnidial development and mycoparasitism of *Ampelomyces quisqualis* parasitic on Erysiphales. Transactions of the Mycological Society of Japan, 21(3): 329–338.

Hassan, S. and Mathesius, U. (2012). The role of flavonoids in root–rhizosphere signalling: Opportunities and challenges for improving plant–microbe interactions. Journal of Experimental Botany, 63(9): 3429–3444.

Hiroyuki, K., Satoshi, T., Shun-ichi, T., Yoshihara, T., Sakamura, S., Shimanuki, T., ... and Tajimi, A. (1989). New fungitoxic sesquiterpenoids, chokols AG, from stromata of Epichloe typhina and the absolute configuration of chokol E. Agricultural and Biological Chemistry, 53(3): 789–796.

Hmiri, S., Rahouti, M., Habib, Z., Satrani, B., Ghanmi, M. and El Ajjouri, M. (2011). Evaluation of the antifungal potential of essential oils of *Mentha pulegium* and *Eucalyptus Camaldulensis* in the biological control of fungi responsible for the deterioration of apples in storage. Bulletin of the Royal Society of Cork Sciences.

Horn, W.S., Simmonds, M.S.J., Schwartz, R.E. and Blaney, W.M. (1995). Phomopsichalasin, a novel antimicrobial agent from an endophytic Phomopsis sp. Tetrahedron, 51(14): 3969–3978.

Hu, Y., Zhang, J., Kong, W., Zhao, G. and Yang, M. (2017). Mechanisms of antifungal and anti-aflatoxigenic properties of essential oil derived from turmeric (*Curcuma longa* L.) on *Aspergillus flavus*. Food Chemistry, 220: 1–8.

Hughes, B.G. and Lawson, L.D. (1991). Antimicrobial effects of *Allium sativum* L. (garlic), *Allium ampeloprasum* L. (elephant garlic), and *Allium cepa* L.(onion), garlic compounds and commercial garlic supplement products. Phytotherapy Research, 5(4): 154–158.

Huzaifa, U., Labaran, I., Bello, A.B.and Olatunde, A. (2014). Phytochemical screening of aqueous extract of Garlic (*Allium sativum*) bulbs. Rep. Opinion. 6: 1–4.

Hwang, E.I., Yun, B.S., Kim, Y.K., Kwon, B.M., KIM, H.G., Lee, H.B., ... and Kim, S.U. (2000). Phellinsin A, a novel chitin synthases inhibitor produced by Phellinus sp. PL3. The Journal of Antibiotics, 53(9): 903–911.

Ibrahim, N. and Kebede, A. (2020). *In vitro* antibacterial activities of methanol and aqueous leave extracts of selected medicinal plants against human pathogenic bacteria. Saudi Journal of Biological Sciences, 27(9): 2261–2268.

Iida, W., Hirano, K., Amemiya, Y. and Mita, Y. (1983). Influence of soil moisture on the incidence of damping-off of cucumber caused by *Fusarium oxysporum* f. sp. *cucumerinum* and *Pythium aphanidermatum*. Technical Bulletin of the Faculty of Horticulture Chiba University, (31): 81–86.

Islam, S.M.M., Masum, M.M.I. and Fakir, M.G.A. (2009). Prevalence of seed-borne fungi in sorghum of different locations of Bangladesh. Scientific Research and Essay, 4(3): 175–179.

Ismaili, R., Lamiri, A. and Moustaid, K. (2014). Etude de l'activité antifongique des huiles essentielles de trois plantes aromatiques marocaines (study of the antifungal activity of essential oils of three moroccan aromatic plants). International Journal of Innovation Science and Research, 12(2): 2351–8014.

Ito, S.I., Nagata, A., Kai, T., Takahara, H. and Tanaka, S. (2005). Symptomless infection of tomato plants by tomatinase producing *Fusarium oxysporum* formae speciales nonpathogenic on tomato plants. Physiological and Molecular Plant Pathology, 66(5): 183–191.

Iwasa, T., Yamamoto, H. and Shibata, M. (1970). Studies on validamycins, new antibiotics. I *Streptomyces hygroscopicus* var. limoneus nov. var., validamycin-producing organism. The Journal of Antibiotics, 23(12): 595–602.

Jarecka, A., Saniewska, A., Bialy, Z. and Jurzysta, M. (2008). The effect of *Medicago arabica*, M. hybrida and M. sativa saponins on the growth and development of *Fusarium oxysporum* Schlecht f. sp. tulipae apt. Acta Agrobotanica, 61(2).

Jean, B. (2009). Pharmacognosie, Phytochimie, Plantes Médicinales (4e éd.). Lavoisier.

Jeong, W., Cha, M.K. and Kim, I.H. (2000). Thioredoxin-dependent hydroperoxide peroxidase activity of bacterioferritin comigratory protein (BCP) as a new member of the thiol-specific antioxidant protein (TSA)/alkyl hydroperoxide peroxidase C (AhpC) family. Journal of Biological Chemistry, 275(4): 2924–2930.

Jiménez-Reyes, M.F., Carrasco, H., Olea, A.F. and Silva-Moreno, E. (2019). Natural compounds: A sustainable alternative to the phytopathogens control. Journal of the Chilean Chemical Society, 64(2): 4459–4465.

Jochum, C.C., Osborne, L.E. and Yuen, G.Y. (2006). Fusarium head blight biological control with *Lysobacter enzymogenes* strain C3. Biological Control, 39(3): 336–344.

Jones, J.D. and Dangl, J.L. (2006). The plant immune system. The Nature, 444(7117): 323–329.

Jones, R.K. (2000). Assessments of Fusarium head blight of wheat and barley in response to fungicide treatment. Plant Disease, 84(9): 1021–1030.

Kagale, S., Marimuthu, T., Thayumanavan, B., Nandakumar, R. and Samiyappan, R. 2004. Antimicrobial activity and induction of systemic acquired resistance in rice by leaf extract of *Datura metel* against *Rhizoctonia solani* and *Xanthomonas oryzae pv. oryzae*. Physiology and Molecular Plant Pathology, 65: 91–1000.

Kalagatur, N.K., Mudili, V., Siddaiah, C., Gupta, V.K., Natarajan, G., Sreepathi, M.H., ... and Putcha, V. L. (2015). Antagonistic activity of *Ocimum sanctum* L. essential oil on growth and zearalenone production by *Fusarium graminearum* in maize grains. Frontiers in Microbiology, 6: 892.

Kalia, P.and Sharma, S.K. (1988). Biochemical genetics of powdery mildew resistance in pea. Theoretical and Applied Genetics, 76(5): 795–799.

Kaur, R., Singh, R.S. and Alabouvette, C. (2007). Antagonistic activity of selected isolates of fluorescent *Pseudomonas* against *Fusarium oxysporum* f. sp. *ciceri*. Asian Journal of Plant Sciences.

Kedia, A., Prakash, B., Mishra, P.K. and Dubey, N.K. (2014). Antifungal and antiaflatoxigenic properties of *Cuminum cyminum* (L.) seed essential oil and its efficacy as preservative in stored commodities. International Journal of Food and Microbiology, (7): 168–169.

Kennedy, D.O. and Wightman, E.L. (2011). Herbal extracts and phytochemicals: Plant secondary metabolites and the enhancement of human brain function. Advances in Nutrition, 2(1): 32–50.

Keukens, E.A., de Vrije, T., Fabrie, C.H., Demel, R.A., Jongen, W.M. and de Kruijff, B. (1992). Dual specificity of sterol-mediated glycoalkaloid induced membrane disruption. Biochimica et Biophysica Acta (BBA)-Biomembranes, 1110(2): 127–136.

Keukens, E.A., de Vrije, T., van den Boom, C., de Waard, P., Plasman, H.H., Thiel, F. and de Kruijff, B. (1995). Molecular basis of glycoalkaloid induced membrane disruption. Biochimica et Biophysica Acta (BBA)-Biomembranes, 1240(2): 216–228.

Khakimov, A.A., Omonlikov, A.U. and Utaganov, S.B.U. (2020). Current status and prospects of the use of biofungicides against plant diseases. GSC Biological and Pharmaceutical Sciences, 13(3): 119-126.

Khalil, H. and Javeed, A. (2019). *In-vitro* investigtions of some chosen botanicals and bio-fungicide on mycelial development and conidial germination of *Cercospora arachidicola* and *Cercosporidium personatum*. *In-vitro*, 15: 3.

Khan, B., Akash, Z., Asad, S., Javed, N., Rajput, N.A., Jabbar, A., ... and Atif, R.M. (2017). Antagonistic potential of *Trichoderma harzianum* against *Fusarium oxysporum* f. sp. *cubense* associated with Panama Wilt of banana. Pakistan Journal of Phytopathology, 29(1): 111–116.

Khan, T.Z., Gill, M.A. and Khan, M.G. (2000). Seed borne fungi of rice from central Punjab and their control. Pakistan Journal of Phytopathology, 12(1): 12–14.

Khoddami, A., Wilkes, M.A. and Roberts, T.H. (2013). Techniques for analysis of plant phenolic compounds. Molecules, 18(2): 2328–2375.

Kim, B.S., Moon, S.S. and Hwang, B.K. (1999). Isolation, antifungal activity, and structure elucidation of the glutarimide antibiotic, streptimidone, produced by *Micromonospora coerulea*. Journal of Agricultural and Food Chemistry, 47(8): 3372–3380.

Kim, H.J., Lee, S.H., Kim, C.S., Lim, E.K., Choi, K.H., Kong, H.G., ... & Moon, B.J. (2007). Biological control of strawberry gray mold caused by *Botrytis cinerea* using *Bacillus licheniformis* N1 formulation. Journal of Microbiology and Biotechnology, 17(3): 438–444.

Kim, J.C., Choi, G.J., Kim, H.J., Kim, H.T., Ahn, J.W. and Cho, K.Y. (2002). Verlamelin, an antifungal compound produced by a mycoparasite, *Acremonium strictum*. The Plant Pathology Journal, 18(2): 102–105.

Klaenhammer, T.R. (1993). Genetics of bacteriocins produced by lactic acid bacteria. FEMS Microbiology Reviews, 12(1-3): 39–85.

Köhl, J., Molhoek, W.M.L., Van der Plas, C.H. and Fokkema, N.J. (1995). Effect of *Ulocladium atrum* and other antagonists on sporulation of *Botrytis cinerea* on dead lily leaves exposed to field conditions. Phytopathology, 85(4): 393–400.

Koldaş, S., Demirtas, I., Ozen, T., Demirci, M.A. and Behçet, L. (2015). Phytochemical screening, anticancer and antioxidant activities of *Origanum vulgare* L. ssp. *viride* (Boiss.) Hayek, a plant of traditional usage. Journal of the Science of Food and Agriculture, 95(4): 786–798.

Kolombet, L.V., Starshov, A.A. and Schisler, D.A. (2005). Greenhouse testing *Trichoderma asperellum* GJS 03-35 and yeast *Cryptococcus nodaensis OH 182.9 as biocontrol agents against Fusarium* head blight of wheat. Mycology and phytopathology, 5: 80–88.

Kolombet, L.V., Zhigletsova, S.K., Kosareva, N.I., Bystrova, E.V., Derbyshev, V.V., Krasnova, S.P. and Schisler, D. (2008). Development of an extended shelf-life, liquid formulation of the biofungicide Trichoderma asperellum. World Journal of Microbiology and Biotechnology, 24: 123–131.

Kolomiyets, E.I., Romanovskaya, T.V. and Zdor, N.A. (2006.) Biopreparations—as an alternative to chemicals. Plant Protection and Quarantine, 10: 18–20.

Koshino, H., Yoshihara, T., Sakamura, S., Shimanuki, T., Sato, T. and Tajimi, A. (1989). Novel C-11 epoxy fatty acid from stromata of *Epichloe typhina* on Phleum pretense. Agriculture Biology and Chemistry, 53: 2527–2528.

Krebs, B., Höding, B., Kübart, S., Workie, M.A., Junge, H., Schmiedeknecht, G. and Hevesi, M. (1998). Use of Bacillus subtilis as biocontrol agent. I. Activities and characterization of *Bacillus subtilis* strains. Journal of Plant Diseases and Protection, 181–197.

Kulisic, T., Radonic, A., Katalinic, V. and Milos, M. (2004). Use of different methods for testing antioxidative activity of oregano essential oil. Food Chemistry, 85(4): 633–640.

Kumar, S. and Kaushik, N. (2012). Metabolites of endophytic fungi as novel source of biofungicide: A review. Phytochemistry Reviews, 11(4): 507–522.

Kumar, S. and Pandey, A.K. (2013). Chemistry and biological activities of flavonoids: An overview. The Scientific World Journal, 2013.

Lahlali, R., Peng, G., McGregor, L., Gossen, B.D., Hwang, S.F. and McDonald, M.R. (2011). Mechanisms of the biofungicide Serenade® (Bacillus subtilis QST713) in suppressing clubroot. Biocontrol Sciences Technology, 21: 1351–1362.

Lambert, R.J.W., Skandamis, P.N., Coote, P.J. and Nychas, G.J. (2001). A study of the minimum inhibitory concentration and mode of action of oregano essential oil, thymol and carvacrol. Journal of Applied Microbiology, 91(3): 453–462.

Lanzotti, V., Bonanomi, G. and Scala, F. (2013). What makes Allium species effective against pathogenic microbes? Phytochemistry Reviews, 12(4): 751–772.

Latha, P., Anand, T., Ragupathi, N., Prakasam, V. and Samiyappan, R. (2009). Antimicrobial activity of plant extracts and induction of systemic resistance in tomato plants by mixtures of PGPR strains and Zimmu leaf extract against Alternaria solani. Biological Control, 50(2): 85–93.

Laurentin, H. (2007). Genetic diversity in sesame (*Sesamum indicum* L.): Molecular markers, metabolic profiles and effect of plant extracts on soil-borne pathogenic fungi. Dissertation.

Lax, A.R., Templeton, G.E. and Meyer, W.L. (1985). Isolation, purification, and biological activity of a self-inhibitor from conidia of *Colletotrichum gloeosporioides*. Phytopathology, 75(4): 386–390.

Lee, J.P., Lee, S.W., Kim, C.S., Son, J.H., Song, J.H., Lee, K.Y., ... and Moon, B.J. (2006). Evaluation of formulations of Bacillus licheniformis for the biological control of tomato gray mold caused by *Botrytis cinerea*. Biological Control, 37(3): 329–337.

Lee, J.Y., Lee, J.Y., Moon, S.S. and Hwang, B.K. (2005). Isolation and antifungal activity of 4-phenyl-3-butenoic acid from *Streptomyces koyangensis* strain VK-A60. Journal of Agricultural and Food Chemistry, 53(20): 7696–7700.

Li, S., Jochum, C.C., Yu, F., Zaleta-Rivera, K., Du, L., Harris, S.D. and Yuen, G.Y. (2008). An antibiotic complex from *Lysobacter enzymogenes* strain C3: Antimicrobial activity and role in plant disease control. Phytopathology, 98(6): 695–701.

Li, Y., Kong, D., Fu, Y., Sussman, M.R. and Wu, H. (2020). The effect of developmental and environmental factors on secondary metabolites in medicinal plants. Plant Physiology and Biochemistry, 148: 80–89.

Liaqat, A., Zahoor, T., Atif Randhawa, M. and Shahid, M. (2019). Characterization and antimicrobial potential of bioactive components of sonicated extract from garlic (*Allium sativum*) against foodborne pathogens. Journal of Food Processing and Preservation, 43(5): e13936.

Lida, W., Hirano, K., Amemiya, Y. and Mita, Y. (1983). Influence of soil moisture on the incidence of damping-off of cucumber caused by *Fusarium oxysporum* f. sp. *cucumerinum* and *Pythium aphanidermatum*. Technical Bulletin of the Faculty of Horticulture Chiba University, 31: 81–6.

Lima, G. and Cicco, V.D. (2006). Integrated strategies to enhance biological control of postharvest diseases. Advances in Postharvest Technologies for Horticultural Crops, 173–194.

Liu, X.Y., Wang, Q., Cui, S.W. and Liu, H.Z. (2008). A new isolation method of β-D-glucans from spent yeast *Saccharomyces cerevisiae*. Food Hydrocolloids, 22(2): 239–247.

Lu, H., Zou, W.X., Meng, J.C., Hu, J. and Tan, R.X. (2000). New bioactive metabolites produced by Colletotrichum sp., an endophytic fungus in *Artemisia annua*. Plant Science, 151(1): 67–73.

Mahajan, M., Kuiry, R. and Pal, P.K. (2020). Understanding the consequence of environmental stress for accumulation of secondary metabolites in medicinal and aromatic plants. Journal of Applied Research on Medicinal and Aromatic Plants, 18: 100255.

Manousi, N., Sarakatsianos, I. and Samanidou, V. (2019). Extraction techniques of phenolic compounds and other bioactive compounds from medicinal and aromatic plants. In Engineering Tools in the Beverage Industry (pp. 283–314). Woodhead Publishing.

Mansour, M.M., Salem, M.Z.M., Khamis, M.H. and Ali, H.M. (2015). Natural durability of *Citharexylum spinosum* and *Morus alba* woods against three mold fungi. BioResources, 10(3): 5330–5344.

Martillo, E.E. and Reuveni, M. (2009). A new potent bio-fungicide for the control of Banana Black Sigatoka. In Phytopathology, 99(6): S80S80.

Masaoka, Y., Kojima, M., Sugihara, S., Yoshihara, T., Koshino, M. and Ichihara, A. (1993). Dissolution of ferric phosphate by alfalfa (*Medicago sativa* L.) root exudates. In Plant Nutrition from Genetic Engineering to Field Practice, 79–82. Springer, Dordrecht.

Masum, M.M.I., Islam, S.M.M. and Fakir, M.G.A. (2009). Effect of seed treatment practices in controlling of seed-borne fungi in sorghum. Scientific Research and Essay, 4(1): 22–27.

Matthies, A. and Buchenauer, H. (2000). Effect of tebuconazole (Folicur®) and prochloraz (Sportale®) treatments on Fusarium head scab development, yield and deoxynivalenol (DON) content in grains of wheat following artificial inoculation with *Fusarium culmorum*. Journal of Plant Diseases and Protection, 33–52.

Mazid, M., Khan, T.A. and Mohammad, F. (2011). Role of secondary metabolites in defense mechanisms of plants. Biology and Medicine, 3(2): 232–249.

Mbili, N., Vries, F., Opara, U.L. and Lennox, C.L. (2015). Chemical composition and antifungal activity of citrus and lemongrass essential oils in combination with cold storage regimes against *Botrytis cinerea*. Proceedings of the IOBC/WPRS Working Group "Integrated Plant Protection in Fruit Crops, Subgroup Pome Fruit Diseases", Stellenbosch, South Africa, 24–28 November 2014, 110: 165–168.

McCallum, J.L., Tsao, R. and Zhou, T. (2002). Factors affecting patulin production by Penicillium expansum. Journal of Food Protection, 65(12): 1937–1942.

McMullen, M., Bergstrom, G., De Wolf, E., Dill-Macky, R., Hershman, D., Shaner, G. and Van Sanford, D. (2012). A unified effort to fight an enemy of wheat and barley: Fusarium head blight. Plant Disease, 96(12): 1712–1728.

McMullen, M., Jones, R. and Gallenberg, D. (1997). Scab of wheat and barley: A re-emerging disease of devastating impact. Plant Disease, 81(12): 1340–1348.

Miller, J.D. (1995). Fungi and mycotoxins in grain: implications for stored product research. Journal of Stored Product Research, 31: 1–6.

Misato, T., Ishii, I., Asakawa, M., Okimoto, Y. and Fukunaga, K. (1959). Antibiotics as protectant fungicides against rice blast (2) The therapeutic action of blasticidin S. Japanese Journal of Phytopathology, 24(5): 302–306.

Moghadam, Z.A., Hosseini, H., Hadian, Z., Asgari, B., Mirmoghtadaie, L., Mohammadi, A. and Javadi, N. H.S. (2019). Evaluation of the antifungal activity of cinnamon, clove, thymes, *Zataria multiflora*, cumin and caraway essential oils against Ochratoxigenic Aspergillus ochraceus. Journal of Pharmaceutical Research International, 26(1): 1–16.

Moghaddam, M., Mehdizadeh, L., Basak, A., Chakraborty, R. and Mandal, S.M. (2016). Essential oil and antifungal Therapy. In Recent Trends in Antifungal Agents and Antifungal Therapy. Springer: Berlin/Heidelberg, Germany, 29–74.

Mohamed, A.A., El-Hefny, M., El-Shanhorey, N.A. and Ali, H.M. (2020). Foliar application of bio-stimulants enhancing the production and the toxicity of *Origanum majorana* essential oils against four rice seed-borne fungi. Molecules, 25(10): 2363.

Moslem, M.A. and El-Kholie, E.M. (2009). Effect of neem (*Azardirachta indica* A. Juss) seeds and leaves extract on some plant pathogenic fungi. Pakistan Journal of Biological Sciences, 12(14): 1045–1048.

Moyo, B., Masika, P.J. and Muchenje, V. (2012). Antimicrobial activities of *Moringa oleifera* Lam leaf extracts. African Journal of Biotechnology, 11(11): 2797–2802.

Muchembled, J., Deweer, C., Sahmer, K. and Halama, P. (2018). Antifungal activity of essential oils on two *Venturia inaequalis* strains with different sensitivities to tebuconazole. Environmental Science and Pollution Research, 25(30): 29921–29928.

Mukesh, S., Vipul, K., Mohammad, S., Sonika, P. and Anuradha, S. (2016). Trichoderma-a potential and effective bio fungicide and alternative source against notable phytopathogens: A review. African Journal of Agricultural Research, 11(5): 310–316.

Mukherjee, S., Das, P. and Sen, R. (2006). Towards commercial production of microbial surfactants. Trends in Biotechnology, 24(11): 509–515.

Nabigol, A. and Morshedi, H. (2011). Evaluation of the antifungal activity of the Iranian thyme essential oils on the postharvest pathogens of strawberry fruits. African Journal of Biotechnology, 10(48): 9864–9869.

Naboulsi, I., Aboulmouhajir, A., Kouisni, L., Bekkaoui, F. and Yasri, A. (2018). Plants extracts and secondary metabolites, their extraction methods and use in agriculture for controlling crop stresses and improving productivity: A review. Academia Journal of Medicinal Plants, 6: 223–240.

Nagar, P.K., Iyer, R.I. and Sircar, P.K. (1982). Cytokinins in developing fruits of *Moringa pterigosperma* Gaertn. Physiologia Plantarum, 55(1): 45–50.

Nagar, P.K., Iyer R.I. and Sircar, P.K. (2006). Cytokinins in developing fruits of *Moringa pterigosperma* Gaertn. Physiology of Plant, (55): 45–50.

Nally, M.C., Pesce, V.M., Maturano, Y.P., Assaf, L.R., Toro, M.E., De Figueroa, L.C. and Vazquez, F. (2015). Antifungal modes of action of Saccharomyces and other biocontrol yeasts against fungi isolated from sour and grey rots. International Journal of Food Microbiology, 204: 91–100.

Nashwa, S.M. and Abo-ElyouSr, K.A. (2013). Evaluation of various plant extracts against the early blight disease of tomato plants under greenhouse and field conditions. Plant Protection Science, 48(2): 74–79.

Nastić, N., Švarc-Gajić, J., Delerue-Matos, C., Barroso, M.F., Soares, C., Moreira, M.M. and Radojković, M. (2018). Subcritical water extraction as an environmentally-friendly technique to recover bioactive compounds from traditional Serbian medicinal plants. Industrial Crops and Products, 111: 579–589.

Neergaard, P. (1970, December). Seed pathology of rice disease problems. In Proc. of International Symposium on Plant Pathology. New Delhi, 69–81.

Nikhil, B. and Sahu, R.K. (2014). Evaluation of some fungicides, botanicals and essential oils against the fungus *Colletotrichum falcatum* causing red rot of sugarcane. The Bioscan, 9(1): 175–178.

Obire, O.and Anyanwu, E.C. (2009). Impact of various concentrations of crude oil on fungal populations of soil. International Journal of Environmental Science & Technology, 6(2): 211–218.

Olea, A.F., Bravo, A., Martínez, R., Thomas, M., Sedan, C., Espinoza, L. and Carrasco, H. (2019). Antifungal activity of eugenol derivatives against *Botrytis cinerea*. Molecules, 24(7): 1239.

Omuketi, W.E. (2020). Diversity, Antagonistic Potential of Endophytes, Phytochemicals and Antimicrobial Activity of Selected Agroforestry Trees Against Xanthomonas Campestris Pv. Musacearum and Cercospora Zeae-Maydis. PhD diss., Maseno University, 2020.

Ongena, M. and Jacques, P. (2008). Bacillus lipopeptides: Versatile weapons for plant disease biocontrol. Trends in Microbiology, 16(3): 115–125.

Oniha, M., Eni, A., Akinnola, O., Omonigbehin, E.A., Ahuekwe, E.F. and Olorunshola, J.F. (2021). *In vitro* antifungal activity of extracts of *moringa oleifera* on phytopathogenic fungi affecting *Carica papaya*. Open Access Macedonian Journal of Medical Sciences, 9(A): 1081–1085.

Osbourn, A. (1996). Saponins and plant defence—a soap story. Trends in Plant Science, 1(1): 4–9.

Oxenham, S.K., Svoboda, K.P. and Walters, D.R. (2005). Antifungal activity of the essential oil of basil (*Ocimum basilicum*). Journal of Phytopathology, 153(3): 174–180.

Park, J.Y., Kim, S.H., Kim, N.H., Lee, S.W., Jeun, Y.C. and Hong, J.K. (2017). Differential inhibitory activities of four plant essential oils on in vitro growth of *Fusarium oxysporum* f. sp. *fragariae* causing Fusarium wilt in strawberry plants. The Plant Pathology Journal, 33(6): 582.

Park, T.H., Vleeshouwers, V.G., Hutten, R.C., van Eck, H.J., van der Vossen, E., Jacobsen, E. and Visser, R.G. (2005). High-resolution mapping and analysis of the resistance locus Rpi-abpt against *Phytophthora infestans* in potato. Molecular Breeding, 16(1): 33–43.

Patel, S. and Saraf, M. (2017). Biocontrol efficacy of *Trichoderma asperellum* MSST against tomato wilting by *Fusarium oxysporum* f. sp. *lycopersici*. Archives of Phytopathology and Plant Protection, 50(5-6): 228–238.

Paulitz, T.C. and Bélanger, R.R. (2001). Biological control in greenhouse systems. Annual Review of Phytopathology, 39(1): 103–133.

Pereira, M.C., Vilela, G.R., Costa, L.M.A.S., Silva, R.F.D., Fernandes, A.F., Fonseca, E.W.N.D. and Piccoli, R.H. (2006). Inibição do desenvolvimento fúngico através da utilização de óleos essenciais de condimentos. Ciência e Agrotecnologia, 30: 731–738.

Pezet, R. and Pont, V. (1990). Ultrastructural observations of pterostilbene fungitoxicity in dormant conidia of *Botrytis cinerea* pers. Journal of Phytopathology, 129(1): 19–30.

Pieterse, C.M.J., Van Wees, S.C.M., Ton, J., Van Pelt, J.A. and Van Loon, L.C. (2002). Signalling in rhizobacteria-induced systemic resistance in *Arabidopsis thaliana*. Plant Biology, 4(05): 535–544.

Pieterse, C.M. and Loon, V.L.C. (1999). Salicylic acid-independent plant defence pathways. Trends in Plant Science, 4(2): 52–58.

Pieterse, C.M., Zamioudis, C., Berendsen, R.L., Weller, D.M., Van Wees, S.C. and Bakker, P.A. (2014). Induced systemic resistance by beneficial microbes. Annual Review of Phytopathology, 52: 347–375.

Pinto, E., Pina-Vaz, C., Salgueiro, L., Gonçalves, M.J., Costa-de-Oliveira, S., Cavaleiro, C., ... and Martinez-de-Oliveira, J. (2006). Antifungal activity of the essential oil of *Thymus pulegioides* on Candida, Aspergillus and dermatophyte species. Journal of Medical Microbiology, 55(10): 1367–1373.

Placinta, C.M., D'Mello, J.F. and Macdonald, A.M.C. (1999). A review of worldwide contamination of cereal grains and animal feed with Fusarium mycotoxins. Animal Feed Science and Technology, 78(1-2): 21–37.

Pohl, C.H., Kock, J.L. and Thibane, V.S. (2011). Antifungal free fatty acids: A review. Science Against Microbial Pathogens: Communicating Current Research and Technological Advances, 3: 61–71.

Polo, K.J.J., Campos, H.L.M., Olivera, C.C., Nakayo, J.L.J. and Flores, J.W.V. (2021). Biofungicide for the control of *Botrytis Cinerea* and *Fusarium Oxysporum*: A laboratory study. Chemical Engineering Transactions, 87: 517–522.

Prakash, B., Kedia, A., Mishra, P.K. and Dubey, N.K. (2015). Plant essential oils as food preservatives to control moulds, mycotoxin contamination and oxidative deterioration of agri-food commodities–Potentials and challenges. Food Control, 47: 381–391.

Price, K.R., Johnson, I.T., Fenwick, G.R. and Malinow, M.R. (1987). The chemistry and biological significance of saponins in foods and feeding stuffs. Critical Reviews in Food Science & Nutrition, 26(1): 27–135.

Punja, Z.K. (1985). The biology, ecology, and control of *Sclerotium rolfsii*. Annual Review of Phytopathology, 23(1): 97–127.

Qessaoui, R., Amarraque, A., Lahmyed, H., Ajerrar, A., Mayad, E.H., Chebli, B., ... and Bouharroud, R. (2020). Inoculation of tomato plants with rhizobacteria suppresses development of whitefly *Bemisia tabaci* (Gennadius) (Hemiptera: Aleyrodidae): Agro-ecological application. Plos One, 15(4): e0231496.

Qian, G.L., Hu, B.S., Jiang, Y.H. and Liu, F.Q. (2009). Identification and characterization of *Lysobacter enzymogenes* as a biological control agent against some fungal pathogens. Agricultural Sciences in China, 8(1): 68–75.

Qnais, E.Y., Elokda, A.S., Abu Ghalyun, Y.Y. and Abdulla, F.A. (2007). Antidiarrheal Activity of the Aqueous extract of *Punica granatum* (Pomegranate) peels. Pharmaceutical Biology, 45(9): 715–720.

Rajput, N.A., Atiq, M., Javed, N., Ye, Y.H., Zhao, Z., Syed, R.N. and Dou, D. (2018). Antimicrobial effect of Chinese medicinal plant crude extracts against *Rhizoctonia solani* and *Pythium aphanidermatum*. Fresenius Environmental Bulletin, 27: 3941–3949.

Rajput, N.A., Pathan, M.A., Mubeen Lodhi, A., Daolong D. and Rajput S. (2011). Effect of neem (*Azadirachta indica*) products on seedling growth of shisham dieback. African Journal of Microbiology Research, 5(27): 4937–4945.

Rakh, R.R., Raut, L.S., Dalvi, S.M. and Manwar, A.V. (2011). Biological control of *Sclerotium rolfsii*, causing stem rot of groundnut by Pseudomonas cf. monteilii 9. Recent Research in Science and Technology, 3(3): 26–34.

Ramos, A.R., Falcao, L.L., Barbosa, G.S., Marcellino, L.S. and Gander, E.S. (2007). Neem (*Azadirachta indica* A. Juss) components; candidates for the control of *Crinipellis perniciosa* and Phytophthora spp. Microbiological Research, 162: 238–243.

Rangiah, K. and Gowda, M. (2019). Method to quantify plant secondary metabolites: Quantification of neem metabolites from leaf, bark, and seed extracts as an example. pp. 21–30. *In*: M. Gowda, A. Sheetal and C. Kole (eds.). The Neem Genome. Compendium of Plant Genomes. Springer International Publishing.

Reuveni, M., Barbier, M. and Viti, A.J. (2020). Essential tea tree oil as a tool to combat black Sigatoka in banana. Outlooks on Pest Management, 31(4): 180–186.

Rezvyakova, S., Eremin, L., Matveychuk, P. and Mitina, E. (2021). The influence of biofungicide and chemical fungicides on the manifestation of diseases and the yield of soybeans. In E3S Web of Conferences, 247: 01046. EDP Sciences.

Rhayour, K., Bouchikhi, T., Tantaoui-Elaraki, A., Sendide, K. and Remmal, A. (2003). The mechanism of bactericidal action of oregano and clove essential oils and of their phenolic major components on *Escherichia coli* and *Bacillus subtilis*. Journal of Essential Oil Research, 15(5): 356–362.

Roddick, J.G., & Drysdale, R.B. (1984). Destabilization of liposome membranes by the steroidal glycoalkaloid α-tomatine. Phytochemistry, 23(3): 543–547.

Rodgers, P.B. (1989). Potential of biological control organisms as a source of antifungal compounds for agrochemical and pharmaceutical product development. Pesticide Science, 27(2): 155–164.

Rodrigues, L., Banat, I.M., Teixeira, J. and Oliveira, R. (2006). Biosurfactants: Potential applications in medicine. Journal of Antimicrobial Chemotherapy, 57(4): 609–618.

Roger, F. and Keinath, A. (2010). Biofungicides and Chemicals for Managing Diseases in Organic Vegetable Production. Clemson, SC: Clemson University Cooperative Ext. Information Leaflet, 88.

Romeo, L., Iori, R., Rollin, P., Bramanti, P. and Mazzon, E. (2018). Isothiocyanates: An overview of their antimicrobial activity against human infections. Molecules, 23(3): 624.

Rosca C.O., Parvu, M., Vlase, L. and Tamas, M. (2007). Antifungal activity of Aloe vera leaves. Fitoterapia, 78(3): 219–222.

Rowayshed, G.H., Abd-Elhameed, A.A., Abd-Elghany, M.E.A., Shahat, A.A. and Younes, O.A.A. (2014). Effective chemical compounds and antibacterial activities of marjoram leaves, teucrium leaves and fennel fruits essential oils. Middle East Journal of Applied Sciences, 4: 637–647.

Ryder, M.H., Yan, Z., Terrace, T.E., Rovira, A.D., Tang, W. and Correll, R.L. (1999). Use of Bacillus isolated in China to suppress take-all and rhizoctonia root rot, and promote seedling growth of glasshouse-grown wheat in Australian soils. Soil Biology and Biochemistry, 31: 19–29.

Sadeghi, M., Zolfaghari, B., Senatore, M. and Lanzotti, V. (2013). Antifungal cinnamic acid derivatives from Persian leek (*Allium ampeloprasum* Subsp. Persicum). Phytochemistry Letters, 6(3): 360–363.

Sahashi, N. and Shishiyama, J. (1986). Increased papilla formation, & major factor of induced resistance in the barley - Erysiphegraminis t.sp. kordei system. Canadian Journal of Botany, 64: 2178–21S1.

Saks, Y. and Barkai-Golan, R. (1995). Aloe vera gel activity against plant pathogenic fungi. Postharvest Biology and Technology, 6(1-2): 159–165.

Salem, M.Z., Hamed, S.A.E.K.M. and Mansour, M. (2019). Assessment of efficacy and effectiveness of some extracted bio-chemicals as bio-fungicides on Wood. Drvna Industrija, 70(4): 337–350.

Salem, M.Z., Mansour, M.M. and Elansary, H.O. (2019). Evaluation of the effect of inner and outer bark extracts of Sugar Maple (*Acer saccharum* var. *saccharum*) in combination with citric acid against the growth of three common molds. Journal of Wood Chemistry and Technology, 39(2): 136–147.

Sales, M.D.C., Costa, H.B., Fernandes, P.M.B., Ventura, J.A. and Meira, D.D. (2016). Antifungal activity of plant extracts with potential to control plant pathogens in pineapple. Asian Pacific Journal of Tropical Biomedicine, 6(1): 26–31.

Sangoyoni, T.E. (2004). Post-Harvest Fungal Deterioration of Yam (*Dioscorea rotundata* Poir) and its control. Ph.D. International Institute of Tropical Agriculture. Ibadan, Nigeria, 179 pp.

Saravanakumar, D., Ciavorella, A., Spadaro, D., Garibaldi, A. and Gullino, M. L. (2008). *Metschnikowia pulcherrima* strain MACH1 outcompetes *Botrytis cinerea, Alternaria alternata* and *Penicillium expansum* in apples through iron depletion. Postharvest Biology and Technology, 49(1): 121–128.

Saravanakumar, D., Spadaro, D., Garibaldi, A. and Gullino, M.L. (2009). Detection of enzymatic activity and partial sequence of a chitinase gene in *Metschnikowia pulcherrima* strain MACH1 used as post-harvest biocontrol agent. European Journal of Plant Pathology, 123(2): 183–193.

Sata, N., Matsunaga, S., Fusetani, N., Nishikawa, H., Takamura, S. and Saito, T. (1998). New antifungal and cytotoxic steroidal saponins from the bulbs of an elephant garlic mutant. Bioscience, Biotechnology, and Biochemistry, 62(10): 1904–1911.

Savary, S., Willocquet, L., Pethybridge, S.J., Esker, P., McRoberts, N.and Nelson, A. (2019). The global burden of pathogens and pests on major food crops. Nature Ecology & Evolution, 3(3): 430–439.

Schäfer, H. and Wink, M. (2009). Medicinally important secondary metabolites in recombinant microorganisms or plants: progress in alkaloid biosynthesis. Biotechnology Journal: Healthcare Nutrition Technology, 4(12): 1684–1703.

Scherm, H., Ngugi, H.K., Savelle, A.T. and Edwards, J.R. (2004). Biological control of infection of blueberry flowers caused by Monilinia vaccinii-corymbosi. Biological Control, 29(2): 199–206.

Schmiedeknecht, G., Bochow, H. and Junge, H. (1998). Use of *Bacillus subtilis* as biocontrol agent. II. Biological control of potato diseases. Journal of Plant Diseases and Protection, 376–386.

Schönbeck, F. and Schlösser, E. (1976). Preformed substances as potential protectants. In Physiological Plant Pathology, 653–678. Springer, Berlin, Heidelberg.

Selmanoglu, G.and Koçkaya, E.A. (2004). Investigation of the effects of patulin on thyroid and testis, and hormone levels in growing male rats. Food and Chemical Toxicology, 42(5): 721–727.

Šernaitė, L., Rasiukevičiūtė, N., Dambrauskienė, E., Viškelis, P. and Valiuškaitė, A. (2020). Biocontrol of strawberry pathogen *Botrytis cinerea* using plant extracts and essential oils. Zemdirbyste Agric, 107: 147–152.

Shao, X., Cheng, S., Wang, H., Yu, D. and Mungai, C. (2013). The possible mechanism of antifungal action of tea tree oil on *Botrytis cinerea*. Journal of Applied Microbiology, 114(6): 1642–1649.

Sharanappa, R. and Vidyasagar, G.M. (2013). Anti-Candida activity of medicinal plants. A review. International Journal of Pharmacy and Pharmaceutical Sciences, 5: 9–16.

Sharga, B.M.and Lyon, G.D. (1998). *Bacillus subtilis* BS 107 as an antagonist of potato blackleg and soft rot bacteria. Canadian Journal of Microbiology, 44(8): 777–783.

Sharma, A., Sharma, N.K., Srivastava, A., Kataria, A., Dubey, S., Sharma, S. and Kundu, B. (2018). Clove and lemongrass oil based non-ionic nanoemulsion for suppressing the growth of plant pathogenic *Fusarium oxysporum* f. sp. *lycopersici*. Industrial Crops and Products, 123: 353–362.

Shen, Y., Yang, H., Chen, J., Liu, D.and Ye, X. (2013). Effect of waxing and wrapping on phenolic content and antioxidant activity of citrus during storage. Journal of Food Processing and Preservation, 37(3): 222–231.

Shiraishi, T., Yamaoka, N. and Kunoh, H. (1989). Association between increased phenylalanine ammonialyase activity and cinnamic acid synthesis and the induction of temporary inaccessibility caused by *Erysiphe graminis* primary germ tube penetration of the barley leaf. Physiological and Molecular Plant Pathology, 34(1): 75–83.

Shitan, N. (2016). Secondary metabolites in plants: Transport and self-tolerance mechanisms. Bioscience, Biotechnology, and Biochemistry, 80(7): 1283–1293.

Shittu, L., Bankole, M., Ahmed, T., Aile, K., Akinsanya, M., Bankole, M., ... and Ashiru, O.A. (2006). Differential antimicrobial activity of the various crude leaves extracts of *Sesame radiatum* against some common pathogenic micro-organisms. Available at SSRN 3017601.

Shokouhi, D. and Seifi, A. (2020). Organic extracts of seeds of Iranian Moringa peregrina as promising selective biofungicide to control *Mycogone perniciosa*. Biocatalysis and Agricultural Biotechnology, 30: 101848.

Sikkema, J.A.N., de Bont, J.A. and Poolman, B. (1995). Mechanisms of membrane toxicity of hydrocarbons. Microbiological Reviews, 59(2): 201–222.

Silva, J.L., Souza, P.E., Monteiro, F.P., Freitas, M.L.O., Júnior, S. and Belan, L.L. (2014). Antifungal activity using medicinal plant extracts against pathogens of coffee tree. Revista Brasileira de Plantas Medicinais, 16: 539–544.

Silva, N.C.C. and Fernandes Júnior, A.J.J.O.V.A. (2010). Biological properties of medicinal plants: A review of their antimicrobial activity. Journal of Venomous Animals and Toxins including Tropical Diseases, 16: 402–413.

Sipahi, R.E., Castell-Perez, M.E., Moreira, R.G., Gomes, C. and Castillo, A. (2013). Improved multilayered antimicrobial alginate-based edible coating extends the shelf life of fresh-cut watermelon (*Citrullus lanatus*). LWT-Food Science and Technology, 51(1): 9–15.

Smyth, T., Ramachandran, V.N.and Smyth, W.F. (2009). A study of the antimicrobial activity of selected naturally occurring and synthetic coumarins. International Journal of Antimicrobial Agents, 33(5): 421–426.

Soković, M., Stojković, D., Glamočlija, J., Ćirić, A., Ristić, M. and Grubišić, D. (2009). Susceptibility of pathogenic bacteria and fungi to essential oils of wild Daucus carota. Pharmaceutical Biology, 47(1): 38–43.

Soliman, M.M., Hamid, O.M.A., Maqsood, H.A., Aziza, S., El Sanosi, Y. and Ragab, O.A.(2015). Insulin-mimetic effects of cinnamon extract in wistar rats. Lucr. Stiint. Ser. Med. Vet., 55: 61–68.

Spadaro, D. and Droby, S. (2016). Development of biocontrol products for postharvest diseases of fruit: The importance of elucidating the mechanisms of action of yeast antagonists. Trends in Food Science & Technology, 47: 39–49.

Spadaro, D., Vola, R., Piano, S. and Gullino, M.L. (2002). Mechanisms of action, efficacy and possibility of integration with chemicals of four isolates of the yeast *Metschnikowia pulcherrima* active against postharvest pathogens on apples. Postharvest Biology and Technology, 24: 123–134.

Stanghellini, M.E. and Phillips, J.M. (1975). *Pythium aphanidermatum*: its occurrence and control with pyroxychlor in the Arabian desert at Abu Dhabi. Plant Disease Reporter, 59(7): 559–563.

Stein, T. (2005). *Bacillus subtilis* antibiotics: Structures, syntheses and specific functions. Molecular Microbiology, 56(4): 845–857.

Strobel, G.A., Miller, R.V., Martinez-Miller, C., Condron, M.M., Teplow, D.B. and Hess, W.M. (1999). Cryptocandin, a potent antimycotic from the endophytic fungus *Cryptosporiopsis* cf. quercina. Microbiology, 145(8): 1919–1926.

Sturrock, C.J., Woodhall, J., Brown, M., Walker, C., Mooney, S.J. and Ray, R.V. (2015). Effects of damping-off caused by *Rhizoctonia solani* anastomosis group 2-1 on roots of wheat and oil seed rape quantified using X-ray Computed Tomography and real-time PCR. Frontiers in Plant Science, 6: 461.

Sudantha, I.M. and Suwardji, S. (2021, July). Trichoderma biofungicides formulations on shallot growth, yield and fusarium wilt disease resistance. In IOP Conference Series: Earth and Environmental Science, 824(1): 012032. IOP Publishing.

Sugiyama, A. and Yazaki, K. (2014). Flavonoids in plant rhizospheres: Secretion, fate and their effects on biological communication. Plant Biotechnology, 31(5): 431–443.

Sugiyama, A., Yamazaki, Y., Yamashita, K., Takahashi, S., Nakayama, T. and Yazaki, K. (2016). Developmental and nutritional regulation of isoflavone secretion from soybean roots. Bioscience, Biotechnology, and Biochemistry, 80(1): 89–94.

Sun, Y., Wang, Y., Han, L.R., Zhang, X. and Feng, J.T. (2017). Antifungal activity and action mode of cuminic acid from the seeds of *Cuminum cyminum* L. against *Fusarium oxysporum* f. sp. *niveum* (FON) causing Fusarium wilt on watermelon. Molecules, 22(12): 2053.

Susanto, A. (2002). Biological control of *Ganoderma boninense* pat., the causal agent of basal stem rot disease of oil palm. AGRIS.

Tajkarimi, M.M., Ibrahim, S.A.and Cliver, D.O. (2010). Antimicrobial herb and spice compounds in food. Food Control, 21(9): 1199–1218.

Takeuchi, T., Hikiji, T., Nitta, K., Yamazaki, S., Abe, S., Takayama, H. and Umezawa, H. (1957). Biological studies on kanamycin. The Journal of Antibiotics, 10(3): 107–114.

Talreja, T. (2010). Screening of crude extract of flavonoids of *Moringa oleifera* against bacterial and fungal pathogen. Journal of Phytology, 2(11).

Tasleem-uz-Zaman, K., Gill, M.A. and Khan, M.G. (2000). Seed borne fungi of rice from central Punjab and their control. Pakistan Journal of Phytopathology, 12(1): 12–14.

Tata, S., Donli, P.O., Mohammed, F.K., Peter, A. and Ahmed, A. (2018). Comparative studies on the effect of aqueous and methanolic extracts of some botanicals on growth and sporulation of *Colletotrichum graminicola*. International Journal of Innovation Research, 5: 1–7.

Tava, A. and Avato, P. (2006). Chemical and biological activity of triterpene saponins from Medicago species. Natural Product Communications, 1(12): 1934578X0600101217.

Tekauz, A., McCallum, B., Ames, N. and Fetch, J.M. (2004). Fusarium head blight of oat—current status in western Canada. Canadian Journal of Plant Pathology, 26(4): 473–479.

Thambugala, K.M., Daranagama, D.A., Phillips, A.J., Kannangara, S.D. and Promputtha, I. (2020). Fungi vs. fungi in biocontrol: An overview of fungal antagonists applied against fungal plant pathogens. Frontiers in Cellular and Infection Microbiology, 10: 604923.

Tian, C., Wang, M., Shen, C. and Zhao, C. (2012). Accuracy mass screening and identification of phenolic compounds from the five parts of *Abutilon theophrasti* Medic. by reverse-phase high-performance liquid chromatography-electrospray ionization-quadrupoles-time of flight-mass spectrometry. Journal of Separation Science, 35(5-6): 763–772.

Tiwari, A.K., Kumar, K., Razdan, V.K. and Rather, T.R. (2004). Mass production of Trichoderma viride on indigenous substrates. Annals of Plant Protection Sciences, 12(1): 71–74.

Tripathy, B., Satyanarayana, S., Abedulla Khan, K., Raja, K. and Tripathy, S. (2017). Preliminary phytochemical screening and comparison study of *in vitro* antioxidant activity of selected medicinal plants. International Journal of Pharmacy and Life Sciences, 8: 5598–5604.

Tzortzakis, N.G., Tzanakaki, K. and Economakis, C.D. (2011, November). Effect of origanum oil and vinegar on the maintenance of postharvest quality of tomato. In: Food and Nutritional Science, (2): 974–982.

Üstüner, T., Kordali, Ş. and Bozhüyük, A.U. (2018). Herbicidal and fungicidal effects of *Cuminum cyminum*, *Mentha longifolia* and *Allium sativum* essential oils on some weeds and fungi.

Van der Ent, S., Verhagen, B.W., Van Doorn, R., Bakker, D., Verlaan, M.G., Pel, M.J., ... and Pieterse, C.M. (2008). MYB72 is required in early signaling steps of rhizobacteria-induced systemic resistance in Arabidopsis. Plant Physiology, 146(3): 1293–1304.

Vijayan, M. (1989). Studies on early blight of tomato caused by Alternaria solani (Ellis and Martin) Jones and Grout. M.Sc. (Ag.) Thesis, Tamil Nadu Agricultural University, Coimbatore, India, 106.

Vinodhini, R. and Narayanan, M. (2008). Bioaccumulation of heavy metals in organs of fresh water fish *Cyprinus carpio* (Common carp). International Journal of Environmental Science & Technology, 5(2): 179–182.

Vitullo, D., Lima, G., Castoria, R., De Curtis, F., Maiuro, L. and De Cicco, V. (2008). Biochemical and physical interactions among *Fusarium oxysporum*, biocontrol agents and tomato plant roots. In International Congress of Plant Pathology (Vol. 90, No. 2, Supplement, pp. S2–372).

Vukic, Milan, Slavica and Grujić, Bozana Odzaković. (2017). Application of edible films and coatings in food production. In Advances in Applications of Industrial Biomaterials. Springer, Cham, 121–138.

Wade, T.C. and Baker, T.B. (1977). Opinions and use of psychological tests: A survey of clinical psychologists. American Psychologist, 32(10): 874.

Walker, G.M. and White, N.A. (2017). Introduction to fungal physiology. Fungi: Biology and Applications, 1–35.

Walker, V., Couillerot, O., Von Felten, A., Bellvert, F., Jansa, J., Maurhofer, M., Bally, R., Moenne-Loccoz, Y. and Comte, G. (2012). Variation of secondary metabolite levels in maize seedling roots induced by inoculation with Azospirillum, Pseudomonas and Glomus consortium under field conditions. Plant Soil, 356: 151–163.

Wang, C., Zhang, J., Chen, H., Fan, Y. and Shi, Z. (2010). Antifungal activity of eugenol against *Botrytis cinerea*. Tropical Plant Pathology, 35: 137–143.

Wang, Y., Sun, Y., Zhang, Y., Zhang, X. and Feng, J. (2016). Antifungal activity and biochemical response of cuminic acid against *Phytophthora capsici* Leonian. Molecules, 21(6): 756.

White, T.J., Bruns, T., Lee, S.J.W.T. and Taylor, J. (1990). Amplification and direct sequencing of fungal ribosomal RNA genes for phylogenetics. PCR Protocols: A Guide to Methods and Applications, 18(1): 315–322.

Wicklow, D.T., Roth, S., Deyrup, S.T. and Gloer, J.B. (2005). A protective endophyte of maize: Acremonium zeae antibiotics inhibitory to *Aspergillus flavus* and *Fusarium verticillioides*. Mycology Research, 109: 610–618.

Wokocha, R.C. and Okereke, V.C. (2005). Fungitoxic activity of extracts of some medicinal plants on *Sclerotium rolfsii*, causal organism of the basal stem rot diseases of tomato. Nigerian Journal of Plant Protection, 22: 106–110.

Wu, H.S., Wang, Y., Zhang, C.Y., Bao, W., Ling, N., Liu, D.Y. and Shen, Q.R. (2009). Growth of *in vitro Fusarium oxysporum* f. sp. *niveum* in chemically defined media amended with gallic acid. Biological Research, 42(3): 297–304.

Wu, X.Z., Cheng, A.X., Sun, L.M. and Lou, H.X. (2008). Effect of plagiochin E, an antifungal macrocyclic bis (bibenzyl), on cell wall chitin synthesis in *Candida albicans* 1. Acta Pharmacologica Sinica, 29(12): 1478–1485.

Wu, Y.M., Wang, Z.W., Hu, C.Y.and Nerín, C. (2018). Influence of factors on release of antimicrobials from antimicrobial packaging materials. Critical Reviews in Food Science and Nutrition, 58(7): 1108–1121.

Wynn, J.P., Kendrick, A. and Ratledge, C. (1997). Sesamol as an inhibitor of growth and lipid metabolism in *Mucor circinelloides* via its action on malic enzyme. Lipids, 32(6): 605–610.

Xie, Y., Huang, Q., Wang, Z., Cao, H.and Zhang, D. (2017). Structure-activity relationships of cinnamaldehyde and eugenol derivatives against plant pathogenic fungi. Industrial Crops and Products, 97: 388–394.

Xie, Y., Yang, Z., Cao, D., Rong, F., Ding, H. and Zhang, D. (2015). Antitermitic and antifungal activities of eugenol and its congeners from the flower buds of *Syzgium aromaticum* (clove). Industrial Crops and Products, 77: 780–786.

Xue, A.G., Voldeng, H.D., Savard, M.E., Fedak, G., Tian, X. and Hsiang, T. (2009). Biological control of fusarium head blight of wheat with *Clonostachys rosea* strain ACM941. Canadian Journal of Plant Pathology, 31(2): 169–179.

Yasmeen, A., Basra, S.M.A., Ahmad, R. and Wahid, A. (2012). Performance of late sown wheat in response to foliar application of *Moringa oleifera* Lam. leaf extract. Chilean Journal of Agricultural Research, 72(1): 92–97.

Yazdani, F., Mafi, M., Farhadi, F., Tabar-Heidar, K., Aghapoor, K., Mohsenzadeh, F. and Darabi, H.R. (2005). Supercritical CO_2 extraction of essential oil from clove bud: Effect of operation conditions on the selective isolation of eugenol and eugenyl acetate. Zeitschrift für Naturforschung B, 60(11): 1197–1201.

Ye, R.W., Tao, W., Bedzyk, L., Young, T., Chen, M. and Li, L. (2000). Global gene expression profiles of *Bacillus subtilis* grown under anaerobic conditions. Journal of Bacteriology, 182(16): 4458–4465.

Yu, Y.B. and Yang, S.F. (1979). Auxin-induced ethylene production and its inhibition by aminoethyoxyvinylglycine and cobalt ion. Plant Physiology, 64(6): 1074–1077.

Zhang, J., Howell, C.R. and Starr, J.L. (1996). Suppression of Fusarium colonization of cotton roots and Fusarium wilt by seed treatments with *Gliocladium virens* and *Bacillus subtilis*. Biocontrol Science and Technology, 6(2): 175–188.

Zhang, Z. and Yuen, G.Y. (1999). Biological control of *Bipolaris sorokiniana* on tall fescue by *Stenotrophomonas maltophilia* strain C3. Phytopathology, 89(9): 817–822.

Zhang, Z. and Yuen, G.Y. (2000). The role of chitinase production by *Stenotrophomonas maltophilia* strain C3 in biological control of *Bipolaris sorokiniana*. Phytopathology, 90(4): 384–389.

Zhang, Z., Yuen, G.Y., Sarath, G. and Penheiter, A.R. (2001). Chitinases from the plant disease biocontrol agent, Stenotrophomonas maltophilia C3. Phytopathology, 91(2): 204–211.

Zhao, J., Zhou, L., Wang, J., Shan, T., Zhong, L., Liu, X. and Gao, X. (2010). Endophytic fungi for producing bioactive compounds originally from their host plants. Current Research, Technology and Education Tropics in Applied Microbiology and Microbial Biotechnology, 1: 567–576.

Zimand, G., Elad, Y. and Chet, I. (1996). Effect of *Trichoderma harzianum* on *Botrytis cinerea* pathogenicity. Phytopathology, 86(11): 1255–1260.

Znini, M., Cristofari, G., Majidi, L., El Harrak, A., Paolini, J. and Costa, J. (2013). *In vitro* antifungal activity and chemical composition of *Warionia saharae* essential oil against 3 apple phytopathogenic fungi. Food Science and Biotechnology, 22(1). 113–119.

Chapter 9

How to Survey and Select Promising Biofungicides?

Nasreen Musheer,[1,*] *Anam Choudhary,*[1] *Arshi Jamil,*[1] *Rabiya Basri,*[2] *Mohd Majid Jamali,*[3] *Sajjad Khan*[4] and *Sabiha Saeed*[1]

1. Introduction

Biofungicides are derivatives of natural resources like microorganisms, plants and their metabolites, extracts, etc., to control pests or pathogens and to improve crop production without causing a toxic effect on the environment and human health (Kumar and Singh, 2014; Archana et al., 2022). The biological fungicides produced could control a wide range of pathogens' infestation in crop plants at a reasonable cost. Biofungicides of better quality with improved shelf life are more eco-friendly than chemical fungicides (Kubheka et al., 2020). According to "Environmental Protection Agency (EPA) of the United States of America (2012)", cyanobacteria, bacteria and fungi (Guo et al., 2017); algae, bryophytes and pteridophytes (Aulakh et al., 2019); higher plants including gymnosperms and angiosperms could be used in the synthesis of biofungicides (Guha et al., 2005; Joshi and Sati, 2012; Commisso et al., 2021). The EPA also promotes safety and regulations of biofungicides for wide acceptability. Besides crop protection potential, biofungicides improve plant growth and develop resistance against abiotic and biotic stresses (Madbouly and Abdelbacki, 2017; Macena et al., 2020). Biofungicides are now widely used in the agricultural sector to promote organic farming as well as incorporated with Integrated Disease Management (IDM) practices. The persistentce of chemical residue in agricultural products has increased awareness amongst farmers to improve the quality of crop yield and ensure food security globally (Gupta, 2018). Increasing consumer demand for residues-free food has also changed market regulations and made biofungicides cost-effective and a good alternative to conventional synthetic fungicides (Kumar and Singh, 2014).

[1] Plant Protection, Agricultural Sciences, Aligarh Muslim University, Aligarh.
[2] Agricultural Sciences, ITM University, Gwalior.
[3] Life Science, Glocal University.
* Corresponding author: musheernasreen@gmail.com

Biofungicides offer multiple effects such as a target specific, antifeedant, deterrent, toxicant, fungicidal and repellent effects against phytopathogens. Considering the positive impact of biofungicides on crop protection and production, researchers are now focused for seeking and introducing novel biofungicides to be commercialized at a large scale (Haider et al., 2020). Biocontrol agents' and phytoextracts offer different mechanisms to protect host plants against phytopathogens, including direct and indirect antagonisms (Santos et al., 2021).

Many scientists have been introducing biological derivatives to control pests/pathogens to protect crops, increase food production, enhance traditional organic farming and ensure food security for decades (Anuagasi, 2017; Subba and Mathur, 2022). Therefore, antifungal bioactive metabolites of biocontrol agents and plants are being isolated and extracted by conventional and non-conventional methods for the development and formulations of biofungicides.

For the development and formulation of biofungicides, screening the antifungal potential of microbes, plant products and their metabolites is essential. The commercialization of biofungicides is dependent on the ecological niche associated with interactions, mode of interactions, formulation efficacy and stability, target specificity, trade cost, residual risk and regulatory laws (Kohl et al., 2011; Raymaekers et al., 2020). An effective biofungicide should be inexpensive, endurable, compatible with the agrofarming system and eco-friendly. The selection of biological material to formulate biofungicides is based on four criteria: (1) antagonistic microbes and plant products are genetically stable; (2) the formulation process should be reasonable ; (3) compatibility with other components of IDM practice; and (4) non-pathogenic to humans.

There are primarily two types: microbial derived biofungicides (which include bacteria and fungi) and plant-based biofungicides, which comprise algae, lichen, bryophytes and angiosperms (Subba and Mathur, 2022). Microbial biocontrol agents (BCAs) may be fungal or bacterial strains, which are usually isolated from the phyllosphere, rhizosphere or endosphere to benefit the plants in different ways. Whereas, plant extracts are prepared using different parts of the plants such as leaves, fruits, bulbs, rhizomes, etc. Sometimes, microbial or plant-based biofungicides are found ineffective under field conditions. To overcome this problem, such biofungicides could be combined with other components of an integrated disease management module and act synergistically against pathogens under field conditions (Trejo-Raya et al., 2021; Pirttila et al., 2021). Many studies have reported that the plant secretes volatile and non-volatile essential oil compounds that are potent in limiting the fungal growth of pathogens (Iqbal et al., 2020; Kumar et al., 2021b).

This chapter describes the importance of biofungicides and needs the antimicrobial potential screening of various microbial strains and plant products. Accordingly, the mode of action of different classes of biofungicides against plant pathogens as well as their new beneficial traits involved in plant health improvement will be explained. Further, the chapter extends the knowledge of effective biofungicide formulation and commercialization for broad spectrum application in agricultural cropping systems. Innovative ideas required in the identification of new efficient members of microorganisms and phytochemicals to be used as biofungicides in order to solve the fungicidal resistant and residual problems are also enlighted.

2. Screening and formulation of biological products for novel biofungicides

To produce an effective biofungicide, one should ensure the stability and antagonistic potential of biological material under varied conditions of the environment. The isolation of plant material and biological control agents and their metabolites need to be based on the antagonistic stability in natural ecosystems, including tropical soils, disease suppressive soils and pathogen-associated ecological niche adaption (either phylloplane or rhizoplane). For isolation and identification of plants as well as microbial metabolites, appropriate analytical methods such as Gas Chromatography-Mass Spectrometry (GC-MS) and Liquid Chromatography-Mass Spectrometry (LC–MS) are required (Maia et al., 2018; Srivastava and Singh, 2020; Luh Suriani et al., 2020). Publicly shared libraries enable access to the selection of biological control microorganisms or natural compounds that can be used for biofungicide formulation in reference to research institutes, organizations or companies, e.g., the National Center for Biotechnological Information (NCBI). The antagonistic activity is next screened against the target pathosystem or niche. They are used for biofungicide assays after positive antagonistic tests. Several types of biofungicide screenings have been adopted in the production of effective biofungicides against pathogens (Raymaekers et al., 2020). Usually biocontrol action against pathogens is screened through an *in vitro* "dual culture assay" where cultural growth of fungal plant pathogens and biocontrol agents form inhibition zones by repressing the mycelial growth of a pathogen (Bedine Boat et al., 2020). Following this, the culture filtrate assay is used to identify the volatiles of biocontrol agents that inhibit pathogen growth. Furthermore, if used on phylloplane surfaces, the microbial biofungicide formulation is exposed to ultraviolet A and B radiation multiple times during incubation (Kohl et al., 2011; Sartori et al., 2020) to ensure its effectiveness under harsh environmental conditions. Other techniques, such as microtiter plates methods, use of liquid biofungicide medium to examine the antifungal potential of bioactive metabolites and essential oils in reducing spore germination and inhibiting mycelial growth. To determine the Minimum Inhibitory Concentration (MIC) of a biofungicide at (EC50) that could inhibit less than 50% of total fungal growth are also used (Perina et al., 2019). Recently, Jethva et al. (2020) analyzed the biofungicides' effects on the pathogen's respiration activity using a resazurin assay. Moreover, Raymaekers et al. (2020) studied the indirect antagonism of biofungicides assisted by phenotypic-or physiological-based methods. The phenotypic-based biofungicide assays measure the disease severity and development of symptoms (like rot, wilt, scorching, blight, mildew, etc.). Conversely, the physiochemical-based analyses are characterized by quantitative plant responses like accumulation of oxygen free radicals, which causes toxicity in plants and lead to cell death. Noutoshi et al. (2012) examined cellular death because of excess production of reactive oxygen species by using specific fluorescent dyes like dichlorofluorescin diacetate (DCFDA) or evans-blue. Besides controlling phytopathogens, biofungicides stimulate plant growth and yield by increasing production of IAA and gibberellins (Nandhini et al., 2018; Gholami et al., 2019) and improving soil fertility that enhances plant nutrient uptake, especially P and K (Upamanya et al., 2020). Sufyan et al. (2020) reported that PGPR bioprimed seeds of Cicer arietinum were noticed with higher root length, fresh-dry mass and total biomass than uninoculated seeds. Similarly, if seed priming is done with aqueous extracts of plants and algal extracts, it causes higher plant vigour and primary metabolite synthesis of photopigments, glucose and amino acids in multiple plants. For example, *Acacia nilotica* and *Sapindus mukorossi* extract improve peanut, chickpea, okra and sunflower plants' vigour (Raf et al., 2015), maize plants treated with *Toddalia asiatica* (Aiyaz et al., 2015), Oryza sativa with *Cymbopogon citratus* (Naveenkumar et al., 2017) and Vigna radiata with *Sargassum polyphyllum, Turbinaria ornata, Gelidiopsis* sp., Padina tetrastomatica, Gracilaria corticata (Sarkar et al., 2018) and tomatoes with *Cinnamomum zeylanicum* (Kowalska et al., 2020).

After screening, the development of a biofungicide formulation considers a number of important factors, including efficient microorganisms and carriers material (such as adjuvants, suppressants, buffering systems and coating compounds) with the highest level of environmental tolerance (Sachdev et al., 2018). Many studies have reported the antagonistic potential of plant extracts and fungal biocontrol agents in the suppression of plant pathogens under *in vitro* and *in vivo* conditions (Nandhini et al., 2018; Gholami et al., 2019; Pavithra et al., 2020). For instance, the combination of *Piper caninum* and *P. betel* var. *nigra* leaves extract and rhizobacteria such as *Enterobacter cloacae, Bacillus subtilis* and *Stenotrophomonas maltophilia* inhibited the fungal growth of *Pyricularia oryzae* as well as improved the growth and yield quality of bali rice (Luh Suriani et al., 2020). Ali et al. (2019) determined that priming the seeds with aqueous extracts of garlic improves the seed germination of Solanum melongena by enhancing antioxidative enzymes like SOD and POD as well as modulating reactive oxygen species. For formulations of biofungicides, microbes/phytoextracts should follow the characteristics of the natural compounds' antifungal potential, direct and indirect mechanisms against plant pathogens and be capable of promoting plant growth. In addition to providing the desired efficacy against plant pathogens, industrial manufacturers ensure that the biofungicide manufacturing process is of low-cost, efficient and non-toxic to plants, humans and the environment. The biofungicide producer investigates culture optimization, fermentation and restoration for long-distance transport and deployment. Recently, nanotechnology has been introduced in the production and improvement of biofungicide efficacy like alginate-silica anaoparticles and carbon nanotubes encapsulated with microbial bioactive metabolites (Hersanti et al., 2020; Pour et al., 2019). The commercial use of biofungicide products follows a registration process (Glare et al., 2016; Feldmann et al., 2022), as biofungicides are made of living things that require more precautions in handling, storage and shelf-life than traditional fungicides. Commercially, biofungicides are used to treat seed, soil, planting material and growth media to specifically control the activity of plant pathogenic fungi. When used as a protectant for foliar applications prior to the pathogen infection, they are as effective as chemical fungicides in their action. For instance, biofungicides could be effective to control air-borne and soil-borne plant pathogens, e.g., *Alternaria alternata* and *Sclerotinia sclerotiorum*, respectively (Hu et al., 2020). Biofungicides can be more persistent than conventional fungicides, are often safer to producers and are occasionally less expensive.

3. Categories of biofungicides

Biofungicides are broadly classified into microbial types, including fungi and bacteria (Jaber and Alananbeh, 2018; Myo et al., 2019; Ferrigo et al., 2020; Nian et al., 2021); and plant-derived biofungicides, including algae, lichens, bryophytes and vascular plants (Soliman et al., 2018; Vehapi et al., 2020; Valadbeigi et al., 2014; Commisso et al., 2021; Liu et al., 2020). For the synthesis of biofungicides, the microbial or plant parts are extracted (Falade et al., 2017) and suspensions are made (Sujarit et al., 2020; Eakjamnong et al., 2021) by using various solvents such as water, ethanol, methanol, acetone, ethyl acetate, etc. (Wei et al., 2021).

3.1 Microbial derived biofungicides

Several species of beneficial microbes (bacteria, fungi, viruses or protozoans) producing antifungal metabolites have been identified to be used in biofungicide synthesis (Kumar and Singh, 2014; Eakjamnong et al., 2021). The microbial biocontrol agents offer multitrophic interactions to control disease without causing damage or side-effects to plants, the environment or beneficial microbes, including humans (Kilic and Akay, 2008). As shown in Table 9.1, the biocontrol agents possess great potential to control plant pathogens and diseases in plants. Antifungal microorganisms have been isolated from plant rhizospheres and plant tissues (Zhou et al., 2007; Srivastava et al., 2016; Jakubiec-Kresniak et al., 2018). The endophytic microbes (which reside inside the host plant tissues) secrete bioactive compounds to develop resistance against diseases and promote plant growth (Segaran and Sathiavelu, 2019). Kumar and Kaushik (2013) reported that endophytic fungi secrete various secondary metabolites such as alkaloids, terpenoids, steroids, isocoumarins, chromones, phenolics and other volatile compounds. Besides secreting antifungal metabolites, microbes employ direct and indirect modes of antagonism against fungal pathogens (Carmona-Hernandez et al., 2019).

Based on the association of microbes with an ecological niche, biofungicides are categorized into three groups: First, phylloplane-microbial associated niche benefits the plant by promoting growth and controlling foliar disease (Wang et al., 2020). For example, *Cladosporium cladosporioides* has controlled the phylloplane pathogen *Pyricularia oryzae* causing rice blast disease by inhibiting conidia germination, appressorium formation and mycelial growth as well as secretes cell-degrading enzymes (Chaibub et al., 2020). Few genera of bacteria (e.g., *Bacillus*, *Pseudomonas*) and yeasts (e.g., *Candida, Pichia*) are able to produce cell-degrading enzymes, antibiotics and other volatile compounds as phylloplane, which help in inducing host resistance and alleviating fruit oxidation especially against post-harvest fruit diseases causing pathogens such as *Botrytis*, *Colletotrichum*, *Monilinia* and *Penicillium* (Dukare et al., 2019). Second, rhizoplane associated microorganisms niche offers competition with other microbes in complex soil niches and produces a wide variety of antimicrobial secondary metabolites (Srivastava and Singh, 2020). Otoguro and Suzuki (2018) studied the bacteria, fungi and endophytic microbes which could be used as biocontrol agents against the grapevine pathogen. The facultative intracellular endophytes exert direct and indirect mechanisms to suppress

Table 9.1: Utilization of biocontrol fungi in restricting the growth of phytopathogenic fungi causing severe diseases in crop plants cited by Subba and Mathur, 2022 with addition of new references.

Biocontrol fungi	Phytopathogenic fungi	Diseases	Host	References
Metarhizium brunneum BIPESCO5 and *Beauveria bassiana* (NATURALIS)	*Fusarium moniliforme* and *F. culmorum*, *F. oxysporum*	crown and root rot	*Capsicum annum* L. (chilli)	Jaber and Alanambeh (2018)
Trichoderma asperellum	*Sclerospora graminicola*, *Fusarium oxysporum* and *Acremonium* sp.	downy mildew	*Pennisetum glaucum L.* (pearl millet)	Nandhini et al. (2018)
Phialocepahala fortinii isolate CKG.I.11	*F. oxysporum* f. sp. *asparagi* and *Asparagus offcinalis*	fusarium disease	*Asparagus officinalis* L. (asparagus)	Surono and Narisawa (2018)
Coprinopsis urticicola M2, *Rhizoctonia zeae* M32 *Fomes fomentarius* M40 and *Rhizoctonia endophytica* M9	*Gaeumannomyces graminis* var. *tritici*	take-all disease	*Triticum aestivum* L. (wheat)	Gholami et al. (2019)
Aureobasidium pullulans	*Trichoderma pleuroticola*, *T. pleuroti*	pleurotus ostreatus disease	*Pleurotus ostreatus* (Oyester mushroom)	Roberti et al. (2019)
Trichoderma harzianum INAT11	*Fusarium verticilloids* and *Fusarium graminearum*	fusarium ear rot	*Zea mays* L. (maize)	Ferrigo et al. (2020)
Penicillium olsonii ML37, *Anthracocyctis fcocculosa* F63P, Anthracoyctis *focculosa* P1P1 and *Sarocladium strictum* C113L	*Fusarium graminearum*	wheat head blight	*Triticum aestivum* L. (wheat)	Rojas et al. (2020)
Trichoderma harzianum	*F. oxysporum*	moler disease	*Allium* sp.	Rusita and Sasongko (2020)
T. harzianum QTYC77	*Fusarium oxysporum* f. sp. *cucumerium*	wilt	*Cucumis sativus* L. (cucumber)	Zhang et al. (2020a)
Talaromyces tratensis KUFA0091	*Alternaria padwickii*, *Bipolaris oryzae* and *Curcularia lunata*	dirty panicle	*Oryza sativa* L. (rice)	Eakjamnong et al. (2021)
Fusarium commune W5	*Fusarium fujikuroi*	bakanae	*Oryza sativa* L. (rice)	Saito et al. (2021)
T. harzianum and *T. asperellum*	*Colletorichum truncatum*	anthracnose	*Capsicum annuum* L. (chilli pepper)	Yadav et al. (2021)
T. atroviride	*Fusarium pseudograminearum* and *Rhizoctonia cerealis*	Fusarium crown rot	*Triticum aestivum* L. (wheat)	Sui et al. (2022)

Table 9.1 contd. ...

...Table 9.1 contd.

Biocontrol fungi	**Phytopathogenic fungi**	**Diseases**	**Host**	**References**
Trichoderma species	*Fusarium* spp., *Sclerotinia* spp., *Pythium* spp., *Rhizoctonia* spp., *Botrytis* spp.,	Soil borne disease	Leguminous plants	Al-Ameri and Rammadan (2022)
T. hamatum SG18 and *T. koningiopsis* SG6	*Phytophthora xcambivora*	Ink disease	*Castanea sativa* L. (Sweet chestnut)	Frascella et al. (2022)
Trichoderma asperellum	*Fusarium verticillioides* and *Ustilago maydis*	Ear rot and corn smut	*Zea mays* L. (maize)	Cuervo-Parra et al. (2022)

the pathogens (Backer et al., 2018; Salma and Jogen, 2011). Several endophytic fungi such as *Acremonium*, *Chaetomium*, *Cladosporium*, *Gliocladium*, *Paraconiothyrium*, *Sarocladium* and *Trichoderma* have been described as biocontrol of phytopathogenic fungi (De Silva et al., 2019; Gama et al., 2020; Madbouly and Abdel-Wareth, 2020). *Trichoderma* is well known amongst biocontrol agents, which resides in the rhizoplane and is exploited as fungal biofungicides. Several studies have been reported that *Trichoderma* exert multiple biocontrol mechanisms to control fungal phytopathogens including mycoparasitism, antibiosis, competition for nutrients, induced defence response (Adnan et al., 2019; Zhang et al., 2016). It provides significant biofertilizer potential in promoting the plant growth (Hang et al., 2022). Therefore, different species of *Trichoderma* genus such as *asperellum*, *atroviride*, *harzianum* and *virens* are being used extensively in the management of plant diseases, either as seed or soil treatments (Woo et al., 2014). Presently, *Trichoderma* species are used extensively preparing commercial biofungicides throughout global markets (Meher et al., 2020). The third category of associated niche includes fungus-infecting viruses (mycoviruses) which are used to control different fungal pathogens (Dobbs et al., 2021). De Miccolis et al., (2022) reported a new putative ssDNA mycovirus belonging to the family Parvoviridae which has controlled the pre- and post-harvest yield losses caused by *Monilinia fructicola* in Serbia. Recently, Wang et al. (2022) isolated Fusarium oxysporum mymonavirus (FoMyV1) ssRNA from fungus *Fusarium oxysporum* and used against phytopathogenic fungi *Sclerotinia sclerotiorum* and *Botrytis cineaea*. Zhou et al., (2021) stated that chrysovirus 1 (PtCVI) alters the endophytic and phytopathogenic traits of *Pestalotiopsis theae*.

Several researchers have reported that yeasts could be used as biological agents to control pathogens (Yuliar et al., 2015; Fu et al., 2016). A few endophytic species of yeasts, e.g., *Aureobasidium*, *Candida*, *Pichia*, *Rhodotorula* and *Saccharomyces* are used as biocontrol fungi (Madbuly and Abdel-Wareth, 2020). Additonally, endophytic bacteria such as *Bacillus amyloliquefaciens*, *B. vellezensis*, *B. subtilis*, *Mesorhizobium*, *Burkholdera*, *Pesudomonas fluorescens*, *Pantoea*, *Rhizobium* and *streptomyces* are quick to develop microbial biofungicides (Cheffi-Azbou et al., 2020; Hassan et al, 2018; Bolivar-Anillo et al., 2020; Nagpal et al., 2020). The Plant

Growth Promoting Rhizobacteria (PGPR) also possesses biofungicidal properties (Sufyan et al., 2020; Sharf et al., 2021; Aquino et al., 2021). Amongst the PGPR, *Pseudomonas* genus is reported as the best biocontrol agent (Panpatte et al., 2016), containing siderophores and derivates of 4-diacetylphloroglucinol and phenazine as well as its combination with chitosan [50–1.5% (v/v)] causes inhibition in the mycelial growth of *A. alternata* and *F. solani* (Trejo-Raya et al., 2021). Panpatte et al. (2016) determined that the *Pseudomonas* possesses both biocontrolling and plant growth promoting characteristics. Other bacteria genera such as *Bacillus*, *Streptomyces*, *Pseudomonas*, etc., have been reported as biocontrolling agents of phytopathogenic fungi in many crops (Miljakovic et al., 2020; Ghadamgahi et al., 2022; Chatterjee et al., 2020). Heo et al. (2022) reported the biological potential of *Burkholderia contaminans* AYoo1 against fusarium wilt and bacterial speck diseases of tomato. Ghadamgahi et al. (2022) showed the antagonism of *P. aeruginosa* strain FG106 against the pathogens namely, *Alternaria alternata*, *Botrytis cinerea*, *Clavibacter michiganensis* subsp. *michiganensis*, *Phytophthora colocasiae*, *P. infestans*, *Rhizoctonia solani* and *Xanthomonas euvesicatoria* pv. *Perforans* of tomato, potato, taro and strawberry. Sharma et al., (2022) determined the *Pseudomonas fluorescens* antagonistic activity against the *Fusarium solani* pathogen that causes root rot disease in okra. Currently, many new species exhibiting antimicrobial potential are being investigated to control fungal pathogens causing severe disease in plants.

3.2 Plant-Incorporated Protectant (PIP's)

Some plants have the feature of producing antifungal metabolites that are coded by genes (Gupta and Dikshit, 2010). These days, scientists use genes to modify or incorporate desired traits in a crop plant, especially to develop resistance in a host against a fungal pathogen called plant-incorporated protectant (Kumar and Singh, 2014). For example, scientists have incorporated biodegradable cry protein to develop transgenic Bt cotton that is resistant to cotton ball worm. Maize varieties are incorporated by the gene which codes for Bt pesticidal protein to be resistant to the European corn borer. According to the "European Union, Genetically Modified (GM) food and feed" plant genes code for pesticidal substances that are used in genetic modification or incorporation into other plants. Later, the modification and incorporation of genes into plants may use scientific assessment to confirm no adverse effects on the environment and public health (according to Regulation (EC) 1829/2003). PIPs, according to Koundal and Rajendran (2003), are an economically sound strategy for providing more food, feed and forage to the people of developing countries.

3.3 Plant-derived biofungicides

Plants are a rich source of various extractable primary and secondary bioactive metabolites (Dhaliwal and Arora, 2006). More than 250,000 higher plant species are estimated to be reservoirs of a diverse range of secondary metabolites such as phytoalexins and phytoanticipins, phenol, alkaloids, flavonoids, terpenoids, glycosides, tannins, lignin, fatty acids, saponin, glucosinolates, essential oils, photopigments, organic acids, protease inhibitors and so on (Adebo and Medina-Meza, 2020); derived metabolites exhibit antimicrobial and antioxidant potentials, which may act either directly or indirectly to protect plants against deleterious fungal pathogens (Velu et al., 2018; Lahlou et al., 2019). Algal, lichen and angiosperm antifungal and anti-mycotoxigenic activities, as well as medicinal aromatic herbal plants' antioxidant potential, are important in suppressing fungal phytopathogens such as Fusarium verticillioides, A. flavus and A. ochraceous.Thus, the selection of these categories of plants could be used for the discovery of biofungicides, which are non-phytotoxic, economically viable and alternative to synthetic chemicals (Gupta and Dikshit, 2010; Dikhoba et al., 2019; Schmid et al., 2022). Such plant species are relevant to the management of various plant diseases (Kumar and Singh, 2014; De Corato, 2017). Plant essential oils with varying constituents of terpenoid and phenolic compounds have potent antimicrobial characteristics (Bakkali et al., 2008). The essential oils have a distinct composition and produce naturally highly volatile compounds that can be replicated using steam distillation techniques (Floudas et al., 2021; Raveau et al., 2020). Isolation of higher plants can be used in the development of natural products for agricultural industries in three different ways: fungicides in an unmodified state (crude extracts), 2) help in the synthesis of new complex products that are necessary in 'building blocks', and 3) introduce new modes of biofungicide actions to encounter the problem of synthetic chemical resistant development in pathogenic fungi and bacteria (Zaker, 2016; Shuping and Eloff, 2017). The isolation and extraction of plants' metabolites constitutes maceration of plant material with different solvents like ethanol, methanol, ether, etc., followed by fractionation-alization and purification (Zaynab et al., 2018; Aqil et al., 2010; Aiyaz et al. 2015). As technology improves, new ways to extract bioactive secondary metabolites, like GC–MS and LC–MS, will be created.

Many researchers have demonstrated that the plant metabolites extract could control phytopathogens under *in vitro* and *in vivo* (Shuping and Elof, 2017; Abdelkhalek et al., 2020; Saputri and Utami, 2020; El-Shahir et al., 2022) as given in Table 9.2. For instance, aqueous extracts of plant including *Hibiscus rosa*, *Melia azedarach* and *Cassic fistula* at concentrations 25, 50 and 100% per litre of potato dextrose agar are effective to control the mycelial growth of *Fusarium graminearum* causing Head Blight (FHB) of wheat (Akbar et al., 2022). Abo-Elyousr et al. (2022) determined the alternative control of tomato fusarium wilt disease caused by *Fusarium Oxysporum* f. sp. *Lycopersici* (FOL) using aqueous extracts of *Calotropis procera*. *Pinus wallachiana* leaf extracts showing antifungal activity against *Fusarium oxysporum* f. sp. *cubense* (Ain et al., 2022). The aqueous extracts of *Dianthus caryophyllus, Capsicum annum, Cinnamomum verum* and *Cymbopogon citratus* at three concentrations 2.5, 5 and 10% could control green mould disease of orange caused by *Penicillium digitatum* (Elhelaly et al., 2022). The aqueous seeds extract of *Piper nigrum*, *Xylopia aethiopica, Aframomum melegueta* and fresh leaves of *Cymbopogon citratus* inhibited the mycelial growth of *Fusarium verticillioides*, *Colletotrichum* sp., *Aspergillus niger* and *Aspergillus flavus* (Gyasi et al., 2022). The aqueous extracts of hornwort, sage and thorn apple showed biocontrol activity against *Rhizoctonia solani* (Al-Baldawy et al., 2021). The plants extract of *Phytolacca americana*, *Prunus laurocerasus*, *Rhododendron*

Table 9.2: Phytoextracts and their phytochemicals leading antifungal activity to control fungal phytopathogen.

Phytoextracts	**Phytochemicals**	**Control fungal pathogens**	**Pathogenic diseases**	**References**
Eysenhardtia polystachya L. (leaf extract)	Isoflavones	*Sclerotium cepivorum*	Allium root rot	Rodriguez Guadarrama et al. (2018)
Azadirachta indica L. (neem leaf extract)	Azadirachtin	*Puccinia triticina*	Wheat leaf rust	Shabana et al. (2017)
Syzygium aromaticum L. (clove oil)	Eugenol and caryophyllene (essential oil)	*Fusarium oxysporum* f. sp. *Lycopersici*	Fusarium wilt disease	Sharma et al. (2017)
Melissa officinalis L. (essential oil)	P-mentha- 1,2,3-triol, P-menth-3-en-8-ol, piperitenone oxide and Zpiperitone oxide (essential oil)	*Botrytis cinerea, P. expansum* and *R. stolonifer*	Post-harvest disease of apple	El Ouadi et al. (2017)
Calotropis procera L. (leaf extract)	terpenoids, flavonoids, saponins, steroids and cardiac glycosides	*Alternaria alternata*	Leaf spot disease	Ali et al. (2020); Morsy et al. (2016)
Curcuma longa L. (turmeric rhizome extract)	essential oils and curcumin	*Aspergillus flavus, A. parasiticus, Candida albicans, Fusarium solani, Penecillium expansum*	Root rot and blue mould disease	Song et al. (2020); Narayanan et al., (2020)
Curcuma amada L. (ginger rhizome extract)	alpha-zingiberene, geranial, 6-gingerol and 6-shagoal	*A. flavus, A. parasiticum and* and *Fusarium verticilloids*	Root rot and blue mould disease	Kavithaet al. (2020); Nerilo et al. (2020)
Punica granatum L. (pomegranate fruit and leaf extract)	Tannins and flavanoids	*Alternaria alternata, Aspergillus niger, Fusarium oxysporum, Fusarium graminearum,* and *pythium digitatum*	–	Zhu et al. (2019)
Piper nigrum L. (pepper leaf and fruits extract)	limonene, sabinene and β-caryophyllene	*F. oxysporum* and *A. niger*	tomato rot	Castellanos et al. (2020)

Table 9.2 contd. ...

...Table 9.2 contd.

Phytoextracts	**Phytochemicals**	**Control fungal pathogens**	**Pathogenic diseases**	**References**
Porella platyphylla, *Cinclidotus fontinaloides* and *Anomodon viticulosus* (liverwort extract)		*Botrytis cinerea*	–	Latinovic et al. (2019)
Marsilea minuta (aquatic fern extract)	benzoic acid-4-ethoxy, ethyl ester, monoester of benzoic acid and farnesol acetate	*Aspergillus niger, A. terreus, A. favus* and *Fusarium solani*	–	Sabithira and Udayakumar (2018)
Heterodermia leucomelos L. (thali extract)	Secondary metabolites	*Aspergillus niger, Aspergillus favus, Fusarium oxysporum, Fusarium solani* and *Colletotrichum falcatum*	Mould, Rot, Fusarium wilt Fusarium root rot and red rot of sugarcane	Babiah et al. (2015)
Parmotrema tinctorum L. and *Flavoparmelia caperata* L. (thali extract)	steroids, quinones, terpenoids, peptides, xanthones, sulphur-containing chromenones, etc.	*Fusarium solani*	Ginger rhizome rot	Shivanna and Garampalli (2015); Kellogg et al. (2017)
Ipomoea batatas L. (Purple sweet potato)	carotenoids, polyphenols, ascorbic acid, alkaloids, coumarins and saponins	*Fusarium oxysporum* f. sp. *capsici*	Fusarium fruit rot on chilli	Saputri and Utami (2020)
Ipomoea batatas L. (sweet potato extract)	carotenoids, polyphenols, ascorbic acid, alkaloids, coumarins and saponins	*Ceratocystis fmbriata*	Black rot Sweet potatoes	Liu et al., (2020); Escobar et al. (2022)
Moringa peregrina L. (leaf and flower extract)	phenols and total flavonoids content	*Mycogone perniciosa*	Agaricus bisporus	Shokouhi and Seif (2020); Abhary (2022)
Oryza sativa L. (leaf extract)	total phenolic, total flavonoid, anthocyanin, tannins content, antioxidant activity and phenolic compounds	*Fusarium oxysporum*	Tomato wilt	Wei et al. (2020); El-Beltagi et al. (2022)

Table 9.2 contd. ...

...Table 9.2 contd.

Phytoextracts	Phytochemicals	Control fungal pathogens	Pathogenic diseases	References
Zizyphus spina-christi L. (methanolic leaves and fruits extract)	Gallic acid, ellagic acid, Phenol, 2,5-bis(1,1-dimethylethyl) (40.24%) and Decane, 2-methyl-(18.53%) leaves and fruits quercetin, D-mannonic acid, 2,3,5,6-tetrakis-o-(trimethylsilyl), and γ-lactone (22.72%)	*Alternaria alternata, A. citri* and *A. radicina*	Tomato leaf spot	El-Shahir et al. (2022)
Allium sativum L. (Garlic bulb aqueous extract)	Allicin (diallylthiosulphinate)	*Alternaria solani, Alternaria tenuissima, Alternaria alternata, Fusarium oxysporum, Fusarium sambucinum, Colletotrichum coccodes, Rhizoctonia solani, Phoma exigua, Pectobacterium carotovorum, Streptomyces scabiei*	Potato foliar and soil borne	Sarfraz et al. (2020); Steglinska et al. (2022)
Citrullus colocynthis (Bitter apple leaf and root extract)	Spinasterol, 22,23-dihydrospinastero and 1,1-diphenyl-2-picrylhydrazyl (DPPH)	*Magnaporthe grisea, Rhizoctonia solani, and Phytophthora infostans*	-	Ahmed et al. (2022)
Nannochloropsis sp., *Phaeodactylum tricornutum, Scenedesmus obliquus, Chlorella vulgaris* and *Spirulina* sp. (Algal aqueous extracts)	carotenoid (β-Carotene and astaxanthin) and phenolic	*Sclerotium rolfsii, Rhizoctonia solani, Botrytis cinerea and Alternaria alternata. S. obliquus*	-	Schmid et al., (2022)

Table 9.2 contd. ...

...*Table 9.2 contd.*

Phytoextracts	Phytochemicals	Control fungal pathogens	Pathogenic diseases	References
Cladonia foliacea, Hypotrachyna cirrhata, Leucodermia leucomelos, Platismatia glauca and *Pseudevernia furfuracea* (lichen thalli extracts)	protolichesterinic acid, lecanoric acid and orsellinic acid	*A.spergillus candidus, A. flavus, A. fumigatus, A. nidulans, A. niger, A. ochraceus, A. parasiticus, A. restrictus, A. stellatus and A. ustus*	mould disease	Furmanek et al. (2021)
Ocimum sanctum L. (tulsi leaf aqueous extract)	camphor, eucalyptol and eugenol	*Alternaria alternata*	leaf spot disease in crop plants	Ojha et al. (2021); Yamani et al. (2016)
Allium sativum L. (garlic bulb extract)	Allicin (diallylthiosulphinate)	*Colletotrichum, Colletotrichum gloeosporioides, the oomycete Phytophthora cactorum and the bacterium Xanthomonas fragariae* associated with	strawberry disease	Oladejo and Imami (2022); Steglinska et al. (2022)

ponticum, Trachystemon orientalis, Smilax excelsa were found suppressive against the pathogens, e.g., *Alternaria solani, B. cinerea* and *Rhizoctonia solani* (Onaran et al., 2016). Some studies have also demonstrated the positive responses of algal and lichen extracts to suppress the fungal pathogens (Schmid et al., 2022; Furmanek et al., 2021). Furmanek et al. (2021) investigated the crude alcoholic extracts of lichens thalli such as *Cladonia foliacea, Hypotrachyna cirrhata, Leucomelos leucomelos, Platismatia glauca* and *Pseudevernia furfuracea* which contain protolichesterinic acid that showed strong antifungal potential against *Aspergillus flavus*. Others *C. foliacea, Nephroma arcticum* and *Parmelia sulcata* secrete lecanoric acid to supress the growth of *Aspergillus fumigatus* and *Evernia prunastri, Hypogymnia physodes, Umbilicaria cylindrica* and *Variospora dolomiticola* produce orsellinic acid to inhibit the mycelial growth of *Aspergillus niger*. Similarly, solvent extracts, essential oils and compounds of medicinal plants have been tested against phytopathogenic fungi (Seepe et al., 2021a, b). Medicinal plant extracts having antifungal potential against Fusarium pathogens have been reported in many studies Seepe et al. (2019), Seepe et al. (2021a, 2021b), Pizzolitto et al. (2020), Xiao et al. (2021). Mannai et al. (2021) evaluated the *in vitro* and *in vivo* antifungal activity of *Raphanus raphanistrum* extracts against Fusarium and Pythiaceae, affecting apple and peach seedlings. Phytoextracts of *Azadirachta indica, Allium sativum* and many others significantly inhibit the fungal growth and spore germination (Haider et al., 2020).

4. Mode of action

As shown in Fig. 9.1, the microbial derived biofungicides and how they work are described in terms of how microorganisms fight off plant pathogens (Spadaro et al., 2002).The interaction of biocontrol agents with the host and pathogen results in direct biocontrol action, which includes: a sequential event that leads to pathogen recognition, attack, penetration, coiling and dissolution of the host cell wall and membrane through toxin or enzymes, which ultimately causes host cell death (Gajera et al., 2020; Kohl et al., 2019). Indirect effects lead to resource competition and trigger plant defensive responses (Ram et al., 2018; Crowder and Harwood, 2014).

Among the biocontrol agents, *Trichoderma* is being used extensively to show both direct and indirect mechanisms against *Macrophomina phaseolina*, *Rhizoctonia solani*, *Sclerotinia slcerotiorum* and *Sclerotium rolfsii* (Ram et al., 2018). Moreover, few genera of yeast showed antifungal potential against fungal pathogens. For example, *Metschnikowia pulcherrima* inhibited the mycelial growth of *Alternaria, Aspergillus, Botrytis, Fusarium, Monilinia* and *Penicillium* (Gore-Lloyd et al., 2019; Oro et al., 2014); *Pichia guilliermondii* suppresses the growth of *Botrytis cinerea* (Freimoser et al., 2019; Zhang et al., 2011), *Saccharomycopsis schoenii* control *Candida auris* mycelial growth (Abbey et al., 2019). The biocontrol agent secretes or excretes hydrolytic enzymes and secondary metabolites that interferes in the pathogen growth and multiplication (Hegedus and Marx, 2013; Leiter et al., 2017). For examples, *Bacillus amyloliquefaciens* strain FZB42 and *B. subtilis* AU195 secretes Bacillomycin D toxin against *F. oxysporum, F. graminearum* and *A. flavus*; the *T. virens* produces gliotoxin against *R. solani*; Iturin A of *B. subtilis* QST713 control *Botrytis* sp. and *R. solani* (Hilber- Bodmer et al., 2017; Oro et al., 2014; Saravanakumar et al., 2008; Spadaro et al., 2002). Other antifungal proteins are also secreted by *Penicillium chrysogenum* and *Aspergillus giganteus* to control *B. cinerea, Fusarium* sp., *Cochliobolus carbonum, A. flavus, Puccinia graminis, Puccinia recondita, Erysiphe graminis, Mycosphaerella grisea, A. alternata* (Leiter et al., 2017; Marx et al., 2008; Toth et al., 2020). Mycoparasitism of biocontrol agents accomplished by secreting cell wall degrading hydrolyzing enzymes (CWDEs) such as alpha-1,3-glucans and beta-1,6-glucans (Almeida et al., 2021); chitinases can hydrolyze chitin (Senol et al., 2014); peptidases catalyze the peptide bonds between the amino acids in a protein (Da Silva et al., 2016) and beta-1,3 glucanase and glycosidases hydrolyze the beta-1,3 bonds in glucans (Zhou et al., 2017). The hydrolysis of the fungal pathogen cell wall components breaking the osmotic pressure of the protoplast ensure complete disintegration of fungal mycelia or conidia (Silva et al., 2019). Chitinases causes cell wall lysis of fungal pathogens such as *B. cinerea* (Fernandez-Caballero et al., 2009); *A. alternata, Cladosporium* sp., *F. oxysporum, F. solani* and *R. solani* (Anitha and Rabeeth, 2010; Sharma et al., 2020); *Ceratocystis fimbriata, Verticillium dahlia* and *Ustilaginoidea virens* (Liu et al., 2020). The basic amino acids nature of secondary metabolites favour the interaction and action of biocontrol agents with the negatively charged cell membrane of the pathogen, which improves pathogen cell wall permeability and generates excessive reactive oxygen species that cause cell death (Leiter et al., 2017). Thus, biochemical characteristics of secondary metabolites are considered promising in

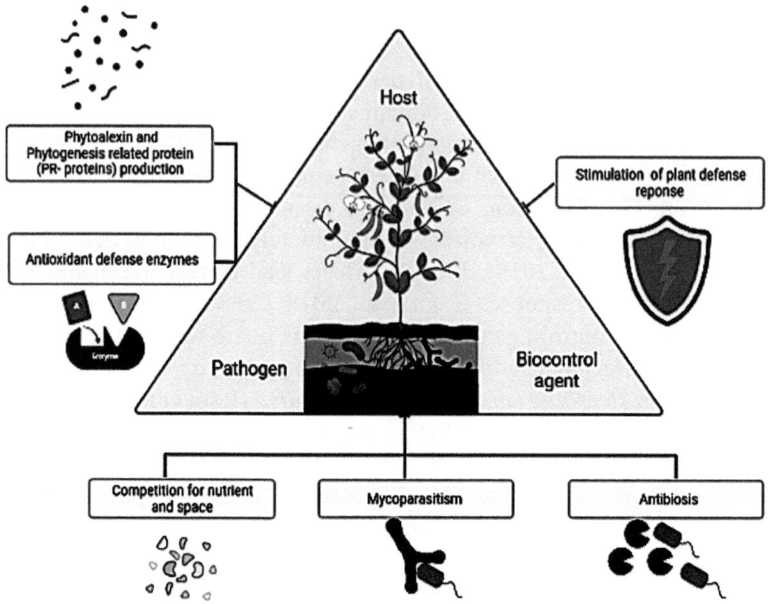

Fig. 9.1: A model of biocontrol agents mechanisms of interaction with the host and pathogens: host-biocontrol interaction stimulates Induced Systemic Resistance (ISR) and sytemic acquired resistance in the plant; biocontrol agents-pathogen interaction control the pathogen growth by providing different mechannisms such as competition where biocontrol agents utilizes nutrients and space more rapidly to colonize faster than the pathogen, mycoparasitism fungal biocontrol agents directly parasitizes the pathogen in a number of steps- recognition by host mycoparasite, secretion of cell wall degrading enzymes, penetration and ultimately cause cell lysis, Antibiosis-biocontrol agents secretes antifungal compounds to control phytopathogenic fungi; host-pathogen interaction elucidates production of phytoalexin, pathogenesis related to protein and antioxidants compounds to further prevent spread of the pathogen infection and also help the plant develop resistance against pathogen.

the production of natural biofungicide formulations. Hashem et al. (2021) developed cost-effective FTB biofungicide using *Trichoderma harzianum* JF419706 and lignocellulosic as a carrier to reduce the root rot disease severity of vegetable crops. This also ensures sustainability to be used alone or in combination with other IDM practices in organic farming and reduces the chemical input in crop management.

Competition is the main biological indirect antagonistic mechanism employed between biocontrol agents and plant pathogenic fungi for high nutrient uptake and availability of space within the same limited environment (Lahlali et al., 2022). Another important indirect antagonistic characteristic of fungal biocontrol agents is the induction of host resistance; this promotes plant growth and improves plant stress tolerance. The microbes can stimulate local or systemic defens ive responses in plants against the target pathogens (Ram et al., 2018; Zhao et al., 2020). Microorganism Associated Molecular Patterns (MAMPs) are also known to trigger defence responses in plants against attacking pathogens (Kohl et al., 2019) by regulating phytohormone-dependent pathways such as Salicylic Acid (SA), Jasmonates (JA), ethylene (ET), Abscisic acid (ABA), auxin (Indole-3-Acetic Acid: IAA) and gibberellins (GA) as shown in Fig. 9.2. Phytohormone profiling plays a role in modulating growth and

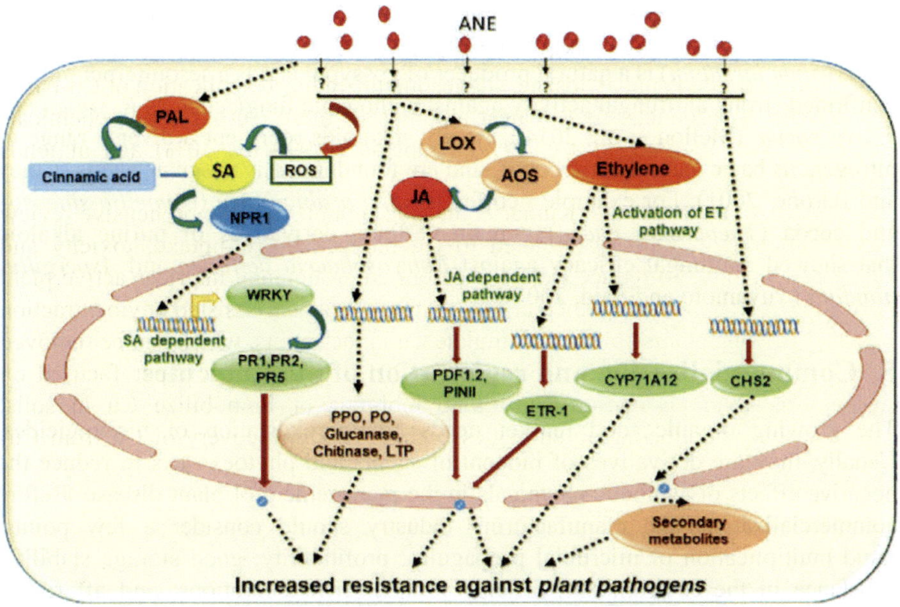

Fig. 9.2: Schematic representation of phytoextract inducing plant defence response against plant pathogens (cited from Shukla et al., 2019).

defence regulatory proteins in response to biotic and abiotic stresses (Adnana et al., 2019). Unlike the biocontrol agents' action mechanisms against pathogens, plant extracts act on host plants as synthetic fungicides. Generally, phytoextracts act as protectants or systemic fungicides, which can improve the efficacy of biofungicides against phytopathogens. The plants' secondary metabolites possesses biofungicidal properties, which are divided into three main groups: phenolic, terpenes and nitrogen compounds based on biosynthetic pathways and mode of action in plants (Amiri et al., 2017; Tripathi and Shukla, 2010; Aiyaz et al., 2015; Naboulsi et al., 2018).

Among the plant's secondary metabolites, phenolic compound functions are a diverse range in plants such as enhancing growth and development (e.g., thickening the cell wall, hormone production and pigmentation, reproduction and defence response against stressors, e.g., osmoregulation, UV protection, anti-herbivory roles and antimicrobial activity). For instance, lignin accumulation in *Pichia membranefaciens* induces resistance against *Rhizopus stolonifera* (Zhang et al., 2020). Phenolic acids are produced in plants by three different biosynthetic pathways: (i) the shikimic/chorizmic acid or succinyl benzoate produced through phenyl propanoid (C_6–C_3) pathoway; (ii) the acetate/malonate or polyketide produced by side-chain-elongated phenyl propanoids pathway (including the large group of flavonoids (C_6–C_3–C_6) and some quinones); and (iii) the acetate/mevalonate produced via aromatic terpenoids, mostly monoterpenes, by dehydrogenation pathway (Mandal et al., 2010). Many studies have determined the antifungal properties of phenolic compounds against the pathogens, such as *P. oryzae* and *Aspergillus parasiticus* (Loran et al., 2022, Lattanzio et al., 2006). Among the three groups, terpenes is the largest class of secondary metabolites that is synthetized in the cytoplasm of plant cells to produce its derivatives of monoterpenoids, sesquiterpenoids, diterpenoids and triterpenoids through the mevalonic acid pathway (Singh and Sharma, 2015). The cotton plant (*Gossypium hirsutum*) is a natural producer of gossypol and is a sesquiterpenoid that confirmed strong antifungal activity against pathogenic fungi, such as *R. solani* and *F. oxysporum* (Mellon et al., 2014). Finally, alkaloids represent a diverse range of nitrogenous basic natural compounds, and are found in many vascular plants (Coley and Barone, 2001). For example , coffee (*Coffea arabica*), tea (*Camellia sinensis*) and cocoa (*Theobroma cacao*) contain caffeine derivatives of purine alkaloid that showed antifungal efficacy against *Botryosphaeria dothidea* and *Aspergillus nidulans* (Yuvamoto and Said, 2007).

5. Commercialization and registration of biofungicides

The growing organic food market needs commercialization of biofungicides. Usually, they are derivatives of biocontrol agents and phytoextracts to reduce the negative effects of synthetic chemicals in the management of plant disease. Before commercialization, the manufacturing industry should consider a few points: rapid multiplication of microbial propagules; profitability; good storage stability; resistance in the environment; efficiency under field conditions; and affordable substrate availability in order to enhance the product quality and market value (Locatelli et al., 2018; Sachdev et al., 2018). Registration systems are then disclosed to confirm the toxicity, production efficiency and safety of the biofungicide products (Desai et al., 2016). The registration of biofungicide products should reveal the following information: systemic name and common name of microbial agents and plants; natural occurrence; morphological description of the microorganism and plant; details of the manufacturing process; mammalian toxicity; environmental toxicity and residual analysis. Thus, registered chemical fungicidal products become expensive to use commercially (Chandler et al., 2011). However, the registration of biofungicidal products is easier than chemical pesticides, which influences industries to develop more products and meets consumer demand globally (Kumar, 2012; Gupta and Dikshit, 2010). But some regulatory bodies had the concept that biofungicides were chemical pesticides made in a laboratory (Chandler et al., 2011), which caused problems for the companies that made them. Another challenge in the line of biofungicide registration requires environmental toxicology assessments that are expensive for certain countries, which limits the market size and consumer demands (Elad, 2001). Additionally, the tedious registration process, the commercialization cost and market time also affect biofungicide companies. For instance, the registration charges include pollution testing of Trichodex® products, which was found economically acceptable at both national and international trade levels (Elad, 2001). Several commercial products derived from microbes and plant extracts have been registered and used both at national and international levels, as summarized in Table 9.3.

It is estimated that North America requires about 40% of global biofungicides (Kumar, 2012). Asian markets of China and India are the world's fastest growing biofungicides countries to meet the demand of organic farming and residue-free crop production. Currently, there are 970 biopesticide products registered through

Table 9.3: Biofungicidal products used to control pathogen and plant diseases.

Biological control agents	**Trade name**	**Target organisms/disease**	**Manufacturer**
Streptomyces lydicus	Actinovate	Botrytis, downy and powdery mildew, *Pythium, Phytophthora, Rhizoctonia*	Novozymes, Denmark
Ulocladium oudemansii U3	Botry stop	*Botrytis cinerea, Sclerotinia sclerotiorum* and *Monilinia* spp.	Bioworks Inc., USA
Bacillus subtillis QST 713	Rhapsody	Powdery mildew, downy mildew, leaf spot, *Rhizoctonia, Pythium, Phytophthora, rust, Erysiphe, Perenospora, Alternaria, Colletotrichum*	Bayer Indian Syntans Limited
Bacillus subtillis GB03	Companion liquid	Leaf spots, Powdery mildew, Black Root Rot, Early Blight Crown Rot, Damping-off Fungus, Grey Mould, Leaf blight Grey Mould, Blight, Root Rot *Pythium* spp., *Rhizocotonia, Pythium, Phytophthora, Aspergilus* spp., *Alternaria* spp., *Botrytis cinerea, Erysiphe cichoracearum, Podosphaera xanthii*	Evergreen grower supply, LLC, India
Coniothryium minitans	Contans WG	*Sclerotinia sclerotiorum, S. minor*	SIPCAMP AGRO USA, Inc.
Bacillus amyloliquefaciens	Double nickel, Triathlon	Botrytis, downy and powdery mildew, Pythium, *Rhizoctonia*, etc.	Certis Group Commercial Organization, Singapore
Agrobacterium radiobacter K84	Galltrol	*Agrobacterium tumefaciens*	AgBioChem USA, Inc.
Streptomyces griseoviridis	Mycostop	*Alternaria, Botrytis, Pythium, Phytophthora, Rhizoctonia*	A.G.Bio-Systems Private Limited, Hyderabad, India
Trichoderma harzianum	Root Shield, Plant shield T-22 planter box	*Cylindrocladium, Fusarium, Pythium, Rhizoctonia, Thielaviopis*	Bioworks Inc., USA Bioglobal India Private Ltd.
Gliocladium catenulatum JII446	Prestop WP	*Botrytis, Fusarium, Pythium, Phytophthora, Rhizoctonia, Verticillium*	Lellemend, Canada
Gliocladium virens GL-21	SoilGard	*Rhizoctonia, Pythium*	Green Thumb Garden Care Private Ltd. Hyderabad

Table 9.3 contd. ...

...*Table 9.3 contd.*

Biological control agents	Trade name	Target organisms/disease	Manufacturer
Ageobacterium radiobacterstrain K 1026	Nogall	*Agrobacterium tumefaciens*	BASF, Australia
Ampelomyces quisquallis isolate M-10	AQ10	Powdery mildew	Ecogen,USA
Pseudomans fuorescens A506	Blight ban A506	Fire blight	Nufarm

the Central Insecticides Board and Registration Committee (CIBRC) government body in India (Mishra et al., 2020). It is estimated that 175 biofungicides with active ingredients are registered and 700 products available in the market, valued at US$ 1.3 billion in 2011and is expected to increase to US$ 3.2 billion in the future (Kumar, 2012). Globally, it is expected that biofungicides market will exceed to the value of US$ 1.6 billion in 2020 to a value of US$ 3.4 billion by 2025 (Biofungicides market, 2020). There were approximately 400 active ingredients of biocontrol agents registered to be manufactured of more than 1250 biofungicides formulations products (USEPA, 2013). In India, among the fungal biofungicides products, only eight types have been registered under Central Insecticidal Board (CIB) Act, 3 & 3 (B) so far (Mishra et al., 2020). In India, more than 100 commercially registered biofungicide and/or growth-promoting products based on various *Trichoderma* spp., are being used (Woo et al., 2014). However, Gupta and Dikshit (2010), reported that Indian farmers had the awareness for wide adaptation of biofungicides in their cropping system instead of chemical fungicides. It is also reported that in India the demand of organic food motivates a good opportunity of biofungicides. In United States, report > 200 microbial biofungicides products are being sold in comparison to 60 products in the European Union. It is notable that US biofungicides market is valued at around US $205 million and expected to be increased approximately US$ 300 million by 2020. Currently, the US biofungicide markets are anticipated in growing over US$ 300 million by 2020. In recent years, analysts expect a higher development of biofungicides in the agriculture sector as many minor and major manufacturers lead commercial production of bio control agents and plant extracts worldwide.

Multi-step requirements in the commercialization of biofungicides:

(a) Isolation of microorganism and plant extracts from the natural ecosystem
(b) Evaluation of bioagent and plant extracts both *in-vitro* and under polyhouse.
(c) Evaluation of microbes and plants extracts as biofertilizers and biofungicides under uncontrolled conditions of field.
(d) Mass production
(e) Formulation
(f) Delivery

(g) Compatibility
(h) Registration and release
(i) substrate used in the formulation.

5.1 Advantages of Biofungicides use in agriculture

Biologically derived biofungicides control the indiscriminate and unscientific use of chemical pesticides to kill or limit the deleterious phytopathogens without harming the host (Srivastava et al., 2016). Biofungicides are generally less toxic and less expensive than chemical pesticides, offer multiple benefits such as being effective against target pathogens, minimal or non-toxic effects, safer towards the human and environmental health and high biodegradability, which increases their acceptability in organic farming to boost crop yields and ensure food safety and security (Gupta and Dikshit, 2010; Kumar and Singh, 2014; Stevenson et al., 2012). Kumar (2012) reported that organic agriculture and public health are two major concerns toward application of biofungicides (Kumar, 2012; Kumar and Singh, 2014) without leaving residual effects on the consumers. It can be used as a component of Integrated Pest Management (IPM) in place of synthetic chemicals. The advantage of using plant-based biofungicides, is that they are eco-friendly, fast-acting, non-pollutive and cost-effective compared to chemical fungicides (Kumar et al., 2021a, b).

Use of biofertilizers in agriculture system has the following benefits:

- Biofertilizers functions as chemical fertilizers.
- Biofertilizers are eco-safe, cost effective and non-pollutive, which help in reducing synthetic agrochemicals loads.
- Microbes in biofertilizers help in direct fixation of atmospheric nitrogen for plants.
- Increases the mineral nutrients availability to plants.
- Improves soil texture and soil fertility.
- Improves plant growth promoting hormones, vitamins and proteins.
- They increase the multiplication rate of beneficial microorganisms in the plant rhizosphere.
- They control and inhibit plant pathogenic fungi.

5.2 Disadvantage

1. High specificity of biofungicides towards target pests/pathogens may necessitate an exact identification of the pest/pathogen.
2. Biofungicides are not as quick in action as chemical fungicides against a pathogen hence, it cannot be effective in a pest/pathogen outbreak.
3. Biofungicides efficacy often differs under the influence of various biotic and abiotic factors.
4. As biofungicides are derived from biological material, one of the major problems is a short shelf life. It can be stored at room temperature up to 6 mon or it may extend up to 2 yr at a cool temperature. However, chemical fungicides have

much higher efficacy and longer shell life during storage with no or less change in their efficacy.

5. Toxicological and environmental impact tests are expensive.
6. Biofungicides pose phytotoxic effects in seed germination, growth and development of a plant (Meepagala et al., 2019; Zhao et al., 2020).
7. Living organisms evolve new races in response of biological, chemical, physical or others control components.
8. The crop specific biofungicides limits a wide acceptability. For example, use of non-specific Rhizobium as a fertilizer will not promote root nodulation or boost crop yield.
9. The effectiveness of microbial fertilizers in improving soil fertility depends on various factors including moisture content, pH, temperature, organic matter and the microbial species.
10. Contamination issue affects the quality of microbial mass production.
11. Long-time exposure under sunlight may inactivate or kill the microbial cell to reproduce as they are light-sensitive.
12. The high cost of commercialization and registration process has limited the production and trade of biofungicides.

6. Conclusion

This chapter concludes that biofungicides derived from microbes and plants could provide good disease control as synthetic fungicides by keeping the following points into consideration: biofungicidal products have strong antifungal potential under exposure to biotic and abiotic stresses; long shell life, target specific, no residual toxicity, cost-effective, accepted in the IDM program and environmentally acceptable without imposing adverse effects on human and animal health. To address the concerns of plant growers and food consumers, this chapter describes biofungicide screening, formulation product development and disease control. Although using biofungicides brings a few challenges, including identification of natural sources having antifungal potential, performance evaluation, registration and formulation, storage and discrepancy. Despite all the challenges, biofungicides are still used to manage plant diseases. Thus, biofungicides demand substantial collaboration with academic, research, industry and government institutions to develop sustainability in the agriculture system. These findings could lead to the development of a new biofungicide, which could compete with the current use of synthetic fungicides. Biofungicides have a significant impact on increasing agricultural productivity and food security by enhancing farmers' livelihoods. This would emphasize the screening of novel biofungicides to meet the current status of farmers. However, the rising demand for biofungicides motivates researchers, research institutes, biotechnologists and major stakeholders to compete with new technologies such as tissue culture and nanotechnology in the production of large quantities of fungicidal plants and other biofungicide compounds. To produce biofungicidal substances in substantial numbers, however, new technologies like tissue culture and nanotechnology are being introduced by research scientists, research institutes, biotechnologists and other key stakeholders. To make biofungicides globally accepted, funding projects, training of researchers and developing specialized research institutes should be engaged in introducing new technologies that help in increasing entrepreneurship and promoting industrial growth. The efficacy of different biofungicides varies under field conditions, especially when disease severity is low to moderate. The best success is most likely to be achieved by combining biofungicides with the integrated disease management components. These characteristics raise environmental awareness and food security concerns. Industrial use of biofungicides has helped developing countries' economies in a large way by increasing export income, creating jobs and reducing poverty.

Suggestions for further studies

With understanding the advantages and disadvantages, this chapter would suggest the use of biofungicides agriculture cropping system by considering the following points: If biofungicides are used in combination of other IDM components, it would overcome the target specific problems:

1. Increasing public and environment health concerns would suggest that further studies should be taken on the impact of biofungicides in enhancing the crop yield and ensuring food security. 2. Policy makers, researchers, government, stakeholders and individuals should be actively involved to tackle the concern of conservation and production of novel plants and biocontrol agents and their development as biofungicides. 3. Regulatory agencies, policy makers and government will develop a model for the registration of biofungicides. 4. New technologies like tissues culture techniques, genetic engineering and other molecular tools should be emphasized and adopted for improving the efficacy of biofungicides formulation. 5. For large scale deployment of biofungicides products, the stakeholders should have collective knowledge of their optimization and adaptation associated with basic safety information data to provide the general public health concern. 6. Formulated biofungicides should have higher efficacy at variable environmental conditions and longer shelf-life than synthetic chemicals. 7. If biofungicides are used in combination of other IDM components, it would overcome the target specific problems. 8. Universal biofungicides development is an important perspective to reduce toxic effects of residues on plants.

References

Abbey, J.A., Percival, D., Abbey, L. et al. (2019). Purification of chitinase enzymes from *Bacillus subtilis* bacteria TV-125, investigation of kinetic properties and antifungal activity against *Fusarium culmorum*. Postharvest Biology and Technology, 14: 1–15.

Abdelkhalek, A., Al-Askar, A.A. and Behiry, S.I. (2020). Bacillus licheniformis strain POT1 mediated polyphenol biosynthetic pathways genes activation and systemic resistance in potato plants against alfafa mosaic virus. Scientific Reports, 10(1): 16120. https://doi.org/10.1038/s41598-020-72676-2.

Abhary, M. (2022). Variation and diversity within wild Moringa peregrina (Forssk.) Fiori; seed size and shape. 10.15413/ajmp.2022.0107.

Abo-Elyousr, K.A.M., Ali, E.F. and Sallam, N.M.A. (2022). Alternative control of tomtao wilt using the aqueous extract of *Calotropis procera*. Horticulturae, 8(3): 197. https://doi.org/10.3390/horticulturae8030197.

Adebo, O.A. and Gabriela Medina-Meza, I. (2020). Impact of fermentation on the phenolic compounds and antioxidant activity of whole cereal grains: A mini review. Molecules, 25(4): 927. https://doi.org/10.3390/molecules25040927.

Adnan, M., Islam, W., Shabbir, A., Khan, K.A., Ghramh, H.A., Huang, Z. et al. (2019). Plant defense against fungal pathogens by antagonistic fungi with Trichoderma in focus. Microbial Pathogenesis, 129: 7–18. https://doi.org/10.1016/j. micpath.2019.01.042.

Ahmed, M., Sajid, A.R., Javeed, A., Aslam, M., Ahsan, T., Hussain, D., Mateen, A., Li, X., Qin, P. and Ji, M. (2022). Antioxidant, antifungal, and aphicidal activity of the triterpenoids spinasterol and 22,23-dihydrospinasterol from leaves of *Citrullus colocynthis* L. Scientific Reports, 12(1): 4910. https://doi.org/10.1038/s41598-022-08999-z. PubMed: 35318417, PubMed Central: PMC8940894.

Ain, Q.U., Asad, S., Ahad, K., Safdar, M.N. and Jamal, A. (2022). Antimicrobial activity of *Pinus wallachiana* leaf extracts against *Fusarium oxysporum* f. sp. *cubense* and analysis of its fractions by HPLC. Pathogens, 11(3). https://doi.org/10.3390/pathogens11030347. PubMed: 35335671, PubMed Central: PMC8953374.

Aiyaz, M., Divakara, S.T., Chandranayaka, S. and Niranjana, S.R. (2015). Efficacy of seed hydropriming with phytoextracts on plant growth promotion and antifungal activity in maize. Int. J. Pest Manag., 61(2): 153–160. https://doi.org/10.1080/09670874.2015.1025116.

Akbar Khan Asma and Nadeem, Azra and Karim, Robina, Khan, Hamzullah and Khan. Muhammad and Ahmad, Zahoor and Khan, Nasrullah and Research, Adaptive and Pakistan [Program]. (2022). Application of aqueous plant extracts and fungicides for management of Fusarium head blight of wheat, 19: 951–956.

Al-Baldawy, M.S.D., Matloob, A.A.A.H. and Almammory, M.K.N. (2021). Effect of plant extracts and biological control agents on *Rhizoctonia solani* Kuhn. IOP Conference Series: Earth and Environmental Science, 735(1): 012079. https://doi.org/10.1088/1755-1315/735/1/012079.

Ali, M., Hayat, S., Ahmad, H., Ghani, M.I., Amin, B., Atif, M.J. and Cheng, Z. (2019). Priming of *Solanum melongena* L. seeds enhances germination, alters antioxidant enzymes, modulates ROS, and improves early seedling growth: Indicating aqueous garlic extract as seed-priming bio-stimulant for eggplant production. Applied Sciences, 9(11): 2203. https://doi.org/10.3390/app9112203.

Ali, M., Hayat, S., Ahmad, H., Ghani, M.I., Amin, B., Atif, M.J. and Cheng, Z. (2019). Priming of *Solanum melongena* L. seeds enhances germination, alters antioxidant enzymes, modulates ROS, and improves early seedling growth: Indicating aqueous garlic extract as seed-priming bio-stimulant for eggplant production. Applied Sciences, 9(11): 2203. https://doi.org/10.3390/app9112203.

Ali, M., Haroon, U., Khizar, M., Chaudhary, H.J. and Munis, M.F.H. (2020). Facile single step preparations of phyto-nanoparticles of iron in Calotropis procera leaf extract to evaluate their antifungal potential against Alternaria alternata. Current Plant Biology, 23: 100157. https://doi.org/10.1016/j.cpb.2020.100157.

Almeida, D.A., Horta, M.A.C., Ferreira Filho, J.A., Murad, N.F. and de Souza, A.P. (2021). The synergistic actions of hydrolytic genes reveal the mechanism of *Trichoderma harzianum* for cellulose degradation. Journal of Biotechnology, 334: 1–10. https://doi.org/10.1016/j.jbiotec.2021.05.001.

Al-Ameri, H. and Rammadan, N. (2022). Biological control mechanisms of *Trichoderma* species to induce resistance in Leguminous plants, 2: 144–150.

Amiri, R., Nikbakht, A., Etemadi, N. and Sabzalian, M.R. (2017). Nutritional status, essential oil changes and water-use efficiency of rose geranium in response to arbuscular mycorrhizal fungi and water deficiency stress. Symbiosis, 73(1): 15–25. https://doi.org/10.1007/s13199-016-0466-z.

Anitha, A. and Rabeeth, M. (2010). Degradation of fungal cell walls of phytopathogenic fungi by lytic enzyme of Streptomyces griseus. African Journal of Plant Science, 4: 61–66.

Anuagasi, C.L11. (2017). The impact of biofungicides on agricultural yields and food security in Africa. International Journal of Agricultural Technology. Anukwuorji. C. a2. and Okereke. C. n3. Okigbo, R. N1, 13(6): 953–978.

Aqil, F., Zahin, M., Ahmad, I., Owais, M., Khan, M.S.A., Bansal, S.S., Farooq, S., Ahmad, I., Owais, M. and Shahid, M. (2010). Antifungal activity of medicinal plant extracts and phytocompounds: A review combating fungal infections. Springer-Verlag, pp. 449–484.

Aquino, J.P.A., Antunes, J.E.L., Bonifácio, A., Rocha, S.M.B., Amorim, M.R., Alcântara Neto, F. and Araujo, A.S.F. (2021). Plant growth-promoting bacteria improve growth and nitrogen metabolism in maize and sorghum. Theoretical and Experimental Plant Physiology, 33(3): 249–260. https://doi.org/10.1007/s40626-021-00209-x.

Archana, H.K., Darshan, G., Thungri, Bashyal, B. and Aggarwal, R. (2022). Biopesticides: A key player in agro-environmental sustainability. 10.1016/B978-0-323-91595-3.00021-5.

Aulakh, M.K., Kaur, N. and Saggoo, M.I.S. (2019). Bioactive phytoconstituents of pteridophytes—A review. Indian Fern Journal, 36: 37–79.

Babiah, P.S., Upreti, D.K. and John, S.A. (2015). Assessment of fungicidal potential of lichen Heterodermia leucomelos (L.) Poelt against pathogenic fungi. Current Research in Environmental and Applied Mycology, 5(2): 92–100. https://doi.org/10.5943/cream/5/2/3.

Backer, R., Rokem, J.S., Ilangumaran, G., Lamont, J., Praslickova, D., Ricci, E., Subramanian, S. and Smith, D.L. (2018). Plant growth-promoting rhizobacteria: Context, mechanisms of action, and roadmap to commercialization of biostimulants for sustainable agriculture. Frontiers in Plant Science, 9: 1473. https://doi.org/10.3389/fpls.2018.01473.

Bakkali, F., Averbeck, S., Averbeck, D. and Idaomar, M. (2008). Biological effects of essential oils—A review. Food and Chemical Toxicology, 46(2): 446–475. https://doi.org/10.1016/j.fct.2007.09.106.

Bedine Boat, M.A., Sameza, M.L., Iacomi, B., Tchameni, S.N. and Boyom, F.F. (2020). Screening, identification and evaluation of Trichoderma spp. for biocontrol potential of common bean damping-off pathogens. Biocontrol Science and Technology, 30(3): 228–242. https://doi.org/10.1080/09583157.2019.1700909.

Bhattacharya, A., Sood, P. and Citovsky, V. (2010). The roles of plant phenolics in defence and communication during *Agrobacterium* and *Rhizobium* infection. Molecular Plant Pathology, 11(5): 705–719. https://doi.org/10.1111/j.1364-3703.2010.00625.x.

Biofungicides market. (2020). https://www.marketsandmarkets.com/Market-Reports/biofungicide-market-8734417.html.

Bizuneh, G.K. (2021). The chemical diversity and biological activities of phytoalexins. Advances in Traditional Medicine, 21(1): 31–43. https://doi.org/10.1007/s13596-020-00442-w.

Bolívar-Anillo, H.J., Garrido, C. and Collado, I.G. (2020). Endophytic microorganisms for biocontrol of the phytopathogenic fungus *Botrytis cinerea*. Phytochemistry Reviews, 19(3): 721–740. https://doi.org/10.1007/s11101-019-09603-5.

Boraste, A., Vamsi, K.K., Jhadav, A., Khairnar, Y., Gupta, N., Trivedi, S., Patil, P., Gupta, G., Gupta, M., Mujapara, A.K. and Joshi, B. (2009). Biofertilizers: A novel tool for agriculture. International Journal of Microbiology Research, 1(2): 23–31. https://doi.org/10.9735/0975- 5276.1.2.23-31.

Carmona Hernandez, S., Reyes-Perez, J.J., Chiquito-Contreras, R.G., Rincon-Enriquez, G., Cerdan-Cabrera, C.R. and Hernandez-Montiel, L.G. (2019). Biocontrol of postharvest fruit fungal diseases by bacterial antagonists: A review Agronomy.

Castellanos, L.M., Olivas, N.A., Ayala-Soto, J., Contreras, C.M.D., Ortega, M.Z., Salas, F.S. and Ochoa, L.H. (2020). *In vitro* and *in vivo* antifungal Activity of clove (Eugenia caryophyllata) and pepper (*Piper nigrum* L.) essential oils and functional extracts against Fusarium oxysporum and *Aspergillus niger* in tomato (*Solanum lycopersicum* L.). International Journal of Microbiology.

Chaibub, A.A., de Sousa, T.P., de Oliveira, M.I.S., Arriel-Elias, M.T., de Araújo, L.G. and de Filippi, M.C.C. (2020). Efficacy of Cladosporium cladosporioides C24G as a multifunctional agent in upland rice in agroecological systems. International Journal of Plant Production, 14(3): 463–474. https://doi.org/10.1007/s42106-020-00097-2.

Chandler, D., Bailey, A.S., Tatchell, G.M., Davidson, G., Greaves, J. and Grant, W.P. (2011 July 12). The development, regulation and use of biopesticides for integrated pest management. Philosophical Transactions of the Royal Society of London. Series B, Biological Sciences, 366(1573): 1987–1998. https://doi.org/10.1098/rstb.2010.0390, PubMed: 21624919, PubMed Central: PMC3130386.

Chatterjee, A., Willett, J.L.E., Nguyen, U.T., Monogue, B., Palmer, K.L., Dunny, G.M. and Duerkop, B.A. (2020). Parallel genomics uncover novel enterococcal-bacteriophage interactions. mBio, 11(2): e03120-19. https://doi.org/10.1128/mBio.03120-19, PubMed: 32127456, PubMed Central: PMC7064774.

Cheffi Azabou, M., Gharbi, Y., Medhioub, I., Ennouri, K., Barham, H., Tounsi, S. and Triki, M.A. (2020). The endophytic strain Bacillus velezensis OEE1: An efficient biocontrol agent against Verticillium wilt of olive and a potential plant growth promoting bacteria. Biological Control, 142. https://doi.org/10.1016/j.biocontrol.2019.104168.

Choudhury, P., Dobhal, S., Srivastava, S. and Saha, Kundu S. (2018). Role of botanical plant extracts to control plant pathogens—a review D. Indian Journal of Agricultural Sciences, 52(4): 341–346. doi 10.18805/IJARe.A-5005.

Coley, P.D. and Barone, J.A. (2001). Ecology of defenses. pp. 426–433. *In*: S.A. Levin (ed.). Encyclopedia of Biodiversity (2nd ed), 2.

Commisso, M., Guarino, F., Marchi, L., Muto, A., Piro, A. and Degola, F. (2021). Bryo-activities: A review on how bryophytes are contributing to the arsenal of natural bioactive compounds against fungi. Plants, 10(2): 203. https://doi.org/10.3390/plants10020203.

Copping, L.G. and Menn, J.J. (2000). Biopesticides: A review of their action, applications and efficacy. Pest Management Science, 56(8): 651–676. https://doi.org/10.1002/1526-4998(200008)56:8<651::AID-PS201>3.0.CO;2-U.

Crowder, D.W. and Harwood, J.D. (2014). Promoting biological control in a rapidly changing world. Biological Control, 75: 1–7. https://doi.org/10.1016/j.biocontrol.2014.04.009.

Cuervo-Parra, J.A., Pérez España, V.H., Zavala-González, E.A., Peralta-Gil, M., Aparicio Burgos, J.E. and Romero-Cortes, T. (2022). Trichoderma Asperellum strains as potential biological control agents against Fusarium verticillioides and Ustilago maydis in maize. Biocontrol Science and Technology, 32(5): 624–647. https://doi.org/10.1080/09583157.2022.2042196.

Das, S., Chaudhari, A.K., Deepika, A.S., Singh, V.K., Dwivedy, A.K. and Dubey, N.K. (2020). Foodborne microbial toxins and their inhibition by plant-based chemicals. Functional and Preservative Properties of Phytochemicals, 6: 165–207. Academic Press.

De Corato, U., Salimbeni, R., De Pretis, A., Avella, N. and Patruno, G. (2017). Antifungal activity of crude extracts from brown and red seaweeds by a supercritical carbon dioxide technique against fruit postharvest fungal diseases. Postharvest Biology and Technology, 131: 16–30. https://doi.org/10.1016/j.postharvbio.2017.04.011.

De Miccolis Angelini, R.M., Raguseo, C., Rotolo, C., Gerin, D., Faretra, F. and Pollastro, S. (2022). The Mycovirome in a worldwide collection of the brown rot fungus Monilinia fructicola. Journal of Fungi, 8(5): 481. https://doi.org/10.3390/jof8050481.

Desai, S., Kumar, G., Amalraj, E., Rao, V. and Peter, A. (2016). Challenges in regulation and registration of biopesticides: An overview. 10.1007/978-81-322-2644-4_19.

De Silva, N., Lumyong, S., Hyde, K.D., Bulgakov, T., Phillips, A.J.L. and Yan, J.Y. (2016). Mycosphere Essays 9: Defining biotrophs and hemibiotrophs. Mycosphere, 7(5): 545–559. https://doi.org/10.5943/mycosphere/7/5/2.

De Silva, N.I., Brooks, S., Lumyong, S. and Hyde, K.D. (2019). Use of endophytes as biocontrol agents. Fungal Biol. Rev., 33(2): 133–148.DOI : 10.1016/j.fbr.2018.10.001.

Dhaliwal, G.S. and Arora, R. (2006). Integrated Pest Management (2nd ed.). Kalyani Publishers.

Dikhoba, P.M., Mongalo, N.I., Elgorashi, E.E. and Makhafola, T.J. (2019). Antifungal and antimycotoxigenic activity of selected South African medicinal plants species. Heliyon, 5(10): e02668. https://doi.org/10.1016/j.heliyon.2019.e02668.

Dobbs, E., Deakin, G., Bennett, J., Fleming-Archibald, C., Jones, I., Grogan, H. and Burton, K. (2021). Viral interactions and pathogenesis during multiple viral infections in Agaricus bisporus. mBio, 12(1): e03470-20. https://doi.org/10.1128/mBio.03470-20.

Dukare, A.S., Paul, S., Nambi, V.E., Gupta, R.K., Singh, R., Sharma, K. and Vishwakarma, R.K. (2019). Exploitation of microbial antagonists for the control of postharvest diseases of fruits: A review. Critical Reviews in Food Science and Nutrition, 59(9): 1498–1513. https://doi.org/10.1080/10408398.2017.1417235.

Eakjamnong, W., Keawsalong, N. and Dethoup, T. (2021). Novel ready-to-use dry powder formulation of Talaromyces tratensis KUFA0091 to control dirty panicle disease in rice. Biological Control, 152. https://doi.org/10.1016/j.biocontrol.2020.104454.

Elad, Y. (2001). Trichodex: Commercialization of *Trichoderma harzianum*. pp. 45–50. *In*: P. Jarvis (ed.). Agro Report, Biopesticides: Trends and Opportunities, T39 – A Case Study. PJB Publications.

El-Beltagi, H.S., Mohamed, H.I., Aldaej, M.I., Al-Khayri, J.M., Rezk, A.A., Al-Mssallem, M.Q., Sattar, M.N. and Ramadan, K.M.A.-Khayri, Jameel & Rezk, Adel & Al-Mssallem, Muneera & Ramadan, Khaled and Sattar, Naeem. (2022). Production and antioxidant activity of secondary metabolites in Hassawi rice (Oryza sativa L.) cell suspension under salicylic acid, yeast extract, and pectin elicitation. *In Vitro* Cellular and Developmental Biology – Plant. https://doi.org/10.1007/s11627-022-10264-x.

Elhelaly, S., Allam, A.Y. and El-Shennawy, M. (2022). Effect of some aqueous plant extracts on controlling orange, green mold disease and quality of "Washington" navel orange. Alexandria Science Exchange Journal, 43(1): 141–149. https://doi.org/10.21608/asejaiqjsae.2022.226129.

El Ouadi, Y., Manssouri, M., Bouyanzer, A., Majidi, L., Bendaif, H., Elmsellem, H., Shariati, M.A., Melhaoui, A. and Hammouti, B. (2017). Essential oil composition and antifungal activity of Melissa officinalis originating from north-Est Morocco, against postharvest phytopathogenic fungi in apples. Microbial Pathogenesis, 107: 321–326. https://doi. https://doi.org/10.1016/j.micpath.2017.04.004.

El-Shahir, A.A., El-Wakil, D.A., Abdel Latef, A.A.H. and Youssef, N.H. (2022). Bioactive compounds and antifungal activity of leaves and fruits methanolic extracts of Ziziphus spina-christi L. Plants, 11(6): 746. https://doi.org/10.3390/plants11060746.

Escobar-Puentes, A.A., Palomo, I., Rodríguez, L., Fuentes, E., Villegas-Ochoa, M.A., González-Aguilar, G.A., Olivas-Aguirre, F.J. and Wall-Medrano, A. (2022). Sweet potato (Ipomoea batatas L.) phenotypes: From agroindustry to health effects. Foods, 11(7): 1058. https://doi.org/10.3390/foods11071058.

Falade, M.J., Enikuomehin, O.A., Borisade, O.A. and Aluko, M. (2017). Control of Cowpea (*Vigna unguiculata* L. Walp.) diseases with intercropping of Maize (*Zea mays* L.) and spray of plant extracts. Journal of Advances in Microbiology, 7(4): 1–10. https://doi.org/10.9734/JAMB/2017/38156.

Feldmann, F., Jehle, J., Bradáčová, K. and Weinmann, M. (2022). Biostimulants, soil improvers, bioprotectants: Promoters of bio-intensification in plant production. Journal of Plant Diseases and Protection, 129(4): 707–713. https://doi.org/10.1007/s41348-022-00567-x.

Fernandez-Caballero, C., Romero, I., Goñi, O., Escribano, M.I., Merodio, C. and Sanchez-Ballesta, M.T. (2009). Characterization of an antifungal and cryoprotective class I chitinase from table grape berries (Vitis vinifera Cv. Cardinal). Journal of Agricultural and Food Chemistry, 57(19): 8893–8900. https://doi.org/10.1021/jf9016543.

Ferrigo, D., Mondin, M., Ladurner, E., Fiorentini, F., Causin, R. and Raiola, A. (2020). Effect of seed biopriming with *Trichoderma harzianum* strain INAT11 on Fusarium ear rot and Gibberella ear rot diseases. Biological Control, 147. https://doi.org/10.1016/j.biocontrol.2020.104286.

Floudas, D., Binder, M., Riley, R., Barry, K., Blanchette, R.A., Henrissat, B., Martínez, A.T., Otillar, R., Spatafora, J.W., Yadav, J.S., Aerts, A., Benoit, I., Boyd, A., Carlson, A., Copeland, A., Coutinho, P.M., de Vries, R.P., Ferreira, P., Findley, K., Foster, B., Gaskell, J., Glotzer, D., Górecki, P., Heitman, J., Hesse, C., Hori, C., Igarashi, K., Jurgens, J.A., Kallen, N., Kersten, P., Kohler, A., Kües, U., Kumar, T.K., Kuo, A., LaButti, K., Larrondo, L.F., Lindquist, E., Ling, A., Lombard, V., Lucas, S., Lundell, T., Martin, R., McLaughlin, D.J., Morgenstern, I., Morin, E., Murat, C., Nagy, L.G., Nolan, M., Ohm, R.A., Patyshakuliyeva, A., Rokas, A., Ruiz-Dueñas, F.J., Sabat, G., Salamov, A., Samejima, M., Schmutz, J., Slot, J.C., St John, F., Stenlid, J., Sun, H., Sun, S., Syed, K., Tsang, A., Wiebenga, A., Young, D., Pisabarro, A., Eastwood, DC., Martin, F., Cullen, D., Grigoriev, I.V. and Hibbett, D.S. (2012). The Paleozoic origin of enzymatic lignin decomposition reconstructed from 31 fungal genomes. Science, 336(6089): 1715–1719. https://doi.org/10.1126/science.1221748.

Frascella, A., Sarrocco, S., Mello, A., Venice, F., Salvatici, C., Danti, R., Emiliani, G., Barberini, S. and Della Rocca, G. (2022). Biocontrol of Phytophthora xcambivora on Castanea sativa: Selection of Local Trichoderma spp. Isolates for the management of ink Disease. Forests, 13(7): 1065. https://doi.org/10.3390/f13071065.

Freimoser, F.M., Rueda-Mejia, M.P., Tilocca, B. and Migheli, Q. (2019). Biocontrol yeasts: Mechanisms and applications. World Journal of Microbiology and Biotechnology, 35(10): 154. https://doi.org/10.1007/s11274-019-2728-4.

Fu, S.F., Sun, P.F., Lu, H.Y., Wei, J.Y., Xiao, H.S., Fang, W.T., Cheng, B.Y. and Chou, J.Y. (2016). Plant growth-promoting traits of yeasts isolated from the phyllosphere and rhizosphere of Drosera spatulata Lab. Fungal Biology, 120(3): 433–448. https://doi.org/10.1016/j.funbio.2015.12.006.

Furmanek, Ł., Czarnota, P. and Seaward, M.R.D. (2021). The effect of lichen secondary metabolites on Aspergillus fungi. Archives of Microbiology, 204(1): 100. https://doi.org/10.1007/s00203-021-02649-0.

Gachango, E., Kirk, W., Schafer, R. and Wharton, P. (2012). Evaluation and comparison of biocontrol and conventional fungicides for control of postharvest potato tuber diseases. Biological Control, 63(2): 115–120. https://doi.org/10.1016/j.biocontrol.2012.07.005.

Gajera, H.P., Hirpara, D.G., Savaliya, D.D. and Golakiya, B.A. (2020). Extracellular metabolomics of Trichoderma biocontroller for antifungal action to restrain Rhizoctonia solani Kuhn in cotton. Physiological and Molecular Plant Pathology, 112: 101547. https://doi.org/10.1016/j. pmpp.101547.

Gama, DdS., Santos, Í.A.F.M., De Abreu, L. Md, Medeiros, F.H. Vd, Duarte, W.F. and Cardoso, P.G. (2020). Endophytic fungi from Brachiaria grasses in Brazil and preliminary screening of Sclerotinia sclerotiorum antagonists. Scientia Agricola, 77(3). https://doi.org/10.1590/1678-992x-2018-0210.

Ghadamgahi, F., Tarighi, S., Taheri, P., Saripella, G.V., Anzalone, A., Kalyandurg, P.B., Catara, V., Ortiz, R. and Vetukuri, R.R. (2022). Plant growth-promoting activity of Pseudomonas aeruginosa FG106 and its ability to act as a biocontrol agent against potato, tomato and taro pathogens. Biology, 11(1). https://doi.org/10.3390/biology11010140, PubMed: 35053136, PubMed Central: PMC8773043.

Gholami, M., Amini, J., Abdollahzadeh, J. and Ashengroph, M. (2019). Basidiomycetes fungi as biocontrol agents against take all disease of wheat. Biological Control, 130: 34–43. https://doi. https://doi.org/10.1016/j.biocontrol.2018.12.012.

Glare, T., Gwynn, R. and Moran-Diez, M. (2016). Development of biopesticides and future opportunities. 10.1007/978-1-4939-6367-6_16.

Gore-Lloyd, D., Sumann, I., Brachmann, A.O., Schneeberger, K., Ortiz-Merino, R.A., Moreno-Beltrán, M., Schläfli, M., Kirner, P., Santos Kron, A., Rueda-Mejia, M.P., Somerville, V., Wolfe, K.H., Piel, J., Ahrens, C.H., Henk, D. and Freimoser, F.M. (2019). Snf2 controls pulcherriminic acid biosynthesis and antifungal activity of the biocontrol yeast Metschnikowia pulcherrima. Molecular Microbiology, 112(1): 317–332. https://doi.org/10.1111/mmi.14272.

Guha, P., Ghosh, M. and Radhanath, G.K. (2005). Antifungal activity of the crude extracts and extracted phenols from gametophytes and sporophytes of two species of Adiantum. Taiwania, 50: 272–283. https://doi.org/10.6165/tai.2005.50(4).272.

Gupta, P.K. (2018). Toxicity of fungicides. pp. 569–580. *In*: R.C. Gupta (ed.). https://doi, Veterinary Toxicology: Basic and Clinical Principles. https://doi.org/10.1016/B978-0-12-811410-0.00045-3. Academic Press.

Gupta, S. and Dikshit, A.K. (2010). Biopesticides: An ecofriendly approach for pest control. Journal of Biopesticides, 3(1): 186–188.

Guo, S.Y., Liu, W.X., Han, L.F. and Chen, J.Z. (2017). Antifungal activity of lichen extracts and usnic acid for controlling the saprolegniasis. International Journal of Environmental and Agriculture Research, 3(5): 43–47. https://doi.org/10.25125/agriculture-journal-IJOEAR-APR-2017-35.

Gyasi, E., Kotey, D.A., Adongo, B.A., Adams, F.K., Owusu, E.O. and Mohammed, A. (2022). Management of Major seed-borne fungi of Cowpea (*Vigna unguiculata* (L.) Walp) with four selected botanical extracts. Advances in Agriculture, 2022: 1–8. https://doi.org/10.1155/2022/3125240.

Haider, E.K., Muhammad, A.M., Ashraf, W., Yaseen, S., Bilal, M., Akbar, U., Fatima, I. and Shahid, T. (2020). Efficacy of phytoextracts against fungal disease of vegetable. Journal of Biodiversity and Environmental Sciences, 16,71 82.

Halama, P. and Van Haluwin, C. (2004). Antifungal activity of lichen extracts and lichenic acids. BioControl, 49(1): 95–107. https://doi.org/10.1023/B:BICO.0000009378.31023.ba.

Hang, X., Meng, L., Ou, Y., Shao, C., Xiong, W., Zhang, N., Liu, H., Li, R., Shen, Q. and Kowalchuk, G.A. (2022). *Trichoderma*-amended biofertilizer stimulates soil resident *Aspergillus* population for joint plant growth promotion. npj Biofilms and Microbiomes, 8(1): 57. https://doi.org/10.1038/s41522-022-00321-z.

Hashem, M., Mostafa, Y.S., Alamri, S., Abbas, A.M. and Eid, E.M. (2021). Exploitation of agro-industrial residues for the formulation of a new active and cost effective biofungicide to control the root rot of vegetable crops. Sustainability, 13(16): 9254. https://doi.org/10.3390/su13169254.

Hassan, Y.I., Trofimova, D., Samuleev, P., Miah, M.F. and Zhou, T. (2018). Chapter 16. Omics approaches in enzyme discovery and engineering. pp. 297–322. *In*: D. Barh and V. Azevedo (eds.). Omics Technologies and Bio-engineering, 2. Elsevier.

Hegedüs, N. and Marx, F. (2013). Antifungal proteins: More than antimicrobials? Fungal Biology Reviews, 26(4): 132–145. https://doi.org/10.1016/j.fbr.2012.07.002.

Heo, A.Y., Koo, Y.M. and Choi, H.W. (2022). Biological control activity of plant growth promoting rhizobacteria Burkholderia contaminans AY001 against tomato fusarium wilt and bacterial speck diseases. Biology, 11(4): 619. https://doi.org/10.3390/biology11040619.

Hersanti, D., Djaya, L., Ilahiyyat, I. and Ruhyaman, R.R. (2020). Ability of *Trichoderma harzianum* in carbon fiber and silica nano particles formulation to control *Fusarium oxysporum in vitro*. IOP Conference Series: Earth and Environmental Science, 458(1). https://doi.org/10.1088/1755-1315/458/1/012016.

Hilber-Bodmer, M., Schmid, M., Ahrens, C.H. and Freimoser, F.M. (2017). Competition assays and physiological experiments of soil and phyllosphere yeasts identify *Candida subhashii* as a novel antagonist of filamentous fungi. BMC Microbiology, 17(1): 4. https://doi.org/10.1186/s12866-016-0908-z.

Hu, Z., Guo, J., Da Gao, B. and Zhong, J. (2020). A novel mycovirus isolated from the plant-pathogenic fungus Alternaria dianthicola. Archives of Virology, 165(9): 2105–2109. https://doi.org/10.1007/s00705-020-04700-9.

Iqbal, A., Irshad, G., Naz, F., Ghuffar, S., Khursheed, R., Bashir, A., Mustafa, M.Z., Mehmood, A. and Hassan, N.U. (2020). Characterization of Fusarium solani, causal agent of pea wilt and its bio-management. Pakistan Journal of Phytopathology, 32(2): 265–271. https://doi.org/10.33866/10.33866/phytopathol.030.02.0534.

Isabelle, B., Philippe, P., Harry, O.L., Lucienne, D. and Gladys, L.M. (2012). Insecticidal and antifungal chemicals produced by plants: A review. Environmental Chemistry Letters. Springer Verlag, 10(4): 325–347. ff10.1007/s10311-012-0359-1ff. ffhal-01767269f.

Itelima, J.U., Bang, W.J. and Onyimba, I.A. (2018). A review: biofertilizer; a key player in enhancing soil fertility and crop productivity. J. Microbiol. Oj, E. Biotechnology Reports, 2(1): 22–28.

Jaber, L.R. and Alananbeh, K.M. (2018). Fungal entomopathogens as endophytes reduce several species of Fusarium causing crown and root rot in sweet pepper (*Capsicum annuum* L.). Biological Control, 126: 117–126. https://doi.org/10.1016/j.biocontrol.2018.08.007.

Jakubiec-Krzesniak, K., Rajnisz-Mateusiak, A., Guspiel, A., Ziemska, J. and Solecka, J. (2018). Secondary metabolites of actinomycetes and their antibacterial, antifungal and antiviral properties. Polish Journal of Microbiology. Polish Society of Microbiologists, 67(3): 259–272. https://doi.org/10.21307/pjm-2018-048.

Jethva, K.D., Bhatt, D.R. and Zaveri, M.N. (2020). Antimycobacterial screening of selected medicinal plants against Mycobacterium tuberculosis H37Rv using agar dilution method and the microplate resazurin assay. International Journal of Mycobacteriology, 9(2): 150–155. https://doi.org/10.4103/ijmy.ijmy_60_20.

Joseph, B., Dar, M.A. and Kumar, V. (2008). Bioefficacy of plant extracts to control Fusarium solanif. sp. melongenae incitant of brinjal wilt. Global Journal of Biotechnology and Biochemistry, 3: 56–59.

Joshi, S. and Sati, S.C. (2012). Antifungal potential of gymnosperms: A review. 10.13140/RG.2.1.3408.0084.

Kavitha, K., Vijaya, N., Krishnaveni, A., Arthanareeswari, M., Rajendran, S., AlHashem, A. and Subramania, A. (2020). Nanomaterials for antifungal applications. Nanotoxicity, 385–398.

Keeffe, E.O., Hughes, H., McLoughlin, P., Tan, S.P. and Carthy, N.Mc. (2019). Methods of analysis for the *in vitro* and *in vivo* determination of the fungicidal activity of seaweeds: A mini review. Journal of Applied Phycology, Kellogg. Jeonju University,.

Kellogg, J.J. and Raja, H.A. (2017). Endolichenic fungi: A new source of rich bioactive secondary metabolites on the horizon. Phytochemistry Reviews, 16(2): 271–293. https://doi.org/10.1007/s11101-016-9473-1.

Kiliç, A. and Akay, M.T. (2008). A three generation study with genetically modified Bt corn in rats: Biochemical and histopathological investigation. Food and Chemical Toxicology, 46(3): 1164–1170. http://www.ncbi.cim.nih.gov/pubmed/18191319.Accessed Retrieved July 24 2014. https://doi.org/10.1016/j.fct.2007.11.016.

Köhl, J., Postma, J., Nicot, P., Ruocco, M. and Blum, B. (2011). Stepwise screening of microorganisms for commercial use in biological control of plant-pathogenic fungi and bacteria. Biological Control, 57(1): 1–12. https://doi.org/10.1016/j.biocontrol.2010.12.004.

Kohl, J., Kolnaar, R. and Ravensberg, Willem, J. (2019). Mode of action of microbial biological control agents against plant diseases: Relevance beyond efficacy. Frontiers in Plant Science, 10. DOI=10.3389/fpls.2019.00845.

Kordali, S., Usanmaz, A., Cakir, A., Komaki, A. and Ercisli, S. (2016). Antifungal and herbicidal effects of fruit essential oils of four *Myrtus communis* genotypes. Chemistry and Biodiversity, 13(1): 77–84. https://doi.org/10.1002/cbdv.201500018.

Koundal, K.R. and Rajendran, P. (2003). Plant insecticidal proteins and their potential for developing transgenic resistant to insect pests. Indian Journal of Biotechnology, 2: 110–120.

Kowalska, J., Tyburski, J., Krzymińska, J. and Jakubowska, M. (2020). Cinnamon powder: An *in vitro* and *in vivo* evaluation of antifungal and plant growth promoting activity. European Journal of Plant Pathology, 156(1): 237–243. https://doi.org/10.1007/s10658-019-01882-0.

Kubheka, S.F., Tesfay, S.Z., Mditshwa, A. and Magwaza, L.S. (2020). Evaluating the efficacy of edible coatings incorporated with Moringa leaf extract on postharvest of "Maluma" avocado fruit quality and its biofungicidal effect. HortScience, 55(4): 410–415. https://doi.org/10.21273/HORTSCI14391-19.

Kumar, S. (2012). Biopesticides: A need for food and environmental safety. Journal of Biofertilizer and Biopesticide, 3: 107–113.

Kumar, S. and Kaushik, N. (2012). Metabolites of endophytic fungi as novel source of biofungicide: A review. Phytochemistry Reviews, 11(4): 507–522. https://doi.org/10.1007/s11101-013-9271-y.

Kumar, S. and Singh, A. (2014). Biopesticides for integrated crop management: Environmental and regulatory aspects. Journal of Biofertilizer and Biopesticide, 5: 121–129, 6202, 1000e121. http://doi.org/10.4172155.

Kumar, D., Arya, S.K., Shamim, M., Srivastava, D., Tyagi, M. and Singh, K.N. (2021a). *In-vitro* and *in-vivo* activity of essential oils against postharvest pathogen *Colletotrichum musae* of banana. J. of Postharvest and Technology, 09: 53–63.

Kumar, M., Charishma, K., Sahu, K.P., Sheoran, N., Patel, A., Kundu, A. and Kumar, A. (2021b). Rice leaf associated Chryseobacterium species: An untapped antagonistic favobacterium displays volatile mediated suppression of rice blast disease. Biological Control, 161: 2021.104703. https://doi.org/10.1016/j.biocontrol.2021.104703.

Lahlali, R., Ezrari, S., Radouane, N., Kenfaoui, J., Esmaeel, Q., El Hamss, H., Belabess, Z. and Barka, E.A. (2022). Biological control of plant pathogens: A global perspective. Microorganisms, 10(3): 596. https://doi.org/10.3390/microorganisms10030596.

Lahlou, Y., Rhandour, Z., Amraoui, B.E. and Bamhaoud, T. (2019). Screening of antioxidant activity and the total polyphenolic contents of six medicinal Moroccan's plants extracts. Journal of Materials and Environmental Science, 10: 133–1348.

Latinovic, N., Sabovljević, M.S., Vujicic, M., Latinovic, J. and Sabovljevic, A. (2019). Bryophyte extracts suppress growth of the plant pathogenic fungus Botrytis cinerea. Botanica Serbica, 43(1): 9–12. https://doi.org/10.2298/BOTSERB1901009L.

Lattanzio, V., Lattanzio, V., Cardinali, A. and Imperato, F. (2006). Role of phenolics in the resistance mechanisms of plants against fungal pathogens and insects. Phytochemistry, 661: 23–67.

Law, J.W., Ser, H., Khan, T.M., Chuah, L., Pusparajah, P., Chan, K., Goh, B. and Lee, L. (2017). The potential of Streptomyces as biocontrol agents against the rice blast fungus, *Magnaporthe oryzae* (*Pyricularia oryzae*) Front Microbiol., 8(3): 1–10. DOI=10.3389/fmicb.2017.00003.

Leiter, É., Gáll, T., Csernoch, L. and Pócsi, I. (2017). Biofungicide utilizations of antifungal proteins of filamentous ascomycetes: Current and foreseeable future developments. BioControl, 62(2): 125–138. https://doi.org/10.1007/s10526-016-9781-9.

Leontiev, R., Hohaus, N., Jacob, C., Gruhlke, M.C.H. and Slusarenko, A.J. (2018). A comparison of the antibacterial and antifungal activities of thiosulfinate analogues of allicin. Scientific Reports, 8(1): 6763. https://doi.org/10.1038/s41598-018-25154-9, PubMed: 29712980, PubMed Central: PMC5928221.

Liu, M., Gong, Y., Sun, H., Zhang, J., Zhang, L., Sun, J., Han, Y., Huang, J., Wu, Q., Zhang, C. and Li, Z. (2020). Characterization of a Novel chitinase from sweet potato and its fungicidal effect against *Ceratocystis fmbriata*. Journal of Agricultural and Food Chemistry, 68(29): 7591–7600. https://doi.org/10.1021/acs.jafc.0c01813.

Locatelli, G.O., dos Santos, G.F., Botelho, P.S., Finkler, C.L.L. and Bueno, L.A. (2018). Development of *Trichoderma* sp. formulations in encapsulated granules (CG) and evaluation of conidia shelf-life. Biological Control, 117: 21–29. https://doi.org/10.1016/j.biocontrol.2017.08.020.

Loi, M., Paciolla, C., Logrieco, A. and Mule, G. (2020). Plant bioactive compounds in pre- and post-harvest management for aflatoxins reduction. Frontiers in Microbiology, 11.

Lorán, S., Carramiñana, J.J., Juan, T., Ariño, A. and Herrera, M. (2022). Inhibition of *Aspergillus parasiticus* growth and aflatoxins production by natural essential oils and phenolic acids. Toxins, 14(6): 384. https://doi.org/10.3390/toxins14060384.

Luh Suriani, N., Ngurah Suprapta, D., Nazir, N., Made Susun Parwanayoni N., Agung Ketut Darmadi, A., Andya Dewi, D., Sudatri, N.W., Fudholi, A., Sayyed, R.Z., Syed, A., Elgorban, A.M., Bahkali, A.H., El Enshasy, H.A. and Dailin, D.J. (2020). A mixture of piper leaves extracts and rhizobacteria for sustainable plant growth promotion and bio-control of blast pathogen of organic bali rice. Sustainability, 12: 8490. https://doi.org/10.3390/su12208490.

Macena, A.M.F., Kobori, N.N., Mascarin, G.M., Vida, J.B. and Hartman, G.L. (2020). Antagonism of Trichoderma-based biofungicides against Brazilian and North American isolates of Sclerotinia sclerotiorum and growth promotion of soybean. BioControl, 65(2): 235–246. https://doi.org/10.1007/s10526-019-09976-8.

Madbouly, A.K. and Abdelbacki, A.M. (2017). Biocontrol of certain soilborne diseases and promotion of growth of Capsicum annuum using biofungicides. Pakistan Journal of Botany, 49: 371–378.

Madbouly, A.K. and Abdel-Wareth, M.T. (2020). The use of Chaetomium taxa as biocontrol agents. *In* Recent Developments on Genus Chaetomium; Springer: Cham, Switzerland, pp. 251–266.

Mahmud, A.A., Upadhyay, S.K., Srivastava, A.K. and Bhojiya, A.A. (2021). Biofertilizers: A Nexus between soil fertility and crop productivity under abiotic stress. Current Research in Environmental Sustainability, 3: 1–14. https://doi.org/10.1016/j.crsust.2021.100063.

Maia, N., da, C., Souza, P.N. et al. 2018. Fungal endophytes of Panicum maximum and Pennisetum purpureum: Isolation, identification, and determination of antifungal potential. Revista Brasileira de Zootecnia, 47. https://doi.org/10.1590/rbz4720170183.

Mandal, S.M., Chakraborty, D. and Dey, S. (2010). Phenolic acids act as signaling molecules in plant-microbe symbioses. Plant Signaling and Behavior Apr., 5(4): 359–368. https://doi.org/10.4161/psb.5.4.10871. Epub April 7 2010. PubMed: 20400851, PubMed Central: PMC2958585.

Mannai, S., Benfradj, N., Karoui, A., Salem, I.B., Fathallah, A., M'Hamdi, M. and Boughalleb-M'Hamdi, N. (2021). Analysis of chemical composition and *in vitro* and *in vivo* antifungal activity of Raphanus raphanistrum extracts against Fusarium and Pythiaceae, affecting apple and peach seedlings. Molecules, 26(9): 2479. https://doi.org/10.3390/molecules26092479.

Martínez-Romero, D., Serrano, M., Bailén, G., Guillén, F., Zapata, P.J., Valverde, J.M., Castillo, S., Fuentes, M. and Valero, D. (2008). The use of a natural fungicide as an alternative to preharvest synthetic fungicide treatments to control lettuce deterioration during postharvest storage. Postharvest Biology and Technology, 47(1): 54–60. https://doi.org/10.1016/j.postharvbio.2007.05.020.

Marx, F., Binder, U., Leiter, E. and Pócsi, I. (2008). The Penicillium chrysogenum antifungal protein PAF, a promising tool for the development of new antifungal therapies and fungal cell biology studies. Cellular and Molecular Life Sciences, 65(3): 445–454. https://doi.org/10.1007/s00018-007-7364-8.

Meher, J., Rajput, R.S., Bajpai, R., Teli, B. and Sarma, B.K. (2020). Trichoderma: A globally dominant commercial Biofungicide. *In*: C. Manoharachary, H.B. Singh and A. Varma (eds.). *Trichoderma*: Agricultural Applications and Beyond. Soil Biology, 61. https://doi.org/10.1007/978-3-030-54758-5_9. Springer.

Meepagala, K.M., Clausen, B.M., Johnson, R.D., Wedge, D.E. and Duke, S.O. (2019). A phytotoxic and antifungal metabolite (pyrichalasin H) from a fungus infecting Brachiaria eruciformis (signal grass). Journal of Agricultural Chemistry and Environment, 8: 115–128. https://doi.org/10.4236/jacen.2019.83010.

Mellon, J.E., Dowd, M.K., Beltz, S.B. and Moore, G.G. (2014). Growth inhibitory effects of gossypol and related compounds on fungal cotton root pathogens. Letters in Applied Microbiology, 59(2): 161–168. https://doi.org/10.1111/lam.12262.

Miljaković, D., Marinković, J. and Balešević-Tubić, S. (2020). The significance of Bacillus spp. in disease suppression and growth promotion of field and vegetable crops. Microorganisms, 8(7): 1037. https://doi.org/10.3390/microorganisms8071037.

Mishra, J., Dutta, V. and Arora, N.K. (2020). Biopesticides in India: Technology and sustainability linkages. 3 Biotech. May, 10(5): 210. https://doi.org/10.1007/s13205-020-02192-7. Epub April 24 2020. PubMed: 32351868, PubMed Central: PMC7181464

Morsy, N., Al Sherif, E.A. and Abdel-rassol, T.M.A. (2016). Phytochemical analysis of Calotropis procera with antimicrobial activity investigation. Main Group Chemistry, 15(3): 267–273. https://doi.org/10.3233/MGC-160206.

Mukesh, S., Vipul, K., Mohammad, S., Sonika, P. and Anuradha, S. (2016). Trichoderma—A potential and effective bio fungicide and alternative source against notable phytopathogens: A review. African Journal of Agricultural Research, 11(5): 310–316. https://doi.org/10.5897/AJAR2015.9568.

Myo, E.M., Liu, B., Ma, J., Shi, L., Jiang, M., Zhang, K. and Ge, B. (2019). Evaluation of *Bacillus velezensis* NKG-2 for bio-control activities against fungal diseases and potential plant growth promotion. Biological Control, 134: 23–31. https://doi.org/10.1016/j.biocontrol.2019.03.017.

Naboulsi, I., Aboulmouhajir, A., Kouisni, L., Bekkaoui, F. and Yasri, A. (2018). Plants extracts and secondary metabolites, their extraction methods and use in agriculture for controlling crop stresses and improving productivity: A review. Acad. Journal of Medicinal Plants, 6(8): 223–240. DOI: 10.15413/ajmp.2018.0139.

Nagpal, S., Sharma, P., Sirari, A. and Gupta, R.K. (2020). Coordination of *Mesorhizobium* sp. and endophytic bacteria as elicitor of biocontrol against Fusarium wilt in chickpea. European Journal of Plant Pathology, 158(1): 143–161. https://doi.org/10.1007/s10658-020-02062-1.

Nandhini, M., Rajini, S.B., Udayashankar, A.C., Niranjana, S.R., Lund, O.S., Shetty, H.S. and Prakash, H.S. (2018). Diversity, plant growth promoting and downy mildew disease suppression potential of cultivable endophytic fungal communities associated with pearl millet. Biological Control, 127: 127–138. https://doi.org/10.1016/j.biocontrol.2018.08.019.

Narayanan, V.S., Muddaiah, S., Shashidara, R., Sudheendra, U.S., Deepthi, N.C. and Samaranayake, L. (2020). Variable antifungal activity of curcumin against planktonic and biofilm phase of different candida species. Indian Journal of Dental Research, 31(1): 145–148. https://doi.org/10.4103/ijdr.IJDR_521_17.

Naveenkumar, R., Muthukumar, A., Sangeetha, G. and Mohanapriya, R. (2017). Developing eco-friendly biofungicide for the management of major seed borne diseases of rice and assessing their physical stability and storage life. Comptes Rendus Biologies, 340(4): 214–225. https://doi.org/10.1016/j.crvi.2017.03.001.

Nerilo, S.B., Romoli, J.C.Z., Nakasugi, L.P., Zampieri, N.S., Mossini, S.A.G., Rocha, G.H.O., Gloria, E. Md, Abreu Filho, B. Ad and Machinski, Jr., M. (2020). Antifungal activity and inhibition of aflatoxins production by *Zingiber officinale* Roscoe essential oil against *Aspergillus flavus* in stored maize grains. Ciência Rural, 50(6): e20190779, 11. https://doi.org/10.1590/0103-8478cr20190779.

Nian, J., Yu, M., Bradley, C.A. and Zhao, Y. (2021). Lysobacter enzymogenes strain C3 suppresses mycelium growth and spore germination of eight soybean fungal and oomycete pathogens and decreases disease incidences. Biological Control, 152. https://doi.org/10.1016/j.biocontrol.2020.104424.

Noutoshi, Y., Ikeda, M., Saito, T., Osada, H. and Shirasu, K. (2012). Sulfonamides identified as plant immune-priming compounds in high-throughput chemical screening increase disease resistance in Arabidopsis thaliana. Frontiers in Plant Science, 3: 245. https://doi.org/10.3389/fpls.2012.00245.

Ojha, S., Goyal, M. and Chittoriya, D. (2021). Efficacy of Ocimum sanctum against Alternaria causing diseases on medicinal plants. International Journal of Scientific Research in Biological Sciences, 8: 60–64.

Oladejo, O. and Imani, J. (2022). Inhibitory effect of CUSTOS, a formulated allium-based extract, on the growth of some selected plant pathogens. International Journal of Plant Biology, 13(2): 44–54. https://doi.org/10.3390/ijpb13020006.

Onaran, A. (2016). Antifungal activity of some plant extracts against different plant pathogenic fungi. International Journal of Advances in Agricultural and Environmental Engineering 3. Sağlam, H.D., Materials, A.P., 284–287.

Ons, L., Bylemans, D., Thevissen, K. and Cammue, B.P.A. (2020). Combining biocontrol agents with chemical fungicides for integrated plant fungal disease control. Microorganisms, 8(12): 1930. https://doi.org/10.3390/microorganisms8121930.

Oro, L., Ciani, M. and Comitini, F. (2014). Antimicrobial activity of *Metschnikowia pulcherrima* on wine yeasts. Journal of Applied Microbiology, 116(5): 1209–1217. https://doi.org/10.1111/jam.12446.

Otoguro, M. and Suzuki, S. (2018). Status and future of disease protection and grape berry quality alteration by microorganisms in viticulture. Letters in Applied Microbiology, 67(2): 106–112. https://doi.org/10.1111/lam.13033.

Panpatte, D.G., Jhala, Y.K., Shelat, H.N. and Vyas, R.V. (2016). Pseudomonas fluorescens: A promising biocontrol agent and PGPR for sustainable agriculture. In Microbial inoculants in sustainable agricultural productivity, 1. Research, Australia.

Pavithra, G., Bindal, S., Rana, M. and Srivastava, S. (2020). 54.62. Role of endophytic microbes against plant pathogens: A review. Asian Journal of Plant Sciences, 19: 54–62. https://doi.org/10.3923/ajps.

Perina, F.J., de Andrade, C.C.L., Moreira, S.I., Nery, E.M., Ogoshi, C. and Alves, E. (2019). Cinnamomun zeylanicum oil and trans-cinnamaldehyde against Alternaria brown spot in tangerine: Direct effects and induced resistance. Phytoparasitica, 47(4): 575–589. https://doi.org/10.1007/s12600-019-00754-x.

Perspectives. (2016). D.P. Singh, H.B. Singh and R. Prabha (eds.). Springer, pp. 257–270.

Pirttila, A.M., Mohammad, P.T.H., Baruah, N. and Koskimaki, J.J. (2021). Biofertilizers and biocontrol agents for agriculture: How to identify and develop new potent microbial strains and traits. Microorganisms, 13: 9(4): 817. https://doi.org/10.3390/microorganisms9040817.

Pizzolitto, R.P., Jacquat, A.G., Usseglio, V.L., Achimón, F., Cuello, A.E., Zygadlo, J.A. and Dambolena, J.S. (2020). Quantitative-structure activity relationship study to predict the antifungal activity of essential oils against Fusarium verticillioides. Food Control, 108: 106836. https://doi.org/10.1016/j.foodcont.2019.106836.

Pour, M.M., Saberi-Riseh, R., Mohammadinejad, R. and Hosseini, A. (2019). Nano-encapsulation of plant growth-promoting rhizobacteria and their metabolites using alginate-silica nanoparticles and carbon nanotube improves UCB1 pistachio micropropagation. Journal of Microbiology and Biotechnology, 29(7): 1096–1103. https://doi.org/10.4014/jmb.1903.03022.

Raf, H., Dawar, S. and Zaki, M.J. (2015). Seed priming with extracts of Acacia nilotica (L.) Willd. ex Delile and Sapindus mukorossi (l) plant parts in the control of root rot fungi and growth of plants. Pakistan Journal of Botany, 47: 1129–1135.

Ram, R.M., Keswani, C., Bisen, K. et al. (2018). Biocontrol technology: Eco-friendly approaches for sustainable agriculture. *In*: D. Barh and V. Azevedo (eds.). Omics Technologies and Bioengineering, 2: Towards Improving Quality of Life. Elsevier, Inc.

Ramón, Romero Gomez, S., Feregrino-Perez, A., Rodriguez-Guadarrama, Hugo and Guevara-Gonzalez. (2018). Antifungal activity of mexican endemic plants on agricultural phytopathogens: A review, 1–11. https://doi.org/10.1109/CONIIN.2018.8489793.

Raveau, R., Fontaine, J. and Lounès-Hadj Sahraoui, A. (2020). Essential oils as potential alternative biocontrol products against plant pathogens and weeds: A review. Foods, 9(3). https://doi.org/10.3390/foods9030365.

Raymaekers, K., Ponet, L., Holtappels, D., Berckmans, B. and Cammue, B.P.A. (2020). Screening for novel biocontrol agents applicable in plant disease management—A review. Biological Control, 144: 104240. https://doi.org/10.1016/j.biocontrol.2020.104240.

Roberti, R., Di francesco, A., Innocenti, G. and Mari, M. (2019). Potential for biocontrol of Pleurotus ostreatus green mould disease by Aureobasidium pullulans De Bary (Arnaud). Biological Control, 135. 10.1016/j.biocontrol.2019.04.016.

Rodriguez-Guadarrama, H., Guevara-Gonzalez, R., Romero Gomez, S. and Feregrino-Perez, A. (2018). Antifungal activity of Mexican endemic plants on agricultural phytopathogens: A review, 1–11. 10.1109/CONIIN.2018.8489793.

Rojas, E.C., Birgit, J., Hans, J.L.J., Meike, A.C.L., Pilar, E., Yuwei, D. and David, B.C. (2020). Selection of fungal endophytes with biocontrol potential against Fusarium head blight in wheat. Biological Control, 144. https://doi.org/10.1016/j.biocontrol.2020.104222.

Rusita, I. and Sasongko, H. (2020). Effectivity of Trichoderma harzianum as bio-fungicide against moler disease and bio-stimulator of shallot growth. Journal of Agri-Food Science and Technology, 1: 12. 10.12928/jafost.v1i1.1941.

Sabithira, G. and Udayakumar, R. (2018). Antibacterial and antifungal activities of leaf and stem of Marsilea minuta L. against selected microbial pathogens. Journal of Applied Biology and Biotechnology, 6, 60612: 71–78. https://doi.org/10.7324/JABB.2018.

Sachdev, S., Singh, A. and Singh, R.P. (2018). Optimization of culture conditions for mass production and bio-formulation of Trichoderma using response surface methodology. 3 Biotech, 8(8): 360. https://doi.org/10.1007/s13205-018-1360-6.

Saito, H., Sasaki, M., Nonaka, Y., Tanaka, J., Tokunaga, T., Kato, A., Thuy, T.T.T., Vang, L.V., Tuong, L.M., Kanematsu, S., Suzuki, T., Kurauchi, K., Fujita, N., Teraoka, T., Komatsu, K. and Arie, T. (2021). Spray application of nonpathogenic fusaria onto rice flowers controls bakanae disease (caused by Fusarium fujikuroi) in the next plant generation. Applied and Environment Microbiology Jan, 4, 87(2): e01959-20. https://doi.org/10.1128/AEM.01959-20, PubMed: 33158893, PubMed Central: PMC7783350.

Salma, M. and Jogen, C.K. (2011). A review on the use of biopesticides in insect pest management. International Journal of Advanced Science and Technology, 1: 169–178.

Santos, G.A., Silva, R., Moreira, S., Vicentini, S. and Ceresini, P. (2021). Biofungicides: An eco-friendly approach for plant. Disease Management. https://doi.org/10.1016/B978-0-12-819990-9.00036-6.

Saravanakumar, D., Ciavorella, A., Spadaro, D., Garibaldi, A. and Gullino, M.L. (2008). Metschnikowia pulcherrima strain MACH1 outcompetes Botrytis cinerea, Alternaria alternata and Penicillium expansum in apples through iron depletion. Postharvest Biology and Technology, 49(1): 121–128. https://doi.org/10.1016/j.postharvbio.2007.11.006.

Sarfraz, M., Nasim, M.J., Jacob, C. and Gruhlke, M.C.H. (2020). Efficacy of allicin against plant pathogenic fungi and unveiling the underlying mode of action employing yeast based chemogenetic profiling approach. Applied Sciences, 10(7): 2563. https://doi.org/10.3390/app10072563.

Sarkar, G., Jatar, N., Goswami, P., Cyriac, R., Suthindhiran, K. and Jayasri, M.A. (2018). Combination of diferent marine algal extracts as biostimulant and biofungicides. Journal of Plant Nutrition, 41(9): 1163–1171. https://doi.org/10.1080/01904167.2018.1434201.

Saputri, D.D. and Utami, A.W.A. (2020). The potency of purple sweet potato (Ipomoea batatas) leaf extract as biofungicide for controlling Fusarium fruit rot on chili. Journal of Agriculture and Applied Biology, 1(1): 1–8. https://doi.org/10.11594/jaab.01.01.01.

Sartori, M., Bonacci, M., Barra, P., Fessia, A., Etcheverry, M., Nesci, A. and Barros, G. (2020). Studies on possible modes of action and tolerance to environmental stress conditions of different biocontrol agents of foliar diseases in Maize. Agricultural Sciences, 11(6): 552–566. https://doi.org/10.4236/as.2020.116035.

Schmid, B., Coelho, L., Schulze, P.S.C., Pereira, H., Santos, T., Maia, I.B., Reis, M. and Varela, J. (2022) Antifungal properties of aqueous microalgal extracts. Bioresource Technology Reports. Peter and Pereira, 18: 101096. https://doi.org/10.1016/j.biteb.2022.101096.

Seepe, H.A., Amoo, S.O., Nxumalo, W. and Adeleke, R.A. (2019). Antifungal activity of medicinal plant extracts for potential management of Fusarium pathogens. Research on Crops, 20: 399–406.

Seepe, H.A., Lodama, K.E., Sutherland, R., Nxumalo, W. and Amoo, S.O. (2020). *In vivo* antifungal activity of South African medicinal plant extracts against Fusarium pathogens and their phytotoxicity evaluation. Plants, 9(12): 1668. https://doi.org/10.3390/plants9121668.

Seepe, H.A., Nxumalo, W. and Amoo, S.O. (2021a). Natural products from medicinal plants against phytopathogenic Fusarium species: Current research endeavours, challenges and prospects. Molecules, 26(21): 6539. https://doi.org/10.3390/molecules26216539.

Seepe, H.A., Ramakadi, T.G., Lebepe, C.M., Amoo, S.O. and Nxumalo, W. (2021b). Antifungal activity of isolated compounds from the leaves of Combretum Erythrophyllum (Burch.) Sond. and Withania somnifera (L.) dunal against Fusarium pathogens. Molecules, 26(16): 4732. https://doi.org/10.3390/molecules26164732.

Segaran, G. and Sathiavelu, M. (2019). Fungal endophytes: A potent biocontrol agent and a bioactive metabolites reservoir. Biocatalysis and Agricultural Biotechnology, 21:, 101284. https://doi.org/10.1016/j.bcab.2019.101284.

Senol, M., Nadaroglu, H., Dikbas, N. and Kotan, R. (2014). Purification of Chitinase enzymes from Bacillus subtilis bacteria TV-125, investigation of kinetic properties and antifungal activity against Fusarium culmorum. Ann. Clin. Microbiol. Antimicrob., 12; 13: 35. doi: 10.1186/s12941-014-0035-3. PMID: 25112904; PMCID: PMC4236515.

Shabana, Y.M., Abdalla, M.E., Shahin, A.A., El-Sawy, M.M., Draz, I.S. and Youssif, A.W. (2017). Efficacy of plant extracts in controlling wheat leaf rust disease caused by Puccinia triticina. Egyptian Journal of Basic and Applied Sciences, 4(1): 67–73. https://doi.org/10.1016/j.ejbas.2016.09.002.

Sharf, W., Javaid, A., Shoaib, A. and Khan, I.H. (2021). Induction of resistance in chili against *Sclerotium rolfsii* by plant-growth-promoting rhizobacteria and *Anagallis arvensis*. Egyptian Journal of Biological Pest Control, 31(1). https://doi.org/10.1186/s41938-021-00364-y.

Sharma, A., Rajendran, S., Srivastava, A., Sharma, S. and Kundu, B. (2017). Antifungal activities of selected essential oils against Fusarium oxysporum f. sp. lycopersici 1322, with emphasis on Syzygium aromaticum essential oil. Journal of Bioscience and Bioengineering, 123(3): 308–313. https://doi.org/10.1016/j.jbiosc.2016.09.011.

Sharma, S., Kumar, S., Khajuria, A., Ohri, P., Kaur, R. and Kaur, R. (2020). Biocontrol potential of chitinases produced by newly isolated Chitinophaga sp. S167. World Journal of Microbiology and Biotechnology, 36(6): 90. https://doi.org/10.1007/s11274-020-02864-9.

Sharma, H., Haq, M.A., Koshariya, A.K., Kumar, A., Rout, S. and Kaliyaperumal, K. (2022) "*Pseudomonas fluorescens*" as an antagonist to control okra root rotting fungi disease in plants. Journal of Food Quality, 2022: 1–8. https://doi.org/10.1155/2022/5608543.

Shivanna, R. and Garampalli, R.H. (2015). Evaluation of fungistatic potential of lichen extracts against Fusarium solani (Mart.) Sacc. causing Rhizome rot disease in Ginger. Journal of Applied Pharmaceutical Science, 5: 067–072. https://doi.org/10.7324/JAPS.2015.501012.

Shokouhi, D. and Seifi, A. (2020). Organic extracts of seeds of Iranian Moringa peregrina as promising selective biofungicide to control Mycogone perniciosa. Biocatalysis and Agricultural Biotechnology, 30: 101848. https://doi.org/10.1016/j.bcab.2020.101848.

Shukla, P., Mantin, E., Adil, M., Bajpai, S., Critchley, A. and Prithiviraj, B. (2019). Ascophyllum nodosum-based biostimulants: Sustainable applications in agriculture for the stimulation of plant growth, stress tolerance, and disease management. Frontiers in Plant Science, 10. 10.3389/fpls.2019.00655.

Shuping, D.S.S. and Eloff, J.N. (2017). The use of plants to protect plants and food against fungal pathogens: A review. African Journal of Traditional, Complementary, and Alternative Medicines, 14(4): 120–127. https://doi.org/10.21010/ajtcam.v14i4.14.

Silva, R.N., Monteiro, V.N., Steindorff, A.S., Gomes, E.V., Noronha, E.F. and Ulhoa, C.J. (2019). Trichoderma/pathogen/plant interaction in pre-harvest food security. Fungal Biology, 123(8): 565–583. https://doi.org/10.1016/j.funbio.2019.06.010.

Singh, B. and Sharma, R.A. (2015). Plant terpenes: Defense responses, phylogenetic analysis, regulation and clinical applications. 3 Biotech., 5(2): 129–151. https://doi.org/10.1007/s13205-014-0220-2.

Soliman, A.S., Ahmed, A.Y., Abdel-Ghafour, S.E., El-Sheekh, M.M. and Sobhy, H.M. (2018). Antifungal bioefficacy of the red algae Gracilaria confervoides extracts against three pathogenic fungi of cucumber plant. Middle East J. Appl. Sci., 8(3): 727–735.

Song, L., Zhang, F., Yu, J., Wei, C., Han, Q. and Meng, X. (2020). Antifungal effect and possible mechanism of curcumin mediated photodynamic technology against Penicillium expansum. Postharvest Biology and Technology, 167: 111234. https://doi.org/10.1016/j.postharvbio.2020.111234.

Spadaro, D., Vola, R., Piano, S. and Gullino, M.L. (2002). Mechanisms of action and efficacy of four isolates of the yeast Metschnikowia pulcherrima active against postharvest pathogens on apples. Postharvest Biology and Technology, 24(2): 123–134. https://doi.org/10.1016/S0925-5214(01)00172-7.

Srivastava, S. and Singh, V.P. (2020). Characterization of biocontrol microorganisms from the rhizoplane of Decalepis arayalpathra and screening of secondary metabolites. Vegetos, 33(2): 352–359. https://doi.org/10.1007/s42535-020-00115-8.

Srivastava, M., Kumar, V., Shahid, M., Pandey, S. and Sing, A. (2016). Trichoderma—A potential and effective bio fungicide and alternative source against notable phytopathogens: A review, 11(5): 310–316, DOI: 10.5897/AJAR2015.9568.

Steglińska, A., Bekhter, A., Wawrzyniak, P., Kunicka-Styczyńska, A., Jastrząbek, K., Fidler, M., Śmigielski, K. and Gutarowska, B. (2022). Antimicrobial activities of plant extracts against *Solanum tuberosum* L. *phytopathogens*. Molecules, 27(5). https://doi.org/10.3390/molecules27051579, PubMed: 35268680, PubMed Central: PMC8911893.

Subba, R. and Mathur, P. (2022). Functional attributes of microbial and plant based biofungicides for the defense priming of crop plants. Theoretical and Experimental Plant Physiology. https://doi.org/10.1007/s40626-022-00249-.

Sufyan, M., Tahir, M.I., Haq, M.I.U., Hussain, S. and Saeed, M. (2020). Effect of seed bio priming with rhizobacteria against root associated pathogenic fungi in chickpea. Pakistani Journal of Phytopathology, 32: 89–96. https://doi.org/10.33866/phyto pathol.032.01.0567.

Sujarit, K., Pathom-Aree, W., Mori, M., Dobashi, K., Shiomi, K. and Lumyong, S. (2020). Streptomyces palmae CMU-AB204T, an antifungal producing-actinomycete, as a potential biocontrol agent to protect palm oil producing trees from basal stem rot disease fungus, *Ganoderma boninense*. Biological Control, 148. https://doi.org/10.1016/j.biocontrol.2020.104307.

Sui, L., Li, J., Philp, J., Yang, K., Wei, Y., Li, H., Li, J., Li, L., Ryder, M., Toh, R., Zhou, Y., Denton, M.D., Hu, J. and Wang, Y. (2022). *Trichoderma atroviride* seed dressing influenced the fungal community and pathogenic fungi in the wheat rhizosphere. Scientific Reports, 12(1): 9677. https://doi.org/10.1038/s41598-022-13669-1.

Surono, Surono and Narisawa, Kazuhiko. (2018). The inhibitory role of dark septate endophytic fungus Phialocephala fortinii against Fusarium disease on the Asparagus officinalis growth in organic source conditions. Biological Control, 121. 10.1016/j.biocontrol.2018.02.017.

Stevenson, P.C., Nyirenda, S.P., Mvumi, B., Sola, P., Kamanula, J.F., Sileshi, G.W. and Belmain, S.R. (2012). Pesticidal plants: A Viable Alternative Insect Pest Management Approachfor Resource-poor Farming in Africa.Biopesticides in Environment and Foodsecurity. http://www.scientificpub.com. Scientific Publishing. ISBN: 978-81-7233-797-1.

Tahir, M.I., Sufyan, M., Haq, M.I.U., Hussain, S. and Saeed, M. (2020). Effect of seed bio priming with rhizobacteria against root associated pathogenic fungi in chickpea. Pakistan Journal of Phytopathology, 32(1): 89–96. https://doi.org/10.33866/phytopathol.032.01.0567.

Tóth, L., Boros, É., Poór, P., Ördög, A., Kele, Z., Váradi, G., Holzknecht, J., Bratschun-Khan, D., Nagy, I., Tóth, G.K., Rákhely, G., Marx, F. and Galgóczy, L. (2020). The potential use of the *Penicillium chrysogenum* antifungal protein PAF, the designed variant PAFopt and its g-core peptide Pgopt in plant protection. Microbial Biotechnology, 13(5): 1403–1414. https://doi.org/10.1111/1751-7915.13559.

Trejo-Raya, A.B., Rodríguez-Romero, V.M., Bautista-Baños, S., Quiroz-Figueroa, F.R., Villanueva-Arce, R. and Durán-Páramo, E. (2021). Effective *in vitro* control of two phytopathogens of agricultural interest using cell-free extracts of Pseudomonas fluorescens and chitosan. Molecules, 26(21): 6359. https://doi.org/10.3390/molecules26216359.

Tripathi, P. and Shukla, A.K. (2010). Exploitation of botanicals in the management of phytopathogenic and storage fungi. Management of fungal Plant Pathogens, 36–50.

United states environmental protection agency (USEPA). (2013). Regulating Biopesticides. https://www3.epa.gov/pesticides/chem_search/ppls/072431-00001-20131113.pdf.

Upamanya, G.K., Bhattacharyya, A. and Dutta, P. (2020). Consortia of entomo-pathogenic fungi and bio-control agents improve the agro-ecological conditions for brinjal cultivation of Assam. 3 Biotech., 10(10): 450. https://doi.org/10.1007/s13205-020-02439-3.

Valadbeigi, T., Bahrami, A.M. and Shaddel, M. (2014). Antibacterial and antifungal activities of different lichens extracts. J. Med. Microbiol. Infec. Disorders, 2(2): 71–75.

Vashist, H. and Jindal, A. (2012). Antimicrobial activities of medicinal plants–review. Int. J. Res. Pharm. Biomed. Sci., 10(4): 222–230. DOI: 10.4236/jbise.2014.714107.

Vehapi, M., Koçer, A.T., Yılmaz, A. and Özçimen, D. (2020). Investigation of the antifungal effects of algal extracts on apple-infecting fungi. Archives of Microbiology, 202(3): 455–471. https://doi.org/10.1007/s00203-019-01760-7.

Velu, G., Palanichamy, V. and Rajan, A.P. (2018). Phytochemical and pharmacological importance of plant secondary metabolites in modern medicine. pp. 135–156. *In*: S.M. Roopan and G. Madhumitha (eds.). Agriculturists. Bioorganic Phase in Natural Food: An Overview. Springer International Publishing.

Wang, J., Li, C., Song, P., Qiu, R., Song, R., Li, X., Ni, Y., Zhao, H., Liu, H. and Li, S. (2022). Molecular and biological characterization of the first Mymonavirus identified in Fusarium oxysporum. Frontiers in Microbiology, 13 Apr 21; 13: 870204. https://doi.org/10.3389/fmicb.2022.870204, PubMed: 35531277, PubMed Central: PMC9069137.

Wang, M., Xue, J., Ma, J., Feng, X., Ying, H. and Xu, H. (2020). Streptomyces lydicus M01 regulates soil microbial community and alleviates foliar disease caused by alternaria alternata on cucumbers. Frontiers in Microbiology, 11. DOI=10.3389/fmicb.2020.00942.

Wei, H., Wang, Y., Jin, Z., Yang, F., Hu, J. and Gao, M.T. (2021). Utilization of straw-based phenolic acids as a biofugicide for a green agricultural production. Journal of Bioscience and Bioengineering, 131(1): 53–60. https://doi.org/10.1016/j.jbiosc.2020.09.007.

Woo, S.L., Ruocco, M., Vinale, F., Nigro, M., Marra, R., Lombardi, N., Pascale, A., Lanzuise, S., Manganiello, G. and Lorito, M. (2014). Trichoderma-based products and their wdespread use in agriculture. Open Mycology Journal, 8(1): 71–126. https://doi.org/10.2174/1874437001408010071.

Xiao, Y., Liu, Z., Gu, H., Yang, F., Zhang, L. and Yang, L. (2021). Improved method to obtain essential oil, asarinin and sesamin from Asarum heterotropoides var. mandshuricum using microwave-assisted steam distillation followed by solvent extraction and antifungal activity of essential oil against Fusarium spp. Industrial Crops and Products, 162: 113295. https://doi.org/10.1016/j.indcrop.2021.113295.

Yadav, M., Dubey, M.K. and Upadhyay, R.S. (2021). Systemic resistance in chilli pepper against anthracnose (Caused by *Colletotrichum truncatum*) induced by *Trichoderma harzianum*, *Trichoderma asperellum* and *Paenibacillus dendritiformis*. Journal of Fungi, 7(4): 307. https://doi.org/10.3390/jof7040307.

Yamani, H.A., Pang, E.C., Mantri, N. and Deighton, M.A. (2016 May 17). Antimicrobial activity of Tulsi (*Ocimum tenuiflorum*) essential oil and their major constituents against three species of bacteria. Frontiers in Microbiology, 7: 681. https://doi.org/10.3389/fmicb.2016.00681, PubMed: 27242708, PubMed Central: PMC4868837.

Yuliar, N.Y.A., Nion, Y.A. and Toyota, K. (2015). Recent trends in control methods for bacterial wilt diseases caused by Ralstonia solanacearum. Microbes and Environments, 30(1): 1–11. https://doi.org/10.1264/jsme2.ME14144.

Yuvamoto, P.D. and Said, S. (2007). Germination, duplication cycle and septum formation are altered by caffeine, caffeic acid and cinnamic acid in Aspergillus nidulans. Microbiology, 76: 735–738.

Zaker, M. (2016). Natural plant products as eco-friendly fungicides for plant diseases control—A review. Agriculturists, 14(1): 134–141. https://doi.org/10.3329/agric.v14i1.29111.

Zaynab, M., Fatima, M., Abbas, S., Sharif, Y., Umair, M., Zafar, M.H. and Bahadar, K. (2018). Role of secondary metabolites in plant defense against pathogens. Microbial Pathogenesis, 124: 198–202. https://doi.org/10.1016/j.micpath.2018.08.034.

Zhang, D., Spadaro, D., Garibaldi, A. and Gullino, M.L. (2011). Potential biocontrol activity of a strain of pichia guilliermondii against grey mold of apples and its possible modes of action. Biological Control, 57(3): 193–201. https://doi.org/10.1016/j.biocontrol.2011.02.011.

Zhang, S., Sun, F., Liu, L., Bao, L., Fang, W., Yin, C. and Zhang, Y. (2020a). Dragonfy-associated Trichoderma harzianum QTYC77 is not only a potential biological control agent of Fusarium oxysporum f. sp. cucumerinum but also a source of new antibacterial agents. J. Agric. Food Chem., 68: 14161–14167. https://doi.org/10.1021/acs.jafc.0c05760.

Zhang, T., Jin, Y., Zhao, J.H., Gao, F., Zhou, B.J., Fang, Y.Y. and Guo, H.S. (2016). Host-induced gene silencing of the target gene in fungal cells confers effective resistance to the cotton wilt disease pathogen Verticillium dahliae. Molecular Plant, 9(6): 939–942.

Zhang, X., Wu, F., Gu, N., Yan, X., Wang, K., Dhanasekaran, S., Gu, X., Zhao, L. and Zhang, H. (2020). Postharvest biological control of Rhizopus rot and the mechanisms involved in induced disease resistance of peaches by pichia membranefaciens. Postharvest Biology and Technology, 163: 111146. https://doi.org/10.1016/j.postharvbio.2020.111146.

Zhao, C.N., Yao, Z.L., Yang, D., Ke, J., Wu, Q.L., Li, J.K. and Zhou, X.D. (2020). Chemical constituents from Fraxinus hupehensis and their antifungal and herbicidal activities. Biomolecules, 10(1): 74. https://doi.org/10.3390/biom10010074.

Zhou, J., Li, Z., Wu, J., Li, L., Li, D., Ye, X., Luo, X., Huang, Y., Cui, Z.and Cao, H. (2017). Functional analysis of a novel b-(1,3)-glucanase from Corallococcus sp. strain EGB containing a fascin-like module. Applied and Environmental Microbiology, 83(16): 1–13. https://doi.org/10.1128/AEM.01016-17.

Zhou, L., Li, X., Kotta-Loizou, I., Dong, K., Li, S., Ni, D., Hong, N., Wang, G. and Xu, W.A. (2021). A mycovirus modulates the endophytic and pathogenic traits of a plant associated fungus. ISME Journal, 15(7): 1893–1906. https://doi.org/10.1038/s41396-021-00892-3.

Zhu, C., Lei, M., Andargie, M., Zeng, J. and Li, J. (2019). Antifungal activity and mechanism of action of tannic acid against Penicillium digitatum. Physiological and Molecular Plant Pathology, 107: 46–50. https://doi.org/10.1016/j.pmpp.2019.04.009.

Chapter 10

Bioremediation of Fungicide-contaminated Environment

Josef Jampílek[1,2,*] and *Katarína Kráľová*[3]

1. Introduction

Fungicides are chemical compounds or biological organisms able to kill fungal pathogens or their spores, that are used in the fight against plant diseases (Frost, 2015a,b,c). In the management of fungal infections, besides inorganic and organic fungicides, biofungicides, i.e., fungicides based on microorganisms or natural products can be used (Lilai et al., 2022; Qessaoui et al., 2022; Subba and Mathur, 2022). Fungicides can be applied as (i) contact fungicides acting only on plant surfaces; (ii) localized penetrants, which are absorbed by leaves, pass short distances within the leaf and kill fungal pathogens on plant surfaces and inside treated leaves; (iii) acropetal penetrants, which can penetrate plants through roots, shoots and leaves and can inhibit fungi on and in treated plant surfaces and inside plant parts that lie above the treated surface; (iv) systemic fungicides, which are absorbed into the xylem and phloem and advances up and down in plants, and this kill the fungi on and in treated plant surfaces and inside plant parts occurring above or below the exposed surfaces (Rouabhi, 2010). Fungicides can function as site-specific inhibitors targeting individual sites within the fungal cell or multi-site inhibitors that target many different sites in each fungal cell (Lucas et al., 2015; Hermann and Stenzel, 2019).

[1] Department of Analytical Chemistry, Faculty of Natural Sciences, Comenius University, Ilkovičova 6, Bratislava 842 15, Slovakia.
[2] Department of Chemical Biology, Faculty of Science, Palacky University Olomouc, Šlechtitelů 27, 78371 Olomouc, Czech Republic.
[3] Institute of Chemistry, Faculty of Natural Sciences, Comenius University, Ilkovičova 6, Bratislava 842 15, Slovakia.
Email: kata.kralova@gmail.com
* Corresponding author: josef.jampilek@gmail.com

Site-specific fungicides bind to specific protein targets resulting in disruption of key cellular processes of fungal pathogens, mutation and enhanced expression of a single gene might result in the development of resistance (Lucas et al., 2015). On the other hand, site-non-specific fungicides affect numerous cellular processes influencing the fitness of pathogens and resistance development to such fungicides that is associated with changes in many genes in pathogen genomes (Hawkins and Fraaije, 2018).

Widespread use and frequently also overuse of fungicides in agriculture noticeably contribute to the contamination of soils and water bodies, posing a serious risk for non-target organisms (Zubrod et al., 2019). Moreover, some synthetic fungicides were found to show endocrine-disrupting (Draskau et al., 2019; Elsharkawy et al., 2019; Draskau and Svingen, 2022; Huang et al., 2022), carcinogenic (Silva et al., 2014; Goyal et al., 2018), teratogenic (Li et al., 2016; Battistoni et al., 2019), genotoxic (Ali et al., 2021; Castro et al., 2022) and neurotoxic (Kang et al., 2021) effects, as well as cardiac developmental toxicity (Wang et al., 2021) and the use of several harmful fungicides was banned in many countries, including the European Union. The risk posed by agrochemicals, including fungicides to non-target organisms and human health requires their thorough remediation from environmental matrices, whereby a bioremediation approach using bacteria, microalgae, fungi and plants for the degradation/removal of fungal pathogens is preferred (Aimeur et al., 2016; Raffa and Chiampo, 2021; Randika et al. 2022).

The main *in situ* techniques of microbial remediation of pesticide-contaminated soils are (i) natural attenuation (contaminant removal by microorganisms present in the contaminated soil); (ii) biostimulation (addition of optimized doses of nutrients stimulating the growth of indigenous microorganisms); (iii) bioaugmentation (soil supplementation with microbial strains or enzymes able to effectively degrade contaminants); (iv) bioventing (promoting of the growth of contaminant-degrading microorganisms occurring in soil via the introduction of air or O_2 and nutrients into the unsaturated zone of soil); (v) biosparging (enhancing of O_2 concentration and stimulation of contaminant-degrading microorganisms by injected air under pressure into the saturated soil zone) (Raffa and Chiampo, 2021). As microbial biostimulants can also serve beneficial bacteria and fungi, that can ameliorate the growth, nutrition and yields of plants as well as their tolerance against stresses (Tanveer et al., 2022).

Disruption of soil structure also affects the structure of the microbial community and thus, degradation of fungicides in field studies using undisturbed soil might be faster compared to sieved soil applied in laboratory studies (Hand et al., 2021). Recent progress in fungal-based bioremediation (mycoremediation) of contaminated environment was presented by AI-Jawhari (2022) and the potential of endophytic fungi to be used in bioremediation was described by Krishnamurthy and Naik (2017) and Sharma and Kumar (2021).

Living organisms in the aquatic environment are increasingly endangered by pollutants entering rivers, lakes and seas particularly due to anthropogenic activity. For environment-friendly removal of agrochemicals, drugs, heavy metals and excess nutrients without the application of toxic chemicals, green algae or cyanobacteria

(Hussein et al., 2017; Pan et al., 2018; Pacheco et al., 2020; Avila et al., 2021; Touliabah et al., 2022; Verasoundarapandian et al., 2022), macroalgae (Girardi et al., 2014; Mazur et al., 2017; Navarrete et al., 2019; Rodriguez-Rojas et al., 2019; Jampílek and Kráľová, 2021) or macrophytes (Dosnon-Olette et al., 2009; Anand et al., 2019; Brunhoferova et al., 2021; Rempel et al., 2021; Skufca et al., 2021; Seenivasagan et al., 2022) can be used. For removal of toxic contaminants from agricultural soils, besides microbial remediation, terrestrial plants can be used (Gautam et al., 2017; Romeh, 2017; Katsoula et al., 2020; You et al., 2021; Ghazaryan et al., 2022; Zand and Mühling, 2022;).

The main *in situ* techniques of microbial remediation of pesticide-contaminated soils and phytoremediation techniques used for removal of synthetic and metal-based pesticides are shown in Fig. 10.1.

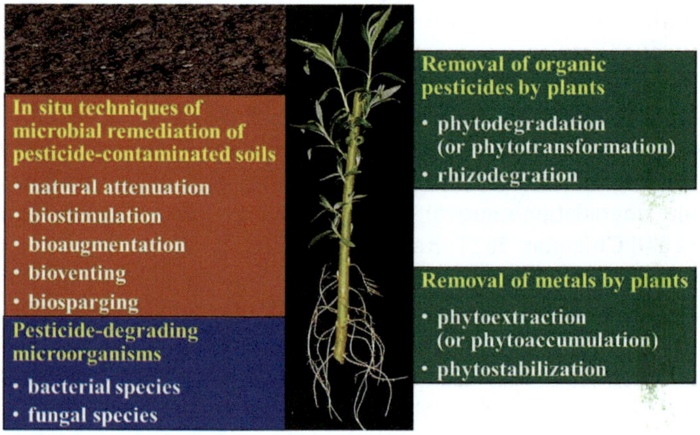

Fig. 10.1: Main *in situ* techniques of microbial remediation of pesticide-contaminated soils and phytoremediation techniques used for removal of synthetic and metal-based pesticides.

According to the Food and Agriculture Organization (FAO) of the United Nations, plant diseases resulting in immense loss of crop production cost the global economy approximately US$220 billion each year (FAO, 2019). The global fungicides market is projected to grow from US$18.43 billion in 2021 to US$25.81 billion by 2028 exhibiting a compound annual growth rate (CAGR, the mean annual growth rate of an investment over a specified period longer than one year) of 4.93% in the forecast period (Fortune Business Insight, 2022). The United States shared about 35% of the fungicide market, followed by Europe and China; both have a share of over 40%. For Asia Pacific countries, that in 2019 shared 2019 38.2% of this market in the on-coming period the fastest growth is predicted (Grand View Research, 2020).

In European countries in 2017 32,642,764 kilograms of bactericides and fungicides were sold (Grand View Research, 2020). Fungicide sales in the European Union represent > 40% of total pesticide sales, whereby ca. 60% of all fungicides is synthetic fungicides. High application of fungicides achieving > 90% of all pesticides is typical for regions with numerous vineyards, whereas in vineyards there is the highest use of inorganic Cu-based fungicides (Ballabio et al., 2018; Panagos et al., 2018), and therefore, to avoid undesirable toxic effects on non-target organisms, the limits for Cu fungicide use corresponding to 4 kg/ha per year were introduced by EU (EFSA, 2020). Biofungicides formulated from living organisms will exhibit the fastest CAGR of 6.7% in terms of volume from 2020 to 2027 (Grand View Research, 2020).

Triantafyllidis et al. (2022) based on fungicide index (FI; kg/ha) corresponding to fungicide active substances used on a hectare of agricultural land found that during the first 20 yr of this millennium the mean FI in 27 countries of the European Union was 0.79 kg/ha. However, in southern countries, the mean FI value was more than one order higher than in Northern ones (1.34 kg/ha vs. 0.11 kg/ha). Considering the use of inorganic fungicides, their use in southern countries was threefold higher than in northern ones (1.69 kg/ha vs. 0.63 kg/ha), whereby the usage of organic fungicides achieved higher average values in western and southern Europe (0.54 and 0.51 kg/ha, respectively). The highest use of fungicides was found in Malta, Cyprus, Italy and Portugal, while the lowest one was in Estonia, Sweden, Latvia and Finland.

As mentioned above, overuse of fungicidal agents contaminating soils and aquatic ecosystems is a global problem resulting not only in significantly lower yields of crops but the persistence of these agrochemicals in environmental matrices may pose a threat to non-target organisms. This chapter presents an up-to-date overview of bioremediation of soils and water bodies contaminated with most commonly used synthetic fungicides via microbial remediation, detoxification of fungicides by laccases, bioremediation of fungicide-contaminated aqueous environment using various types of algae and removal of synthetic organic and inorganic fungicides from the environment using macrophytes and terrestrial plants. The mechanisms of degradation and degradation products of important fungicides, beneficial effects of bioaugmentation using microbial strains effectively able to degrade fungicides and fungicide-degrading microbial strains immobilized on biochar as well as application of chelate-forming agents on bioremediation processes are briefly discussed.

2. Microbial remediation of synthetic fungicides

2.1 Azoles fungicides

Azoles are the most common antifungal agents. According to the central 5-membered heterocycle, they can be divided into imidazoles, triazoles and thiazoles, see Fig. 10.2. In the case of imidazoles and triazoles, the heterocycle has to contain nitrogen atoms in position 3 (imidazole) or position 4 (triazoles). Structures of the below-mentioned azole fungicides are illustrated in Fig. 10.3. These positions of nitrogens are crucial because they chelate iron in the enzyme 14α-demethylase (inhibiting the conversion of lanosterol to ergosterol) and thus cause the formation of an incomplete fungal cell wall. Thiazoles, that are used only as antifungal drugs (abafungin) and not as agricultural fungicides, inhibit not only fungal 14α-demethylase but also sterol 24-C-methyltransferase (which again causes depletion of ergosterol in the fungus membrane) (FRAC Code List, 2022; Jampilek,

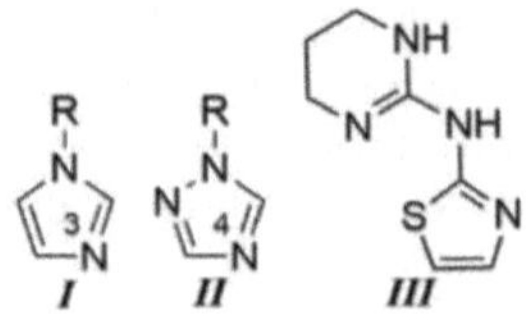

Fig. 10.2: General structures of antifungal azoles: imidazoles (*I*), triazoles (*II*) and thiazoles (*III*).

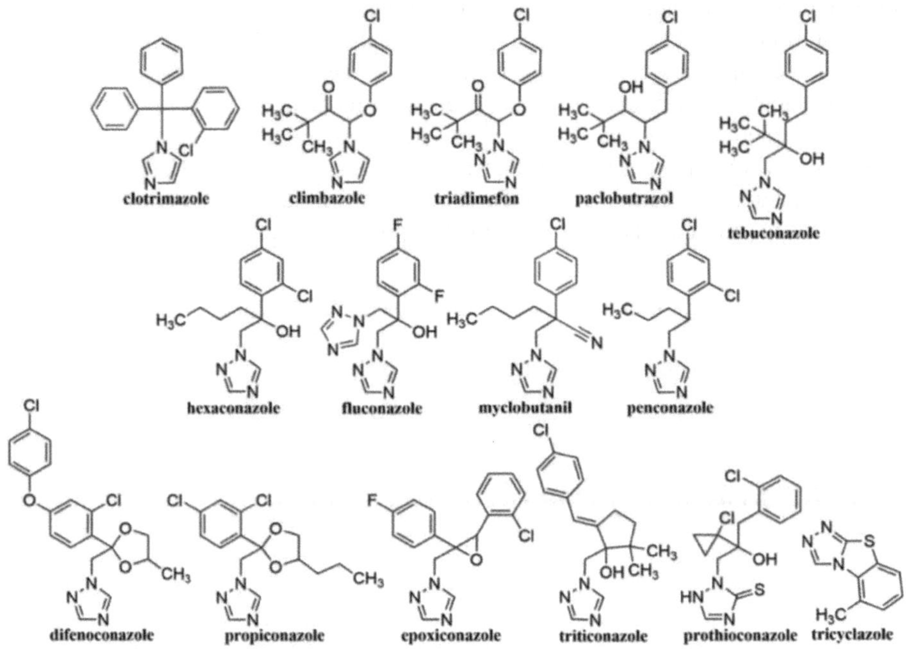

Fig. 10.3: Structures of described azole fungicides.

2016a,b, 2022). The impact of triazole fungicides on the structure of the microbial communities and activities of enzymes occurring in soil was comprehensively overviewed by Roman et al. (2021). The azole fungicides showing toxic effects on algae and fish species are widely used for treatments of fungal infections in humans because of their resistance to microbial degradation. Therefore, for their degradation photolysis under ultraviolet irradiation is preferable (Chen and Ying, 2015). For example, combined treatment using vacuum UV (185 nm) and UVC (254 nm) irradiation can degrade tebuconazole (TEB) in drinking water at the μg/L−mg/L levels, providing transformation products with reduced toxicity compared to the parent fungicide (Del Puerto et al., 2022).

Pacholak et al. (2022) investigated biodegradation of four fungicides, clotrimazole (CLZ), climbazole (CLB), epoxiconazole (EPO) and fluconazole (FLZ) using activated sludge and two Gram-negative bacterial strains, *Stenotrophomonas maltophilia* AsPCl2.3 and *Pseudomonas monteilii* LB2, isolated from the same sludge. In this experiment after 1 mon 71% CLB, 50.0% CLZ and 15.1% EPO were removed, while elimination of FLZ was not observed. During 1 mon *S. maltophilia* AsPCl2.3 was able to remove 21.5% CLB and 16.7% CLZ, while *P. monteilii* LB2 degraded 42% of CLB and 13% of CLZ; degradation of EPO and FLZ by both studied bacterial strains was not detected. On the other hand, except FLZ, complete degradation of tested fungicides providing final degradation product imidazole or triazole was observed using Fenton process and UV irradiation. Cai et al. (2021) investigating biodegradation of CLB and FLZ in activated sludge under aerobic conditions observed the degradation of CLB (half-life ($t_{1/2}$): 5.3 d), which adsorbed to solid sludge, while only 30% of FLZ (remaining predominantly in the liquid) was removed within 77 d of incubation. Whereas no biodegradation products of FLZ were detected, 10 biodegradation products of CLB were identified, whereby the degradation pathway involved oxidative dehalogenation, side chain oxidation and azole ring loss. CLB was biotransformed by whole cells of *Trichoderma harzanium* and *T. asperellum*, by which using the more efficient fungus, *T. harzianum*, even a 91% CLB degradation rate was achieved. During CLB degradation carbonyl reduction into alcohol was observed (Manasfi et al., 2020). Kryczyk-Poprawa et al. (2019) reported that mycelia of *Lentinula edodes* are suitable for *in vitro* degradation of bifonazole (powder and cream) and CLZ (powder) antimycotics. In both antimycotics the degradation process particularly affected the imidazole moiety. Vacuum UV irradiation also efficiently degraded CLZ, whereby eight less toxic transformation products were formed predominantly via hydroxylation in the phenyl ring and/or imidazole group followed by ring opening or loss of the imidazole moiety (Goncalves et al., 2021). Co-culture of bacterial strains *Hydrogenophaga eletricum* 5AE and *Methylobacillus* sp. 8AE efficiently biodegraded EPO and defluorinated approximately 80% of this fungicide in 28 d; whereas *H. eletricum* was assumed to be responsible for defloration, *Methylobacillus* sp. played a likely accessory catabolic role. Considering that in the original bacterial consortium these two bacterial strains were present in a lesser amount, it can be suggested that the removal of EPO via degradation can be performed by less abundant strains in the bacterial community (Alexandrino et al., 2021). After an enrichment period of 6 mon the microbial consortia enriched from estuarine sediment and agricultural soil, which consisted mostly of *Pseudomonas*, *Ochrobactrum* and *Comamonas* spp., eliminated fluorinated fungicides, EPO and fludioxonil in co-metabolic conditions and showed good biodegradation also without a co-substrate; they degraded up to 10 mg/L of EPO or fludioxonil in 3 wk, when elimination of fungicides was faster than their defluorination (Alexandrino et al., 2020). Slow-release fungicide preparations containing EPO, TEB and azoxystrobin, which were fabricated as pellets and granules using biodegradable poly(3-hydroxybutyrate) as a matrix and natural fillers such as peat, wood flour and clay, ensuring a gradual and sustained release of fungicides, showed *in vitro* antifungal effect against *Fusarium verticillioides in vitro*. At supplementation of these preparations to the soil, the degradation of fungicides depended predominantly on the type of formulation, while the impact of the filler type was only minor. Accelerated accumulation of fungicides in the soil at the application of granules was associated with their faster disintegration (Volova et al., 2019).

With increased treatment frequency the microbial functional diversity of the soil exposed to TEB showed an inhibition–recovery–stimulation trend. Initial slow degradation of TEB during repeated treatments showed a decrease, but high TEB concentrations strikingly retarded the degradation. Exposure to fungicide resulted in a considerable reduction of soil microbial biomass and bacterial community diversity and the effect was reinforced with multiple treatments. As TEB-degrading bacterial genera, *Methylobacterium*, *Burkholderia*, *Hyphomicrobium* and *Dermacoccus* were identified, and their relative abundance in treated soils showed an increase by 42.1–34687.1% compared to the control. However, high fungicide concentration reduced the relative abundances of *Methylobacterium* but stimulated those of the other three above-mentioned bacterial genera (Han et al., 2021). Helicur 250 EW fungicidal preparation containing an active ingredient TEB, which was sprayed on leaves of *Horderum vulgare* plants cultivated in soil, affected soil bacterial community compared to control soil. In contrast to *Proteobacteria*, which occurred as predominant taxon in both soils, *Ramlibacter tataounensis*, *Azospirillum palatum* and *Kaistobacter terrae* were detected only in TEB-contaminated soil; *Bacillus arabhattai*, *B. soli* and *B. simplex* was detected exclusively in the control soil. Fungicidal preparation inhibited not only the activities of all soil enzymes (except arylsulfatase) but also the growth and development of treated plants (Bacmaga et al., 2020). The TEB-degrading strain *Alcaligenes faecalis* WZ-2 immobilized on wheat straw-derived biochar performed faster degradation of TEB in soil resulting in reduced $t_{1/2}$ compared to the application of non-immobilized bacterial cells (13.3 d vs. 40.8–18.7 d), although using biochar alone somewhat delayed degradation of TEB in soil. Whereas a dose of 10 mg/kg exhibited an adverse impact on activities of soil enzymes such as urease, dehydrogenase and invertase, the bacterial strain immobilized on biochar contributed to faster fungicide degradation of TEB and was able to restore native soil microbial enzyme activities along with the composition of microbiome community (Sun et al., 2020). In the culture the *Serratia marcescens* strain B1 degraded 94.05% of 200 mg TEB/L in 8 hr but at an initial fungicide concentration of 300 mg/L, only 64.11% degradation was observed. Whereas in control soil contaminated with 200 mg/L TEB only 70.42% degradation was observed in 30 d, inoculation with *S. marcescens* B1 at concentration 3×10^7 CFU/g dry soil increased it to 96.46%. Moreover, greenhouse and field experiments suggested the ability of this bacterial strain to remove the residues of TEB in contaminated soil by *Brassica rapa* plants (Wang et al., 2018a). *Bacterial strains Bacillus* sp. 29B3 and *Bacillus* sp. 3B6 isolated from cloud water were able biotransform 100 μM TEB from 50 to 60%. However, in contrast to *Bacillus* sp. 29B3, also showing higher fungicide degrading efficiency at reduced TEB concentration, *Bacillus* sp. 3B6 degraded only 50% fungicide, independently of its concentration, which was associated with faster degradation of the (+)-(*S*)-enantiomer compared to the (−)-(*R*)-enantiomer. From three identified metabolites, the metabolite formed via dehydration of the hydroxylated metabolite was present only at low levels and was not detected with *Bacillus* sp. 29B3. Since commercial formulations contain racemic mixture of TEB, at the assessment of environmental risk it would be necessary to consider the eventual chiral dependence of persistence, biodegradation, bioaccumulation and toxicity (Youness et al., 2018). It should be noted that the adverse impact of TEB on the proliferation of soil microorganisms and the activity of soil enzymes can be mitigated using biostimulating substances such as bird droppings, but the addition of the biostimulant did not improve the yield of spring barley plants; the addition of biostimulants also promoted TEB degradation (Bacmaga et al., 2019).

Zhao et al. (2021) investigated the impact of chiral enantiomers of paclobutrazol (PBZ) fungicide also acting as a plant growth regulator on the structure of the microbial community in soil and found that *R*-isomer ($t_{1/2}$: 50 d) was less persistent than *S*-isomer ($t_{1/2}$: 80 d). The enantiomers greatly affected the soil microbial community and *Pseudomonas* and *Mycobacterium* were likely responsible for the fungicide degradation. The enantioselective impact was reflected mainly in the relative abundance of *Fusarium* sp. The higher effects of tested enantiomers on the structure of the fungal community than on the structure of the bacterial community were associated with the fungicidal activity of PBZ. Considerable impact on the microbial networks showed *R*-PBZ, and therefore *S*-PBZ can be recommended to be used as a plant growth regulator and fungicide suppressing adverse impact on soil microbial communities (Zhao et al., 2021). In paddy simulation models the soils and sediments containing 195 μg/kg and 31.3 μg/kg triadimefon (TDM) showed prolonged t_{12} after elimination of the resident microbial community (86.6–115.5 d vs. 4–28.9 d) suggesting that the removed bacterial community was largely responsible for fungicide degradation. In contrast to *Enterobacter*, *Sphingomonas* and *Xanthomonas* were identified as crucial TDM degrading bacterial genera, which participate in sustaining the resilience of soils and sediments to TDM pollution (Fan et al., 2021). When *S*-TDM and *R*-TDM enantiopure were incubated in different soils with different pH values, considerable enantiomerization among the soils was observed. Whereas the alkaline soil showed accelerated conversion compared to neutral soil, in acidic soil no conversion was observed and *Arthrobacter* and *Halomonas* spp showed higher abundance in alkaline soil than in neutral soil after application of fungicides, while in acidic soil their abundance was diminished. In addition, for converting *S*-TDM to *R*-TDM and *R*-TDM to *S*-TDM in alkaline and neutral soil *Arthrobacter* and *Halomonas* were found to be important (Talab et al., 2018). Using racemic triticonazole (TTC) with initial concentrations ranging from 25.1 to 93.1 ng/g, enantioselective degradation in pear, peach and jujube was observed after 2 hr, 10 d and 3 d, respectively, whereby degradation of *R*-(−)-TTC in pear was slower ($t_{1/2}$: 5.02 d) than that of *S*-(+)-TTC ($t_{1/2}$: 2.01 d) and the racemic triticonazole residues in tested fruits were lower than the maximum residue level set by European Union (Guo et al., 2022).

In non-sterilized soils accelerated difenoconazole (DFZ) degradation was observed compared to sterilized soils. With increasing concentrations of DFZ gradual decrease in soil bacterial community diversity was observed and a decrease in the soil bacterial community network complexity was estimated as well. Some of the 57 operational taxonomic units representing a core bacterial community were involved in DFZ degradation is soil (Zhang et al., 2021a). Surfactin biosurfactant producing bacterium strain (BTKU3) isolated from contaminated soil, that was identified as *Lysinibacillus* sp., was reported to degrade up to 9.1 μg/mL (91%) DFZ after 3 d of incubation at 35°C (Satapute and Jogaiah, 2022). Whereas after 3 mon 99% removal of DFZ was observed in planted as well as unplanted soils (control), an amendment of 5% biochar to soils reduced the degradation to 88% in planted and 83% in unplanted soil. In the presence of biochar, the uptake of the fungicide was considerably lessened, for example, by 45% at the application of 5% biochar, and this reduction was associated with improved sorption and microbial degradation of DFZ in soils supplemented with biochar. Moreover, in biochar-treated planted soils the abundance of *Sphingomonadaceae* and *Pseudomonadaceae* showed an increase of 18 and 63%, respectively, suggesting that soil chemical properties modified by biochar resulted in the increase of DFZ-degrading bacteria, and eventual reduction of fungicide bioavailability by biochar cannot be excluded as well (Cheng et al., 2017). *Sphingomonas* isolate S11 producing surfactin degraded 15.13% propiconazole (PCZ) after 48 hr and inhibited the development of *Fusarium culmorum* colonies *in vitro* by 48.80%, and was able to reduce the deoxynivalenol content of wheat grain inoculated with *F. culmorum* 22-fold (Wachowska et al., 2020). Immobilized cells of *Burkholderia* sp. strain BBK-9 tested for the degradation of PCZ; 20 μg/mL) utilized up to 19.2 μg /mL of the fungicide after 96 hr at 30°C and pH 7 (Satapute et al., 2018).

Sphingobacterium multivorum strain B-3 isolated from sewage, activated sludge and soil was able to degrade 85.6% hexaconazole (HXZ) using an initial inoculum concentration of 0.4 g/L and initial concentration of HXZ of 50 mg/L in 6 d, and it degraded 45.6% HXZ in 60 d in natural soil. Three degradation products, namely 2-(2,4-dichlorophenyl)-1-(1*H*-1,2,4-triazol-1-yl)hexane-2,5-diol, 2-(2,4-dichlorophenyl)hexane-1,2-diol and 1*H*-1,2,4-triazole (see Fig. 10.4) were identified and up-regulation of aldehyde dehydrogenase, monooxygenase, RND transporters and ABC transporters genes was observed (An et al., 2020).

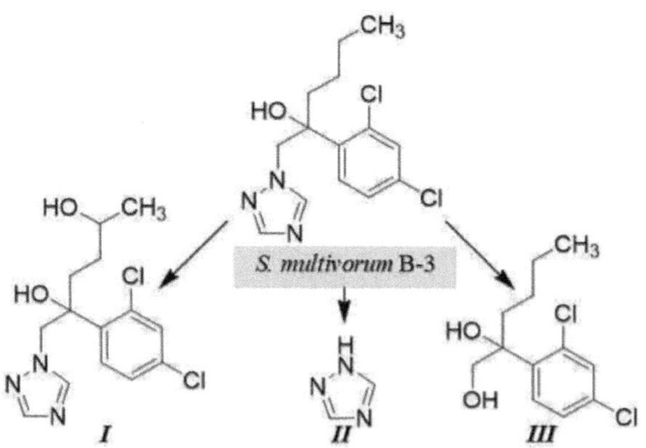

Fig. 10.4: Degradation products of hexaconazole by *Sphingobacterium multivorum* B-3: 2-(2,4-dichlorophenyl)-1-(1*H*-1,2,4-triazol-1-yl)hexane-2,5-diol (*I*), 1*H*-1,2,4-triazole (*II*) and 2-(2,4-dichlorophenyl)hexane-1,2-diol (*III*).

Among *Candida tropicalis*, *Enterobacter cloacae* and *Pseudomonas aeruginosa* microbial strains able to degrade prothioconazole (PTC), *P. aeruginosa* was found to be the most efficient with degradation efficiency up to 93.32%. The strain cultures produced overall 62 different PTC degradation metabolites, by which the main products prothioconazoledesthio, prothioconazole-dechloropropyl and oxidizing PTC were detected. However, transformation products such as C_9H_7NO, $C_{10}H_{17}N_7$ and $C_{12}H_{13}ClN_2O$ were only detected using incubation with *Enterobacter cloacae* (Shi et al., 2020). The comparison of soil microbial communities in plots planted with lowbush blueberries treated with fungicides PTC or chlorothalonil with untreated control plots showed that exposure to PTC resulted in an enhanced abundance of *Clavaria sphagnicola*; the bacterial genus *Rudaea* also showed considerably higher abundance in soils of treated plots. Bacterial degradation of fungicides was accompanied by an increasing abundance of enzymes (S)-2-haloacid dehalogenase and haloacetate dehalogenase assisting in the degradation of Cl-containing organic substances. Moreover, remarkable differences in the abundance of some enzymes associated with soil nutrient cycles (nitronate monooxygenase, xanthine dehydrogenase) were observed in fungicide-treated plots compared to untreated ones (Lloyd et al., 2021).

In batch reactors *Sphingomonas* sp. NJUST37 completely degraded 100 mg/L tricyclazole within 102 hr, whereby biodegradation started with monooxygenation and was followed by triazole ring cleavage, decyanation reaction, hydration reaction, deamination and dihydroxylation up to final mineralization of transformation products. Moreover, when a pilot-scale powdered activated carbon treatment tank was used for the remediation of real fungicide wastewater inoculated with *Sphingomonas* sp. NJUST37, the removal efficiency of tricyclazole achieved even 90% (Wu et al., 2018a). Using inoculation of activated sludge with *Sphingomonas* sp. NJUST26 and *Sphingomonas* sp. NJUST37, bacterial strains degrading 1*H*-1,2,4-triazole and tricyclazole, practically complete removal of both fungicides from wastewater in the bioreactor was achieved, accompanied by toxicity reduction. The inoculated fungicide-biodegrading strains altered the structure of the microbial community, although after long-term operation their dominance was lost (Wu et al., 2018b).

It should be noted that the application of γ-radiation of 1.2 kGy achieved complete mineralization of penconazole and myclobutanil (MCB) in an aqueous solution (Saadaoui et al., 2021) and about 90% of the primary degradation of fluconazole (FLZ) was observed within 1 hr by optimized photo-Fenton removal of the fungicide (Frankowski et al., 2021).

2.2 Benzimidazoles

Benzimidazole fungicides are effective on ascomycetes and basidiomycetes and can be applied to cereals, fruits, vegetables, vines and are also used in post-harvest crop handling. Their mechanism of action is inhibition of β-tubulin polymerization, i.e., they kill cells during division. The best known benzimidazole fungicides include carbendazim (CBZ) and thiabendazole (TBZ). Benomyl and thiophanate-methyl are

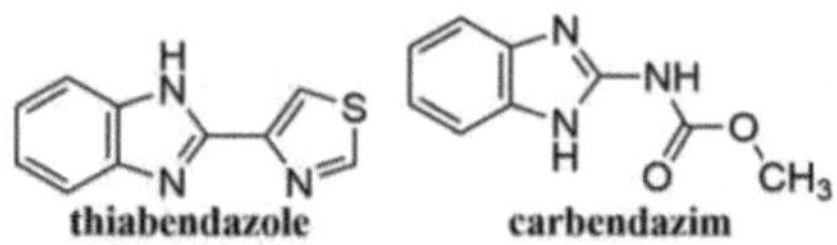

Fig. 10.5: Structures of described benzimidazoles.

also used, but they are activated by conversion to carbendazim (FRAC Code List 2022; Zhou et al., 2016). The structures of the described benzimidazole fungicides are shown in Fig. 10.5.

The bacterial consortium consists predominantly of α-(*Shinella*, *Oligotropha*, *Sphingomonas*), β-(*Methylobacillus*, *Methilibium*, *Hydrogenophaga*) and γ-*Proteobacteria* (*Pseudomonas*, *Hydrocarboniphaga*) and a *Sphingomonas* strain was able to degrade 750 mg/L of TBZ in an aqueous media and up to 500 mg/kg TBZ in soil, showing good degradation capacity at pH 4.5–7.5 and in a temperature range from 15°C to 37°C via mineralization of the benzoyl ring of TBZ benzimidazole moiety; high degradation capacity was also observed in the presence of other fungicides (*ortho*-phenylphenol and diphenylamine) (Perruchon, et al., 2017a). A bacterial consortium composed of α-, β- and γ-*Proteobacteria* with *Sphingomonas* phylotype B13 identified as crucial TBZ degrading strain degraded TBZ via the cleavage of the imidazole moiety, releasing thiazole-4-carboxamidine and the benzoyl moiety transformed to catechol, that might be consumed by the bacterial consortium (Perruchon et al., 2017b).

Carbendazim (CBZ) is a systemic broad-spectrum fungicide used in agriculture against various fungal diseases, however, it is an endocrine-disrupting compound, and therefore it is important to remove it from environmental matrices (Singh et al., 2016). The half-life of CBZ depends on the nature of the soil and can range from several days to 1 yr (Singh et al., 2016). Sunlight radiation can not ensure sufficient CBZ degradation (Salunkhe et al., 2014). The degradation of CBZ in soils can be affected by biochar addition. While in the presence of 0.5 or 5% of sewage sludge-derived biochar produced at 700°C in the soil the degradation of CBZ was pronouncedly suppressed and increased biochar amendment of 10% resulted in remarkably faster CBZ degradation, in soil without amendment eight metabolites of CBZ were identified (Huang et al., 2020).

A microbial consortium consisting of *Streptomyces albogriseolus* and *Brevibacillus borstelensis* strains exhibited ca. 97 and 98% CBZ degradation in 12 hr and 20 hr, respectively, when the initial fungicide concentration of 30 μg/mL was used. Both tested bacterial strains were resistant to CBZ toxicity and grew even at a high fungicide concentration of 500 μg/mL (Arya and Sharma, 2016). Bacterial strains isolated from the agricultural soils identified as *Streptomyces* sp. CB1, *Bacillus subtilis* CB2, *Pseudomonas aeruginosa* CB3 and *Rhizobium leguminosarum* CB were found to degrade CBZ. The higher fungicide-degrading efficiency was shown by *P. aeruginosa* CB3. Whereas the addition of Cu^{2+} ions contributed to higher microbial degradation of CBZ, supplementation of Fe^{2+} ions did not affect it. On the other hand, in the presence of humic acid enhanced and extended growth of bacterial isolates was observed, although the impact on CBZ microbial degradation was modest (Singh et al., 2019). Co-incubation of bacterium *Brevibacillus panacihumi* C17 isolated from agrochemical waste effluent in India with 100 mg/L CBZ for 36 hr at pH 7.0 and applied speed of 180 rpm resulted in 87.25% degradation of CBZ. The fungicide was biosorbed on the microbial outer cell surface utilizing adsorption energy and for describing the bio-sorption process Dubinin–Kaganer–Raduskevich isotherm model was the most suitable. At application of *Brevibacillus* sp. C17 strain the Chemical Oxygen Demand (COD) of the agrochemical waste effluent (880 mg/L) was reduced by 85% within 50 hr of batch treatment. It was shown that the bacterial treatment resulting in CBZ degradation considerably reduced the toxicity of the real effluent and aqueous CBZ solution (Kanjilal et al., 2018).

Chuang et al. (2021) investigated the impact of CBZ and its degrading strain, *Rhodococcus qingshengii* strain djl-6, on the composition, and diversity of soil bacterial and fungal communities under laboratory conditions. At a treatment lasting 2 wk, CBZ showed an adverse effect on bacterial communities and diminished the amount of dominant fungal phylum *Ascomycota*. The presence of CBZ enhanced the connectivity and modularity of microbial co-occurrence networks, thereby contributing to the augmented complexity of soil microbial co-occurrence networks. The strain djl-6 was found to be suitable for bioremediation of CBZ-polluted soils; in the degradation of CBZ strains of *Arthrobacter*, *Bacillus*, *Brevundimonas*, *Lysinibacillus*, *Massilia*, *Mycobacterium*, *Paenibacillus* and *Pseudarthrobacter* also might be involved. CBZ-degrading bacterium strain dj1-11 identified as *Rhodococcus erythropolis*, that was isolated from fungicide-contaminated soil samples, efficiently degraded 1000 mg/L CBZ in Minimal Salts Medium (MSM) at 28°C *in vitro* (333.33 mg/L per day), whereby the addition of NH_4NO_3 (12.5 mM) resulted in remarkable enhancement of the degradation. As degradation metabolites 2-aminobenzimidazole (2-AB) and 2-hydroxybenzimidazole (2-HB) were estimated and *mheI* gene cloned from chromosomal DNA of djl-11 exhibited 99% sequence homology to the *mheI* gene from *Nocardioides* sp. SG-4G (Zhang et al., 2013). Gram-positive, rod-shaped bacterium *Rhodococcus* sp. D-1 showing similarity to *Rhodococcus erythropolis* degraded 98.20% of 200 ppm CBZ within 5 d into primary metabolites 2-AB and 2-HB (see Fig. 10.6); 2-AB was subsequently gradually degraded to CO_2 and H_2O by other enzymes (Bai et al., 2017). Based on genomic analysis, researchers presumed expression of a new kind of enzyme in *Rhodococcus* sp. D-1 capable to hydrolyze CBZ. As rhamnolipid, that is considered a bacterial surfactant, can affect cell transport mainly by modifying cell surface hydrophobicity (Zhong et al., 2016a), its impact on CBZ biodegradation was investigated as well. The presence of rhamnolipid supported CBZ biodegradation in a dose-dependent manner and using a dose corresponding to its Critical Micelle Concentration (CMC; 50 ppm) the highest CBZ biodegradation (97.33%) within 2 d was observed; a triple rhamnolipid dose (150 ppm) on the beginning of treatment remarkably inhibited CBZ biodegradation, which achieved only 0.01% within 2 d but subsequently stimulated CBZ biodegradation up to 99.26% after 5 d. Hence, suitable modification of cell surface hydrophobicity and zeta potential by CBZ along with CBZ emulsification facilitated direct uptake of fungicide and its

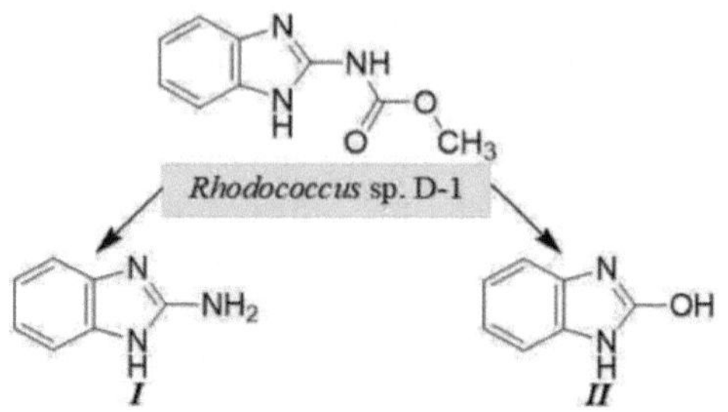

Fig. 10.6: Degradation products of carbendazim by *Rhodococcus* sp. D-1: 2-aminobenzimidazole (*I*) and 2-hydroxybenzimidazole (*II*).

subsequent biodegradation. *Mycobacterium* strain SD-4 showing 98.6% similarity to *Mycobacterium aromaticivorans* JCM 16368^T used CBZ as the only carbon and nitrogen sources for its growth and degraded CBZ to less toxic primary metabolite 2-AB via hydrolysis by CBZ-hydrolyzing esterase (MHEI), which was then converted to 2-HB and 2-HB, subsequently degraded through the ring cleavage. This strain degraded 90% of 50 mg/L CBZ after incubation lasting 72 hr with a mean degradation rate of 0.63 mg/L. The researchers optimized the codon of *mheI* gene to achieve its soluble expression in *E. coli*. As a key amino acid affecting MheI activity, cystein was identified and based on the site-directed mutation experiment, a substantial role of Cys16 and Cys222 in CBZ hydrolysis by MheI was validated (Zhang et al., 2017). Lei et al. (2017) cloned the CBZ hydrolyzing enzyme gene from *Microbacterium* sp. strain djl-6F and heterologously expressed it in *Escherichia coli* BL21. The MHEI-6F protein was found to catalyze direct hydrolysis of CBZ to 2-AB, showing optimum catalytic activity at 45°C and pH 7.0. The presence of metal ions reduced the activity of MHEI-6F, while sodium dodecyl sulfate exhibited a strong inhibitory effect. As catalytic groups essential for CBZ hydrolysis Cys16 and Cys222 of MHEI-6F were identified.

In the investigation of CBZ biodegradation using *Agaricus bisporus* Spent Mushroom Substrate (SMS) and its dominating microbes, it was found that 17.45% CBZ removal was observed with B-1 bacterial isolate broth culture at 30°C in an experiment lasting 6 d. Using sterilized CBZ mixed with *A. bisporus* SMS at 100 μg/g of SMS (control), after 15 d 11.90% CBZ degradation was observed, which increased to 33.50% at the application of mixed inoculum of *Trichoderma* sp. and *Aspergillus* sp., suggesting the potential of SMS to be used in bioremediation of CBZ polluted sites (Ahlawat et al., 2010). Incubation of CBZ with *Trichoderma* spp., *Trichoderma harzianum*, *T. viride* and *T. atroviride* for 5 d resulted in 85, 47 and 21% degradation of the fungicide (Sharma et al., 2016a). A single culture of *Aspergillus niger* degraded 69.66 and 99.96% of CBZ (4 ppm) after 10 and 20 d, respectively, while using the mixed cultures containing *A. niger*, *Exerohilum* sp. and *Fusarium* sp. 98.34% of CBZ was degraded after 10 d. On the other hand, 92.14 and 55.74% degradation after 10 d of incubation was observed in the mixed culture of *Exerohilum* sp. and *Fusarium* sp. exposed to double and tree fold CBZ concentrations (8 and 12 ppm) (Raheem et al., 2021). The bioremediation of CBZ can also be achieved using *Aspergillus versicolor* fungus. However, using an augmented microbial consortium containing bacterial species such as *Pseudomonas*, *Klebsiella* species or *Bacillus subtilis* and *A. versicolor*, as much as 94.4% degradation of CBZ was observed (Rajpal et al., 2023). Bacterial isolate CBW identified as *Pseudomonas* sp. was able to remove 87.1 and 99.1% of CBZ initial concentrations of 1.0 and 10.0 mg/L after 3 d ; pH 7.0 was found to be optimal for CBZ degradation, that to some extent became greater with increasing temperature. During degradation as primary CBZ metabolite, 2-AB was detected, which was subsequently transformed to 2-HB, 1,2-diaminobenzene, catechol and CO_2 (Fang et al., 2010). Alvarado-Gutierrez et al. (2017) investigating the kinetics of CBZ removal by an acclimated microbial consortium in a horizontal tubular biofilm reactor observed the highest instantaneous removal rates in its first section, whereby *Klebsiella*, *Stenotrophomonas* and *Flavobacterium* strains identified in the biofilm were found to degrade the fungicide in axenic culture. After 20 d of incubation at 30°C, pH 7.0, and 120 rpm the CBZ-degrading *Achromobacter* sp. strain GB61 isolates from polluted agricultural soil biodegraded 76.2% CBZ, whereby $t_{1/2}$ of 7.3 d and maximum specific degradation rate (q_{max}) of 0.122 per day were estimated; as biodegradation products 2-AB, 2-HB and benzimidazole were identified (Bhandari et al., 2021).

Comammox bacterium *Nitrospira inopinata* was found to biotransform fenhexamid fungicide, asulam herbicide as well as mianserin and ranitidine drugs, and also biotransformed CBZ when supplied with ammonia but not nitrite as the energy source. Ammonia monooxygenase of *N. inopinata* showed a much higher substrate (ammonia) promiscuity than that of *Nitrososphaera gargensis* and *Nitrosomonas nitrosa* Nm90. At biotransformation of CBZ hydroxylation at the aromatic ring occurred (Han et al., 2019). Murillo-Zamora et al. (2017) investigated the simultaneous removal of CBZ, TEB, metalaxyl (MLX), triadimenol, edifenphos, EPO and fenbuconazole fungicides by a biomixture containing coconut fiber, compost and soil at a volume ratio 45:13:42, which was bioaugmented with *Trametes versicolor*. Whereas the bioaugmented matrix efficiently removed CBZ, edifenphos and MLX, the removal of triazole fungicides failed. Nine days after the application of fungicides their effective detoxification was observed, and it was found that even non-bioaugmented biomixture can be successfully used for the removal of fungicides except triazoles.

2.3 *Strobilurin fungicides*

Strobilurins are benzene or pyrimidine derivatives of methacrylic acid that block the Q_o site of cytochrome b. These Q_o inhibitors, that obstruct mitochondrial respiration by blocking electron transport in mitochondria, mostly contact pesticides with long-degrading time (FRAC Code List, 2022; Feng et al., 2020a). The structures of the strobilurins are shown in Fig. 10.7.

The enrichment culture XS19 containing *Pseudomonas* (69.8%), *Sphingobacterium* (21.2%), *Delftia* (6.3%) and *Achromobacter* (1.6%) bacterial strains enriched from soil was able to degrade 50 mg/L of azoxystrobin (AZO), picoxystrobin, trifloxystrobin, kresoxim-methyl, pyraclostrobin (PYR) and enestroburin strobilurin fungicides at 50 mg/L within 8 d via hydrolysis of their

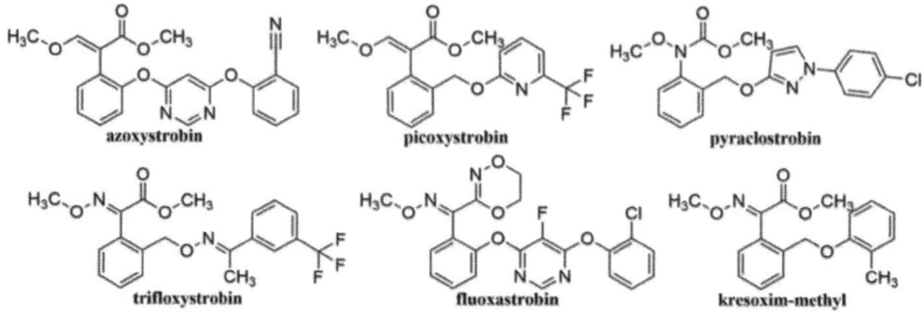

Fig. 10.7: Structures of described strobilurins.

methyl ester group. The crucial role in the degradation of strobilurin fungicides played carboxylesterases in *Pseudomonas* and *Sphingobacterium*, whereby in the presence of bis-*p*-nitrophenyl phosphate, a carboxylesterase inhibitor, the strobilurin-degrading ability of XS19 was suppressed (Wang et al., 2022). Isolated strains *Cupriavidus* sp. and *Rhodanobacter* sp., which degraded AZO added as the only carbon source, were also able to degrade other strobilurin fungicides such as trifloxystrobin, PYR and kresoxim-methyl, although for successful degradation an additional nitrogen source was required. On the other hand, after supplementation of a further carbon source, fungicide degradation was strongly diminished (by ca. 50%). The degradation of AZO in soil was accompanied by modification of the structure of the bacterial community (Howell et al., 2014). The impact of AZO on the microbial and enzymatic activity of AZO-contaminated soil was investigated by Bacmaga et al. (2015). Increasing doses of fungicide (2–50 mg/kg soil dry mass) suppressed the growth of organotrophic bacteria, fungi and actinomycetes along with altering microbial biodiversity. AZO inhibited activities of soil enzymes including dehydrogenases, catalase (CAT), urease, acid phosphatase and alkaline phosphatase, whereby the highest resistance to the impact of AZO exposure exhibited dehydrogenases. The researchers isolated from soil containing 22.50 mg/kg AZO four bacterial species from the genus *Bacillus* and five fungal species from the genus *Aphanoascus*. Feng et al. (2020a) presented a review paper that focused on current progress in strobilurin degradation and analyzed the effectiveness of strobilurin-degrading microbes and microbial consortia, which have the potential to be used in the bioremediation of fungicide-polluted environments. A novel esterase gene, strH, responsible for the de-esterification of strobilurin fungicides by generating the corresponding parent acid was cloned in the newly isolated strain *Hyphomicrobium* sp. strain DY-1. The estimated catalytic efficiencies (K_{cat}/K_m) of StrH were as follows: 2.41 ± 0.19 × 10^{-2} μM^{-1}. s^{-1} for AZO, 2.94 ± 0.02 μM^{-1}. s^{-1} for PYR, 4.64 ± 0.05 μM^{-1}. s^{-1} for picoxystrobin, and 196.32 ± 2.30 μM^{-1}. s^{-1} for trifloxystrobin, by which the higher activity for trifloxystrobin was observed at 50°C and pH 7.0. De-esterification of fungicides was accompanied by their detoxification and alleviation of the adverse impact on algal growth (Jiang et al., 2021).

In laboratory batch experiments using Spodosols amended with 2, 25 and 50 mg AZO/kg soil reduced activities of urease, invertase and phosphatase were observed, that can affect nutrient cycling and carbon utilization. Even the lowest applied fungicide dose of 2 mg/kg inhibited activities of hydrolytic enzymes urease, invertase and phosphatase after 35 d, in contrast to oxidoreductase enzyme CAT, the activity of which was stimulated. AZO also modified relative abundances of soil microbial communities. Approximately 50% of the bacterial populations were represented by *Streptomyces*, *Amycolatopsis* and *Sphingomonas*. Whereas treatment with 2 and 25 mg/kg AZO resulted in increased relative abundances of *Streptomyces* compared to control, exposure to 50 mg/kg soil was accompanied by higher relative abundances of *Amycolatopsis* compared to the control, and the abundance of *Sphingomonas* decreased with increasing AZO doses (Wang et al., 2020). Hocinat and Boudemagh (2016) isolated nine actinomycetes strains showing homology to *Nocardia* sp., and *Streptomyces* sp., that was able to degrade 500 mg/L of Ortiva fungicidal preparation, with AZO as the active ingredient within 21 d at 30°C when it was supplied as the only carbon source. As degradation product of AZO produced by *Bacillus licheniformis* strain TAB7 methyl (*E*)-3-amino-2-(2-{[6-(2-cyanophenoxy) pyrimidin-4-yl]oxy}phenyl)acrylate and *Z*-isomers of AZO and azoxystrobin amine as the metabolites of (*E*)-AZO (see Fig. 10.8) were identified by Mpofu et al. (2021), suggesting that *B. licheniformis* strain TAB7 was able to contribute to the enzymatic isomerization of (*E*)-AZO to (*Z*)-AZO; the AZO amine was found to be less toxic than AZO. *Ochrobactrum anthropi* SH14, a bacterial strain, which utilizes AZO as the only carbon source, was isolated from polluted soil Feng et al. (2020b). This bacterial strain was able to degrade 86.3% of 50 μg AZO/mL in a MSM within 5 d, achieving a maximum specific degradation rate of 0.6122 d^{-1}; the estimated inhibition constant was 188.4680 μg/mL. Aleksova et al. (2021) investigated the impact of Quadris® fungicidal preparation containing AZO as the active ingredient on carbon sources and bacterial catabolic profiles using doses of up to 35.0 mg AZO/kg dry loamy sand soil. It was found that a dose of 2.90 mg/kg dry soil (corresponding to the field recommended concentration)

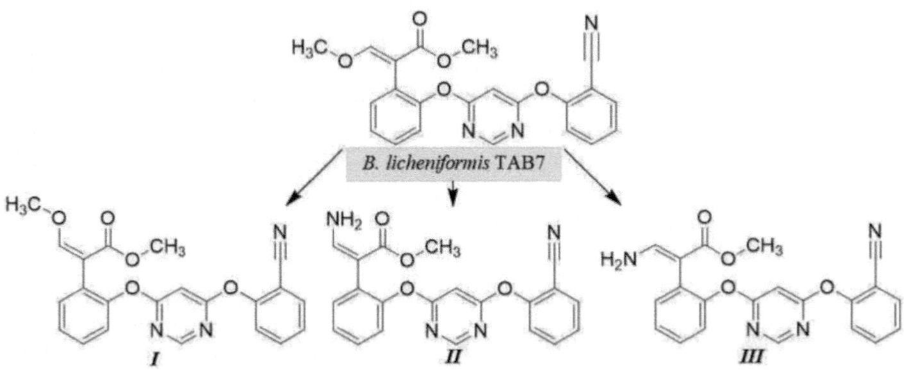

Fig. 10.8: Degradation products of azoxystrobin by *Bacillus licheniformis* TAB7: *Z*-isomers of AZO (*I*), methyl (*Z*)-3-amino-2-(2-{[6-(2-cyanophenoxy)pyrimidin-4-yl]oxy}phenyl)acrylate (*II*) and methyl (*E*)-3-amino-2-(2-{[6-(2-cyanophenoxy)pyrimidin-4-yl]oxy}phenyl)acrylate (*III*).

modified mainly the low-available biolog carbon sources, while doses ranging from 14.65 to 35.00 mg/kg dry soil also functioned on medium and highly utilizable carbon sources. Linear correlation between AZO soil residues and used carbon sources were not observed and it was presumed that fungicidal preparation impacted bacterial catabolic profiles in investigated soils via soil acidification and modifying soil nutrient pool.

A comparison of AZO and PBZ degradation in relatively undisturbed soil during field studies and sieved soil in a laboratory experiment showed that AZO degradation rates expressed by median degradation time (deg. $t_{1/2}$) were 34–37 d and did not depend on applied soil. On the other hand, pronouncedly accelerated degradation of PBZ was observed in undisturbed cores (deg. $t_{1/2}$: 63 d) compared to sieved soil (deg. $t_{1/2}$: 255 d), whereby the moisture level did not affect the rate of degradation. On the other hand, disruption of soil structure greatly affected the microbial community structure (Hand et al., 2020).

Based on the evaluation of PYR degradation efficiency of bacteria from orange cultivation plots in a liquid nutrient medium, the five most efficient strains *Bacillus* sp. CSA-13, *Paenibacillus alvei* CBMAI2221, *Bacillus* sp. CBMAI2222, *Bacillus safensis* CBMAI2220 and *Bacillus aryabhattai* CBMAI2223 were selected. Using these bacterial strains in consortia, their effects, whether synergistic or antagonistic, depended on the used combination of bacteria. The researchers suggested a biodegradation pathway with 15 identified metabolites, including 1-(4-chlorophenyl)-1*H*-pyrazol-3-ol. The reduction of PYR in soil by the native microbiome can be enhanced using bioaugmentation process via inoculation of soil with an efficient bacterial consortium (Birolli et al., 2020). During degradation of AZO, the main intermediate compounds illustrated in Fig. 10.9 were formed and subsequently transformed into non-persistent metabolites. Using *O. anthropi* SH14 strain 57.2, 76.6, 78.7, 88.5 and 89.4% degradation of fluoxastrobin, picoxystrobin, trifloxystrobin. PYR and kresoxim-methyl were also achieved. Application of this bacterial strain for bioaugmentation of AZO-polluted soils resulted in considerably accelerated fungicide degradation manifested in reduced AZO half-life by 95.7 and

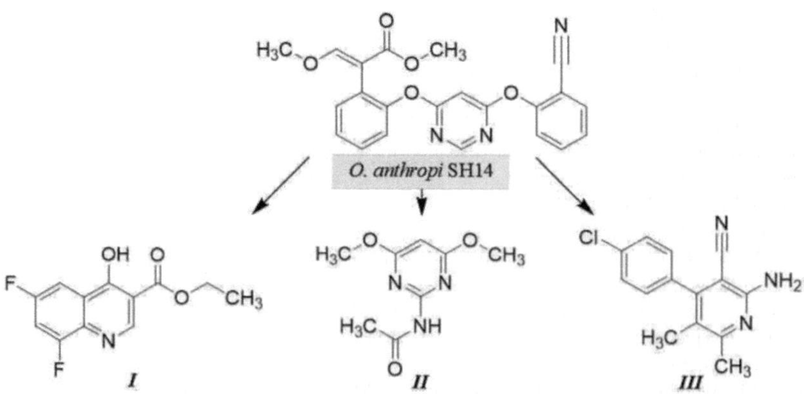

Fig. 10.9: Degradation products of azoxystrobin by *Ochrobactrum anthropi* SH14: ethyl 6,8-difluoro-4-hydroxyquinoline-3-carboxylate (*I*), *N*-(4,6-dimethoxypyrimidin-2-yl)acetamide (*II*) and 2-amino-4-(4-chlorophenyl)-5,6-dimethylpyridine-3-carbonitrile (*III*).

65.6 d in sterile and non-sterile soils compared to control soils missing bacterial treatment (Feng et al. 2020b).

2.4 Other fungicides

Malhotra et al. (2021) analyzed published findings related to the environmental fate, toxicity, metabolic routes, related genes and enzymes as well as evolutionary mechanisms associated with the degradation of carbamate pesticides and discussed the possibilities enabling improved degradation efficiency, including the use of microbial consortia for efficient degradation and metabolic engineering. The structures of the compounds described here are shown in Fig. 10.10.

Pseudomonas putida QTH3 was able to degrade pentachloronitrobenzene (PCNB) achieving 49.84% degradation in 35 d and as degradation products catechol, 2,3,5,6-tetrachloroaniline, 2,3,4,5-tetrachloroaniline, 2,3,4,5,6-pentachloroaniline and pentachlorothioanisole (see Fig. 10.11) were identified. At incubation with 100 mg/L PCNB for 30 min, the degradation rate of the intracellular enzyme achieved 44.73% compared to 8.93% observed for the extracellular enzyme, suggesting that the intracellular enzyme is primarily responsible for fungicide degradation (Wang et al., 2019a). *In-situ* remediation of PCNB-contaminated soil planted with *Panax notoginseng* using soil inoculation with 0.15 kg/m^2 and 0.30 kg/m^2 of solid inoculum of *Cupriavidus* sp., YNS-85 reduced fungicide soil concentration by 50.3 and 74.2%, respectively, within 1 yr compared to the uninoculated control. Although plant biomass, physicochemical properties of the soil and diversity of the soil microbial community did not differ much between inoculated and control groups, reduced soil CAT activity was observed at treatment with *Cupriavidus* sp. YNS-85; on the other hand, fluorescein diacetate esterase activity depended on the applied dose of inoculum (Zhang et al., 2020b). Alfalfa plants also effectively accumulated and degraded PCNB fungicide. In the planted alfalfa, soil PCNB

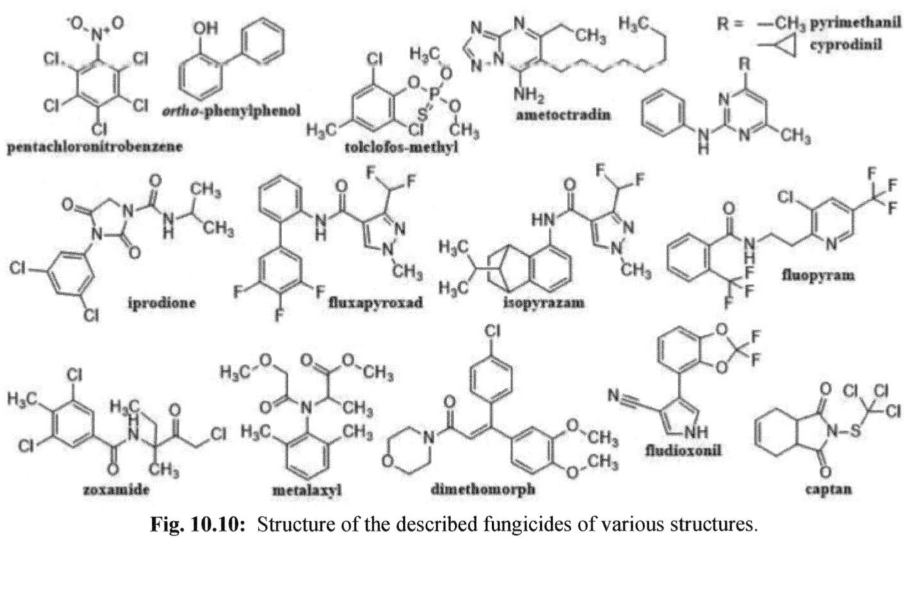

Fig. 10.10: Structure of the described fungicides of various structures.

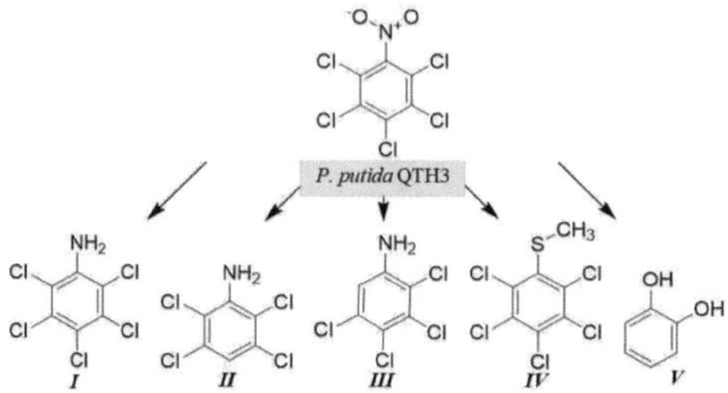

Fig. 10.11. Degradation products of pentachloronitrobenzene by *Pseudomonas putida* QTH3: 2,3,4,5,6-pentachloroaniline (*I*), 2,3,5,6-tetrachloroaniline (*II*), 2,3,4,5-tetrachloroaniline (*III*), pentachlorothioanisole (*IV*), catechol (*V*).

degradation rate achieved 66.26−77.68% within 20 d compared to 48.34% in non-planted soil, in which the fungicide residues also greatly exceeded those observed in planted soil. The presence of PCNB in soil affected the activities of soil enzymes such as urease, polyphenol oxidase, alkaline phosphatase and acid phosphatase; the activities of these enzymes showed an increase following planting with *Medicago sativa* (Li and Yang, 2013). *Sphingomonas haloaromaticamans* strain P3 isolated from soil originating from a wastewater disposal site exhibited fast degradation of *ortho*-phenylphenol (OPP) using it as an energy source. However, the degrading effectiveness of these bacteria depended on the external supply of amino acids or on the co-occurrence of other bacteria that were not involved in fungicide degradation. For example, in the presence of TBZ and diphenylamine *S. haloaromaticamans* P3 proved to metabolize up to 150 mg/L of fungicide within 1 wk in a wide range of temperatures (4−37°C) and pH (4.5−9) (Perruchon et al., 2016). Perruchon et al. (2017c) performed genomic, proteomic and transcription analysis of a *Sphingomonas haloaromaticamans* strain degrading OPP and identified two orthologous operons encoding the orthocleavage of benzoic acid (ben/cat), while proteomic analysis estimated 13 up-regulated catabolic proteins when *S. haloaromaticamans* was growing on OPP and/or benzoic acid. Based on transcriptomics crucial role in the transformation of OPP by *S. haloaromaticamans* played catabolic operons located in the 92-kb scaffold and flanked by transposases and via heterologous expression, and the researchers also isolated a flavin-dependent monoxygenase belonging to mostly up-regulated proteins in the cells growing on the fungicide. The amino acid auxotrophy of fungicide-degrading bacteria can restrain their bioremediation capacity. Supplementation of OPP-degrading *Sphingomonas haloaromaticamans* strain with isoleucine, phenylalanine, tyrosine and methionine prevented its degrading capacity only with methionine, and the bacterium was not able to accomplish de novo biosynthesis of vitamin B12 due to the absence of genes for the construction of the corrin ring, and for its growth and preservation of degrading capacity supply of vitamin B12 at concentrations corresponding to those occurring in the environment (i.e., 0.1 ng/mL) was necessary (Perruchon et al., 2020). Mobile genetic elements of sewage sludge-isolated *Pseudomonas nitroreducens* HBP, able to completely degrade the OPP fungicide, were described by Carraro et al. (2020). Co-inoculation of the culture of *Sphingomonas* sp. 224, known to degrade tolclofos-methyl, a phospholipid biosynthesis inhibitor and organophosphorus fungicide, with biofilm-forming bacterium *Bacillus* sp. E5 ameliorated biofilm formation, achieving about 90% degradation efficiency within 48 hr, which exceeded that observed with the application of *Sphingomonas* sp. 224 alone; with the application of consortium biofilm higher degradation of tolclofos-methyl was also observed *in situ*, using soil (Kwak et al., 2013).

In an *in vitro* experiment, the degradation of ametoctradin by soil-derived microbial consortia achieved 81% in 72 hr, while the half-life of fungicide contained in soil was 2 wk. However, a correlation of degradation with the increased relative abundance of *Burkholderiales* was observed not only in *ex vivo* soils but also *in vitro* in the packed-bed microbial bioreactor (Whittington et al., 2020). Gram-negative bacterial strain showing 99% homology to *Acinetobacter* sp., isolated from cyprodinil-contaminated soil degraded this fungicide into two polar metabolites, see Fig. 10.12. According to researchers, this cyprodinil-degrading strain might have gene coding of amine hydrolase and hydroxylase and imines hydrolysis and monohydroxyl substitution on the benzene ring as a biodegradation pathway was proposed. Whereas the monoclonal strain showed good degradation activity at an initial fungicide concentration of 200 mg/L (degradation after incubation of 2 wk), mixed bacterium degraded cyprodinil at a concentration of 800 mg/L (Chen et al., 2018). The investigation of the impact of pyrimethanil (PYM), a fungicide inhibiting methionine biosynthesis, on soil microbial community under altered rainfall regimes (2 and 8 wk after its application) showed that rainfall extremes modified the impact of PYM on the structure of soil microbial community structure but did not affect PYM effects on soil functions, including enzyme activities, potential nitrification and BIOLOG carbon substrate utilization (Ng et al., 2014).

Paenarthrobacter sp. strain YJN-5 started to degrade iprodione (IPD) via hydrolysis of a fungicide amide bond to *N*-(3,5-dichlorophenyl)-2,4-dioxoimidazolidine, see Fig. 10.13. For this step, the ipaH gene, encoding a

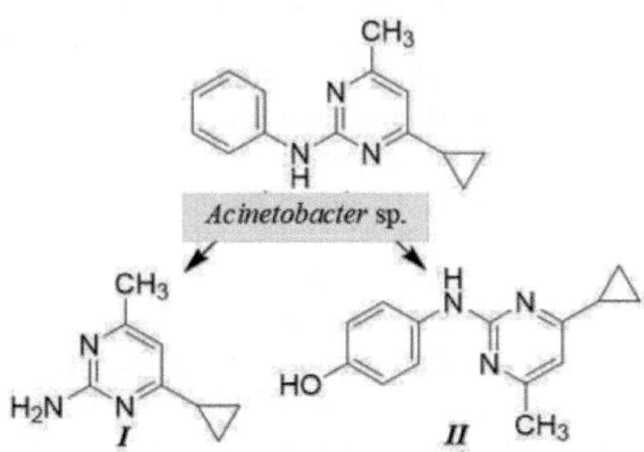

Fig. 10.12: Degradation products of cyprodinil by *Acinetobacter* sp.: 4-cyclopropyl-6-methylpyrimidin-2-amine (*I*), 4-[(4-cyclopropyl-6-methylpyrimidin-2-yl)amino]phenol (*II*).

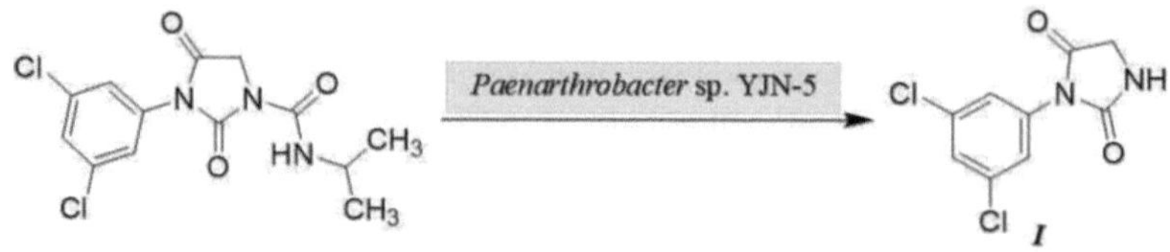

Fig. 10.13: Degradation of iprodione by strain *Paenarthrobacter* sp. YJN-5: *N*-(3,5-dichlorophenyl)-2,4-dioxoimidazolidine (*I*).

novel amidase sharing 40% similarity with an indoleacetamide hydrolase from *Bradyrhizobium diazoefficiens* USDA 1, was found to be responsible. This ipaH gene containing the Ser–Ser–Lys motif characteristic for members of the amidase family and by replacement of Lys82, Ser157 and Ser181 with alanine complete loss of enzymatic activity was observed, suggesting only this gene is inevitable for starting the degradation of IPD (Yang et al., 2018). Treatment of pepper plants by four applications of IPD, whether in soil or sprayed on leaves, resulted in faster pesticide degradation manifested with shorter half times of degradation in 4th treatment compared to the first one (0.48 vs. 1.23 d in soil and 5.95 vs. 365 d on leaves), whereby exposure to IPD considerably modified the composition of the epiphytic and soil bacterial and fungal communities. *Paenarthrobacter* strains able to degrade IPD, which were isolated from soil and phyllosphere, hydrolyzed the fungicide to 3,5-dichloraniline via the formation of 3,5-dichlorophenyl-carboxamide and 3,5-dichlorophenylurea-acetate (Katsoula et al., 2020). Consortium consisting of an *Arthrobacter* sp., strain C1 and an *Achromobacter* sp., strain C2, capable to transform IPD to 3,5-dichloroaniline, was isolated by Campos et al. (2017a). *Arthrobacter* sp., strain C1 exhibited fast degradation of IPD (DT_{50} = 2.3 hr) and its metabolite 3,5-dichlorophenylurea-acetate (DT_{50} = 2.9 h) in mineral salts medium indicating that isopropylamine, formed by hydrolysis of IPD and glycine formed during hydrolysis of 3,5-dichlorophenylurea-acetate were utilized as C and N sources; another metabolite of IPD, 3,5-dichlorophenyl-carboxamide, was degraded by *Arthrobacter* sp., strain C1 only in nutrient-rich medium. Similarly, *Achromobacter* sp., strain C2 transformed IPD and its metabolites only in a nutrient-rich medium. As the *Arthrobacter* sp., strain C1 was able to degrade vinclozolin (structurally similar to IPD) and partially propanil, but not procymidone and phenylureas, it can be stated that degradation was affected by substituents of the carboxamide moiety (Campos et al., 2017b). Campos et al. (2015) earlier reported that *Arthrobacter* sp., strain C1 exhibited good degradation of IPD and formation of 3,5-dichloraniline at pH 7.5 and a temperature of 35°C; effective IPD degradation occurred also in the presence of a mixture of pesticides, that can be present in on-farm biobed system. On the other hand, *Achromobacter* sp., strain C2 can slowly co-metabolize IPD with *Arthrobacter* sp., strain C1. Two bacterial cocultures of *Providencia stuartii* JD and *Brevundimonas naejangsanensis* J3 effectively degraded dimethachlon, IPD and procymidone to simple degradation metabolites. Cocultures of these two bacterial strains immobilized on charcoal-alginate-chitosan carrier degraded dimethachlon (20.25 kg a.i./ha), IPD (50 kg a.i./ha) and procymidone

(7.50 kg a.i./ha) in the field brunisolic soils by 96.74, 95.02 and 96.27%, respectively, what was also manifested in lower half-lives of 1.53, 1.59 and 1.57 d compared to free cocultures (3.60, 4.03 and 3.92 d) and natural dissipation (21.33, 20.51 and 20.09 d). Hence, fast bioremediation of soils polluted with dicarboximide fungicide can be achieved by utilizing synergetic degradation effects of immobilized bacterial cocultures (Zhang et al., 2021b).

Fluxapyroxad affected microbial activity, community structure and functional diversity. It inhibited substrate-induced respiration and microbial biomass on day 15, and later stimulated them. Within the first 15 d, the biomass of both Gram-positive and Gram-negative bacteria was reduced, but after 1 and 2 mon the ratio of Gram-negative to Gram-positive bacteria showed an increase. On the other hand, fluxapyroxad reduced fungal biomass and the ratio of fungi to bacteria was diminished as well (Wu et al., 2015). Fungicide isopyrazam exhibited the highest adsorption (Kd_{ads} up to 23 μg/mL) in soil containing a considerable content of organic matter (0.73%) and minimum half-lives estimated in hydrolysis, photolysis and biodegradation experiments were 16.7, 4.3, and 19.7 d, respectively (Gul et al., 2021b). Ahmad and Gul (2020) investigated the degradation of 10 mg/L isopyrazam by *Aspergillus flavus*, *Penicillium chrysogenum*, *Aspergillus niger*, *Aspergillus terreus*, *Aspergillus fumigatus* fungal strains and *Xanthomonas axonopodis* and *Pseudomonas syringae* bacterial strains for 35 d. The highest biodegradation efficiency of 86 and 80%, and the lowest half-life time of 21 and 28.1 d showed *P. syringae* and *X. axonopodis*, whereby transformation products illustrated in Fig. 10.14 were identified. The degradation efficiency and half-life time of other studied fungal strains were as follows: *A. terreus* 11%/56.8 d; *P. chrysogenum* 18%/44.7 d; *A. fumigatus* 21%/40.7 d; *A. niger* 18%/39.6d; *A. s flavus* 30%/32.6 d (Ahmad and Gul, 2020). The effectiveness of selected fungal and bacterial strains to degrade fluopyram (FPM), a new succinate dehydrogenase inhibitor fungicide, increased in the following order: *Aspergillus fumigatus* (24.2%,) < *Escherichia coli* (82.7%) < *Streptococcus pyogenes* (89.8%,) < *Aspergillus niger* (90.7%,) < *Aspergillus flavus* (91.3%,) < *Aspergillus terreus* (95.4%) < *Streptococcus pneumoniae* (99.3%) (Naeem and Ahmad, 2022). During incubating of FPM with rat and human liver microsomes its phase I metabolites such as benzamide, benzoic acid, 7-hydroxyl, 8-hydroxyl, 7,8-dihydroxylfluopyram, lactam fluopyram, pyridyl acetic acid and Z/E-olefin FPM as well as imide, hydroxyl lactam and 7-hydroxyl pyridyl acetic acid oxidative metabolites were detected, suggesting that oxidation by dechlorination is one of the crucial metabolism mechanisms of this fungicide (Mekonnen et al., 2018). At investigation of zoxamide biodegradation by *Escherichia coli*, *Streptococcus pyogenes* and *Streptococcus pneumoniae* degradation products including 2-(3,5-dichloro-4-methylphenyl)-4-ethyl-4-methyl-4*H*-1,3-oxazin-5(6*H*)-one, 3,5-dichloro-*N*-(3-hydroxy-1-ethyl-1-methyl-2-oxopropyl)-4-methylbenzamide and 3,5-dichloro-4-methylbenzamide (see Fig. 10.15) were identified. The most efficient bacterial biodegradation of zoxamide (28%) was observed with *E. coli*; the half-life of fungicide related to its degradation was found to be 42.5 d for *E. coli,* 58.7 d for *S. pyogenes*, and

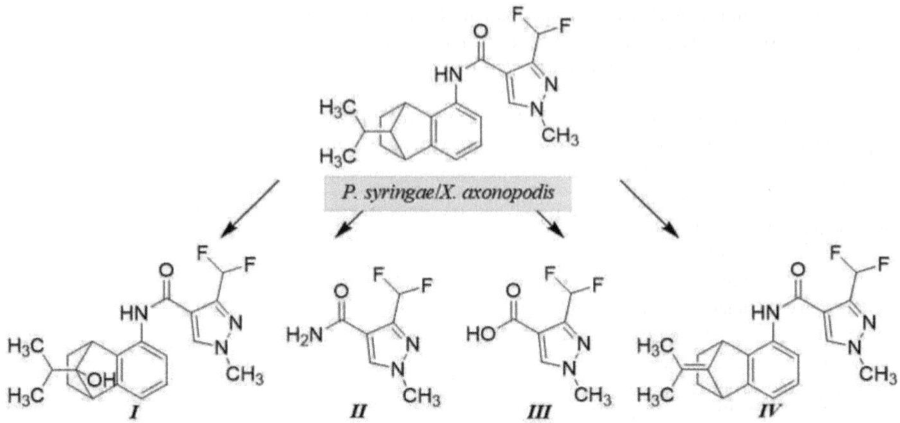

Fig. 10.14: Degradation products of isopyrazam by *Pseudomonas syringae* and *Xanthomonas axonopodis*: 3-(difluoromethyl)-*N*-[(9*S*)-9-hydroxy-9-(propan-2-yl)-1,2,3,4-tetrahydro-1,4-methanonaphthalen-5-yl]-1-methyl-1*H*-pyrazole-4-carboxamide (*I*), 3-(difluoromethyl)-1-methyl-1*H*-pyrazole-4-carboxamide (*II*), and 3-(difluoromethyl)-1-methyl-1*H*-pyrazole-4-carboxylic acid (*III*), 3-(difluoromethyl)-1-methyl-*N*-[9-(propan-2-ylidene)-1,2,3,4-tetrahydro-1,4-methanonaphthalen-5-yl]-1*H*-pyrazole-4-carboxamide (*IV*).

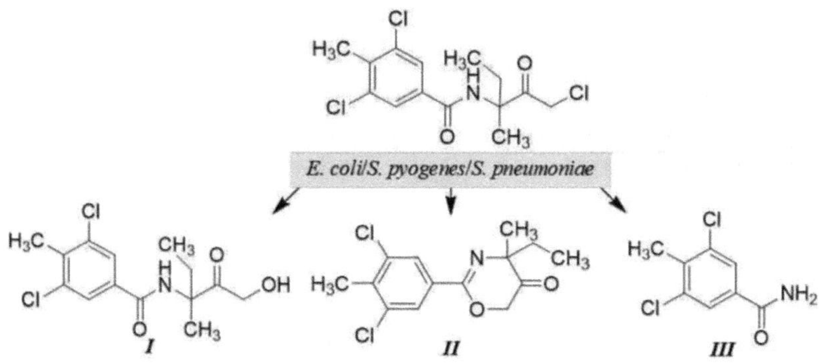

Fig. 10.15: Degradation products of zoxamide by *Escherichia coli*, *Streptococcus pyogenes* and *Streptococcus pneumoniae* strains: 3,5-dichloro-*N*-(3-hydroxy-1-ethyl-1-methyl-2-oxopropyl)-4-methylbenzamide (*I*), 2-(3,5-dichloro-4-methylphenyl)-4-ethyl-4-methyl-4*H*-1,3-oxazin-5(6*H*)-one (*II*), and 3,5-dichloro-4-methylbenzamide (*III*).

67.9 d for *S. pneumoniae* (Ahmad et al., 2020). Zoxamide sorption depended on the physical and chemical characteristics of soils and in standard batch equilibrated experiment adsorption coefficients ranging from 2.5 to 25 μg/mL were estimated. Using hydrolytic, biodegradation and photolytic experiments for zoxamide in the pedosphere minimum half-life of 17.6, 15.7 and 4.3 d, respectively, was observed, whereas the highest half-life achieved 21.9, 22.2 and 7.4 d, respectively (Gul et al., 2021a). The successive application of metalaxyl (MLX) affected soil microbial community structure. Increasing applied times of MLX resulted in its faster degradation, while a rise in bacteria populations, suppression of soil fungi and transient promoting the impact on actinomycetes was observed when the fungicide was supplemented for the second time; with the third MLX application, only the fungal population after treatment was affected in 2 wk (Wang et al., 2019b). The effects of multiple applications of MLX at the recommended field rate and double of recommended field rate on the soil microbial community were also investigated earlier by Wang et al. (2015). *Bacillus cereus* WL08 isolated from soils contaminated with dimethomorph (DMM) immobilized on a system prepared using bamboo charcoal and sodium alginate exhibited higher fungicide-degrading efficiency compared to free *B. cereus* WL08 (96.88 vs. 66.95%; pH 7.0; 30°C) at initial DMM concentration of 50 mg/L within 72 hr, and in a continuous reactor system was able to remove 85.61% of fungicide for 1 mon, whereby the influent concentration ranged from 50 to 100 mg/L. Using immobilized WL08 also reduced the half-life of fungicide in the field soil contaminated with DMM (Zhang et al., 2020a).

Fludioxonil (FLU)-contaminated wastewater treated under microaerophilic conditions in an immobilized cell bioreactor stepwise diminished the Hydraulic Retention Times (HRTs) from 10 to 3.9 d and 12 transformation products were identified. FLU degradation started by hydroxylation and carbonylation of the pyrrole moiety and breaking of the oxidized cyanopyrrole ring at the HN−C bond. 2,2-Difluoro-2*H*-1,3-benzodioxole-4-carboxylic acid transformation product validated the decyanation and deamination of the fungicide and 2,3-dihydroxybenzoic acid metabolite suggested its defluorination. HRT shortening was accompanied by reduced α-diversity and pronounced changes in the β-diversity; the microbial community consisting of bacterial taxa suitable to perform an effective degradation and/or aerobic denitrification (e.g., *Clostridium*, *Oligotropha*, *Pseudomonas* and *Terrimonas*) differed from the composition of the bacterial community originating from the initial activated sludge system (Mavriou et al., 2021a). Using an up-flow immobilized cell bioreactor for treatment of FLU-polluted wastewater the fungicide removal efficiency achieved 95.4%, where by 94.0% mineralization of organic-F was observed. From the start-up to the end of FLU treatment a 51.2-fold enhancement of the relative abundance of *Empedobacter*, *Sphingopyxis* and *Rhodopseudomonas* was observed (Mavriou et al., 2021b). Soil bacterium *Planomicrobium flavidum* strain EF was found to resist captan up to 2000 ppm and in MSM medium supplemented with captan and exhibited higher growth than in the absence of captan. Shaking conditions allowed about 77.5% utilization of captan by *P. flavidum* within 2 hr of growth, and after 24 hr of growth only 0.8% of captan was not removed (Mohamed and Mostafa, 2018). *Bacillus circulans* using captan as the only source of carbon and energy degraded it at first by hydrolysis to *cis*-1,2,3,6-tetrahydrophthalimide, which further degraded to phthalic acid (Fig. 10.16) by a protocatechuate pathway (Megadi et al., 2010).

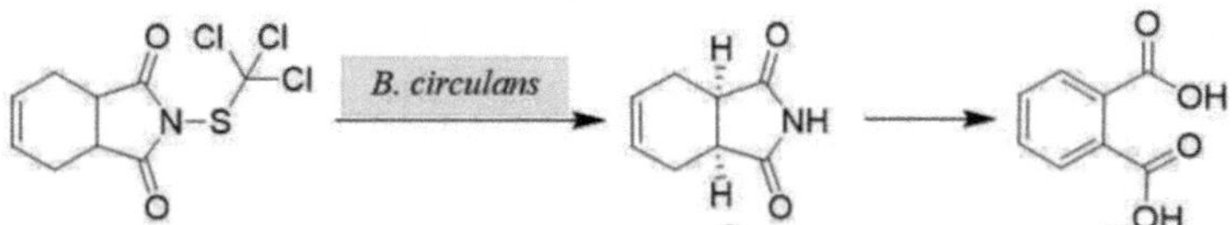

Fig. 10.16: Degradation products of captan by strain *Bacillus circulans*: *cis*-1,2,3,6-tetrahydrophthalimide (*I*) and phthalic acid (*II*).

3. Detoxification of fungicides by laccases

Extracellular and/or cell-free enzymes as well as purified or partially purified enzymes can also be used for the detoxification of contaminants. Moreover, enzymatic effectiveness can be ameliorated via DNA engineering to design excellent bioremediators or using single-enzyme nanoparticles, in which each enzyme molecule is surrounded by a hybrid organic/inorganic polymer network (Bhandari and Sharma, 2021). Laccases (benzenediol: oxygen oxidoreductase, multicopper oxidases), which belong to polyphenoloxidases, can catalyze mono-electronic oxidation of phenolic substrates if atmospheric molecular oxygen is present as their co-substrate (Kumar et al., 2014; Bilal et al., 2019; Janusz et al., 2020). Moreover, in the presence of mediator systems, laccase can also degrade non-phenolic molecules (Sathishkumar et al., 2019). Consequently, it can be also used for bioremediation of wastewaters contaminated with fungicides such as malachite green (MG; 4-{[4-(dimethylamino) phenyl](phenyl) methylidene}-*N*,*N*-dimethylcyclohexa-2,5-dien-1-iminium chloride) (Wang et al., 2018b; Mao et al., 2021; Thoa et al., 2022).

Crude extracellular laccase derived from a fungal strain *Fusarium oxysporum* HUIB02 was able to remove $\geq$ 80% MG after 20 hr of treatment; the addition of Cu^{2+} ions stimulated MG degradation and in the presence of mediators up to 99% degradation of MG was achieved. The efficient MG degradation was observed with MG concentration $\leq$ 100 mg/L, while at a tenfold higher MG concentration it decreased (Thoa et al., 2022). The extracellular yellow laccase from white rot fungus *Trametes* sp. F1635 showed decolorization of MG > 60%, where the best mediators violuric acid and acetosyringon (3,5-dimethoxy-4-hydroxyacetophenon) were estimated (Wang et al., 2018b). As CotA protein present in the spores of *Bacillus* strains confers spore laccase activity, Cheng et al. (2021) cloned and overexpressed CotA-laccase from *Bacillus subtilis* in *Escherichia coli*. Using CotA-laccase up to 90–94% decolorization of MG at freshwater and saline conditions was observed suggesting powerful degradation of the fungicide. Thermostable laccase Ghlac Mut2 variant from *Geothermobacter hydrogeniphilus* (Ghlac), which was cloned and expressed in *E. coli* showed stability in pH range from 4.0 to 8.0 and half-life at 50°C and 60°C of 0.6 hr and 9.8 hr, respectively, caused practically 100% decolorization of 100 mg/L MG in 3 hr at 70°C, whereby the degradation products were not toxic to bacteria and maize plants (Mao et al., 2021).

Removal of MG using laccase immobilized biochar (Pandey et al., 2022a), nanocellulose synthesized from an agro-waste of quinoa husks (Ariaeenejad et al., 2022), magnetic dialdehyde cellulose (Qiao et al., 2022) or graphene oxide (Zhou et al., 2022) resulting in enhanced degradation efficiency was also reported. In isolated treatments with crude laccase extract from *Pleurotus dryinus* grown on municipal biosolid the fungicide degrading efficiency decreased as follows: kresoxim-methyl (96%), >PYR (91%) >CBZ (59%) >FLU (56%) PYM (29%) while using of commercial laccase >90% fungicide removal rate was observed for kresoxim-methyl, pyraclostrobin, CBZ, TBZ and IPD. The catalytic effect observed in a mixture of related compounds was comparable with the laccase mediator effect (Vaithyanathan et al., 2022).

4. Bioremediation of fungicide-contaminated aqueous environment using algae

Microalgae, which are characterized by rapid biomass growth and CO_2 fixation (Gerotto et al., 2020; Kráľová and Jampílek, 2021), can be successfully used for the removal of emerging metal and organic contaminants (Miazek and Brozek-Pluska, 2019; Manikandan et al., 2021). During degradation, microalgae utilize organic contaminants as the primary source of carbon and nutrients, which are indispensable for their growth and development (Haripriyan et al., 2022). Superior biosorption, biodegradation and transformation capacities of cyanobacteria and microalgae determine their use for removal of hazardous pollutants from industrial effluents and agricultural discharges to prevent contamination of the water bodies (Gehlot et al., 2022). Cyanobacteria can sequester toxic metals by biosorption and bioaccumulation, and they are also able to degrade various xenobiotics to nontoxic forms (Pandey et al., 2022b). Algal biomass can be subsequently used for biodiesel production (El-Sheekh et al., 2018; Ismail et al., 2020). Bioremediation of wastewater containing organic contaminants by microalgae and cyanobacteria and utilization of algal biomass is shown in Fig. 10.17.

Cyanobacterium *Nostoc muscorum* and green alga *S. obliquus* exposed to captan (15 mg/L or 30 mg/L) for 1 wk accumulated it intracellularly, whereby *S. obliquus* degraded it with high efficiency to phthalic acid and 1,2,3,6-tetrahydrophthalimide,

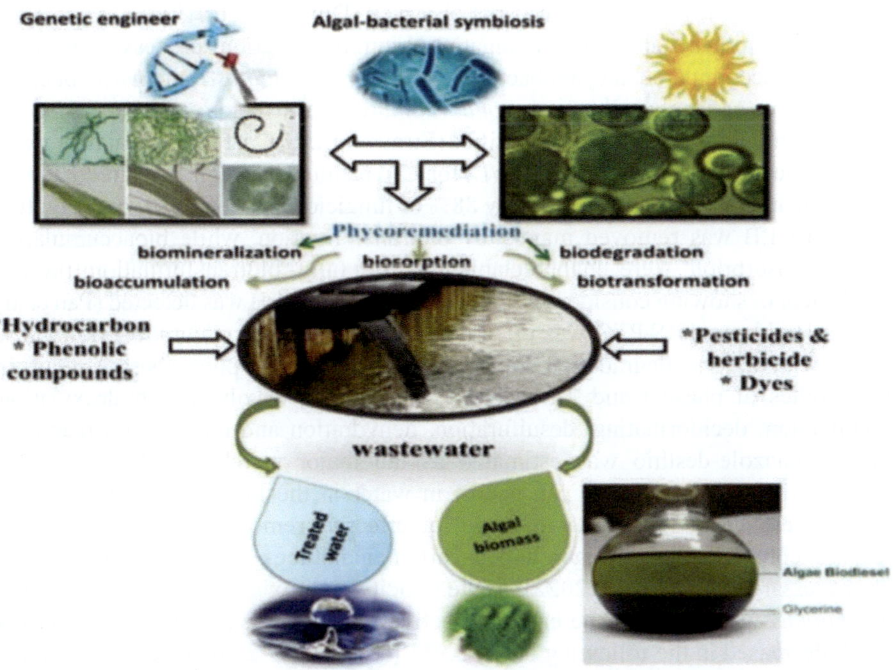

Fig. 10.17. Bioremediation of wastewater containing organic contaminants by microalgae and cyanobacteria and utilization of algal biomass. Adapted from Touliabah et al. (2022). Copyright 2022 MDPI.

i.e., metabolites, which are eco-friendly compounds. *S. obliquus* was more tolerant to captan treatment, which was reflected in lesser impairment of algal growth and lesser reduction of chlorophyll (Chl) content and activities of enzymes phosphoenolpyruvate carboxylase (PEPC) and ribulose-1,5-bisphosphate carboxylase/oxygenase (Rubisco), that are indispensable for photosynthetic processes. Strong oxidative damage induced by higher captan concentration manifested by pronouncedly increased levels of H_2O_2, malondialdehyde (MDA), NADPH oxidase and protein peroxidation was observed particularly in *N. muscorum*. In the studied algal species, glutathione-*S*-transferase enzyme and antioxidant defense system were involved in captan detoxification. Due to the superior antioxidant defense system of *S. obliquus* compared to *N. muscorum* this green alga shows potential to be used for phycoremediation of captan-contaminated aqueous environment (Hamed et al., 2022).

For removal of CBZ lyophilized green alga *Chlorella thermophila* was also investigated showing improved bioadsorption of the fungicide with increasing pH; the maximum bioadsorption was observed at pH 10 (Karkala et al., 2022). Toxicity of CBZ to freshwater diatom *Navicula* sp. was manifested by reduced growth of the diatom (25 h IC_{50}: 2.18 mg/L), while high acute toxicity of the fungicide was associated with its rapid accumulation in diatom cells. Due to the biotransformation of accumulated CBZ within algal cells via hydroxylation, methylation, decarboxylation demethylation and deamination reactions, the algal growth rate was recovered after 72 hr; however, at exposure to > 0.5 mg/L CBZ the enhancement of low Chl levels was not observed (Ding et al., 2019).

On the other hand, at investigating biotransformation processes of azole fungicides accumulated in cyanobacteria *Microcystis aeruginosa* and *Synechococcus* sp., as well as green alga *Chlamydomonas reinhardtii* using batch experiments *in vitro* no transformation products were identified (Stravs et al., 2017). Exposure of freshwater microalgae S. *obliquus* to CLB (2 mg/L) for 12 d resulted in a significant reduction of algal density and Ch content, whereby 88% of fungicide was removed. In the culture solution CLB was removed mainly by biotransformation, while bioaccumulation and bioadsorption were inappreciable; as the only biotransformation product CLB-alcohol showing considerably lower toxicity than CLB was detected (Pan et al., 2018). Highly toxic *S*-PTC was effectively biodegraded by *Chlorella pyrenoidosa* along with abiotic degradation including hydrolysis and photolysis. Among 14 metabolites of phase I and 2 metabolites of phase II involved in hydroxylation, methylation, dechlorinating, desulfuration, dehydration and conjugation reactions, prothioconazole-desthio was estimated as the major metabolite. However, this particularly toxic metabolite can persist in water in the long term, independently of the presence of *C. pyrenoidosa* and for effective removal of prothioconazole-desthio cobalt coated nitrogen-doped carbon nanotubes and peroxymonosulfate are necessary (Zhang et al., 2022). Acute toxicity and toxic effects of MCB against *S. obliquus* were found to be enantioselective; the estimated 96 hr EC_{50} values in mg/L decreased in the following order: 3.951 ((−)-MCB) >2.760 (rac-MCB) >2.128 ((+)-MCB). Different impact of MCB enantiomers on the levels of photosynthetic pigments and activities of antioxidant enzymes was observed as well. In addition, the degradation of MCB enantiomers in *S. obliquus* was also studied. Inoculation with *S. obliquus* resulted in faster fungicide degradation compared to the uninoculated medium, whereby at a concentration of 3 mg/L the degradation of (−)-MCB was faster than that of (+)-MCB (Cheng et al., 2013).

Exposure of cyanobacterial species *Anabaena laxa* and *Nostoc muscorum* to *R*-MLX (10 and 25 mg/L) resulted in a dose-dependent decrease in algal growth, photosynthetic pigment content and activities of PEPC and Rubisco, whereby *A. laxa* showed higher tolerance to the fungicide. Similarly, oxidative damage caused by *R*-MLX to *A. laxa* was less evident than in *N. muscorum* due to pronounced induction of non-enzymatic antioxidants such as polyphenols, flavonoids, tocopherols and glutathione (GSH), and increased activities of antioxidant enzymes (peroxidase, glutathione peroxidase, glutathione reductase and glutathione-*S*-transferase. Moreover, *A. laxa* accumulated more *R*-MLX and might be used for bioremediation of *R*-MLX-contaminated aqueous environment (Hamed et al., 2020). *Scenedesmus obliquus* and *S. quadricauda* freshwater microalgae exposed to 600 μg/L of DMM or PYM for 4 d were able to remove 10 and 24% of these fungicides. Higher fungicide removal by algal cells was observed with *S. quadricauda* compared to *S. obliquus* achieving 1.1 multiple for dimethomorp and 1.53 multiple for pyrimethanil (Dosnon-Olette et al., 2010).

As mentioned above, algae also can effectively remove metals/metalloids, including Cu-based fungicides from the contaminated aqueous environment. The growth of microalgae *S. obliquus*, *Chlorella fusca*, *Ankistrodesmus braunii*, *Chlorella saccharophila* and *Leptolyngbya JSC-1* exposed to 20 ppm Cu was not affected and Cu removal from water by the algae achieved 99.9, 99.3, 97, 96.7, and 96%, respectively. With prolonging the exposure up to 12 d, only low leakage of nucleic acid and protein was observed. The ability of tested algae to remove Cu from soils was lower, namely 73, 75, 71, 70, and 68%, respectively, for *S. obliquus*, *C. fusca*, *A. braunii*, *C. saccharophila* and *Leptolyngbya JSC-1*. In both treatments, the highest Cu removal was obtained with *S. obliquus* (Zada et al., 2022). Extracellular Polymeric Substances (EPS) produced by cyanobacteria species *Nostoc* sp., N27P72 and *Nostoc* sp., FB71 showed good adsorption capacity for Cu^{2+}, whereby using maltose as a carbon source increased the production of EPS as well as enhanced protein and carbohydrate levels achieved, which contributes to the improved ability of cyanobacteria for metal absorbance. Hence, EPS might be used in pesticide bioremediation instead of synthetic and abiotic flocculants (Ghorbani et al., 2022). For Cu removal from a Cu-contaminated aqueous environment macroalgae are also suitable. Filamentous macroscopic green algae *Microspora* sp., occurring in acidic (pH 4.47–5.75) and metal-enriched discharges from a copper mine in India were able to accumulate 1.74–3.66% Cu suggesting the potential of this alga to be used for remediation of Cu-polluted aqueous environment (Equeenuddin et al., 2021). Batch and fixed-bed studies focused on the trapping of Cu^{2+} ions by raw biomass of brown marine macro-algae *Ascophyllum nodosum* showed that *A. nodosum* utilizes cation exchanger properties for Cu removal from an aqueous environment, whereby –COOH and $-SO_3$ are crucial functional groups responsible for the binding of cations. The reuse of the algal biomass in multiple cycles can be ensured by its regeneration using 0.1 M $CaCl_2$ at pH 8.0 (Mazur et al., 2017). Brown macroalga *Pelvetia canaliculata* also functioned as a natural cation exchanger releasing light metals (Na^+, K^+, Mg^{2+} and Ca^{2+}) at sequestration of Cu^{2+} and Zn^{2+} ions by raw macroalgal biomass; at pH 4.0, practically all binding sites were occupied by Cu^{2+} and Zn^{2+} ions reaching maximum uptake capacities about 2.4 mEq/g. The amount of weak acid COOH groups on the biomaterial surface was 1.5-fold higher than that of strong acid $-SO_3$ groups (Girardi et al., 2014). Red macroalgae *Gracilaria lemaneiformis* and *G. lichenoides* exposed to 0−500 μg/L Cu accumulated more Cu^{2+} on the extracellular side (cell wall) than in the cytoplasm and chloroplasts, cell nucleus, mitochondria and ribosomes. Higher content of insoluble polysaccharide in the cell wall of *G. lichenoides* stimulated the extracellular Cu^{2+}-binding and was able to prevent metal toxicity. However, compared to *G. lichenoides*, oxidative stress induced by Cu^{2+} caused stronger oxidative damages in *G. lemaneiformis* resulting in reduced algal growth and contents of photosynthetic pigments and impaired photosynthetic activity as well as increased Reductive Oxygen Species (ROS) and MDA accumulation and electrolyte leakage due to insufficient activities of antioxidant enzymes (Huang et al., 2013).

Whereas no detectable amounts of GSH and comparable levels of phytochelatins (PCs) PC2 were detected in both *Ulva compressa* algae coming from control sites and Cu-polluted seawater sites of northern Chile, enhancement of PC4 and transcript levels of *UcMT1* and *UcMT2* was observed in algae from the Cu-polluted sites. Cu content in *U. compressa* from Cu-polluted sites was more than one order higher compared to algae of control sites (260 and 272 μg Cu/g d.m. vs. 20 μg Cu/g d.m.); algae of Cu-polluted sites showed intracellular Cu accumulation and electrodense nanoparticles containing Cu in the chloroplasts (Espinoza et al., 2021).

5. Remediation of the fungicide-contaminated environment using plants

Phytoremediation is a cost-effective environment-friendly method, which can also use plants to remediate contaminated environmental matrices via extraction, immobilization (stabilization), degradation and volatilization of contaminants (Masarovičová and Kráľová, 2018; Kráľová and Jampílek, 2022a; Tonelli et al., 2022).

5.1 Remediation of synthetic fungicides

Phytodegradation (or phytotransformation) and rhizodegradation (or phytostimulation) are phytoremediation techniques enabling the breakdown of organic contaminants. Whereas in the rhizodegration process the degradation of contaminants occurs in the rhizosphere by root-associated microorganisms, the exudates produced by the plant stimulate activities of microbial enzymes resulting in improved degradation of contaminants (Dominguez et al., 2020; Sivaram et al., 2020; Anum et al., 2022), in phytodegradation process the contaminants are degraded within the plant by the metabolic process utilizing plant enzymes such as peroxidases, phenol oxidases, esterases or nitroreductases, ensuring the transformation of parent contaminant to more polar metabolites, which subsequently conjugate with plant biomolecules forming less toxic metabolites, which are then incorporated into plant tissues (He et al., 2017; Qu et al., 2018; Takkar et al., 2022; Tonelli et al., 2022). For removal of organic and metal contaminants from stormwater runoff and industrial wastewaters Constructed Wetlands (CWs), i.e., artificial wetlands acting as biofilters can be used. Whereas in the filter bed usually sand and gravel are present, the microorganisms colonizing the plant organs of sown vegetation degrade organic contaminants (Lv et al., 2016b; Lyu et al., 2018; Parlakidis et al., 2021). Constructed wetland with a horizontal subsurface flow is shown in Fig. 10.18. The historical development of CWs for wastewater treatment, including their classification, was presented by Vymazal (2022).

The phytoremediation efficiency of four wetland plants *Typha latifolia*, *Phragmites australis*, *Iris pseudacorus* and *Juncus effusus* for imazalil and TEB was investigated by Lv et al. (2016a). The researchers found that plants cultivated in hydroponium in the presence of 10 mg fungicide per liter removed after 24 hr exposure 25–41% TEB and 46–96% imazalil. The accumulation of both tested fungicides in plant tissue achieved 2.8–14.4% fungicide amounts added in hydroponium at the beginning of the experiment, indicating that besides uptake of fungicides the plants were also able to metabolize them. Moreover, the translocation of fungicides and their degradation in the plants were enantioselective, but changes in the enantiomeric fraction of fungicides were not observed in hydroponic solutions. The efficiency of TEB removal in unsaturated and saturated CWs as well as in unplanted CWs and CWs planted with *T. lahfolia*, *P. australis*, *I. pseudacorus*, *J. effusus* and *Berula erecta* was investigated by Lyu et al (2018). Considerably higher fungicide removal was observed in unsaturated CWs (removal rate constant: 2.6–10.9 cm per day) than in saturated CWs (removal rate constant: 1.7–7.9 cm per day). Similarly, higher TEB removal was observed in planted CWs compared to unplanted ones, which was reflected in removal rate constants 3.1–10.9 cm per day and 1.7–2.6 cm per day, respectively. Hence, TEB removal was affected primarily by the system design, hydraulic loading rate and the presence of

Fig. 10.18: Constructed wetland with a horizontal subsurface flow (Roseč, Czech Republic). Photo Jan Vymazal. Adapted from Vymazal (2022). Copyright 2022 MDPI.

plants, but also positively correlated with dissolved O_2 and removal of all nutrients. On the other hand, TEB removal occurred primarily via biodegradation and metabolization inside the plants after its uptake by the plant, which was supported by low levels of TEB sorption to the substrate (0.7–2.1%) and plant phytoaccumulation achieving 2.5–12.1%. Saturated CWs planted with *T. latifolia*, *P. australis*, *I. pseudacorus*, *J. effusus* and *Berula erecta* exhibited pronouncedly higher removal efficiencies of imazalil and TEB compared to unplanted controls only in summer. The removal of fungicides was primarily due to the high rate of metabolization in the saturated CW mesocosms and correlated with the rate of evapotranspiration and the removal of macronutrients (nirogen and phosphorus) during summer and with the dissolved oxygen/oxygen saturation during winter. Besides degradation of the uptaken fungicides inside the plant tissue plant-stimulated microbial degradation in the bed substrate can also play a role, whereby nitrifying bacteria can actively contribute to the biodegradation of fungicides (Lv et al., 2016b). Using buffer zones reduction of field-scale losses of fungicides from subsurface drainage waters to surface waters can be achieved. At the investigation of sorption processes of fungicides boscalid, prochloraz and TEB on various substrates it was found that the potential retention of fungicides decreased as follows: straw >> sediments > soils, while the adsorption capacity of prochloraz was much higher than that of TEB and boscalid. Thus, by using plants which can enhance soil-sediment sorption capacity and straw filters, the sorption processes can be stimulated and improved efficiency of ponds and ditches can contribute to reduced contamination of surface water (Valee et al., 2014). In 45-meter long experimental stream mesocosms > 90% reduction of runoff related peak concentrations of TEB and trifloxystrobin fungicides and indoxacarb and thiacloprid insecticides applied as pesticide mixture was observed (Elsaesser et al., 2013).

PCZ, imazalil and TBZ supplemented to flow-through stream mesocosms (45 m length, 0.4 m width, 0.26 m water depth, discharge 1 L/s), that were planted with the submerged macrophyte *Elodea nuttallii* (Planch.) achieved a higher reduction of fungicides by 20–25% compared to the unvegetated mesocosm, whereby the macrophytes retained about 7–10% of the initially applied amounts of fungicides (Stang et al., 2013).

Wetland plant species *Pontederia cordata*, *J. effusus* and *Sagittaria latifolia* planted in sandy soil containers placed in AZO or imidacloprid treated water for 2 mon exhibited a higher reduction of AZO concentration by 51.7, 24.9 and 28.7% than was observed in non-planted containers. Hence, fast-growing wetland plants, which are tolerant to environmental conditions and require low maintenance, show the potential to be used for the remediation of synthetic fungicides (McKnight et al., 2022). On the other hand, acute exposure to a commercial formulation AMISTAR® containing AZO as an active ingredient was found to alter antioxidant enzymes and damage the aquatic macrophyte *Myriophyllum quitense*, causing oxidative impairment at high concentrations, including DNA damage at doses of 100 and 500 g/L (Garanzini et al., 2019). Considerable lipid and DNA damage along with remarkable inhibition of CAT and peroxidase in *M. quitense* at exposure to 50 and 100 μg/L AZO were reported earlier by Garanzini and Menone (2015). Exposure of maize roots to AZO for 96 hr resulted in accumulation of fungicide within the plant and gradual detoxification of compounds or conjugates also appeared. The crucial detoxified metabolite was the methyl ester hydrolysis product (AZO free acid) followed by GSH conjugate and its derivative lacking the glycine residue from the GSH as well as the glycosylated form of AZO, which was present in a small amount (Dionisio et al., 2019). Two and four weeks after spraying of *Lactuca sativa* plants with AZO, seven transformation products formed by hydrolysis, reduction, hydroxylation, photoisomerization and hydrolytic cleavage of ether bonds of the parent fungicide molecule were identified (Gautam et al., 2017). Under salinity stress, AZO accumulated in roots of *Plantago major* within 2 d more than under cold stress or natural conditions. In roots and leaves AZO carboxylic acid was formed via hydrolysis of methyl ester. In *P. major* roots under cold and salt conditions two metabolites, 2-(6-hydroxypyrimidin-4-yloxy) benzonitrile and AZO benzoic acid were detected, which were not present under natural conditions, but in the period 4–12 d of exposure were detected in leaves in all the treatments. Higher H_2O_2 and proline levels were estimated in the shoots of stressed plants compared to control plants; high NaCl levels of 100 mM considerably increased peroxidase activity and reduced Chl contents compared to control plants (Romeh, 2017). The comparison of AZO accumulating capacity of *Glycin max*, *P. major* and *Helianthus annuus* plants cultivated in AZO-contaminated soil showed that after 10 d of exposure the accumulated AZO amounts in roots decreased as follows: 25.32 mg/kg (*G. max*) > 20.62 mg/kg (*P. major*) > 18.29 mg/kg (*H. annus*), while maximal AZO amounts in the leaves were observed after 10 d of exposure in *P. major* (15.03 mg/kg) and *H. annuus* (9.8 mg/kg) and after 3 d of exposure in *G. max* (3.12 mg/kg). Thus, AZO accumulated more in the roots of plants than in their leaves (/Romeh, 2015). Stormwater wetlands designed mainly for flood protection can also diminish concentrations and remove loads of runoff-related pesticides into downstream aquatic ecosystems. A stormwater wetland was reported to reduce up to 100% of AZO, cymoxanil and cyprodinil fungicides in the runoff from a vineyard catchment during two successive periods of pesticide application and complete removal of dissolved kresoxym methyl and tetraconazole was observed as well (Maillard et al., 2012).

In *Rosa roxburghii* plants sprayed with PYR the degradation of fungicide followed the first-order kinetics and degradation $t_{1/2}$ achieved 3.86–5.95 d in soil and 6.20–7.79 d in *R. roxburghii*, while the terminal residues of fungicide were 0.105–3.153 mg/kg in soil and 0.169–1.236 mg/kg in *R. roxburghii* (Han et al., 2022).

Campos et al. (2017b) to maximize the effectiveness of biobeds for IPD dissipation used bioaugmentation strategy via inoculation of biobeds with *Arthrobacter* strain Cl degrading IPD and/or rhizosphere-assisted biodegradation using ryegrass and investigated alterations in the activity and composition of the microbial community. In bioaugmented biobeds faster biodegradation of the parent fungicide to 3,5-dichlorophenyl-2-carboxamide, 3,5-dichlorophenylurea acetate and 3,5-dichloroaniline metabolites were observed and $t_{1/2}$ under bioaugmented or rhizosphere-assisted treatment was considerably shorter than that of control

(3.4 vs. 9.5 d). Bioaugmentation did not show a remarkable impact on the abundance of α-Proteobacteria and Firmicutes; rhizosphere-assisted treatments resulted in the transient promotion of acid phosphatase and β-glucosidase, and increased hydrolytic activity observed at inoculation of biobeds with *Arthrobacter* Cl correlated with the hydrolysis of IPD. The impact of *Lolium perenne* rhizosphere on the dissipation of IPD fungicide and changes in the microbial community in a biopurification system was studied by Diez et al. (2017). The *L. perenne* rhizosphere was found to enhance IPD dissipation and the presence of plants promoted peroxidase activity. Whereas a comparable gradual increase in bacterial abundance with and without vegetal cover was observed during the experiment, after 40 d abundance of fungal species was reduced but treatment with fungicide led to its increase. In soils co-contaminated with Cd and IPD planting with *Medicago sativa* decreased soil concentration of both contaminants and increased the abundance of Bacteroidetes and Nitrospirae as well as the Cd amount phytoextracted by plants. In addition, using biochar amendment to soil improved plant biomass and Cd phytoextraction efficiency (Zhang et al., 2019).

Parlakidis et al. (2021) investigated the removal of FPM from water coming from cleaning of pesticide spraying equipment using horizontal subsurface flow CWs; the loading of CWs with fungicide-contaminated water occurred daily from December 2019 to January 2021. Whereas unplanted CW removed only 25.09% FPM, CW with planted *P. australis* removed 36.10% FPM. Using bioaugmentation with plant growth-promoting rhizobacteria the removal was enhanced to 70.67%, while application of gravel and zeolite as porous media resulted in 62.06% removal of FPM. On the other hand, CW planted with *T. latifolia* removed 59.98% FPM. Hence, besides plant species, the fungicide removal was mostly affected by bioaugmentation and application of zeolite presence in porous media.

The original biobed system comprises a clay layer at the bottom, a biomixture layer and a grass layer on top or it can be a specially excavated pit filled with a mixture of straw. For depuration of fungicide-contaminated effluents from various agro-food processing industries biobeds, which can remove 93.1−99.9% fungicides from the agro-industrial effluents can be used. In contrast to lipophilic fluxapyroxad, that was largely retained in the biobed, MLX-M and carboxin characterized with higher polarity were greatly dissipated. Biobeds promoted microbial communities, which were not affected by treatment with fungicides; this can be associated with microaerophilic conditions on water saturation of biobeds reflected in a pronounced increase in the abundance of facultative or strict anaerobes like *Chloroflexi*/*Anaerolinae*, *Acidibacter* and *Myxococcota* (Papazlatani et al., 2022). The addition of wood waste-derived biochar to soil after 3 wks reduced the concentrations of *R*-and *S*-MLX enantiomers and their *R*- and *S*-MLX acid metabolites by 57.7−86.3 and 13.3−32.5%, respectively, whereby lower uptake was due to reduced bioavailability of used fungicide enantiomers. The presence of biochar resulted in lower *R*-MLX amounts accumulated in *Lactuca sativa* plants compared to soil indicating that enantioselective uptake of MLX was diminished. Moreover, in the presence of biochar an increase in soil bacterial diversity was observed with an elevated abundance of *Luteimonas*, *Methylophilus* and *Hydrogenophaga*, i.e., microorganisms able to perform enantioselective degradation of fungicide in soil (You et al., 2021). Earlier relatively higher amounts of *S*-MLX enantiomer in *Solanum lycopersicum* and *Cucumis sativus* plants were observed due to faster degradation of (–)-*R*-MLX enantiomer to (–)-*R*-MLX acid metabolite (Li et al., 2013). *Cannabis sativa* L. exposed for 7 d to MLX-M (2–100 μg/mL) removed 67–94% of fungicide from water, whereby its content in plant dry biomass ranged from 106 to 3861 μg/g. The removal of the fungicide was primarily due to the fungicide plant uptake and transformation (Loffredo et al., 2021).

Lemna minor was able to remove 10–40% DMM applied in form of Forum® fungicidal preparation after 96 hr of exposure, achieving higher fungicide removal than at application of the pure ingredient (Megateli et al., 2013). DMM and isoproturon fungicides activate cytochrome P_{450} multienzyme family in an aquatic plant, i.e., *Elodea canadensis*, and detoxification of dimetomorph by this aquatic plant likely occurred via hydroxylation and subsequent glucosylation, possibly yielding soluble as well as cell wall-bound residues (Dosnon-Olette et al., 2011). In a non-waterproofed surface flow CW treated with a mixture of DMM, glyphosate (herbicide) and imidacloprid (insecticide) by simulating a single rain event, the average dissipation of 50% of the applied doses of pesticides was within 1 wk and calculated $t_{1/2}$ in the water phase of the wetland for imidacloprid and DMM were 20.6 and 12.0 d, respectively (Braschi et al., 2022).

CW mesocosms were reported to effectively reduce the levels of chlorothalonil in simulated stormwater runoff (Sherrard et al., 2004).

5.2 Remediation of inorganic fungicides

After absorption by roots, the metals, which are non-biodegradable contaminants, can be accumulated in roots and subsequently translocated into the aerial part of plants; these can be then harvested ensuring definitive removal of metal from the environment. This phytoremediation technique is known as phytoextraction (or phytoaccumulation) (Masarovičová et al., 2010; Sheoran et al., 2016; Ali et al., 2022). On the other hand, migration of metal contaminants by wind and water erosion, leaching and dispersion of the soil can be prevented by their adsorption on the surface of the plant roots, complexation with organic acids produced by roots, precipitation in slightly soluble metal forms or allocation in the root tissues using phytostabilization technique (Radziemska et al., 2017; Schachtschneider et al., 2017; Gavrilescu, 2022; Muthusamy et al., 2022). Under metal stress plants detoxify heavy metals via chelation by specific ligands and sequestration, whereby important chelators are phytochelatins and metallothioneins; plants which do not belong to hyperaccumulators of heavy metals, largely sequester them in root vacuoles (Sharma et al., 2016b; Gavrilescu, 2022). For remediation of environmental matrices contaminated with toxic metals various plant species, including trees, crops, macrophytes or medicinal and aromatic plants can be used (Masarovičová and Kráľová, 2018; Ali et al., 2022; Cherian and Joseph, 2022; Kráľová and Jampílek, 2022b). Sources of toxic metals and their foliar and root uptake in plants are shown in Fig. 10.19.

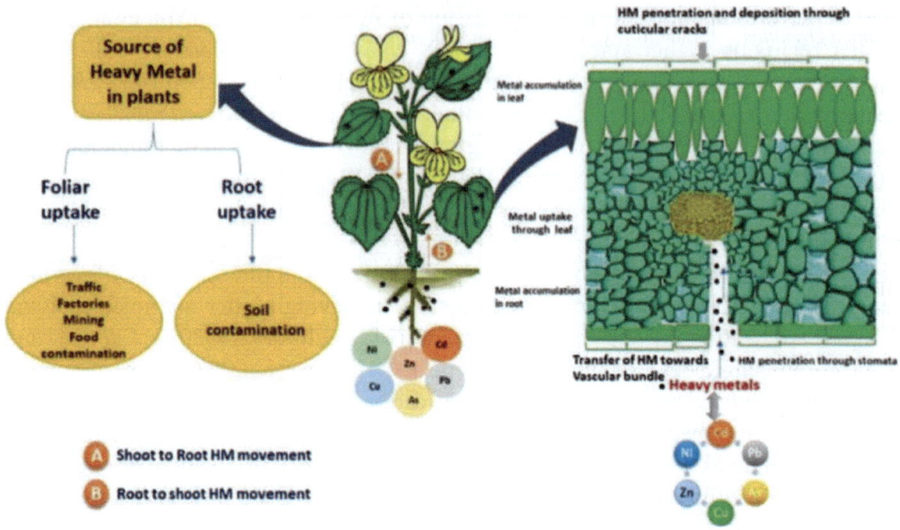

Fig. 10.19: Sources of toxic metals and their foliar and root uptake in plants. Adapted from Hasan et al. (2019). Copyright 2019 MDPI.

Copper is a micronutrient, which is indispensable for normal plant development and growth. As it is a cofactor of many enzymes and a component of plastocyanin, the carrier of photosynthetic electron transport in photosystem II, Cu deficiency has a negative effect on plant growth, while low Cu concentrations have a beneficial growth-stimulating effect on plants. On the other hand, higher Cu concentrations whether in bulk or nanoform, are phytotoxic (Masarovičová et al., 2014; Kráľová et al., 2019, 2021; Kráľová and Jampílek, 2022c). In addition to the application of Cu in fertilizers, it is used as a fungicide in treating plant diseases which can contribute to soil contamination with Cu (Tamm et al., 2022).

Tamm et al. (2022) evaluated the use of Cu-based fungicides in organic agriculture in 12 European countries (Belgium, Bulgaria, Denmark, Estonia, France, Germany, Hungary, Italy, Norway, Spain, Switzerland and the UK) and found that in organic agriculture 3258 t of Cu are used per year, which corresponds to 52% of the permitted annual dosage, suggesting the reduced application of Cu-based plant-protection products; the greatest amounts of these Cu-based agrochemicals were used for olives (1263 t/y, 39%), grapevines (990 t/y, 30%) and almonds (317 t/y, 10%).

Shabbir et al. (2020) and Kumar et al. (2021) published comprehensive review papers summarizing findings related to Cu bioavailability, uptake, toxicity and tolerance in plants. To reduce excess Cu from environmental matrices active plant extraction, chelate-assisted remediation and microorganism-assisted phytoextraction are suitable using plants able to accumulate Cu in the shoots, which can be removed simply (Mackie et al., 2012). Cornu et al. (2017) in a review paper focused on mechanisms enabling microorganisms to mobilize or immobilize Cu in soils, also described bioremediation strategies such as *ex situ* recovery of Cu from Cu-contaminated solids, *in situ* bioimmobilization of Cu and bioaugmentation-assisted phytoextraction for effective Cu removal from soils.

In viticulture copper fungicides have been used against downy mildew plant disease since 1885 when Bordeaux mixture (mixture of $CuSO_4$ with lime) was introduced by Millardet (1885) for the control of *Plasmopara viticola.* As Cu-based antimicrobial compounds are phytotoxic and exhibit an adverse impact on soil microorganisms and their accumulation in the soil is not desirable, 2002 regulation 473/2002 in the European Union imposed several restrictions related to their use (Commission Regulation, 2002; Lamichhane et al., 2018). Testing of plant species for remediation of Cu-contaminated soils (Inceptisol and Entisol) in vineyards showed that *Lolium multiflorum* was able to reduce efficaciously Cu availability in the soil, and *Cyperus compressus* and *Chrysanthemum leucanthemum* also exhibited phytostabilization potential in winter. However, during summer the tested species did not accumulate and stabilize Cu (Melo et al., 2021). Pietrini et al. (2019) investigating phytoremediation potential of *Arundo donax* L. plants, biomass which can be utilized for bioenergy purposes, in a semi-hydroponic mesocosm experiment found that the plants tolerated up to 300 ppm of Cu, whereby their biomass production was not affected. However, treatment with 150 and 300 ppm Cu showed a negative impact on Chl content and photosynthetic electron transport and caused imbalances in leaf concentrations of Fe and Zn. While during the experiment Cu concentration in control plants were ca. 13−16 mg/kg d.m., at treatment with 150 and 300 ppm Cu, it achieved approximately 37 and 29 mg/kg d.m. at day 7 but only just about 18 and 15 mg/kg d.m. at day 28. Maize roots accumulated considerably higher Cu concentrations from Cu-contaminated soil than shoots suggesting phytostabilization potential of *Zea mays* plants. Increasing Cu concentrations in soils supported Cu translocation from root to shoot, whereby under exposure to 200 mg/kg soil maize plants accumulated up to 5210 μg Cu per pot (Zand and Mühling, 2022). *Artemisia absinthium* perennial herb was able to accumulate high Cu concentration in its roots without translocation into the aerial plant part and can be used for Cu phytostabilization in soils contaminated with this metal. However, using citric and malic acids chelating compounds combined with NH_4NO_3, which acts as plant growth stimulating agent, Cu bioavailability along with its translocation into shoots can be improved, and *A. absinthium* plants can also be used for Cu phytoextraction from contaminated soils (Ghazaryan et al., 2022). In Chilean hardwoods quillay and espino exposed to excess Cu (50 and 100 μM) for 6 mon, Cu accumulated predominantly in roots but it was also estimated in tissues outside the periderm. However, while Cu did not affect the growth of quillay, the growth of espino was simulated (Milla-Moreno et al., 2022). The impact of three types of biochars on Cu availability and Cu uptake by *Brassica juncea* plants was investigated during three successive growth cycles using doses of 30 and 60 t/ha. Application of coconut husk biochar did not affect available soil Cu but in the two last growth cycles, it enhanced shoot Cu uptake by 117 and 38%, respectively. On the other hand, orange bagasse biochar applied at a dose of 60 t/ha, while diminishing Cu availability, was not able to effectively reduce the Cu uptake, however with the application of sewage sludge biochar at a dose of 60 t/ha shoot Cu uptake was doubled suggesting that

Cu availability was not affected. Consequently, it can be stated that orange bagasse biochar can be used for Cu immobilization, while sewage sludge biochar has the potential to be used for Cu removal from Cu-contaminated soils (Gonzaga et al., 2018).

In chelate-assisted phytoremediation of metals, the plants take up metals mobilized into the soil solution by chelators in bioavailable forms, whereby the application of rapidly biodegraded chelating agents might be preferred as an alternative to synthetic chelates (Randelovic et al., 2022). *Brassica napus* plants cultivated in a hydroponic solution containing 0.5 μmol Cu/L accumulated in their shoots and roots 195.1 and 7009.1 mg Cu/kg d.m., respectively, corresponding to a translocation factor of 0.028, and considering the actual dry mass of these plant organs, a portion of Cu allocated in shoots was 27.6%. Based on greater Cu accumulation in the roots compared to shoots it can be presumed that the plants adopted an exclusion mechanism by preventing Cu transport into the shoots, enabling them to tolerate the Cu-induced toxicity (Peško and Kráľová, 2013). In *Matricaria recutita* plants cultivated hydroponically in the presence of 12–60 μmol/L $CuSO_4$ bioconcentration factors related to the accumulated metal amount in roots (BCF_{root}) and shoots (BCF_{shoot}) varied in the range 399–570 and 3.8–11.5, respectively. At co-application of equimolar ethylenediaminetetraacetic acid (EDTA) concentration the corresponding BCF_{root} showed a decrease (110–305), in contrast to BCF_{shoot}, which considerably increased (34–39). Due to excellent stimulation of Cu translocation into the shoot by EDTA, in the shoots of *M. recutita* treated with 24 and 60 μmol/L $CuSO_4$ and equimolar EDTA the fraction of Cu related to its total amount accumulated by the plant was 45 and 59%, respectively, while this fraction achieved only 5.2 and 4.8% without EDTA application (Kráľová and Masarovičová, 2008).

As-hyperaccumulator *Pteris vittata* plants accumulated more Cu (1355 mg/kg) than As (487 mg/kg); Cu was predominantly allocated in roots and in fronds only 0.6–18% Cu was estimated. Treatment with Cu resulted in decreased K and enhanced P concentrations in the roots of *P. vittata*. Moreover, high Cu levels were found to reduce As uptake by plants (Xu et al., 2022).

In addition to copper, inorganic fungicides also include sulfur and mercury-based compounds. Sulfur is considered the oldest fungicide, which was already used in ancient times in Greece (about 1000 yr BC) and was also mentioned by Homer (Tweedy, 1981; Williams and Cooper, 2004); sulfur is also one of the most important macronutrients for photosynthetic organisms. It usually does not damage the plants, but a high concentration of S can inhibit the uptake of some other nutrients. The most abundant form of inorganic S in nature, sulfate, is assimilated in both algae and plants in chloroplasts to sulfide (S^{2-}), where for SO_4^{2-} reduction activation of SO_4^{2-} to adenosine 5′-phosphosulphate is necessary, that is subsequently reduced by a GSH-dependent reductase. In vascular plants, adenosine 5′-phosphosulphate reductase is considered the control point in the pathway of SO_4^{2-} reduction (Giordano et al., 2016). The first organic product generated from S, cysteine, is a precursor of S-containing biologically active compounds such as GSH and methionine (Li et al., 2019). Treatment of leaf surfaces and stems with a sulfur spray can be used to prevent the development of plant diseases such as powdery mildew. It is presumed that S can disrupt the metabolisms of harmful fungi, which results in inhibition of their spreading and under contact exposure the spores will be killed. The Australian project Strategic Use of Sulfur in Integrated Pest and Disease Management (IPM) Programs for Grapevines focused on strategies to be applied for the optimum use of sulfur in vineyards (Emmett, 2003). Since S occurs naturally in nature, in environmental matrices it can be involved the natural sulfur cycle (Brillon, 2022).

Mercury-based fungicides, whether they contain Hg^{2+} or organomercurial fungicides show high toxicity and can endanger human health, and therefore their use is banned in many countries. Induced phytoextraction of Hg using thiosulfates and other sulfur-containing compounds, aminopolycarboxylic acids, low-molecular-weight organic acids and enzymes, which can considerably improve uptake of Hg by plants, was overviewed by Makarova et al. (2022). Phytoremediation techniques suitable for Hg removal from Hg-polluted soils, including the use of transgenic plants able to detoxify Hg contaminants, were reported by Malik et al. (2022). Hg-resistant bacteria, which have the *mer* operon system, can bioremediate Hg via volatilization by converting both reactive inorganic and organic Hg compounds to volatile and monoatomic forms. The bacterial cellular mechanism of Hg resistance is associated with smaller Hg uptake, extracellular sequestration and bioaccumulation (Priyadarshanee et al., 2022). *Cupriavidus metallidurans* MSR33 was found to reduce Hg^{2+} under aerobic and anaerobic conditions. In the presence of O_2 after 24 hr exposure, about 97% Hg^{2+} removal was observed compared to 71% estimated at anaerobic conditions, indicating the deciding role of O_2 in the remediation process (Bravo et al., 2020). The resistance mechanism of *C. metallidurans* MSR33 to Hg^{2+} and organomercurial compounds is shown in Fig. 10.20. Hg^{2+} is reduced to Hg^0 by

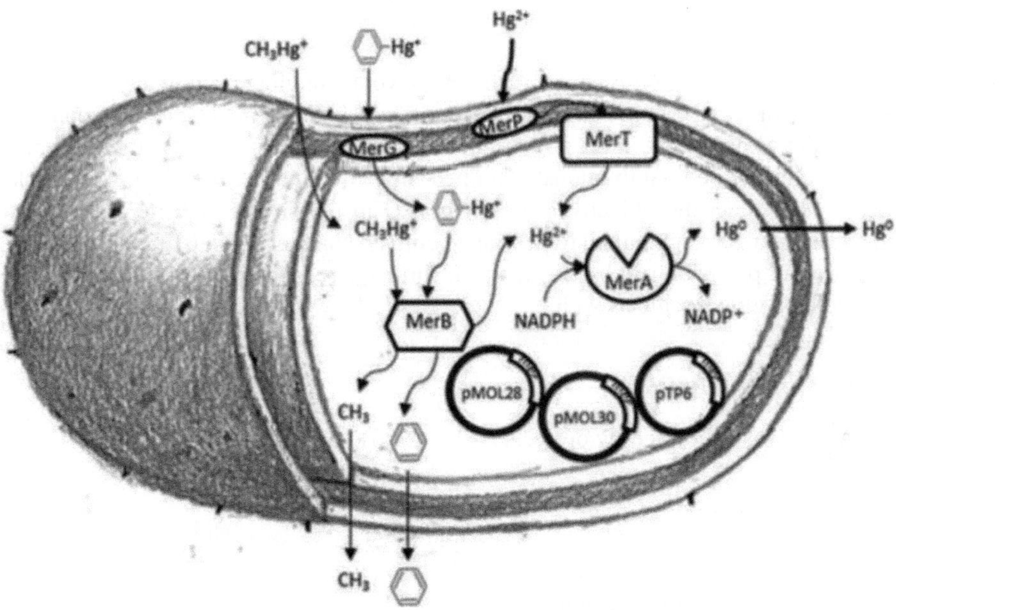

Fig. 10.20: Resistance mechanism of *Cupriavidus metallidurans* MSR33 to Hg^{2+} and organomercurial compounds. Adapted from Bravo et al. (2020). Copyright 2020 MDPI.

mercuric reductase MerA utilizing the reducing power of NADPH. MerB catalyzes the protonolysis of the C–Hg bond in organomercurials, MerP and MerT proteins are involved in the transport of Hg^{2+} inside the cell, while the periplasmic protein MerG is involved in the importing of phenylmercury.

6. Conclusions

Current climate change accompanied by increasing temperature and changing precipitation will affect the geographic distribution of crops, which along with the large rise of harmful fungal phytopathogens results in severe losses in agricultural production. Consequently, the production of sufficient healthy food for the increasingly growing human population can be endangered. However, the majority of synthetic fungicides used in the fight against plant diseases exhibit toxic effects on non-target organisms and their persistence in environmental matrices is undesirable, and their access into the food chain should also be prevented. Therefore, it is necessary to identify versatile microorganisms able to degrade a variety of fungicides to less toxic metabolites together with the utilization of algae, macrophytes and terrestrial plants for bioremediation of fungicide-contaminated soils and water bodies. In addition to supplementing effective fungicide-degrading bacterial/fungal strains or their consortia to the soil, it is desirable to reduce runoff losses of fungicides from crop fields with grass hedges, vegetated buffers, buffer strips and constructed wetlands or by using algae for bioremediation of fungicide-contaminated wastewaters. The use of genetically engineered microorganisms and plants or the application of immobilized fungicide-degrading enzymes like lactase may contribute to the reduction of fungicide contamination. Since fungicide-degradation products are usually less toxic compared to the parent fungicides, eventually during biodegradation, their total mineralization will be accomplished, the adverse impact on consumable crops and human health can be eliminated. Although several fungicides showing endocrine-disrupting or carcinogenic activity were already banned in developed countries, their residues persisting in the soil might be removed and their export to developing countries might be unambiguously stopped. In EU countries, it is desirable to take measures to meet the objectives set by the EU in the Green Deal in 2019, a new environmental strategy aimed to neutralize climate change by 2030, including a 50% reduction in pesticide use by 2030, and a similarly responsible approach would be needed on a global scale as well.

Acknowledgement

This work was supported by projects APVV-22-0133 and VEGA 1/0116/22.

Abbreviations

2-AB	(2-aminobenzimidazole)
AZO	(azoxystrobin)
BCF	(bioconcentration factor)

CAT	(catalase)
CBZ	(carbendazim)
Chlorophyll	(Chl)
CLB	(climbazole)
CLZ	(clotrimazole)
DFZ	(difenoconazole)
DMM	(dimethomorph)
EDTA	(ethylenediaminetetraacetic acid)
EPO	(epoxiconazole)
FLU	(fludioxonil)
FLZ	(fluconazole)
FPM	(fluopyram)
GSH	(glutathione)
2-HB	(2-hydroxybenzimidazole)
HXZ	(hexaconazole)
IPD	(iprodione)
MCB	(myclobutanil)
MDA	(malondialdehyde)
MG	(malachite green)
MLX	(metalaxyl)
MSM	(minimal salts medium)
PBZ	(paclobutrazol)
PCNB	(pentachloronitrobenzene)
PCZ	(propiconazole)
PEPC	(phosphoenolpyruvate carboxylase)
PTC	(prothioconazole)
PYM	(pyrimethanil)
PYR	(pyraclostrobin)
ROS	(reductive oxygen species)
Rubisco	(ribulose-1,5-bisphosphate carboxylase/oxygenase)
SMS	(spent mushroom substrate)
TBZ	(thiabendazole)
TDM	(triadimefon)
TEB	(tebuconazole)
TTC	(triticonazole)

References

Ahlawat, O.P., Gupta, P., Kumar, S., Sharma, D.K. and Ahlawat, K. (2010). Bioremediation of fungicides by spent mushroom substrate and its associated microflora. Indian J. Microbiol., 50: 390–395.

Ahmad, K.S. and Gul, P. (2020). Fungicide isopyrazam degradative response toward extrinsically added fungal and bacterial strains. J. Basic Microbiol., 60: 484–493.

Ahmad, K.S., Sajid, A., Gul, M.M. and Ali, D. (2020). Effective remediation strategy for xenobiotic zoxamide by pure bacterial strains, *Escherichia coli*, *Streptococcus pyogenes*, and *Streptococcus pneumoniae*. Biomed. Res. Int., 2020: 5352427.

Al-Jawhari, I.F.H. (2022). Recent advancements in mycoremediation. pp. 145–161. *In*: Suyal, D.C. and Soni, R. (eds.). Bioremediation of Environmental Pollutants. Springer, Cham.

Aimeur, N., Tahar, W., Meraghni, M., Meksem, N. and Bordjiba, O. (2016). Bioremediation of pesticide (mancozeb) by two *aspergillus* species isolated from surface water contaminated by pesticides. J. Chem. Pharma. Sci., 9: 2668–2670.

Aleksova, M., Kenarova, A., Boteva, S., Georgieva, S., Chanev, C. and Radeva, G. (2021). Effects of increasing concentrations of fungicide Quadris(R) on bacterial functional profiling in loamy sand soil. Arch. Microbiol., 203: 4385–4396.

Alexandrino, D.A.M., Mucha, A.P., Almeida, C.M.R. and Carvalho, M.F. (2020). Microbial degradation of two highly persistent fluorinated fungicides—epoxiconazole and fludioxonil. J. Hazard. Mater., 394: 122545.

Alexandrino, D.A.M., Mucha, A.P., Tomasino, M.P., Almeida, C.M.R. and Carvalho, M.F. (2021). Combining culture-dependent and independent approaches for the optimization of epoxiconazole and fludioxonil-degrading bacterial consortia. Microorganisms, 9: 2109.

Ali, D., Ibrahim, K.E., Hussain, S.A. and Abdel-Daim, M.M. (2021). Role of ROS generation in acute genotoxicity of azoxystrobin fungicide on freshwater snail *Lymnaea luteola* L. Environ. Sci. Pollut. Res., 28: 5566–5574.

Ali, I., Shah, A., Deeba, F., Hussain, H., Yazdan, F., Khan, M.U. and Khan, M.D. (2022). Screening of various *Brassica* species for phytoremediation of heavy metals-contaminated soil of Lakki Marwat, Pakistan. Environ. Sci. Pollut. Res., 29: 37765–37776.

Alvarado-Gutierrez, M.L., Ruiz-Ordaz, N., Galindez-Mayer, J., Santoyo-Tepole, F., Curiel-Quesada, E., Garcia-Mena, J. and Ahuatzi-Chacon, D. (2017). Kinetics of carbendazim degradation in a horizontal tubular biofilm reactor. Bioprocess Biosyst. Eng., 40: 519–528.

An, X.K., Tian, C.Y., Xu, J., Dong, F.S., Liu, X.G., Wu, X.H. and Zheng, Y.Q. (2020). Characterization of hexaconazole-degrading strain *Sphingobacterium multivorum* and analysis of transcriptome for biodegradation mechanism. Sci. Total Environ., 722: 137171.

Anand, S., Bharti, S.K., Kumar, S., Barman, S.C. and Kumar, N. (2019). pp. 89–119. *In*: N. Arora and N. Kumar (eds.). Phytoremediation of Heavy Metals and Pesticides Present in Water Using Aquatic Macrophytes. Phyto and Rhizo Remediation. Microorganisms for Sustainability, Vol 9. Springer, Singapore.

Anum, W., Riaz, U., Murtaza, G., Raza, M.U. and Rahman, A.U. (2022). Sustainable agroecosystems: Recent trends and approaches in phytoremediation and rhizoremediation. *In*: J. Malik and M.R. Goyal (eds.). Bioremediation and Phytoremediation Technologies in Sustainable Soil Management. Apple Academic Press, 20 pp.

Ariaeenejad, S., Motamedi, E. and Salekde, G.H. (2022). Highly efficient removal of dyes from wastewater using nanocellulose from quinoa husk as a carrier for immobilization of laccase. Bioresource Technol., 349: 126833.

Arya, R. and Sharma, A.K. (2016). Bioremediation of carbendazim, a benzimidazole fungicide using *Brevibacillus borstelensis* and *Streptomyces albogriseolus* together. Curr. Pharm. Biotechnol., 17: 185–189.

Avila, R., Peris, A., Eljarrat, E., Vicent, T. and Blanquez, P. (2021). Biodegradation of hydrophobic pesticides by microalgae: Transformation products and impact on algae biochemical methane potential. Sci. Total Environ., 754: 142114.

Bacmaga, M., Kucharski, J. and Wyszkowska, J. (2015). Microbial and enzymatic activity of soil contaminated with azoxystrobin. Environ. Monit. Assess., 187: 615.

Bacmaga, M., Wyszkowska, J. and Kucharski, J. (2019). Biostimulation as a process aiding tebuconazole degradation in soil. J. Soils Sediments, 19: 3728–3741.

Bacmaga, M., Wyszkowska, J. and Kucharski, J. (2020). Response of soil microorganisms and enzymes to the foliar application of Helicur 250 EW fungicide on *Horderum vulgare* L. Chemosphere, 242: 125163.

Bai, N.L., Wang, S., Abuduaini, R., Zhang, M.N., Zhu, X.F. and Zhao, Y.H. (2017). Rhamnolipid-aided biodegradation of carbendazim by *Rhodococcus* sp D-1: Characteristics, products, and phytotoxicity. Sci. Total Environ., 590: 343–351.

Ballabio, C., Panagos, P., Lugato, E., Huang, J.H., Orgiazzi, A., Jones, A., Fernandéz-Ugalde, O., Borelli, P. and Montanarella, L. (2018). Copper distribution in European topsoils: An assessment based on LUCAS soil survey. Sci. Total Environ., 636: 282–298.

Battistoni, M., Di Renzo, F., Menegola, E. and Bois, F.Y. (2019). Quantitative AOP based teratogenicity prediction for mixtures of azole fungicides. Comput. Toxicol., 11: 72–81.

Bhandari, G. and Sharma, M. (2021). Microbial enzymes in the bioremediation of pollutants: Emerging potential and challenges. pp. 75–94. *In*: R. Prasad, S.C. Nayak, R.N. Kharwar and N.K. Dubey (eds.). Mycoremediation and Environmental Sustainability. Fungal Biology. Springer, Cham.

Bhandari, G., Bhatt, P., Gangola, S., Srivastava, A. and Sharma, A. (2022). Degradation mechanism and kinetics of carbendazim using *Achromobacter* sp. strain GB61. Bioremed. J., 26: 150–161.

Bilal, M., Iqbal, H.M.N. and Barceló, D. (2019). Persistence of pesticides-based contaminants in the environment and their effective degradation using laccase-assisted biocatalytic systems. Sci. Total Environ., 695: 133896.

Birolli, W.G., da Silva, B.F. and Rodrigues-Filho, E. (2020). Biodegradation of the fungicide pyraclostrobin by bacteria from orange cultivation plots. Sci. Total Environ., 746: 140968.

Braschi, I., Blasioli, S., Lavrnic, S., Buscaroli, E., Di Prodi, K., Solimando, D. and Toscano, A. (2022). Removal and fate of pesticides in a farm constructed wetland for agricultural drainage water treatment under Mediterranean conditions (Italy). Environ. Sci. Pollut. Res., 29: 7283–7299.

Bravo, G., Vega-Celedón, P., Gentina, J.C. and Seeger, M. (2020). Effects of mercury II on *Cupriavidus metallidurans* strain MSR33 during mercury bioremediation under aerobic and anaerobic conditions. Processes, 8: 893.

Brillon, K. (2022). Sulfur fungicide: Disease and pest control option. https://www.epicgardening.com/sulfur-fungicide/.

Brunhoferova, H., Venditti, S., Schlienz, M. and Hansen, J. (2021). Removal of 27 micropollutants by selected wetland macrophytes in hydroponic conditions. Chemosphere, 281: 130980.

Cai, W.W., Ye, P., Yang, B., Shi, Z.Q., Xiong, Q., Gao, F.Z., Liu, Y.S., Zhao, J.L. and Ying, G.G. (2021). Biodegradation of typical azole fungicides in activated sludge under aerobic conditions. J. Environ. Sci., 103: 288–297.

Campos, M., Perruchon, C., Vasilieiadis, S., Menkissoglu-Spiroudi, U., Karpouzas, D.G. and Diez, M.C. (2015). Isolation and characterization of bacteria from acidic pristine soil environment able to transform iprodione and 3,5-dichloraniline. Int. Biodeterior. Biodegradation, 104: 201–211.

Campos, M., Karas, P.S., Perruchon, C., Papadopoulou, E.S., Christou, V., Menkissoglou-Spiroudi, U., Diez, M.C. and Karpouzas, D.G. (2017a). Novel insights into the metabolic pathway of iprodione by soil bacteria. Environ. Sci. Pollut. Res., 24: 152–163.

Campos, M., Perruchon, C., Karas, P.A., Karavasilis, D., Diez, M.C. and Karpouzas, D.G. (2017b). Bioaugmentation and rhizosphere-assisted biodegradation as strategies for optimization of the dissipation capacity of biobeds. J. Environ. Manage., 187: 103–110.

Carraro, N., Sentchilo, V., Polak, L., Bertelli, C. and van der Meer, J.R. (2020). Insights into mobile genetic elements of the biocide-degrading bacterium *Pseudomonas nitroreducens* HBP-1. Genes, 11: 930.

Castro, M.S., Carvalho Penha, L.C., Torres, T.A., Jorge, M.B., Carvalho-Costa, L.F., Fillmann, G. and Luvizotto-Santos, R. (2022). Genotoxic and mutagenic effects of chlorothalonil on the estuarine fish *Micropogonias furnieri* (Desmarest, 1823). Environ. Sci. Pollut. Res., 29: 23504–23511.

Chen, Z.F. and Ying, G.G. (2015). Occurrence, fate and ecological risk of five typical azole fungicides as therapeutic and personal care products in the environment: A review. Environ. Int., 84: 142–153.

Chen, X.X., He, S., Liu, X.L. and Hu, J.Y. (2018). Biodegradation and metabolic mechanism of cyprodinil by strain *Acinetobacter* sp. from a contaminated-agricultural soil in China. Ecotoxicol. Environ. Saf., 159: 190–197.

Cheng, C., Huang, L., Diao, J.L. and Zhou, Z.Q. (2013). Enantioselective toxic effects and degradation of myclobutanil enantiomers in *Scenedesmus obliquus.* Chirality, 25: 858–864.

Cheng, J.Z., Lee, X.Q., Gao, W.C., Chen, Y., Pan, W.J. and Tang, Y. (2017). Effect of biochar on the bioavailability of difenoconazole and microbial community composition in a pesticide-contaminated soil. Appl. Soil Ecol., 121: 185–192.

Cheng, C.M., Patel, A.K., Singhania, R.R., Tsai, C.H., Chen, S.Y., Chen, C.W. and Dong, C.D. (2021). Heterologous expression of bacterial CotA-laccase, characterization and its application for biodegradation of malachite green. Bioresource Technol., 340: 125708.

Cherian, E. and Joseph, S. (2022). Water hyacinth: A potent source for phytoremediation and biofuel production. *In*: J.T. Puthur and O.P. Dhankher (eds.). Bioenergy Crops, A Sustainable Means of Phytoremediation. CRC Press, Boca Raton, 18 pp.

Chuang, S.C., Yang, H.X., Wang, X., Xue, C., Jiang, J.D. and Hong, Q. (2021). Potential effects of *Rhodococcus qingshengii* strain djl-6 on the bioremediation of carbendazim-contaminated soil and the assembly of its microbiome. J. Hazard. Mater., 414: 125496.

Commission Regulation. (2002). Commission Regulation (EC) No. 473/2002 of 15 March 2002 amending Annexes I, II and VI to Council Regulation (EEC) No. 2092/91 on organic production of agricultural products and indications referring thereto on agricultural products and foodstuffs, and laying down detailed rules as regards the transmission of information on the use of copper compounds. https://www.legislation.gov.uk/eur/2002/473/pdfs/eur_20020473_adopted_en.pdf.

Cornu, J.Y., Huguenot, D., Jezequel, K., Lollier, M. and Lebeau, T. (2017). Bioremediation of copper-contaminated soils by bacteria. World J. Microbiol. Biotechnol., 33: 26.

Del Puerto, O., Goncalves, N.P.F., Medana, C., Prevot, A.B. and Roslev, P. (2022). Attenuation of toxicity and occurrence of degradation products of the fungicide tebuconazole after combined vacuum UV and UVC treatment of drinking water. Environ. Sci. Pollut. Res., 29: 58312–58325.

Diez, M.C., Elgueta, S., Rubilar, O., Tortella, G.R., Schalchli, H., Bornhardt, C. and Gallardo, F. (2017). Pesticide dissipation and microbial community changes in a biopurification system: Influence of the rhizosphere. Biodegradation, 28: 395–412.

Ding, T.D., Li, W. and Li, J.Y. (2019). Toxicity and metabolic fate of the fungicide carbendazim in the typical freshwater diatom *Navicula* species. J. Agric. Food Chem., 67: 6683–6690.

Dionisio, G., Gautam, M. and Fomsgaard, I.S. (2019). Identification of azoxystrobin glutathione conjugate metabolites in maize roots by LC-MS. Molecules, 24: 2473.

Dominguez, J.J.A., Inoue, C. and Chien, M.F. (2020). Hydroponic approach to assess rhizodegradation by sudangrass (*Sorghum x drummondii*) reveals pH- and plant age-dependent variability in bacterial degradation of polycyclic aromatic hydrocarbons (PAHs). J. Hazard. Mater., 387: 121695.

Dosnon-Olette, R., Couderchet, M. and Eullaffroy, P. (2009). Phytoremediation of fungicides by aquatic macrophytes: toxicity and removal rate. Ecotoxicol. Environ. Saf., 72: 2096–2101.

Dosnon-Olette, R., Trotel-Aziz, P., Couderchet, M. and Eullaffroy, P. (2010). Fungicides and herbicide removal in *Scenedesmus* cell suspensions. Chemosphere, 79: 117–123.

Dosnon-Olette, R., Schroeder, P., Bartha, B., Aziz, A., Couderchet, M. and Eullaffroy, P. (2011). Enzymatic basis for fungicide removal by *Elodea canadensis*. Environ. Sci. Polut. Res., 18: 1015–1021.

Draskau, M.K., Boberg, J., Taxvig, C., Pedersen, M., Frandsen, H.L., Christiansen, S. and Svingen, T. (2019). *In vitro* and *in vivo* endocrine disrupting effects of the azole fungicides triticonazole and flusilazole. Environ. Pollut., 255: Part 2, 113309.

Draskau, M.K. and Svingen, T. (2022). Azole fungicides and their endocrine disrupting properties: Perspectives on sex hormone-dependent reproductive development. Front. Toxicol., 4: 883254.

EFSA (European Food Safety Authority). (2020). Modification of the existing maximum residue levels for mandipropamid in kohlrabies and herbs and edible flowers. EFSA Journal, 18: e05958.

Elsaesser, D., Stang, C., Bakanov, N. and Schulz, R. (2013). The landau stream mesocosm facility: Pesticide mitigation in vegetated flow-through streams. Bull. Environ. Contam. Toxicol., 90: 640–645.

Elsharkawy, E.E., El-Nasser, M.A. and Bakheet, A.A. (2019). Mancozeb impaired male fertility in rabbits with trials of glutathione detoxification. Regul. Toxicol. Pharmacol., 105: 86–98.

El-Sheekh, M.M., Abomohra, A., Eladel, H., Battah, M. and Mohammed, S. (2018). Screening of different species of *Scenedesmus* isolated from Egyptian freshwater habitats for biodiesel production. Renew. Energy, 129: 114–120.

Emmett, B. (2003). Strategic Use of Sulphur in Integrated Pest and Disease Management (IPM) Programs for Grapevines. State of Victoria, Department of Primary Industries, Mildura, https://www.langhornecreek.com/wp-content/uploads/2017/08/Sulphur_use_in_IPM_Bob_Emmett_2011.pdf.

Equeenuddin, S.M., Bisoi, K.C. and Barik, C.K. (2021). Natural attenuation of metals by algal mat from acid mine drainage at Malanjkhand copper mine. Arab. J. Geosci., 14: 680.

Espinosa, D., González, A., Pizarro, J., Segura, R., Laporte, D., Rodriguez-Rojas, D., Sáez, C.A. and Moenne, A. (2021). *Ulva compressa* from copper-polluted sites exhibits intracellular copper accumulation, increased expression of metallothioneins and copper-containing nanoparticles in chloroplasts. Int. J. Mol. Sci., 22: 10531.

Fan, X.Y., Fu, Y., Nie, Y.X., Matsumoto, H., Wang, Y., Hu, T.T., Pan, Q.Q., Lv, T.X., Fang, H.D., Xu, H.R., Wang, Y., Ge, H., Zhu, G.N., Liu, Y.H., Wang, Q.W. and Wang, M.C. (2021). Keystone taxa-mediated bacteriome response shapes the resilience of the paddy ecosystem to fungicide triadimefon contamination. J. Hazard. Mater., 417: 126061.

Fang, H., Wang, Y.Q., Gao, C.M., Yan, H., Dong, B. and Yu, Y.L. (2010). Isolation and characterization of *Pseudomonas* sp CBW capable of degrading carbendazim. Biodegradation, 21: 939–946.

FAO. (2019). New standards to curb the global spread of plant pests and diseases. http://www.fao.org/news/story/en/item/1187738/icode/.

Feng, Y.M., Huang, Y.H., Zhan, H., Bhatt, P. and Chen, S.H. (2020a). An overview of strobilurin fungicide degradation: Current status and future perspective. Front. Microbiol., 11: 389.

Feng, Y.M., Zhang, W.P., Pang, S.M., Lin, Z.Q., Zhang, Y.M., Huang, Y.H., Bhatt, P. and Chen, S.H. (2020b). Kinetics and new mechanism of azoxystrobin biodegradation by an *Ochrobactrum anthropi* Strain SH14. Microorganisms, 8: 625.

Fortune Business Insight. (2022). Fungicides Marker Size, Share & Covid-19 Impact Analysis, by Type (Chenmical and Biological), Crop Type (Cereals, Oilseeds & Pulses, Fruits & Vegetables and Others), Application Method (Foliar, Chemigation, Seed Treatment, and Others), and Regional Forecast, 2021–2028. Market Research Report, Report ID:FBI103267. January 2022; https://www.fortunebusinessinsights.com/fungicides-market-103267.

FRAC Code List© 2022. https://www.frac.info/docs/default-source/publications/frac-code-list/frac-code-list-2022--final.pdf?sfvrsn=b6024e9a_2.

Frankowski, R., Platkiewicz, J., Stanisz, E., Grzeskowiak, T. and Zgola-Grzeskowiak, A. (2021). Biodegradation and photo-Fenton degradation of bisphenol A, bisphenol S and fluconazole in water. Environ. Pollut., 289: 117947.

Frost, C. (2015a). Encyclopedia of Fungicides: Volume I (Animal and Plant Disease). Callisto Reference, New York 11375, USA.

Frost, C. (2015b). Encyclopedia of Fungicides: Volume II (Benefits and Drawbacks). Callisto Reference, New York 11375, USA.

Frost, C. (2015c). Encyclopedia of Fungicides: Volume III (Plant Disease Management). Callisto Reference, New York 11375, USA.

Garanzini, D.S. and Menone, M.L. (2015). Azoxystrobin causes oxidative stress and DNA damage in the aquatic macrophyte *Myriophyllum quitense*. Bull. Environ. Contam. Toxicol., 94: 146–151.

Garanzini, D.S., Medici, S., Moreyra, L.D. and Menone, M.L. (2019). Acute exposure to a commercial formulation of Azoxystrobin alters antioxidant enzymes and elicit damage in the aquatic macrophyte *Myriophyllum quitense*. Physiol. Mol. Biol. Plants, 25: 135−143.

Gautam, M., Etzerodt, T. and Fomsgaard, I.S. (2017). Quantification of azoxystrobin and identification of two novel metabolites in lettuce via liquid chromatography-quadrupole-linear ion trap (QTRAP) mass spectrometry. Int. J. Environ. Anal. Chem., 97: 419–430.

Gavrilescu, M. (2022). Enhancing phytoremediation of soils polluted with heavy metals. Curr. Opin. Biotechnol., 74: 21–31.

Geholt, P., Vivekanand, V. and Pareek, N. (2022). Cyanobacterial and microalgal bioremediation: An efficient and eco-friendly approach toward industrial wastewater treatment and value-addition. pp. 343–362. *In*: S. Das and H.R. Dash (eds.). Microbial Biodegradation and Bioremediation, 2nd edn. Techniques and Case Studies for Environmental Pollution, Elsevier.

Gerotto, C., Norici, A. and Giordano, M. (2020). Toward enhanced fixation of CO_2 in aquatic biomass: Focus on microalgae. Front. Energy Res., 8: 213.

Ghazaryan, K.A., Movsesyan, H.S., Minkina, T.M., Nevidomskaya, D.G. and Rajput, V.D. (2022). Phytoremediation of copper-contaminated soil by *Artemisia absinthium*: comparative effect of chelating agents. Environ. Geochem. Health, 44: 1203–1215.

Ghorbani, E., Nowruzi, B., Nezhadali, M. and Hekmat, A. (2022). Metal removal capability of two cyanobacterial species in autotrophic and mixotrophic mode of nutrition. BMC Microbiol., 22: 58.

Giordano, M. and Prioretti, L. (2016). Sulphur and algae: Metabolism, ecology and evolution. pp. 185–209. *In*: M. Borowitzka, J. Beardall and J. Raven (eds.). The Physiology of Microalgae. Developments in Applied Phycology, Vol 6. Springer, Cham, Switzerland.

Girardi, F., Hackbarth, F.V., de Souza, S.M.A.G.U., de Souza, A.A.U., Boaventura, R.A.R. and Vilar, V.J.P. (2014). Marine macroalgae *Pelvetia canaliculata* (Linnaeus) as natural cation exchanger for metal ions separation: a case study on copper and zinc ions removal. Chem. Eng. J., 247: 320–329.

Goncalves, N.P.F., del Puerto, O., Medana, C., Calza, P. and Roslev, P. (2021). Degradation of the antifungal pharmaceutical clotrimazole by UVC and vacuum-UV irradiation: Kinetics, transformation products and attenuation of toxicity. J. Environ. Chem. Eng., 9: 106275.

Gonzaga, M.I.S., Mackowiak, C., de Almeida, A.Q., Wisniewski, A., de Souza, D.F., Lima, I.D. and de Jesus, A.N. (2018). Assessing biochar applications and repeated *Brassica juncea* L. production cycles to remediate Cu contaminated soil. Chemosphere, 201: 278–285.

Goyal, K., Sharma, A., Arya, R., Sharma, R., Gupta, G.K. and Sharma, A.K. (2018). Double edge sword behavior of carbendazim: A potent fungicide with anticancer therapeutic properties. Anti-Cancer Agents Med. Chem., 18: 38–45.

Grand View Research. (2020). Fungicides Market Size, Share & Trends Analysis Report By Product (Inorganic, Biofungicides), By Application (Cereals & Grains, Fruits & Vegetables), By Region, And Segment Forecasts, 2020–2027, Report ID: GVR-4-68038-674-5, May 2020.

Gul, P., Ahmad, K.S., Jaffri, S.B. and Ali, D. (2021a). Lithosphere-stationed fate and eco-detoxification investigation of fungicidal agent Zoxamide possessing chlorinated benzamidic genesis. Int. J. Environ. Sci. Technol., 18: 3127–3142.

Gul, P., Ahmad, K.S., Jaffri, S.B. and Ali, D. (2021b). Variegated pedospheric matrices based pyrzaole fungicide chemico-physical and biological degradation elucidation. Soil Sediment Contam., 30: 998–1024.

Guo, C., Di, S.S., Chen, X.L., Wang, Y.H., Qi, P.P., Wang, Z.W., Zhao, H.Y., Gu, Y.L., Xu, H., Lu, Y.L. and Wang, X.Q. (2022). Evaluation of chiral triticonazole in three kinds of fruits: Enantioseparation, degradation, and dietary risk assessment. Environ. Sci. Pollut. Res., 29: 32855–32866.

Han, P., Yu, Y.C., Zhou, L.J., Tian, Z.Y., Li, Z., Hou, L.J., Liu, M., Wu, Q.L., Wagner, M. and Men, Y.J. (2019). Specific micropollutant biotransformation pattern by the comammox bacterium *Nitrospira inopinata*. Environ. Sci. Technol., 53: 8695–8705.

Han, L.X., Kong, X.B., Xu, M. and Nie, J.Y. (2021). Repeated exposure to fungicide tebuconazole alters the degradation characteristics, soil microbial community and functional profiles. Environ. Pollut., 287: 117660.

Han, L., Wu, Q. and Wu, X.M. (2022). Dissipation and residues of pyraclostrobin in *Rosa roxburghii* and soil under filed conditions. Foods, 11: 669.

Hand, L.H., Marshall, S.J., Dougan, C., Nichols, C., Kende, A., Ritz, K. and Oliver, R.G. (2021). The impact of disturbed soil structure on the degradation of 2 fungicides under constant and variable moisture. Environ. Toxicol. Chem., 40: 2715–2725.

Hamed, S.M., Hassan, S.H., Selim, S., Wadaan, M.A.M., Mohany, M., Hozzein, W.N. and AbdElgawad, H. (2020). Differential responses of two cyanobacterial species to *R*-metalaxyl toxicity: Growth, photosynthesis and antioxidant analyses. Environ. Pollut., 258: 113681.

Hamed, S.M., Okla, M.K., Al-Saadi, L.S., Hozzein, W.N., Mohamed, H.S., Selim, S. and AbdElgawad, H. (2022). Evaluation of the phycoremediation potential of microalgae for captan removal: Comprehensive analysis on toxicity, detoxification and antioxidants modulation. J. Hazard. Mater., 427: 128177.

Haripriyan, U., Gopinath, K.P., Arun, J. and Govarthanan, M. (2022). Bioremediation of organic pollutants: a mini review on current and critical strategies for wastewater treatment. Arch. Microbiol., 204: 286.

Hasan, M.M., Uddin, M.N., Ara-Sharmeen, I.F., Alharby, H., Alzahrani, Y., Hakeem, K.R. and Zhang, L. (2019). Assisting phytoremediation of heavy metals using chemical amendments. Plants, 8: 295.

Hawkins, N.J. and Fraaije, B.A. (2018). Fitness penalties in the evolution of fungicide resistance. Ann. Rev. Phytopathol., 56: 339–360.

He, Y.J., Langenhoff, A.A.M., Sutton, N.B., Rijnaarts, H.H.M., Blokland, M.H., Chen, F., Huber, C. and Schröder, P. (2017). Metabolism of Ibuprofen by *Phragmites australis*: Uptake and phytodegradation. Environ. Sci. Technol., 51: 4576–4584.

Hermann, D. and Stenzel, K. (2019). FRAC mode-of-action classification and resistance risk of fungicides. pp. 589–608. *In*: P. Jeschke, M. Witschel, W. Krämer and U. Schirmer (eds.). Modern Crop Protection Compounds, Volume 3: Insecticides, Third edn. Wiley VCH Verlag GmbH & Co. KGaA.

Hocinat, A. and Boudemagh, A. (2016). Biodegradation of commercial Ortiva fungicide by isolated actinomycetes from the activated sludge. Desalin. Water Treat., 57: 6091–6097.

Howell, C.C., Semple, K.T. and Bending, G.D. (2014). Isolation and characterisation of azoxystrobin degrading bacteria from soil. Chemosphere, 95: 370–378.

Huang, H.Z., Liang, J.S., Wu, X.S., Zhang, H., Li, Q.Q. and Zhang, Q.Y. (2013). Comparison in copper accumulation and physiological responses of *Gracilaria lemaneiformis* and *G. lichenoides* (Rhodophyceae). Chin. J. Oceanol. Limnol., 31: 803–812.

Huang, T., Ding, T.D., Liu, D.H. and Li, J.Y. (2020). Degradation of carbendazim in soil: Effect of sewage sludge-derived biochars. J. Agric. Food Chem., 68: 3703–3710.

Huang, T., Zhao, Y.H., He, J., Cheng, H.G. and Martyniuk, C.J. (2022). Endocrine disruption by azole fungicides in fish: A review of the evidence. Sci. Total Environ., 822: 153412.

Hussein, M., Abdullah, A., Badr El-Din, N. and Mishaqa, E. (2017). Biosorption potential of the microchlorophyte *Chlorella vulgaris* for some pesticides. J. Fertil. Pestic., 8: 1.

Ismail, M.M., Ismail, G.A. and El-Sheekh, M.M. (2020). Potential assessment of some micro- and macroalgal species for bioethanol and biodiesel production. Energ. Source. Part A: Recovery Util. Environ. Eff. doi: 10.1080/15567036.2020.1758853.

Jampílek, J. (2016a). Potential of agricultural fungicides for antifungal drug discovery. Expert. Opin. Drug Dis., 11: 1–9.

Jampílek, J. (2016b). How can we bolster the antifungal drug discovery pipeline? Future Med. Chem., 8: 1393–1397.

Jampílek, J. (2022). Drug repurposing to overcome microbial resistance. Drug Discov. Today, 27: 2028–2041.

Jampílek, J. and Kráľová, K. (2021). Seaweeds as indicators and potential remediators of metal pollution. pp. 51–92. *In*: H.I. Mohamed, H.E.S. El-Beltagi and K.A. Abd-Elsalam (eds.). Plant Growth-Promoting Microbes for Sustainable Biotic and Abiotic Stress Management, Springer.

Janusz, G., Pawlik, A., Świderska-Burek, U., Polak, J., Sulej, J., Jarosz-Wilkołazka, A. and Paszczyński, A. (2020). Laccase properties, physiological functions, and evolution. Int. J. Mol. Sci., 21: 966.

Jiang, W.K., Gao, Q.Q., Zhang, L., Liu, Y.L., Zhang, M.L., Ke, Z.J., Zhou, Y.D. and Hong, Q. (2021). Detoxification esterase StrH initiates strobilurin fungicide degradation in *Hyphomicrobium* sp. Strain DY-1. Appl. Environ. Microbiol., 87: e00103-21.

Kang, J., Bishayee, K. and Huh, S.O. (2021). Azoxystrobin impairs neuronal migration and induces ROS dependent apoptosis in cortical neurons. Int. J. Mol. Sci., 22: 12495.

Kanjilal, T., Panda, J. and Datta, S. (2018). Assessing *Brevibacillus* sp C17: An indigenous isolated bacterium as bioremediator for agrochemical effluent containing toxic carbendazim. J. Water Process. Eng., 23: 174–185.

Karkala, S., Rodrigues, M., Patavardhan, S., D'Souza, L. and Kiran, S. (2022). Chlorophycean micro alga as a potential bioremediant: An investigative study using carbendazima group C carcinogenic fungicide. Asian J. Water Environ. Pollut., 19: 63–69.

Katsoula, A., Vasileiadis, S., Sapountzi, M. and Karpouzas, D.G. (2020). The response of soil and phyllosphere microbial communities to repeated application of the fungicide iprodione: accelerated biodegradation or toxicity? FEMS Microbiol. Ecol., 96: fiaa056.

Kráľová, K. and Masarovičová, E. (2008). EDTA-assisted phytoextraction of copper, cadmium and zinc using chamomille plants. Ecol. Chem. Eng. A, 15: 213–220.

Kráľová, K., Masarovičová, E. and Jampílek, J. (2019). Plant responses to stress induced by toxic metals and their nanoforms. pp. 479–522. *In*: M. Pessarakli (ed.). Handbook of Plant and Crop Stress, 4th ed. CRC Press, Boca Raton.

Kráľová, K. and Jampílek, J. (2021). Impact of metal nanoparticles on marine and freshwater algae. pp. 889–921. *In*: M. Pessarakli (ed.). Handbook of Plant and Crop Physiology, 4th ed. CRC Press, Boca Raton, FL.

Kráľová, K., Masarovičová, E. and Jampílek, J. (2021). Risks and benefits of metal-based nanoparticles for vascular plants. pp. 923–963. *In*: M. Pessarakli (ed.). Handbook of Plant and Crop Physiology, 4th ed. CRC Press, Boca Raton, FL.

Kráľová, K. and Jampílek, J. (2022a). Phytoremediation of environmental matrices contaminated with photosystem II-inhibiting herbicides. pp. 31–80. *In*: S. Siddiqui, M.K. Meghvansi and K.K. Chaudharv (eds.). Pesticides Bioremediation, Springer Nature Switzerland Cham, Switzerland.

Kráľová, K. and Jampílek, J. (2022b). Medicinal and aromatic plant species with potential for remediation of metal(loid)-contaminated soils. pp. 173–236. *In*: T. Aftab (ed.). Sustainable Management of Environmental Contaminants: Eco-friendly Remediation Approaches. Springer Nature.

Kráľová, K. and Jampikek, J. (2022c). Impact of copper-based nanoparticles on economically important plants. pp. 293–339. *In*: K.A. Abd-Elsalam (ed.). Copper Nanostructures: Next-Generation of Agrochemicals for Sustainable Agroecosystems. Elsevier.

Krishnamurthy, Y.L. and Naik, B.S. (2017). Endophytic fungi bioremediation. pp. 47–60. *In*: D. Maheshwari and K. Annapurna (eds.). Endophytes: Crop Productivity and Protection. Sustainable Development and Biodiversity, Vol 16. Springer, Cham.

Kryczyk-Poprawa, A., Zmudzki, P., Maslanka, A., Piotrowska, J., Opoka, W. and Muszynska, B. (2019). Mycoremediation of azole antifungal agents using in vitro cultures of *Lentinula edodes*. 3Biotech, 9: 207.

Kumar, V.V., Sivanesan, S. and Cabana, H. (2014). Magnetic cross-linked laccase aggregates—Bioremediation tool for decolorization of distinct classes of recalcitrant dyes. Sci. Total Environ., 487: 830–839.

Kumar, V., Pandita, S., Sidhu, G.P.S., Sharma, A., Khanna, K., Kaur, P., Bali, A.S. and Setia, R. (2021). Copper bioavailability, uptake, toxicity and tolerance in plants: A comprehensive review. Chemosphere, 262: 127810.

Kwak, Y., Rhee, I.K. and Shin, J.H. (2013). Application of biofilm-forming bacteria on the enhancement of organophosphorus fungicide degradation. Bioremediat. J., 17: 173–181.

Lamichhane, J.R., Osdaghi, E., Behlau, F., Kohl, J., Jones, J.B. and Noel, J. (2018). Thirteen decades of antimicrobial copper compounds applied in agriculture. A review. Agron. Sustain. Dev., 38: 28.

Lei, J., Wei, S.P., Ren, L.J., Hu, S.B. and Chen, P. (2017). Hydrolysis mechanism of carbendazim hydrolase from the strain *Microbacterium* sp djl-6F. J. Environ. Sci., 54: 171–177.

Li, Y.Y. and Yang, H. (2013). Bioaccumulation and degradation of pentachloronitrobenzene in *Medicago sativa*. J. Environ. Manage., 119: 143–150.

Li, Y.B., Dong, F.S., Liu, X.G., Xu, J., Chen, X., Han, Y.T., Cheng, Y.P., Jian, Q. and Zheng, Y.Q. (2013). Enantioselective separation and transformation of metalaxyl and its major metabolite metalaxyl acid in tomato and cucumber. Food Chem., 141: 10–17.

Li, D., Liu, M.G., Yang, Y.S., Shi, H.H., Zhou, J.L. and He, D.F. (2016). Strong lethality and teratogenicity of strobilurins on *Xenopus tropicalis* embryos: Basing on ten agricultural fungicides. Environ. Pollut., 208(Part B): 868−874.

Li, Q., Gao, Y. and Yang, A. (2020). Sulfur homeostasis in plants. Int. J. Mol. Sci., 21: 8926.

Lilai, S.A., Kapinga, F.A., Nene, W.A., Mbasa, W.V. and Tibuhwa, D.D. (2022). The efficacy of biofungicides on cashew wilt disease caused by *Fusarium oxysporum*. Eur. J. Plant Pathol., 163: 453–465.

Lloyd, A.W., Percival, D. and Yurgel, S.N. (2021). Effect of fungicide application on lowbush blueberries soil microbiome. Microorganisms, 9: 1366.

Loffredo, E., Picca, G. and Parlavecchia, M. (2021). Single and combined use of *Cannabis sativa* L. and carbon-rich materials for the removal of pesticides and endocrine-disrupting chemicals from water and soil. Environ. Sci. Pollut. Res., 28: 3601–3616.

Lucas, J.A., Hawkins, N.J. and Fraaije, B.A. (2015). The evolution of fungicide resistance. Adv. Appl. Microbiol., 90: 29–92.

Lv, T., Zhang, Y., Casas, M.E., Carvalho, P.N., Arias, C.A., Bester, K. and Brix, H. (2016a). Phytoremediation of imazalil and tebuconazole by four emergent wetland plant species in hydroponic medium. Chemosphere, 148: 459–466.

Lv, T., Zhang, Y., Zhang, L., Carvalho, P.N., Arias, C.A. and Brix, H. (2016b). Removal of the pesticides imazalil and tebuconazole in saturated constructed wetland mesocosms. Water Res., 91: 126–136.

Lyu, T., Zhang, L., Xu, X., Arias, C.A., Brix, H. and Carvalho, P.N. (2018). Removal of the pesticide tebuconazole in constructed wetlands: Design comparison, influencing factors and modelling. Environ. Pollut., 233: 71–80.

Mackie, K.A., Mueller, T. and Kandeler, E. (2012). Remediation of copper in vineyards—A mini review. Environ. Pollut., 167: 16–26.

Maillard, E., Payraudeau, S., Ortiz, F. and Imfeld, G. (2012). Removal of dissolved pesticide mixtures by a stormwater wetland receiving runoff from a vineyard catchment: an inter-annual comparison. Int. J. Environ. Anal. Chem., 92: 979–994.

Makarova, A.S., Nikulina, E. and Fedotov, P. (2022). Induced phytoextraction of mercury. Sep. Purif. Rev., 51: 174–194.

Malhotra, H., Kaur, S. and Phale, P.S. (2021). Conserved metabolic and evolutionary themes in microbial degradation of carbamate pesticides. Front. Microbiol., 12: 648868.

Malik, S., Prasad, S., Ghoshal, S., Kumari, T. and Shekhar, S. (2022). Plant mediated techniques in detoxification of mercury contaminated soils. *In*: J.A. Malik and M.R. Goyal (eds.). Bioremediation and Phytoremediation Technologied in Sustainable Soil Management, Vol. 3. Apple Academic Press, 32 pp.

Manasfi, R., Chiron, S., Montemurro, N., Perez, S. and Brienza, M. (2020). Biodegradation of fluoroquinolone antibiotics and the climbazole fungicide by *Trichoderma* species. Environ. Sci. Pollut. Res., 27: 2331–23341.

Manikandan, A., Babu, P.S., Shyamalagowri, S., Kamarj, M., Muthukumaran, P. and Aravind, J. (2021). Emerging role of microalgae in heavy metal bioremediation. J. Basic Microbiol., 62: 330–347.

Mao, G., Wang, K., Wang, F., Li, H., Zhang, H., Xie, H., Wang, Z., Wang, F. and Song, A. (2021). An engineered thermostable laccase with great ability to decolorize and detoxify malachite green. Int. J. Mol. Sci., 22: 11755.

Masarovičová, E., Kráľová, K. and Kummerová, M. (2010). Principles of classification of medicinal plants as hyperaccumulators or excluders. Acta Physiol. Plant., 32: 823–829.

Masarovičová, E., Kráľová, K. and Zinjarde, S.S. (2014). Metal nanoparticles in plants. Formation and action. pp. 683–731. *In*: M. Pessarakli (ed.). Handbook of Plant and Crop Physiology, 3rd edn. CRC Press, Boca Raton, FL.

Masarovičová, E. and Kráľová, K. (2018). Woody species in phytoremediation applications for contaminated soils. pp. 319–373. *In*: A.A. Ansari, S.S. Gill, R. Gill, G.R. Lanza and L. Newman (eds.). Phytoremediation, Management of Environmental Contaminants, Vol. 5, Springer International Publishing AG, Cham.

Mavriou, Z., Alexandropoulou, I., Melidis, P., Karpouzas, D.G. and Ntougias, S. (2021a). Bioprocess performance, transformation pathway, and bacterial community dynamics in an immobilized cell bioreactor treating fludioxonil-contaminated wastewater under microaerophilic conditions. Environ. Sci. Pollut. Res., 29: 29597–29612.

Mavriou, Z., Alexandropoulou, I., Melidis, P., Karpouzas, D.G. and Ntougias, S. (2021b). Biotreatment and bacterial succession in an upflow immobilized cell bioreactor fed with fludioxonil wastewater. Environ. Sci. Pollut. Res., 28: 3774–3786.

Mazur, L.P., Pozdniakova, T.A., Mayer, D.A., de Souza, S.M.A.G.U., Boaventura, R.A.R. and Vilar, V.J.P. (2017). Cation exchange prediction model for copper binding onto raw brown marine macro-algae *Ascophyllum nodosum*: batch and fixed-bed studies. Chem. Eng. J., 316: 255–276.

McKnight, A.M., Gannon, T.W. and Yelverton, F. (2022). Phytoremediation of azoxystrobin and imidacloprid by wetland plant species *Juncus effusus*, *Pontederia cordata* and *Sagittaria latifolia.* Int. J. Phytoremediation, 24: 196–204.

Megadi, V.B., Tallur, P.N., Mulla, S.I. and Ninnekar, H.Z. (2010). Bacterial degradation of fungicide captan. J. Agric. Food Chem., 58: 12863–12868.

Megateli, S., Dosnon-Olette, R., Trotel-Aziz, P., Geffard, A., Semsari, S. and Couderchet, M. (2013). Simultaneous effects of two fungicides (copper and dimethomorph) on their phytoremediation using *Lemna minor*. Ecotoxicology, 22: 683–692.

Mekonnen, T.F., Panne, U. and Koch, M. (2018). Prediction of biotransformation products of the fungicide fluopyram by electrochemistry coupled online to liquid chromatography-mass spectrometry and comparison with *in vitro* microsomal assays. Anal. Bioanal. Chem., 410: 2607–2617.

Melo, G.W., Furini, G., Brunetto, G., Comin, J.J., Simao, D.G., Marques, A.C.R., Marchezan, C., Silva, I.C.B., Souza, M. and Soares, C.R. (2021). Identification and phytoremediation potential of spontaneous species in vineyard soils contaminated with copper. Int. J. Phytoremediation, 24: 342–349.

Miazek, K. and Brozek-Pluska, B. (2019). Effect of PHRs and PCPs on microalgal growth, metabolism and microalgae-based bioremediation processes: A review. Int. J. Mol. Sci., 20: 2492.

Milla-Moreno, E., Guy, R.D. and Soolanayakanahally, R.Y. (2022). Enlightening the pathway of phytoremediation: Ecophysiology and X-ray fluorescence visualization of two chilean hardwoods exposed to excess copper. Toxics, 10: 237.

Millardet, P.A. (1885). Traitement du mildiou par le mélange de sulphate de cuivre et de chaux. J. Agric. Prat., 2: 707–710.

Mohamed, E. and Mostafa, F.A. (2018). Captan utilization by a soil bacterium *Planomicrobium flavidum* strain EF. Sains Malays., 47: 85–89.

Mpofu, E., Alias, A., Tomita, K., Suzuki-Minakuchi, C., Tomita, K., Chakraborty, J., Malon, M., Ogura, Y., Takikawa, H., Okada, K., Kimura, T. and Nojiri, H. (2021). Azoxystrobin amine: A novel azoxystrobin degradation product from *Bacillus licheniformis* strain TAB7. Chemosphere, 273: 129663.

Murillo-Zamora, S., Castro-Gutierrez, V., Masis-Mora, M., Lizano-Fallas, V. and Rodriguez-Rodriguez, C.E. (2017). Elimination of fungicides in biopurification systems: Effect of fungal bioaugmentation on removal performance and microbial community structure. Chemosphere, 186: 625–634.

Muthusamy, L., Rajendran, M., Ramamoorthy, K., Narayanan, M. and Kandasamy, S. (2022). Phytostabilization of metal mine tailings—A green remediation technology. pp. 243–253. *In*: V. Kumar, M.P. Shah and S.K. Shahi (eds.). Phytoremediation Technology for the Removal of Heavy Metals and Other Contaminants from soil and Water. Elsevier.

Naeem, H. and Ahmad, K.S. (2022). Fungal and bacterial assisted bioremediation of environmental toxicant (*N*-[2-[3-chloro-5-(trifluoromethyl)-2-pyridinyl] ethyl]-2-(trifluoromethyl) benzamide) holding benzamidic genesis elucidating the eco-friendly strategy. J. Basic Microbiol., 62: 711–720.

Navarrete, A., Gonzalez, A., Gomez, M., Contreras, R.A., Diaz, P., Lobos, G., Brown, M.T., Saez, C.A. and Moenne, A. (2019). Copper excess detoxification is mediated by a coordinated and complementary induction of glutathione, phytochelatins and metallothioneins in the green seaweed *Ulva compressa*. Plant Physiol. Biochem., 135: 423–431.

Ng, E.L., Bandow, C., Proenca, D.N., Santos, S., Guilherme, R., Morais, P.V., Roembke, J. and Sousa, J.P., (2014). Does altered rainfall regime change pesticide effects in soil? A terrestrial model ecosystem study from Mediterranean Portugal on the effects of pyrimethanil to soil microbial communities under extremes in rainfall. Appl. Soil Ecol., 84: 245–253.

Pacheco, D., Rocha, A.C., Pereira, L. and Verdelhos, T. (2020). Microalgae water bioremediation: Trends and hot topics. Appl. Sci., 10: 1886.

Pacholak, A., Burlaga, N., Frankowski, R., Zgola-Grzeskowiak, A. and Kaczorek, E. (2022). Azole fungicides: (Bio)degradation, transformation products and toxicity elucidation. Sci. Total Environ., 802: 149917.

Pan, C.G., Peng, F.J. and Ying, G.G. (2018). Removal, biotransformation and toxicity variations of climbazole by freshwater algae *Scenedesmus obliquus*. Environ. Pollut., 240: 534–540.

Panagos, P., Ballabio, C., Lugato, E., Jones, A., Borrelli, P., Scarpa, S., Orgiazzi, A. and Montanarella, L. (2018). Potential sources of anthropogenic copper inputs to European agricultural soils. Sustainability, 10: 2380.

Pandey, D., Daverey, A., Dutta, K. and Arunachalam, K. (2022a). Bioremoval of toxic malachite green from water through simultaneous decolorization and degradation using laccase immobilized biochar. Chemosphere, 297: 134126.

Pandey, S., Dubey, S.K., Kashyap, A.K. and Jain, B.P. (2022b). Cyanobacteria-mediated heavy metal and xenobiotics bioremediation. pp. 335–350. *In*: P. Singh, M. Fillat and A. Kumar (eds.). Cyanobacterial Lifestyle and its Applications in Biotechnology. Elsevier.

Papazlatani, C.V., Karas, P.A., Lampronikou, E. and Karpouzas, D.G. (2022). Using biobeds for the treatment of fungicide-contaminated effluents from various agro-food processing industries: Microbiome responses and mobile genetic element dynamics. Sci. Total Environ., 823: 153744.

Parlakidis, P., Mavropoulos, T., Vryzas, Z. and Gikas, G.D. (2021). Fluopyram removal from agricultural equipment rinsing water using HSF pilot-scale constructed wetlands. Environ. Sci. Pollut. Res., 29: 29584–29596.

Perruchon, C., Patsioura, V., Vasileiadis, S. and Karpouzas, D.G. (2016). Isolation and characterisation of a *Sphingomonas* strain able to degrade the fungicide *ortho*-phenylphenol. Pest Manag. Sci., 72: 113–124.

Perruchon, C., Pantoleon, A., Veroutis, D., Gallego-Blanco, S., Martin-Laurent, F., Liadaki, K. and Karpouzas, D.G. (2017a). Characterization of the biodegradation, bioremediation and detoxification capacity of a bacterial consortium able to degrade the fungicide thiabendazole. Biodegradation, 28: 383–394.

Perruchon, C., Chatzinotas, A., Omirou, M., Vasileiadis, S., Menkissoglou-Spiroudi, U. and Karpouzas, D.G. (2017b). Isolation of a bacterial consortium able to degrade the fungicide thiabendazole: The key role of a *Sphingomonas* phylotype. Appl. Microbiol. Biotechnol., 101: 3881–3893.

Perruchon, C., Vasileiadis, S., Rousidou, C., Papadopoulou, E.S., Tanou, G., Samiotaki, M., Garagounis, C., Molassiotis, A., Papadopoulou, K.K. and Karpouzas, D.G. (2017c). Metabolic pathway and cell adaptation mechanisms revealed through genomic, proteomic and transcription analysis of a *Sphingomonas haloaromaticamans* strain degrading *ortho*-phenylphenol. Sci. Rep., 7: 6449.

Perruchon, C., Vasileiadis, S., Papadopoulou, E.S. and Karpouzas, D.G. (2020). Genome-based metabolic reconstruction unravels the key role of B12 in methionine auxotrophy of an *ortho*-phenylphenol-degrading *Sphingomonas haloaromaticamans*. Front. Microbiol., 10: 3009.

Peško, M. and Kráľová, K. (2013). Physiological response of *Brassica napus* L. plants to Cu(II) treatment. Proceedings of ECOpole, 7: 155–161.

Pietrini, F., Carnevale, M., Beni, C., Zacchini, M., Gallucci, F. and Santangelo, E. (2019). Effect of different copper levels on growth and morpho-physiological parameters in giant reed (*Arundo donax* L.) in semi-hydroponic mesocosm experiment. Water, 11: 1837.

Priyadarshanee, M., Chatterjee, S., Rath, S., Dash, H.R. and Das, S. (2022). Cellular and genetic mechanism of bacterial mercury resistance and their role in biogeochemistry and bioremediation. J. Hazard. Mater., 423(Pt A): 126985.

Qessaoui, R., Zanzan, M., Ajerrar, A., Lahmyed, H., Boumair, A., Tahzima, R., Alouani, M., Mayad, E.H., Chebli, B., Walters, S.A. and Bouharroud, R. (2022). *Pseudomonas* isolates as potential biofungicides of green mold (*Penicillium digitatum*) on orange fruit. Int. J. Fruit Sci., 22: 142–150.

Qiao, W.C., Zhang, Z.Y., Qian, Y., Xu, L.J. and Guo, H. (2022). Bacterial laccase immobilized on a magnetic dialdehyde cellulose without cross-linking agents for decolorization. Colloids Surf. A Physicochem. Eng., 632: 127818.

Qu, M.J., Li, N., Li, H.D., Yang, T., Liu, W., Yan, Y.P., Feng, X.H. and Zhu, D.W. (2018). Phytoextraction and biodegradation of atrazine by *Myriophyllum spicatum* and evaluation of bacterial communities involved in atrazine degradation in lake sediment. Chemosphere, 209: 439–448.

Radziemska, M., Vaverkova, M.D. and Baryla, A. (2017). Phytostabilization-management strategy for stabilizing trace elements in contaminated soils. Int. J. Environ. Res. Publ. Health, 14: 958.

Raffa, C.M. and Chiampo, F. (2021). Bioremediation of agricultural soils polluted with pesticides: A review. Bioengineering, 8: 92.

Raheem, S.S., Al-Dossary, M.A. and AL-Saad, H.T. (2021). Ability of some fungi to biodegrade carbendazim fungicide. Iraqi J. Agric. Sci., 52: 259–267.

Rajpal, N., Verma, S., Kumar, N., Kumar, N., Lee, J., Kim, K.H., Ratan, J.K. and Divya, N. (2023). Bioremediation of carbendazim and thiamethoxam in domestic greywater using a bioaugmented microbial consortium. Environ. Technol. Innov., 30: 103087.

Randelovic, D., Jakovljevic, K. and Zeremski, T. (2022). Chelate-assisted phytoremediation. pp. 131–154. *In*: V. Pandey (ed.). Assisted Phytoremediation. Elsevier.

Randika, J.L.P.C., Bandara, P.K.G.S.S., Soysa, H.S.M., Ruwandeepika, H.A.D. and Gunatilake, S.K. (2022). Bioremediation of pesticide-contaminated soil: A review on indispensable role of soil bacteria. J. Agric. Sci. Sri Lanka, 17: 19–43.

Rempel, A., Gutkoski, J.P., Nazari, M.T., Biolchi, G.N., Cavanhi, V.A.F., Treichel, H. and Colla, L.M. (2021). Current advances in microalgae-based bioremediation and other technologies for emerging contaminants treatment. Sci. Total Environ., 772: 144918.

Rodriguez-Rojas, F., Celis-Pla, P.S.M., Mendez, L., Moenne, F., Munoz, P.T., Gabriela, L.M., Diaz P, Sanchez-Lizaso, J.L., Brown, M.T., Moenne, A. and Saez, C.A. (2019). MAPK pathway under chronic copper excess in green macroalgae (Chlorophyta): Involvement in the regulation of detoxification mechanisms. Int. J. Mol. Sci., 20: 4546.

Roman, D.L., Voiculescu, D.I., Filip, M., Ostafe, V. and Isvoran, A. (2021). Effects of triazole fungicides on soil microbiota and on the activities of enzymes found in soil: A review. Agriculture, 11: 893.

Romeh, A.A. (2015). Evaluation of the phytoremediation potential of three plant species for azoxystrobin-contaminated soil. Int. J. Environ. Sci. Technol., 12: 3509–3518.

Romeh, A.A.A. (2017). Phytoremediation of azoxystrobin and its degradation products in soil by *P. major* L. under cold and salinity stress. Pestic. Biochem. Physiol., 142: 21–31.

Rouabhi, R. (2010). Introduction and toxicology of fungicides. *In*: O. Carisse (ed.). Fungicides. Intech, Rjeka. https://www.intechopen.com/chapters/12734.

Saadaoui, H., Boujelbane, F., Serairi, R., Ncir, S. and Mzoughi, N. (2021). Transformation pathways and toxicity assessments of two triazole pesticides elimination by gamma irradiation in aqueous solution. Sep. Purif. Technol., 276: 119381.

Salunkhe, V.P., Sawant, I.S., Banerjee, K., Wadkar, P.N., Sawant, S.D. and Hingmire, S.A. (2014). Kinetics of degradation of carbendazim by *B. subtilis* strains: possibility of in situ detoxification. Environ. Monit. Assess., 186: 8599–8610.

Satapute, P.P. and Kaliwal, B.B. (2018). *Burkholderia* sp. strain BBK_9: A potent agent for propiconazole degradation. pp. 87–103. *In*: E.D. Bidoia and R.N. Montagnolli (eds.). Toxicity and Biodegradation testing. Book Series: Methods in Pharmacology and Toxicology, Springer.

Satapute, P. and Jogaiah, S. (2022). A biogenic microbial biosurfactin that degrades difenoconazole fungicide with potential antimicrobial and oil displacement properties. Chemosphere, 286, Part 1: 131694.

Sathishkumar, P., Gu, F.L., Ameen, F. and Palvannan, T. (2019). Fungal laccase mediated bioremediation of xenobiotic compounds. *In*: Y.C. Chang (ed.). Microbial Biodegradation of Xenobiotic Compounds. CRC Press, Boca Raton, 23 pp.

Schachtschneider, K., Chamier, J. and Somerset, V. (2017). Phytostabilization of metals by indigenous riparian vegetation. Water SA, 43: 177–185.

Seenivasagan, R., Karthika, A., Kalidoss, R. and Malik, J.A. (2022). Bioremediation of polluted aquatic ecosystems using macrophytes. pp. 57–79. *In*: J.A. Malik (ed.). Advances in Bioremediation and Phytoremediation for Sustainable Soil Management. Springer, Cham.

Shabbir, Z., Sardar, A., Shabbir, A., Abbas, G., Shamshad, S., Khalid, S., Natasha, Murtaza, G., Dumat, C. and Shahid, M. (2020). Copper uptake, essentiality, toxicity, detoxification and risk assessment in soil-plant environment. Chemosphere, 259: 127436.

Sharma, P., Sharma, M., Raja, M., Singh, D.V. and Srivastava, M. (2016a). Use of *Trichoderma* spp. in biodegradation of Carbendazim. Indian J. Agric. Sci., 86: 891–894.

Sharma, S.S., Dietz, K.J. and Mimura, T. (2016b). Vacuolar compartmentalization as indispensable component of heavy metal detoxification in plants. Plant Cell Environ., 39: 1112–1126.

Sharma, P. and Kumar, S. (2021). Bioremediation of heavy metals from industrial effluents by endophytes and their metabolic activity: Recent advances. Bioresour. Technol., 339: 125589.

Sheoran, V., Sheoran, A.S. and Poonia, P. (2016). Factors affecting phytoextraction: A review. Pedosphere, 26: 148–166.

Sherrard, R.M., Bearr, J.S., Murray-Gulde, C.L., Rodgers, J.H. and Shah, Y.T. (2004). Feasibility of constructed wetlands for removing chlorothalonil and chlorpyrifos from aqueous mixtures. Environ. Pollut., 127: 385–394.

Shi, Y.H., Ye, Z., Hu, P., Wei, D., Gao, Q., Zhao, Z.Y., Xiao, J.J., Liao, M. and Cao, H.Q. (2020). Removal of prothioconazole using screened microorganisms and identification of biodegradation products via UPLC-QqTOE-MS. Ecotoxicol. Environ. Saf., 206: 111203.

Silva, R.C., Barros, K.A. and Pavao, A.C. (2014). Carcinogenicity of carbendazim and its metabolites. Química Nova, 37: 1329–1334.

Singh, S., Singh, N., Kumar, V., Datta, S., Wani, A.B., Singh, D., Singh, K. and Singh, J. (2016). Toxicity, monitoring and biodegradation of the fungicide carbendazim. Environ. Chem. Lett., 14: 317–329.

Singh, S., Kumar, V., Singh, S. and Singh, J. (2019). Influence of humic acid, iron and copper on microbial degradation of fungicide Carbendazim. Biocatal. Agric. Biotechnol., 20: 101196.

Sivaram, A.K., Subashchandrabose, S.R. and Logeshwaran, P. (2020). Rhizodegradation of PAHs differentially altered by C_3 and C_4 plants. Sci. Rep., 10: 16109.

Skufca, D., Kovacic, A., Prosenc, F., Bulc, T.G., Heath, D. and Heath, E. (2021). Phycoremediation of municipal wastewater: Removal of nutrients and contaminants of emerging concern. Sci. Total Environ., 782: 146949.

Stang, C., Elsaesser, D., Bundschuh, M., Ternes, T.A. and Schulz, R. (2013). Mitigation of biocide and fungicide concentrations in flow-through vegetated stream mesocosms. J. Environ. Qual., 42: 1889–1895.

Stravs, M.A., Pomati, F. and Hollender, J. (2017). Exploring micropollutant biotransformation in three freshwater phytoplankton species. Environ. Sci. Process. Impacts, 19: 822–832.

Subba, R. and Mathur, P. (2022). Functional attributes of microbial and plant based biofungicides for the defense priming of crop plants. Theor. Exp. Plant Physiol., 34: 301–333.

Sun, T., Miao, J.B., Saleem, M., Zhang, H.N., Yang, Y. and Zhang, Q.M. (2020). Bacterial compatibility and immobilization with biochar improved tebuconazole degradation, soil microbiome composition and functioning. J. Hazard. Mater., 398: 122941.

Takkar, S., Shandilya, C., Agrahari, R., Chaurasia, A., Vishwakarma, K., Mohapatra, S., Varma, A. and Mishra, A. (2022). Green technology: Phytoremediation for pesticide pollution. pp. 353–375. *In*: V. Kumar, M.P. Shah and S.K. Shahi (eds.). Phytoremediation Technology for the Removal of Heavy Metals and Other Contaminants. Elsevier.

Talab, K.M.A., Yang, Z.H., Li, J.H., Zhao, Y., Omer, S.A.M. and Xiong, Y.B. (2018). The influence of microbial communities for triadimefon enantiomerization in soils with different pH values. Chirality 30: 293–301.

Tamm, L., Thuerig, B., Apostolov, S., Blogg, H., Borgo, E., Corneo, P.E., Fittje, S., de Palma, M., Donko, A., Experton, C., Marín, E.A., Pérez, A.M., Pertot, I., Rasmussen, A., Steinshamn, H., Vetemaa, A., Willer, H. and Herforth-Rahmé, J. (2022). Use of copper-based fungicides in organic agriculture in twelve European countries. Agronomy, 12: 673.

Tanveer, F., Iqrar, I., Shinwari, Z.K. and Gul, I. (2022). Microbial biostimulants and their role in environmental bioremediation. *In*: C.O. Inamuddin, Adetunji, M.I. Ahmed and T. Altalhi (eds.). Microbial Biostimulants for Sustainable Agriculture and Environmental Bioremediation, 1st ed. CRC Press, 12 pp.

Thoa, L.T.K., Thao, T.T.P., Hung, N.B., Khoo, K.S., Quang, H.T., Lan, T.T., Hoang, V.D., Park, S.M., Ooi, C.W., Show, P. and Huy, N.D. (2022). Biodegradation and detoxification of malachite green dye by extracellular laccase expressed from *Fusarium oxysporum*. Waste Biomass Valor., 13: 2511–2518.

Tonelli, F.C.P., Tonelli, F.M.P., Lemos, M.S. and Nunes, N.A.D.M. (2022). Mechanisms of phytoremediation. pp. 37–64. *In*: R.A. Bhat, F.M.P. Tonelli, G.H. Dar and K. Hakeem (eds.). Phytoremediation, Biotechnological Strategies for Promoting Invigorating Environs. Academic Press, Elsevier.

Touliabah, H.E., El-Sheekh, M.M., Ismail, M.M. and El-Kassas, H. (2022). A review of microalgae- and cyanobacteria-based biodegradation of organic pollutants. Molecules, 27: 1141.

Triantafyllidis, V., Kosma, C., Karabagias, I.K., Zotos, A., Pittaras, A. and Kehayias, G. (2022). Fungicides in Europe during the twenty-first century: a comparative assessment using agri-environmental indices of EU_{27}. Water Air Soil Pollut., 233: 52.

Tweedy, B.G. (1981). Inorganic sulfur as a fungicide. pp. 43–68. *In*: F.A. Gunther and J.D. Gunther (eds.). Residue Reviews, Vol 78. Springer, New York, NY.

Vaithyanathan, V.K., Vaidyanathan, V.K. and Cabana, H. (2022). Laccase-driven transformation of high priority pesticides without redox mediators: Towards bioremediation of contaminated wastewaters. Front. Bioeng. Biotechnol., 9: 770435.

Vallee, R., Dousset, S., Billet, D. and Benoit, M. (2014). Sorption of selected pesticides on soils, sediment and straw from a constructed agricultural drainage ditch or pond. Environ. Sci. Pollut. Res., 21: 4895–4905.

Verasoundarapandian, G., Lim, Z.S., Radziff, S.B.M., Taufik, S.H., Puasa, N.A., Shaharuddin, N.A., Merican, F., Wong, C.Y., Lalung, J. and Ahmad, S.A. (2022). Remediation of pesticides by microalgae as feasible approach in agriculture: Bibliometric strategies. Agronomy, 12: 117.

Volova, T., Prudnikova, S., Boyandin, A., Zhila, N., Kiselev, E., Shumilova, A., Baranovskiy, S., Demidenko, A., Shishatskaya, E. and Thomas, S. (2019). Constructing slow-release fungicide formulations based on poly(3-hydroxybutyrate) and natural materials as a degradable matrix. J. Agric. Food Chem., 67: 9220–9231.

Vymazal, J. (2022). The historical development of constructed wetlands for wastewater treatment. Land, 11: 174.

Wachowska, U., Kucharska, K., Pluskota, W., Czaplicki, S. and Stuper-Szablewska, K. (2020). Bacteria associated with winter wheat degrade *Fusarium* mycotoxins and triazole fungicide residues. Agronomy, 10: 1673.

Wang, F.H., Zhu, L.S., Wang, X.G., Wang, J. and Wang, J.H. (2015). Impact of repeated applications of metalaxyl on its dissipation and microbial community in soil. Water Air Soil Pollut., 226: 430.

Wang, X.H., Hou, X.J., Liang, S., Lu, Z.B., Hou, Z.G., Zhao, X.F., Sun, F.J. and Zhang, H. (2018a). Biodegradation of fungicide Tebuconazole by *Serratia marcescens* strain B1 and its application in bioremediation of contaminated soil. Int. Biodeterior. Biodegradation, 127: 185–191.

Wang, S.N., Chen, Q.J., Zhu, M.J., Xue, F.Y., Li, W.C., Zhao, T.J., Li, G.D. and Zhang, G.Q. (2018b). An extracellular yellow laccase from white rot fungus *Trametes* sp. F1635 and its mediator systems for dye decolorization. Biochimie, 148: 46–54.

Wang, Y., Zhang, X.Q., Wang, L., Wang, C.W., Fan, W.X., Wang, M.Q. and Wang, J.M. (2019a). Effective biodegradation of pentachloronitrobenzene by a novel strain *Pseudomonas putida* QTH3 isolated from contaminated soil. Ecotoxicol. Environ. Saf., 182: 109463.

Wang, F.H., Zhou, T.T., Zhu, L.S., Wang, X.G., Wang, J., Wang, J.H., Du, Z.K. and Li, B. (2019b). Effects of successive metalaxyl application on soil microorganisms and the residue dynamics. Ecol. Indic., 103: 194–201.

Wang, X.H., Lu, Z.B., Miller, H., Liu, J.H., Hou, Z.G., Liang, S., Zhao, X.F., Zhang, H. and Borch, T. (2020). Fungicide azoxystrobin induced changes on the soil microbiome. Appl. Soil Ecol., 145: 103343.

Wang, Y.F., Yu, Z.Q., Fan, Z.P., Fang, Y.W., He, L.T., Peng, M.L., Chen, Y.Y., Hu, Z.Y., Zhao, K., Zhang, H.P. and Liu, C.Y. (2021). Cardiac developmental toxicity and transcriptome analyses of zebrafish (*Danio rerio*) embryos exposed to Mancozeb. Ecotoxicol. Environ. Saf., 226: 112798.

Wang, W.J., Zhao, Z., Yan, H., Zhang, H.Y., Li, Q.X. and Liu, X.L. (2022). Carboxylesterases from bacterial enrichment culture degrade strobilurin fungicides. Sci. Total Environ., 814: 152751.

Whittington, H.D., Singh, M., Ta, C., Azcarate-Peril, M.A. and Bruno-Barcena, J.M. (2020). Accelerated biodegradation of the agrochemical ametoctradin by soil-derived microbial consortia. Front. Microbiol., 11: 1898.

Williams, J.S. and Cooper, R.M. (2004). The oldest fungicide and newest phytoalexin—A reappraisal of the fungitoxicity of elemental sulphur. Plant Pathol., 53: 263–279.

Wu, X.H., Xu, J., Liu, Y.Z., Dong, F.S., Liu, X.G., Zhang, W.W. and Zheng, Y.Q. (2015). Impact of fluxapyroxad on the microbial community structure and functional diversity in the silty-loam soil. J. Integr. Agric., 14: 114–124.

Wu, H.B., Shen, J.Y., Jiang, X.B., Liu, X.D., Sun, X.Y., Li, J.S., Han, W.Q., Mu, Y. and Wang, L.J. (2018a). Bioaugmentation potential of a newly isolated strain *Sphingomonas* sp. NJUST37 for the treatment of wastewater containing highly toxic and recalcitrant tricyclazole. Bioresour. Technol., 264: 98–105.

Wu, H.B., Shen, J.Y., Jiang, X.B., Liu, X.D., Sun, X.Y., Li, J.S., Han, W.Q. and Wang, L.J. (2018b). Bioaugmentation strategy for the treatment of fungicide wastewater by two triazole-degrading strains. Chem. Eng. J., 349: 17–24.

Xu, M., Lin, Y., da Silva, E.B., Cui, Q.H., Gao, P., Wu, J. and Ma, L.Q. (2022). Effects of copper and arsenic on their uptake and distribution in As-hyperaccumulator *Pteris vittata*. Environ. Pollut., 300: 118982.

Yang, Z.G., Jiang, W.K., Wang, X.H., Cheng, T., Zhang, D.S., Wang, H., Qiu, J.G., Cao, L., Wang, X. and Hong, Q. (2018). An amidase gene, ipaH, is responsible for the initial step in the iprodione degradation pathway of *Paenarthrobacter* sp strain YJN-5. Appl. Environ. Microbiol., 84: e01150–18.

You, X.W., Suo, F.Y., Yin, S.J., Wang, X., Zheng, H., Fang, S., Zhang, C.S., Li, F.M. and Li, Y.Q. (2021). Biochar decreased enantioselective uptake of chiral pesticide metalaxyl by lettuce and shifted bacterial community in agricultural soil. J. Hazard. Mater., 417: 126047.

Youness, M., Sancelme, M., Combourieu, B. and Besse-Hoggan, P. (2018). Identification of new metabolic pathways in the enantioselective fungicide tebuconazole biodegradation by *Bacillus* sp. 3B6. J. Hazard. Mater., 351: 160–168.

Zada, S., Raza, S., Khan, S., Iqbal, A., Kai, Z., Ahmad, A., Ullah, M., Kakar, M., Fu, P.C., Dong, H.F. and Xueji, Z. (2022). Microalgal and cyanobacterial strains used for the bio sorption of copper ions from soil and wastewater and their relative study. J. Ind. Eng. Chem., 105: 463–472.

Zand, A.D. and Mühling, K.H. (2022). Phytoremediation capability and copper uptake of maize (*Zea mays* L.) in copper contaminated soils. Pollutants, 2: 53–65.

Zhang, X.J., Huang, Y.J., Harvey, P.R., Li, H.M., Ren, Y., Li, J.S., Wang, J.N. and Yang, H.T. (2013). Isolation and characterization of carbendazim-degrading *Rhodococcus erythropolis* djl-11. PLoS One, 8: e74810.

Zhang, Y.K., Wang, H., Wang, X., Hu, B., Zhang, C.F., Jin, W., Zhu, S.J., Hu, G. and Hong, Q. (2017). Identification of the key amino acid sites of the carbendazim hydrolase (MheI) from a novel carbendazim-degrading strain *Mycobacterium* sp SD-4. J. Hazard. Mater., 331: 55–62.

Zhang, M.Y., Wang, J., Bai, S.H., Zhang, Y.L., Teng, Y. and Xu, Z.H. (2019). Assisted phytoremediation of a co-contaminated soil with biochar amendment: Contaminant removals and bacterial community properties. Geoderma, 348: 115–123.

Zhang, C., Li, J.H., Wu, X.M., Long, Y.H., An, H.M., Pan, X.L., Li, M., Dong, F.S. and Zheng, Y.Q. (2020a). Rapid degradation of dimethomorph in polluted water and soil by *Bacillus cereus* WL08 immobilized on bamboo charcoal-sodium alginate. J. Hazard. Mater., 398: 122806.

Zhang, N., Guo, D., Zhu, Y., Wang, X.M., Zhu, L.J., Liu, F., Teng, Y., Christie, P., Li, Z.G. and Lu, Y.M. (2020b). Microbial remediation of a pentachloronitrobenzene-contaminated soil under *Panax notoginseng*: A field experiment. Pedosphere, 30: 563–569.

Zhang, H.P., Song, J.J., Zhang, Z.H., Zhang, Q.K., Chen, S.Y., Mei, J.J., Yu, Y.L. and Fang, H. (2021a). Exposure to fungicide difenoconazole reduces the soil bacterial community diversity and the co-occurrence network complexity. J. Hazard. Mater., 405: 124208.

Zhang, C., Wu, X., Wu, Y.Y., Li, J.H., An, H.M. and Zhang, T. (2021b). Enhancement of dicarboximide fungicide degradation by two bacterial cocultures of *Providencia stuartii* JD and *Brevundimonas naejangsanensis* J3. J. Hazard. Mater., 403: 123888.

Zhang, Z.X., Xie, Y.X., Ye, Y.Z., Yang, Y.L., Hua, R. and Wu, X.W. (2022). Toxification metabolism and treatment strategy of the chiral triazole fungicide prothioconazole in water. J. Hazard. Mater., 432: 128650.

Zhao, H.W., Li, Q.L., Jin, X.T., Li, D., Zhu, Z.Q. and Li, Q.X. (2021). Chiral enantiomers of the plant growth regulator paclobutrazol selectively affect community structure and diversity of soil microorganisms. Sci. Total Environ., 797: 148942.

Zhong, H., Liu, G., Jiang, Y., Brusseau, M.L., Liu, Z., Liu, Y. and Zeng, G.M. (2016). Effect of low-concentration rhamnolipid on transport of *Pseudomonas aeruginosa* ATCC 9027 in an ideal porous medium with hydrophilic or hydrophobic surfaces. Colloids Surf. B: Biointerfaces, 139: 244–248.

Zhou, Y., Xu, J., Zhu, Y., Duan, Y. and Zhou M. (2016). Mechanism of action of the benzimidazole fungicide on *Fusarium graminearum*: Interfering with polymerization of monomeric tubulin but not polymerized microtubule. Phytopathology, 106: 807–813.

Zhou, W.T., Zhang, W.X. and Cai, Y.P. (2022). Enzyme-enhanced adsorption of laccase immobilized graphene oxide for micro-pollutant removal. Sep. Purif. Technol., 294: 121178.

Zubrod, J.P., Bundschuh, M., Arts, G., Brühl, C.A., Imfeld, G., Knäbel, A., Payraudeau, S., Rasmussen, J.J., Rohr, J., Scharmüller, A., Smalling, K., Stehle, S., Schulz, R. and Schäfer, R.B. (2019). Fungicides: An overlooked pesticide class? Environ. Sci. Technol., 53: 3347–3365.

Index

R

S

T

About the Editors

Kamel A. Abd-Elsalam, Ph.D. is currently a Research Professor at the Plant Pathology Research Institute, Agricultural Research Center, Giza, Egypt. Dr. Kamel's research interests include developing, improving and installing plant biosecurity diagnostic tools, understanding and exploiting fungal pathogen genomes and developing eco-friendly hybrid nanomaterials for controlling toxicogenic fungi, plant diseases and agroecosystems applications. He has published 20 books related to nano-biotechnology applications in agriculture and plant protection that were published by the world's major publishing houses (Springer, Tylor Frances and Elsevier). Since 2019, he has served as the Editor-in-Chief of the Elsevier book series, Nanobiotechnology for Plant Protection, he also serves as the Series Editor of the Elsevier book series Applications of Genome Modified Plants and Microbes in Food and Agriculture. He has also participated as an active member of the Elsevier Advisory Panel, giving feedback and suggestions for improvement of Elsevier's products and services since 2020. Kamel Abd-Elsalam has published more than 160 scientific research in international and regional specialized scientific journals with a high impact factor, and has an h-index of 36, i-10 index of 95, with 5142+ citations. He has also served as a Guest Editor for the Journal of Fungi, Plants and Microorganisms, and as a Review, Editor for Frontiers in Genomic Assay Technology and refereed for several reputed journals. In 2014, he was awarded the Federation of Arab Scientific Study Councils Prize for excellent scientific research in biotechnology (fungal genomics) (first ranking). In addition, according to Stanford University's worldwide database rating in 2021, he was listed among the top 2% of the world's most influential scientists by Stanford University and Elsevier. Dr. Kamel earned his Ph.D. in Molecular Plant Pathology from Christian Alberchts University of Kiel (Germany) and Suez Canal University (Egypt), and in 2008, he was awarded a postdoctoral fellowship from the same institution. Dr. Kamel was a visiting associate professor at Mae Fah Luang University in Thailand, the Institute of Microbiology at TUM in Germany, the Laboratory of Phytopathology at Wageningen University in the Netherlands and the Plant Protection Department at Sassari University in Italy and Moscow University in Russia. He was ranked in Top 2% most influential scientist in the world in nanobiotechnology for the year 2020 by Stanford University.

Mousa A. Alghuthaymi is currently a professor, he obtained a Ph.D. in Microbiology from King Saud University in 2013. Dr. Mousa's research interests include the development, improvement and establishment of plant biosecurity diagnostic tools, the understanding and exploitation of fungal pathogen genomes, and the development of eco-friendly hybrid nanomaterials for the control of toxicogenic fungi, plant diseases and agroecosystems applications. The head of the Biology Department at the College of Science and Human Studies in Shaqra University, Saudi Arabia, and a member of the University's Scientific Council since March 2022, and previously worked as the Head of the Chemistry Department between 2016–2018. He has published 14 chapters in 12 books and published about 40 research papers in refereed scientific journals.

Salah M. Abdel-Momen I received my Ph.D. in Plant Pathology from Texas A&M University through a US-AID grant and completed a postdoctoral program at the University of Maryland's Department of Molecular Genetics. I returned to Egypt to work as a researcher at the Plant Pathology Research Institute, Agricultural Research Center, where I helped establish a laboratory for molecular plant pathology with colleagues, which assisted our senior and junior colleagues in manipulating molecular tools and techniques in plant disease diagnosis and plant microbe interaction. As I continued to work at the Agricultural Research Center, I was responsible for a variety of tasks, including conducting plant disease research and overseeing multiple master's theses and Ph.D. dissertations until I was elevated to the level of professor. As a trainee or trainer, I took part in many programs at the American University in Cairo and Chicago State University in the fields of advanced management, report writing, technology evaluation and transfer, and intellectual property rights. This is in addition to other extension and on-farm training programs on various crops in most Egyptian governorates. Such activities provided me with expertise and a broad view of agricultural challenges in Egypt, allowing me to become the editor-in-chief of the Egyptian Journal of Agricultural Research and president of the Egyptian Society of Plant Pathology. This experience also aided my appointment to several administrative positions, including Deputy Director of the Plant Pathology Research Institute for Training and Extension, Vice President of the Agricultural Research Center, President of the Agricultural Research Center, and the Egypt's minister of agriculture and lands reclamation from 2012 to 2013. Serving in these positions exposed and enabled me to get a broad understanding of agricultural problems and output, as well as to participate in plans for developing and sustaining agriculture.

For Product Safety Concerns and Information please contact our EU representative GPSR@taylorandfrancis.com Taylor & Francis Verlag GmbH, Kaufingerstraße 24, 80331 München, Germany

Batch number: 10726519

Printed by Printforce, the Netherlands